Đặng Vũ Đỗ Thiếu Lan

Biology of Human Reproduction

Sculpture in wood by Joan Miró (1931) titled *Man and Woman*.
An evocation of the ambiguities and complexities of human sexuality.
(Reprinted with permission from the Artists Rights Society, New York, NY.)

Biology of
Human Reproduction

Ramón Piñón, Jr.
UNIVERSITY OF CALIFORNIA, SAN DIEGO

University Science Books
Sausalito, California

University Science Books
55D Gate Five Road
Sausalito, CA 94965
www.uscibooks.com

Order Information:
   Phone (703) 661-1572
   Fax (703) 661-1501

Production Manager: *Susanna Tadlock*
Manuscript Editor: *Deena Cloud*
Designer: *Robert Ishi*
Illustrators: *John and Judy Waller*
Compositor: *Wilsted & Taylor Publishing Services*
Printer and Binder: *Maple-Vail Book Manufacturing Group*

This book is printed on acid-free paper.

Library of Congress Cataloging-in-Publication Data

Piñón, Ramón, 1939–
   Biology of human reproduction / Ramón Piñón.
      p. cm.
   Includes bibliographical references and index.
   ISBN 1-891389-12-2
   1. Human reproduction.   I. Title.

   QP251   .P54 2001
   612.6—dc21   2001035350

Printed in the United States of America
10 9 8 7 6 5 4 3 2

*For Kathleen and
all our children*

Abbreviated Table of Contents

Detailed Table of Contents

Part IV   A NEW LIFE   269

CHAPTER 11   Fertilization and Implantation   271

CHAPTER 16   Infertility and Assisted Reproduction   399

CHAPTER 17   Sexually Transmitted Diseases   427

Preface

THIS TEXT HAS DEVELOPED out of my teaching human reproduction for a number of years at two levels, an advanced level for biology majors and an introductory-intermediate level for nonscience students. I first taught human reproduction in the winter quarter of 1973. The field has changed immensely since then. Indeed, most of the information presented in this book was unknown at that time. The field has matured enormously, and there is now a large body of information about human reproduction that is available to experts but has not yet made its way into public consciousness. I take it for granted that most people are interested in reproduction, but the complexity of the field may discourage many from tackling it seriously.

The present text is intended for nonscience majors, which means simply that an extensive background in biology is not presumed. Nevertheless, this is not a text for the completely naive. I do assume that all students have acquired a certain level of knowledge about biology (and perhaps reproduction) in high school, at the university, or through their own reading. Indeed, my experience has been that many college-level students, perhaps because of the inherent interest in the subject, have done a certain amount of reading on their own. I am mindful that many students might find some parts of the text rough going. Nevertheless, my experience has been that a willingness to tackle a complex subject, to think analytically and critically, and to ask questions and not be dismayed with incomplete or unsatisfactory answers are perhaps the most important prerequisites for the course.

My objective is to provide a systematic overview of the biology of human reproduction at what might be called an intermediate level—that is, beyond what is generally covered in introductory general biology texts. This means that we will explore many aspects of human reproduction in a rather detailed way. The material is not necessarily easy, but I believe it is always interesting. An overview of the scope of the text is provided in Chapter 1. As will be evident, the study of human reproduction is a multidisciplinary endeavor bringing together several different fields of research. The spectrum of topics in the text covers essentially the entire range of studies of human reproduction. I have ordered the topics in what I think is a logical manner, but I recognize that not all instructors will agree with me. The first eight chapters form the core of the material and focus on the anatomy, physiology, and endocrinology of the reproductive system. The remaining chapters explore other aspects of human reproduction, and they can be covered in whatever order fits best with the instructor's overall plan. The focus of the text is biology, but in Chapter 19 I have ventured into what to me is the uncharted territory of human sexuality. Here I have limited my discussion to the little that we know about the biological basis of a few selected aspects of human sexuality.

Throughout the text I have sought to emphasize the experimental, observational, and clinical findings that form the basis of our understanding of the different aspects of human reproduction. Such findings are often presented in the form of tables or graphs that have been taken directly from the primary literature. The level of coverage varies from introductory to intermediate and, in some cases, perhaps advanced. For a few topics I have also provided a short historical perspective which I hope may add to an appreciation of how our understanding has developed. I have also tried to conform (with only two exceptions) to new recommendations for anatomical terminology that avoid the use of eponymous terms. In addition, except in a few instances, I have avoided mentioning the names of scientists and clinicians associated with the experiments and observations discussed, although the works that provided illustrations or graphs are cited in the readings at the end of chapters or in the legends to the figures or tables. It was not my intent to provide in this text a comprehensive review citing all contributions. I feel that a text of this type should focus more properly on ideas and insight. The bibliography at the end of each chapter is divided into a general reading list and a list of journal articles that provide more detailed experimental or clinical findings.

I am hopeful that both students and instructors will find something of value in the text. I think the text can be used successfully by instructors with different teaching styles and approaches. For the student, I hope the text is stimulating and thought-provoking. I have also included questions at the end of each chapter that I hope will engage and develop the reader's understanding of the material. Topics that can be explored in more detail through library research are included as well. I hope that the text will form the foundation on which the student can continue to build after leaving the university.

Acknowledgments

I wish to acknowledge the encouragement that my friend Dr. Vernon Ávila provided in the early stages of writing. He graciously read a number of chapters of the first draft of the text. The comments and encouragement of various reviewers are greatly appreciated. I have tried to incorporate many of their suggestions. Special thanks are due Drs. Gregory Erickson, Kenneth Jones, Oliver Jones, Linda Olson, and Dorothy Hollingsworth from the UCSD School of Medicine for providing me with figures that appear in the text; Dr. A. H. Holstein, University of Hamburg, who graciously gave me permission to reprint the scanning electron micrographs of spermatogenesis; and Dr. S. D. Gilman, University of Chicago, for the Chapter 17 frontispiece art. I greatly appreciate the work of three former students, Pamela Hill and Ranee Munaim, for helping with the Index, and Lauren Sakoda, for preparing Figures 19.4 and 19.5. I thank Bruce Armbruster and Jane Ellis at University Science Books for the reception and encouragement they gave me when presented with the first draft of the text. Jackie Vignes deserves special thanks for her unfailing patience and courtesy in helping me prepare the several versions of the text that I have used in my teaching.

Ramón Piñón, Jr.

San Diego, California
January 2002

Biology of Human Reproduction

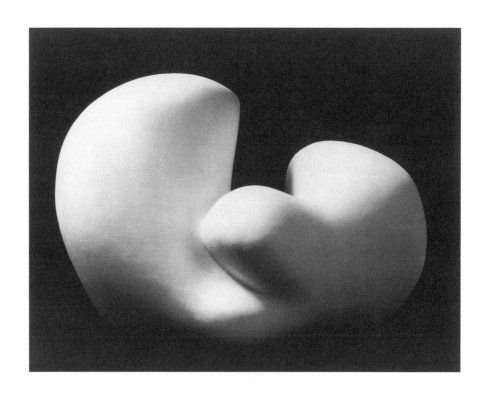

Jean (Hans) Arp, *Human Concretion*, 1935.
(The Museum of Modern Art, New York. Gift of the Advisory Committee.
© 2000 The Museum of Modern Art, New York.)

PART

I

INTRODUCTION

Henri Matisse, *Dance*, 1938.
(© 2001 Succession H. Matisse, Paris/Artists Rights Society [ARS], New York, NY.)

The Study of Human Reproduction

*In the beginning the Author of the human fabric fashioned
two human beings for the conservation of the species in such a
way that the male should furnish the primary principle of the
infant, the female indeed should fitly conceive it and should
nourish the little child arising from this principle as she would
nourish some member of her own body until the child should
become stronger and could be given forth into the air which
surrounds us. Both male and female received instruments
suitable for these functions and peculiar to them alone.
To these organs was imparted so great a power and
attraction of delight in the generative act that the living
creatures are incited by this power, and whether or not
they are young or foolish or devoid of reason, they
fall to the task of propagating the species not
otherwise than if they were the wisest of beings.*

—The Epitome of Andreas Vesalius. 1543

Translated from the Latin by L. R. Lind and with anatomical notes by C. W. Asling.
Massachusetts Institute of Technology Press, Cambridge, 1949, p. 83.

HUMAN INTEREST IN REPRODUCTION is of immense antiquity. We probably would
not be too far wrong if we guessed that curiosity about reproduction is as old as
human consciousness. This interest has extended to reproduction not only of the human species, but of all living organisms. Human societies have bred animals for work
and pleasure, and they have cultivated plants for food and ornamental purposes for
millennia. As empirical enterprises these activities have been remarkably successful.
From it have come all our domesticated animals and an amazing variety of plants. But
until very recently both of these activities were carried out without any real understanding of the nature of reproduction. For most of the history of our species, the multitude of questions that must have been asked by our ancestors—the meaning of sex,
the reason for the two sexes (genders), the relationship between sex and reproduction,
the nature and function of the organs of reproduction, the meaning of menstruation in

3

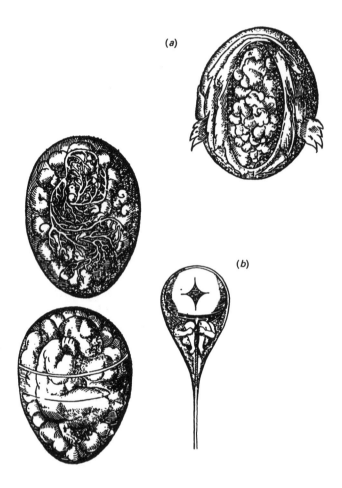

Figure 1.1
Drawings of two highly imaginative models of reproduction. (a) The "ovist" view by Rueff (1554). (b) The "spermist" view by Harsoeker (1694). (University of Glasgow Museum of Art, with permission.)

human females, pregnancy, birth, puberty, the loss of reproductive ability in the female (menopause), the causes of infertility, the nature of the sexual drive (libido)—remained unanswered or, from a modern perspective, were answered in highly fanciful ways. It was not too long ago, for example, that *preformationist* theories, according to which every individual existed fully formed in miniature form in the egg (the "ovist" view) or in the sperm (the "spermist" view), seemed quite reasonable (Fig. 1.1). Fanciful as they might seem to us now, these ideas represented early attempts to understand the nature of reproduction.

The first steps in understanding reproductive processes were taken with the birth of the scientific revolution during the sixteenth century. Classical antiquity, which bequeathed us an immense legacy in history, law, philosophy, literature, and mathemat-

1st step of understanding

ics, left, by comparison, only a meager inheritance in science, especially in the realm of reproduction. It was not until the end of the nineteenth century that the first significant clues about the nature of reproduction were found, and it was not until the middle of the twentieth century that a realistic understanding of the reproductive cycles in mammals, including humans, was achieved. The impressive rate of progress that characterized the latter half of the twentieth century, in contrast to the much more gradual rate during the preceding periods, can be attributed to the vigorous way in which the advances and techniques of the diverse disciplines of endocrinology, physiology, anatomy, genetics, embryology, biochemistry, and molecular biology were brought to bear on the study of reproduction. Reproductive biology at the beginning of the twenty-first century is a complex, multidisciplinary, and sophisticated science. Still, despite the enormous progress of the last 20 years, huge gaps in our understanding remain. You will see as you read these chapters that many important questions are still unanswered. The next 20 years should bring new and exciting insights into the nature of reproduction.

The reproductive biology of a species, understood as the totality of processes and relationships that determine reproductive success, is multidimensional, for it encompasses interactions at many different levels—molecular, cellular, tissue, organ, organismal, and behavioral. Indeed, the reproductive biology of any species can be studied at each of these levels. In this book we will focus primarily on the cellular, tissue, and organ interactions and, in particular, on the mechanisms that regulate and integrate the elements of the reproductive system. Keep in mind that human reproduction cannot be studied in isolation, separate from that of other species. Humans, after all, are primates, mammals, and vertebrates, and our reproductive system retains characteristics of its evolutionary heritage. The essentials of sexual reproduction are more or less universal and relatively easy to comprehend, but in nature, that universal theme has been played out in what almost appears to be an unlimited set of variations. The details of those variations are what makes the study of reproduction so fascinating. Every species has evolved its own quirks and unique features, even if it shares many features with related species. Although the focus in this text is on human reproductive biology, reference will be made on many occasions to information obtained from studies—for example, in the mouse, sheep, and the Rhesus monkey, three species that have been studied intensely. Extrapolation of conclusions from those species to the human case is not always possible. Nevertheless, in many instances, inferences drawn from animal studies will have a certain applicability to humans. Animal studies have provided important insights into human reproductive biology not obtainable by direct human experimentation.

AREAS OF FOCUS

Human reproductive biology encompasses many related areas of study. This division reflects the historical development of the science, and also the distinct questions and problems into which the field naturally divides itself. Exploration of these subdisciplines will form the substance of this book. These different areas of study are grouped in this text under four major themes, briefly summarized below. I hope that this format will be helpful in bringing some coherence to an immense field of study. The subject is demanding, but certainly well worth the effort. Every field of study generates its own special terminology. Reproductive biology is no different. This circumstance should not be viewed as an obstacle, but rather as a way of clarifying concepts and ideas. What follows is an overview of the topics covered in the text.

Part I. Introduction

Chapters 1 and 2 set the stage for the rest of the text. Sex and sexual reproduction, despite their ubiquity, remain mysterious phenomena whose significance continue to be discussed vigorously today by evolutionary biologists. In Chapter 2 we will see how human reproduction fits into the wider context of sexual reproduction. In addition, Chapter 2 includes a brief review of the essentials of the genetic machinery, the basis of all life. We will consider the nature of the chromosome cycle that characterizes sexually reproducing species; the sex chromosomes; and the nature and origin of mutations. Mutations are especially important because they are the source of many reproductive abnormalities and birth defects.

The functioning reproductive system provides an excellent example of the indispensable role that molecules known as *hormones* play in all biological processes. We devote Chapter 3 to a short review of the important classes and properties of hormones and hormone action. The characteristics of the hormones that play major roles in reproduction (discussed in more detail in later chapters) are described.

Part II. The Gonads

The *gonads*—the *ovaries* in females and the *testes* in males—define the irreducible difference between the two sexes. The two sexes also differ anatomically and physiologically in a number of ways. In mammals, including humans, sexual differentiation, which includes the formation of the gonads and the internal and external genitalia, begins very early in embryogenesis and is completed before birth. Our understanding of these processes indicates that male development appears to be "imposed" on a substratum that is essentially programmatically female. We will see in Chapter 4 what is meant when we say that male development is the *induced* pathway, while female development is the *constitutive*, or *default*, pathway. Moreover, we have come to understand how far from "fail-safe" these developmental processes are. Thanks in no small measure to the analysis of the "exceptions," we are now beginning to develop a better understanding of the genes and developmental pathways that are responsible for the fundamental female/male dichotomy.

The ovaries not only produce the egg; they also secrete the steroid sex hormones estrogen and progesterone, which together orchestrate the menstrual cycle. Both are multifunctional hormones that play important roles in many different tissues. The complex nature of ovarian function is considered in Chapter 5. The testis produces sperm at a prodigious rate, and its hormones determine the male body type. What is perhaps not appreciated is that the testis holds the key to evolutionary change. Testicular function is reviewed in Chapter 6.

Part III. The H-P-G Axis

The brain is the final arbiter of reproduction. Specifically, the hypothalamus, the pituitary gland, and certain elements of the central nervous system (CNS) constitute important components of the *hypothalamic-pituitary-gonadal (H-P-G) axis*. It is probably fair to say that the major contribution of twentieth-century reproductive biology was the discovery and elaboration of how the hypothalamus, pituitary, and gonads interact. These three tissues communicate with each other through a few important hormones. The hypothalamus, through its secretion of the hormone *gonado-*

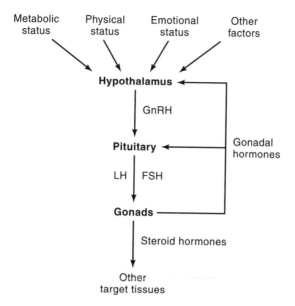

Figure 1.2
Schematic model of the H-P-G axis. The primary organs involved are the hypothalamus, the pituitary, and the gonads. The hypothalamus integrates diverse stimuli resulting from interaction with the environment. These signals regulate the release of the hypothalamic hormone GnRH, which in turn stimulates the production of the gonadotropic hormones LH and FSH from the pituitary. Under LH and FSH regulation, the gonads produce the steroid hormones that are responsible for the male-female differences, and which also regulate the pituitary and hypothalamus.

tropin-releasing hormone (GnRH), stimulates the pituitary to secrete the gonado-tropic hormones, *luteinizing hormone* (LH) and *follicle-stimulating hormone* (FSH). The gonads, primarily under LH and FSH stimulation, produce the sex steroid hormones required for the production of the egg in the female and the sperm in the male; these hormones are also responsible for the somatic differences between the male and female (body shape, distribution of body hair, and perhaps even differences in behavior). The sex steroid hormones also exert feedback effects on both the pituitary and the hypothalamus, thereby regulating the activity of each and forming a self-regulatory network (Fig. 1.2). We will review the nature of the H-P-G axis in Chapter 7.

No one is born reproductively competent. In all mammalian species, a period of growth and development after birth is required before reproductive capability is attained. In primates, and especially in humans, the period between birth and the attainment of sexual maturity is much longer than in other mammalian species. The onset of puberty signals the "reawakening" of the H-P-G axis during the pubertal period. Much is known about the neuroendocrine changes that take place during this period, but the nature of the ultimate signals that initiate the onset of puberty remains elusive (see Chapter 8). Clues about the factors that play a part in this crucial and fascinating aspect of human reproductive biology are slowly beginning to emerge.

Reproductive activity is also known to be influenced by a variety of factors, including nutritional status, physical activity, and emotional states. A unifying way to understand these effects is to consider them as examples of generalized stress. We have come to understand that different neural elements in the CNS function as *sensors* of stress, and in turn modulate the activity of the H-P-G axis. In Chapter 9 we will consider the different types of stress and the way in which each affects the activity of the H-P-G axis.

The effects of aging on reproductive function are complex. The cessation of the reproductive potential in the female, *menopause*, comes when she is still relatively young, with some 30 or more years left to live. Is menopause a pathological condition? Should it be approached as a condition that requires medical attention? How to adjust to the postmenopausal period productively and effectively is an important question for many women today. What are the consequences of the menopause? How should the female deal with the loss of ovarian function? Should she go on estrogen replacement therapy or not? For the male, reproductive aging is less dramatic, but no less real. New studies are helping us understand this critical period in our lives (see Chapter 10).

Part IV. A New Life

The first step in the formation of a new person is fertilization. Fertilization involves the fusion of two highly specialized cells—the egg and the spermatozoon. The reproductive success of any species is measured by the ability to ensure fruitful fertilizations. Fertilization appears to be a precarious affair. The lifetimes of the egg and the spermatozoon are very short, which means that release of both gametes has to be timed so that an egg will be available to be fertilized at the appropriate time. In species characterized by internal fertilization—that is, in which fertilization occurs inside the female—transport of the sperm from the site of sperm deposition to the site of fertilization requires that the environment of the female internal genitalia be regulated rather precisely. The crucial event, fusion of the egg and the sperm, although still far from being completely understood, is considered in Chapter 11.

One of the characteristic features of mammals is that the offspring are born alive after a period of development in the uterus of the mother. In placental mammals (in contrast to marsupial mammals), development of the embryo takes place completely within the maternal uterus and is dependent on the formation of a specialized organ composed of both maternal and fetal components, the *placenta*. The placenta permits the mammalian embryo to develop in a relatively secure and protected environment. *Implantation* describes the process by which the newly formed embryo establishes contact with the uterus and initiates the development of the placenta. Birth usually takes place when the fetus has become mature enough to survive outside the mother. It may be surprising that we still do not understand the factors that initiate labor and birth. Nevertheless, a few clues are emerging. Indispensable to mammalian life is lactation. The development of the mammary gland and its function in producing milk is considered in Chapter 12.

Not all pregnancies are successful. Things can go wrong, and we have to face the loss of an embryo or a fetus, or the tragic consequences of an infant with birth defects. How do we understand such occurrences? What can be done to prevent them? The factors that contribute to fetal loss and birth defects are considered in Chapter 13. Errors specific to the development and maturation of the H-P-G axis also occur and can have drastic consequences for the reproductive system as a whole. A brief review of the types of errors affecting reproductive function is given in Chapter 14.

Part V. Societal Issues

The need or desire to control or limit human fertility and its opposite, the desire to improve the fertility of couples unable to have children, appear to be a constant in human societies. However, only in the latter half of the twentieth century were the twin problems of controlling fertility and infertility approached systematically. Strong imperatives for having children as well as for limiting fertility have probably coexisted in most human societies since ancient times. The introduction of the oral contraceptive pill in the late 1950s was revolutionary, a powerful social force in changing male-female relationships. Newer contraceptive strategies, currently under investigation, are absolutely essential if the world's population is to be stabilized in this century. These different issues are discussed in Chapter 15.

The problems of infertility have weighed heavily on infertile couples—particularly on the woman, since until recently, infertility was attributed to females, and only rarely to males. We now have a much clearer sense of the natural pattern of fertility in the female and the factors that influence changes in fertility. The delay of childbearing that is becoming typical in industrialized societies is not without consequences. The widely held belief that delay will not affect a woman's ability to conceive and carry a baby to term may not be valid. Lifestyle factors, particularly the increasing use and abuse of alcohol, smoking, and drugs, may be contributing to the burden of infertility. Therapy for normal loss of fertility or infertility due to genetic or physiological disturbances is available. Since the birth of the first "test-tube baby" in 1970, assisted reproduction techniques, which have made it possible for many infertile couples to have children, have increased in number and sophistication. The "brave new world" so often alluded to in popular literature is already here, with its impressive successes, but also with a host of questions that societies are trying to define and settle. We will consider these issues in Chapter 16.

It is probably fair to say that we live in a turbulent time when it comes to sexual matters. One consequence of this turbulent time has been the spread of sexually transmitted diseases (STDs). STDs have, in fact, reached epidemic proportions in the United States and in many other parts of the world. The price paid in disease, pregnancy loss, and birth defects is incalculable. We will consider the nature of STDs and review the features of the most important ones in Chapter 17.

Cancers of the reproductive tissues, breast, uterus, and ovaries in the female, and prostate in the male, take a terrific toll in our society. An understanding of the evolutionary context of the risk factors, particularly in the female, is slowly making us realize that the incidence of these cancers could be reduced significantly. Reproductive cancers may be the price we have paid for modernity, but we may not have to pay it indefinitely (see Chapter 18).

Sexuality and reproduction are often thought to be synonymous, but they are in fact quite distinct. The study of the biological basis of human sexuality is in its infancy. In nonprimate mammalian species, sexual behavior tends to be very stereotyped, and some progress has been made in understanding the molecular basis of such instinctive responses. In humans, the separation of innate from learned behavior is especially difficult, and the extent to which innate, nonconscious responses still govern our behavior remains an unresolved problem. The rules that guided our conduct to a large extent 20 years ago no longer seem to apply. No longer is sexual desire, homosexuality, or transsexuality discussed only by psychiatrists or psychotherapists. There is a pressing desire to try to understand the biological roots of human sexuality in all its complexity.

Perhaps only then will it be possible to bridge the gaps that separate our societal discussions about these topics. We are not yet there, but we are making progress by formulating the appropriate questions (see Chapter 19).

Finally, in Chapter 20, we consider two new technologies that may have revolutionary consequences for biology and medicine: *cloning* and the derivation of *embryonic stem cells*. Application of both advances to humans may well have important benefits, but it also raises unprecedented ethical dilemmas because these processes require the generation of human embryos for research purposes. As a society we are not yet certain how to respond to the complex issues that human embryo research raises.

SOURCES OF INFORMATION ON HUMAN REPRODUCTION

Information about human reproduction appears in many forms, including textbooks, reference books, books that discuss the societal and ethical questions arising from advances in reproductive technologies, research journals, and, increasingly, Internet articles. Below, I provide a partial list of general readings selected from the many sources available. Books included are those published in or after 1990. Many other references appear in the readings at the end of each chapter. Research that bears directly or indirectly on aspects of human reproduction discussed in this text is published in many different types of journals. The list below includes the sources that I have found useful.

Textbooks

1. General

Not many textbooks on human reproduction have been published. Those listed here differ in level of difficulty, format, style, and emphasis.

Johnson, M. H., and B. J. Everitt. 1995. *Essential Reproduction*, 4th ed. Blackwell Science, Oxford, UK.

Jones, R. E. 1997. *Human Reproductive Biology*, 2nd ed. Academic Press, San Diego, CA.

Wood, J. W. 1994. *Dynamics of Human Reproduction: Biology, Biometry, Demography*. Aldine de Gruyter, New York, NY.

Yu, H.-S. 1994. *Human Reproductive Biology*. CRC Press, Inc., Boca Raton, FL.

2. Human embryology

Carlson, B. M. 1999. *Human Embryology and Developmental Biology*. Mosby, St. Louis, MO.

Dunstan, G. R. (Ed.). 1990. *The Human Embryo: Aristotle and the Arabic and European Traditions*. University of Exeter Press, Exeter, UK.

England, M. A. 1996. *Life Before Birth*, 2nd ed. Mosby-Wolfe, London, UK.

Larsen, W. J. 1997. *Human Embryology*. Churchill Livingston, New York, NY.

Moore, K. L., *et al.* 1994. *Color Atlas of Clinical Embryology*. Saunders, Philadelphia, PA.

O'Rahilly, R., and F. Müller. 1996. *Human Embryology and Teratology*. Wiley-Liss, New York, NY.

Sadler, T. W. 1995. *Langman's Medical Embryology*, 7th ed. Williams and Wilkins, Baltimore, MD.

Thorogood, R. (Ed.). 1997. *Embryos, Genes, and Birth Defects*. John Wiley, Chichester, NY.

3. Reproductive endocrinology/physiology

Adashi, E. Y., *et al.* (Eds.). 1996. *Reproductive Endocrinology, Surgery, and Technology*. Lippincott-Raven, Philadelphia, PA.

Edwards, R. G., *et al.* (Eds.). 1996. *GnRH Analogues and Reproductive Medicine*. 1996. Oxford University Press, Oxford, UK.

Fauser, B. C. J. M. (Ed.). 1999. *Molecular Biology in Reproductive Medicine*. Parthenon Publishing Group, New York, NY.

Findlay, J. K. (Ed.). 1994. *Molecular Biology of the Female Reproductive System*. Academic Press, San Diego, CA.

Ginsbury, J. (Ed.). 1996. *Drug Therapy in Reproductive Endocrinology*. Oxford University Press, London, UK.

Hildt, E., and S. Graumann (Eds.). 1999. *Genetics in Human Reproduction*. Ashgate, Brookfield, VT.

Loke, Y. W., and A. King. 1995. *Human Implantation: Cell Biology and Immunology.* Cambridge University Press, New York, NY.

Mishell, D. R. (Ed.). 1999. *Reproductive Endocrinology.* Current Medicine, Inc., Philadelphia, PA.

Mitchell, M. E. (Ed.). 1990. *Eicosanoids in Reproduction.* CRC Press, Boca Raton, FL.

Ombelet, W., and A. Vereecken (Eds.). 1995. *Modern Andrology.* Oxford University Press, Oxford, UK.

Yen, S. C., *et al.* (Eds.). 1999. *Reproductive Endocrinology: Physiology, Pathophysiology, and Clinical Management.* Saunders, Philadelphia, PA.

4. Human reproduction: family and society

Boling, P. (Ed.). 1995. *Expecting Trouble: Surrogacy, Fetal Abuse, and New Reproductive Technologies.* Westview Press, Boulder, CO.

Campbell, K. L., and J. W. Wood (Eds.). 1994. *Human Reproductive Ecology: Interactions of Environment, Fertility, and Behavior.* New York Academy of Sciences, New York, NY.

Dolgin, J. L. 1997. *Defining the Family: Law, Technology, and Reproduction in an Uneasy Age.* New York University Press, New York, NY.

Dunaif, G. E., *et al.* (Eds.). 1998. *Human Diet and Endocrine Modulation: Estrogenic and Androgenic Effects.* ILSI Press, Washington, D.C.

Guidelines for Reproductive Toxicity Risk Assessment. 1999. U.S. Environmental Protection Agency, Washington, D.C.

Humber, J. M., and R. F. Almeder (Eds.). 1996. *Reproduction, Technology, and Rights.* Humana Press, Totowa, NJ.

IPPF Charter on Sexual and Reproductive Rights: Vision 2000. International Planned Parenthood Federation, London, UK.

Paul, M. E. (Ed.). 1993. *Reproductive and Developmental Hazards.* U.S. Department of Health and Human Services, Atlanta, GA.

Pollard, I. 1994. *A Guide to Reproduction: Social Issues and Human Concerns.* Cambridge University Press, New York, NY.

Schettler, T., *et al.* 1999. *Generations at Risk: Reproductive Health and the Environment.* MIT Press, Cambridge, MA.

Van Steirteghem, A., *et al.* (Eds.). 1996. *Genetics and Assisted Human Conception.* Oxford University Press, London, UK.

Journals

1. Journals that specialize in different aspects of human reproduction

AIDS
American Journal of Obstetrics and Gynecology
Contraception
Current Opinion in Obstetrics and Gynecology
European Journal of Obstetrics and Gynecology
Fertility and Sterility
Human Reproduction
International Journal of Gynecology and Obstetrics
Journal of Andrology
Journal of Clinical Endocrinology and Metabolism
Journal of Reproduction and Fertility

2. Journals that occasionally publish articles with relevance to human reproduction

American Journal of Human Genetics
Bailliere's Clinical Endocrine Metabolism

British Medical Bulletin
Bulletin of the New York Academy of Sciences
Endocrine and Metabolic Clinics of North America
Endocrine Reviews
Hormone Research
Human Genetics
Journal of the American Medical Association
Lancet
Nature
Nature Genetics
Nature Medicine
New England Journal of Medicine
Proceedings of the National Academy of Sciences (USA)
Quarterly Review of Biology
Osteoporosis International
Postgraduate Medicine
Science
Trends in Endocrinology and Metabolism

3. Journals for nonexperts that occasionally carry articles on human reproduction

American Scientist
BioScience
Discovery
Science News
Scientific American

4. Electronic databases

MEDLINE: National Library of Medicine
Covers medicine, health, and biomedical sciences, and includes journal articles
1966–present. Includes links out to many full-text journal articles at the publishers'
Web sites. The MEDLINE database is available free at various sites on the Internet.

OMIM: Online Mendelian Inheritance in Man
OMIM is the online version of Dr. Victor McKusick's comprehensive source on
genetic disorders, *Mendelian Inheritance in Man.*

BIOETHICSLINE
Part of the service from the National Library of Medicine. Covers the ethical, legal,
and public policy issues surrounding health care and biomedical research. Free to
everyone.

Pablo Picasso, *Splitting of the Cell*, 1935.
An intriguing way of picturing cell division (1 cell into 2) and sexual reproduction
(2 cells into 1). (© 1999 Estate of Pablo Picasso/Artists Rights Society [ARS], New York, NY.)

CHAPTER 2

Sex, Reproduction, and Mutation

Sex is the queen of problems in evolutionary biology.
Perhaps no other natural phenomenon has aroused so much
interest; certainly none has sowed so much confusion.

—G. Bell. 1982

The Masterpiece of Nature: The Evolution and Genetics of Sexuality.
University of California Press, Berkeley, CA.

Endogenous oxidative damage to germ line DNA is likely
to lead to heritable mutations and increased incidence of
birth defects, genetic diseases, and cancer in offspring.

—A. A. Woodall and B. N. Ames. 1997

Nutritional prevention of DNA damage to sperm and consequent risk
reduction in birth defects and cancer in offspring. In *Preventive Nutrition.*
A. Bendich and R. J. Deckelbaum, Eds. Human Press Inc., Totowa, NJ.

W̲E ACCEPT SEX AND REPRODUCTION as part of the natural order of things, and ~~Sex ≠ reproduction~~ we generally consider sex as being necessary for reproduction. In reality, however, they are distinct phenomena. It may come as a surprise that sex has been a problem for biologists for at least 100 years. It is not immediately obvious why this should be so. In this chapter we will try to understand why the meaning and significance of sex continue to be discussed vigorously by evolutionary biologists. We will also see how human reproduction fits into the wider context of sexual reproduction. We will start with a brief review of the essentials of the genetic machinery that lies at the basis of all life. We will consider the nature of the chromosome cycle that characterizes sexually reproducing species and the significance of sex chromosomes. Finally, we will take up the nature and origin of mutations, which are the driving force for evolutionary change. Without them, evolutionary change would be impossible. On the other hand, mutations have a powerful negative impact on human well-being in terms of prenatal mortality, birth defects, reduced viability and fertility, and many different diseases. Hence, understanding the mechanism of mutagenesis and identifying the factors that contribute to the origin of mutations are of special importance.

15

CELLS AND ORGANISMS

Evolution of cells

Cell: basic unit of life [handwritten margin note]

3 billion year [handwritten margin note]

The cell is the basic unit of life. All living organisms are composed of cells. The simplest organisms are unicellular and microscopic. Plants and animals are multicellular organisms made up of millions of cells that are organized into tissues, organs, and organ systems that carry out specialized functions. The earliest fossil evidence for the existence of cellular life dates to over 3 billion years ago. The features of the fossil remains suggest that these very early organisms were **prokaryotic** ("before a nucleus"), as opposed to **eukaryotic** ("having a nucleus"). Eukaryotic cells arose about 1.5 billion years ago (Table 2.1). Eukaryotic cells were so named because they contain a prominent membrane-bound structure called the *nucleus*. In contrast, prokaryotic cells are defined by the absence of a nucleus. Many other differences distinguish eukaryotic from prokaryotic cells. The extent of these differences was not fully realized until the 1960s, when application of the electron microscope to the study of cells revealed the important differences in internal architecture between the two.

3 domains – characterized from each other [handwritten margin note]

All extant organisms are grouped into three domains—**Bacteria, Archaea,** and **Eukarya.** Each domain is characterized by specific features that are absent in the other two domains. Until recently, the relationship between the three domains has been depicted in the form of an evolutionary tree as shown in Fig. 2.1. According to this view, Bacteria branched earliest from the putative common ancestor of all cells. The Archaea and Eukarya emerged from a different branch, and would be considered more closely related to each other than to Bacteria. New data, however, is leading evolutionary biologists to question the validity of this model, in particular the existence of a single common ancestor, and the relationships among the three domains implied by the traditional model. There is general agreement, however, regarding the kingdoms that make up the Eukarya. The earliest eukaryotic organisms were unicellular, and probably not

Table 2.1
Evolution in Outline

Event	Millions of years ago (approximate)
Formation of the Earth	4500
First cellular life: prokaryotes, asexual reproduction	3900–3400
First eukaryotes: mitosis, meiosis, sexual reproduction	1500
First vertebrates	600–500
Amphibians	370
Reptiles	320
Primitive mammals	220
Modern mammals	120
Extinction of dinosaurs and spread of mammals	64
Primitive primates	38–24
Homo erectus	3–1.5
Homo sapiens	0.2

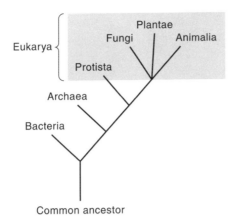

Figure 2.1
The traditional evolutionary tree of the three domains of living organisms—Archaea, Bacteria, and Eukarya. The Bacteria are considered to be the first to emerge from the universal ancestor of all cellular life. The Archaea and Eukarya represent a separate and later branch. Recent comparisons of genes from members of the three domains have raised serious doubts about the validity of the evolutionary relationships depicted in the traditional model. The Eukarya consists of four kingdoms—Protista, which are unicellular organisms, and the multicellular Fungi, Plantae, and Animalia.

much larger than their prokaryotic counterparts. Extant descendants of these early eukaryotes are probably the organisms that now make up the Protista, of which the better-known members are the algae, protozoa, slime molds, and water molds. The Protista probably represent the earliest branch of the Eukarya. Around 600 million years ago, true multicellular organisms began to appear. These include members of the three other eukaryotic kingdoms—the **Fungi**, the **Animalia**, and the **Plantae**. Many species of bacteria and simple eukaryotes can form multicellular aggregates, but such structures are not really equivalent to true multicellular organisms, which are characterized by cell differentiation and specialization that result in the development of tissues and organs.

The molecular biology of cells

Despite the differences in structure and function between various types of cells, they all contain the same four classes of molecules: *carbohydrates*, *lipids*, *proteins*, and *nucleic acids*. Each class of molecules plays a distinct role in the life of a cell. Carbohydrates (sugars and starches) are the primary sources of energy. Lipids are a structurally diverse group of molecules that serve a variety of functions. For example, phospholipids are the structural elements of the membranes found in all types of cells. In many animal cells, another important group of lipids are the steroids, which include cholesterol and all the steroid hormones (estrogens, androgens, progestins, corticosteroids, and mineralocorticoids).

Proteins are composed of long chains of amino acids, or polypeptides, and are often referred to as the workhorses of the cell because of their diverse functions. Many proteins are enzymes, carrying out the many metabolic reactions of the cell. Other pro-

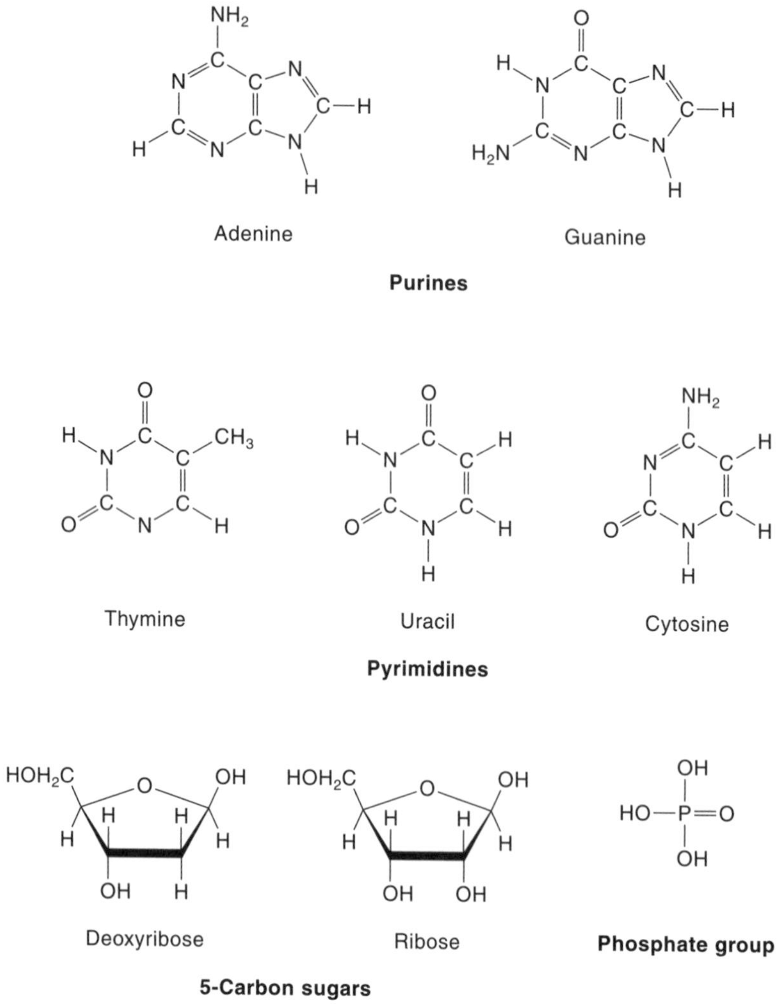

Figure 2.2

Components of nucleic acids—nitrogenous bases, five-carbon sugars (deoxyribose and ribose), and phosphate groups. There are five nitrogenous bases, two purines—adenine and guanine—and three pyrimidines—thymine, uracil, and cytosine.

teins are hormones and hormone receptors (for example, insulin and the insulin receptor) by which the many tissues in multicellular organisms are regulated. Other proteins are transport molecules, carrying minerals, vitamins, and other molecules throughout the body. Other proteins have diverse structural roles (myosin in muscle cells, keratin in skin cells).

The nucleic acids, DNA and RNA, are long-chain molecules made up of linear arrays of nucleotides. A nucleotide consists of a sugar, a phosphate, and a nitrogenous base. Two of these bases are compounds called purines (**A**, adenine, and **G**, guanine), and three are compounds called pyrimidines (**C**, cytosine, **T**, thymine, and **U**, uracil). There are two types of sugars, deoxyribose and ribose (Fig. 2.2)—hence the two

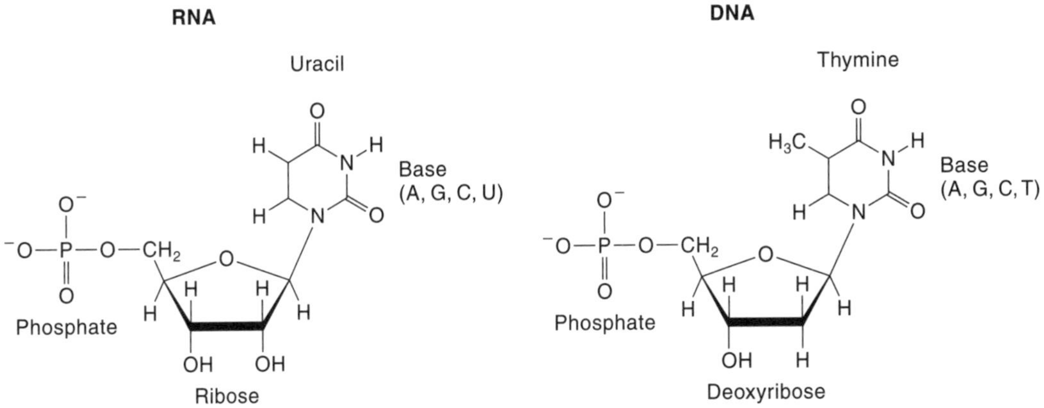

RNA

Uracil

DNA

Thymine

Base
(A, G, C, U)

Phosphate

OH   OH

Ribose

Base
(A, G, C, T)

Phosphate

OH   H

Deoxyribose

Figure 2.3
Structures of DNA and RNA nucleotides. A DNA nucleotide consists of a nitrogenous base
(adenine, guanine, cytosine, or thymine), deoxyribose, and a phosphate group. An RNA
nucleotide consists of a nitrogenous base (adenine, guanine, cytosine, or uracil), ribose, and
a phosphate group.

types of nucleic acid. DNA nucleotides contain deoxyribose, a phosphate, and A, G,
C, or T as the base. RNA nucleotides contain ribose, a phosphate, and A, G, C, or U as
the base (Fig. 2.3).

An RNA molecule is a single-stranded RNA nucleotide chain in which the sugar-
phosphate groups form the "backbone" of the molecule, while the bases are perpen-
dicular to the backbone (Fig. 2.4). In contrast, a DNA molecule is double-stranded

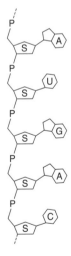

Figure 2.4
Schematic representation of an RNA molecule. The sugar-phosphate groups form the
"backbone" of the nucleotide chain.

consisting of two DNA nucleotide chains bound to each other by chemical bonds between the purine and pyrimidine components of the nucleotide (Fig. 2.5). The two chains are said to be complementary to each other because a T on one chain is always bonded to an A on the other, and a G is always bonded to a C. This base-pairing rule ensures that once the sequence of nucleotides in one chain is known, the sequence in the other chain follows immediately. The two chains are also "antiparallel" to each other because the orientation of the sugar-phosphate bonds are opposite in the two chains (Fig. 2.5).

gene: genetic machinery responsible for cell fxn

The genetic machinery responsible for cell function is fundamentally the same in all cells. The basis for all cellular function is the **gene**, a sequence in a DNA molecule that specifies the linear sequence of amino acids in a protein. The gene is, therefore, an informational molecule, and the information that it carries is coded in the sequence of the DNA nucleotides, T, A, C, and G, taken three at a time. There are 64 possible ways in which the four DNA bases can be grouped as triplets (for example, AAA, TTA,

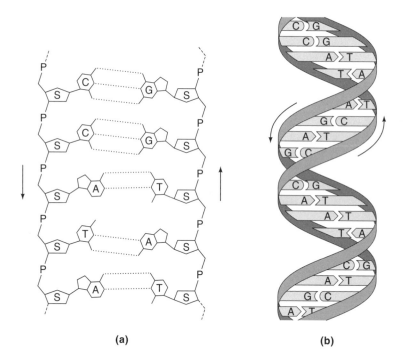

(a)   (b)

Figure 2.5
Schematic representation of a DNA molecule. DNA is a double-stranded, helical molecule.
(a) Each DNA strand is a polynucleotide chain consisting of a linear sequence of nucleotides. The two strands of the helix are held together by chemical bonds between the nucleotides of the two strands. The constraint on the bonds is that adenine (A) always pairs with thymine (T), and guanine (G) with cytosine (C). Because of this constraint, one strand is complementary to the other, meaning that if the nucleotide sequence in one strand is known, the strand in the other can be predicted. Note that the orientation of the sugar-phosphate bonds in the two strands are opposite to each other. As a consequence, the two strands are antiparallel. The informational content of the DNA lies in the sequence of nucleotides. (b) Model of the DNA double helix. The sugar-phosphate backbone is represented by the two helical bands.

ATC, TCG, GCA, etc.). The triplets are referred to as **codons** because they code for the 20 different naturally occurring amino acids found in proteins.

The gene represents stored information that can be *expressed*, a process requiring two steps. In the first step, **transcription**, the sequence of DNA nucleotides is converted into a sequence of RNA nucleotides forming a molecule called **messenger RNA** (mRNA). In the second step, **translation**, the sequence of nucleotides in the mRNA determines the sequence of amino acids in the synthesis of a protein molecule. Expression of a gene, then, represents a flow of genetic information from DNA to RNA to protein.

Transcription is carried out by the enzyme RNA polymerase, together with several other proteins known as *transcription factors* (Fig. 2.6). Translation, also referred to

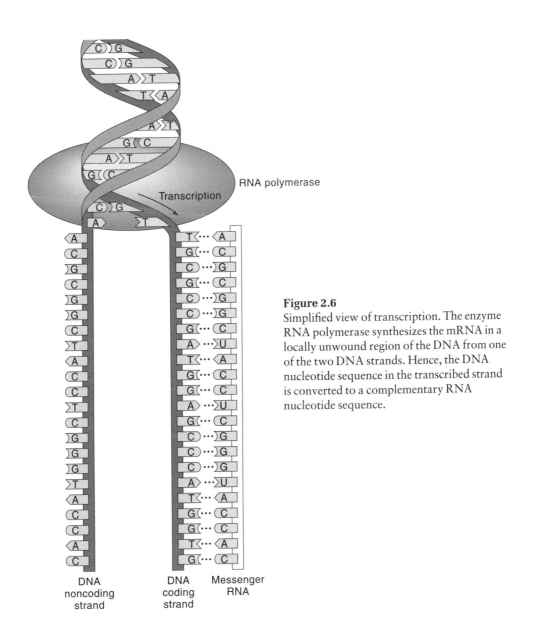

Figure 2.6
Simplified view of transcription. The enzyme RNA polymerase synthesizes the mRNA in a locally unwound region of the DNA from one of the two DNA strands. Hence, the DNA nucleotide sequence in the transcribed strand is converted to a complementary RNA nucleotide sequence.

22

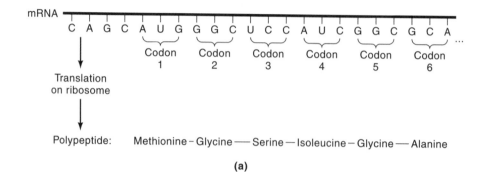

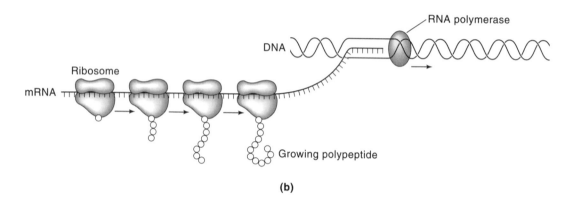

Figure 2.7
Translation. (a) The sequence of codons in the mRNA is converted into the sequence of amino acids in the polypeptide chain. (b) Translation is carried out on the ribosomes. Several ribosomes can attach to one mRNA at the same time. After attaching, the ribosome moves along the length of the mRNA, generating the new polypeptide chain.

as protein synthesis, takes place on the ribosomes. Ribosomes are small, complex structures consisting of two RNA molecules (ribosomal RNA, or rRNA) and about 30 proteins (Fig. 2.7). In prokaryotic cells, translation is directly coupled to transcription so that the mRNA is translated as soon as it is synthesized. In eukaryotes, the two processes are separated physically. The gene is transcribed in the nucleus; the resulting mRNA is transported from the nucleus into the cytoplasm, and translation takes place on ribosomes in the cytoplasm.

The set of proteins that a cell synthesizes at any given time is determined by the set of genes expressed (active) at that time, which in turn depends on the needs of the cell. A complex machinery regulates the gene expression. Expression of some genes may be required on a more-or-less continuous basis, for example, to maintain the metabolic activity needed to keep the cell alive. Such genes tend to be the same in all cells, and they are often referred to as "housekeeping genes." Other genes are expressed only when the need for that gene product arises. In a multicellular organism, tissues can be distinguished from one another by the set of genes they express. For example, a heart muscle cell is different from a liver or brain cell because the spectrum of proteins it synthesizes is different from that of the liver or brain cell.

[handwritten margin notes:]
housekeeping genes: express at the same level in all cells

Tissues → ≠ from each other by the expression of a set of genes.

Genes and chromosomes

Genes are part of larger structures called **chromosomes**. In prokaryotes, the chromosome is a circular double-stranded DNA molecule. The prokaryotic chromosome contains the entire complement of DNA—the **genome**—of that species. Genes are positioned along the length of the chromosome, and the position of each gene is fixed and unique. This means that gene order along the chromosome is preserved, a fact that has been of immense practical value in identifying and isolating genes and the proteins that they encode. The full circumference of the chromosome of the bacterium *E. coli* measures about 1.3 mm, a value many times greater than the longest dimension of the bacterial cell. This implies that the chromosome must be tightly coiled and folded in order to fit inside the cell.

[handwritten margin note: gene position on chromosome is fixed & unique.]

In eukaryotes, the genome is partitioned into linear DNA molecules in an arbitrary, but unique, manner for any given species (Table 2.2). The genomes of mammals, for example, all have about the same amount of DNA, but the number of chromosomes into which the DNA is apportioned varies greatly. In general, closely related species tend to have similar numbers of chromosomes, but there are many exceptions. The Indian muntjac deer is a very close relative of the Chinese muntjac deer, but their chromosome number is quite different (6 versus 46, respectively). On the other hand, dolphins and baleen whales, distantly related, both have 44 chromosomes. The entire DNA of a human cell would extend to a length of about 6 meters if it were organized into a single chromosome.

[handwritten margin note: closely related species tend to have similar # of chromosome (exception occur)]

Eukaryotic chromosomes are immensely complex structures, consisting of a long DNA molecule and different kinds of proteins. The DNA-protein structure is precisely coiled and folded in order to fit into the cell nucleus (Fig. 2.8). The lowest order of coiling involves folding the DNA helix around a core of *histone* proteins forming the nucleosome. The nucleosome structure is then subjected to several orders of folding to yield the final structure. The larger eukaryotic chromosomes of plants and animals can be visualized with the aid of specific stains. The stain molecules bind to the DNA and protein components of the chromosomes in a unique and reproducible manner for each chromosome. The differences in the staining pattern, visualized by optical microscopes, can be used to identify the different chromosomes of an organism. The display

[handwritten margin note: protein / DNA coil & fold 1) histone → nucleosome]

Table 2.2
Chromosome Number in Various Species

Species	Diploid number	Species	Diploid number
Baleen whale	44	Human	46
Cat	38	Indian muntjac deer	6
Chimpanzee	48	Mosquito	6
Chinese muntjac deer	46		
Dog	78		
Dolphin	44	Barley	14
Donkey	62	Corn	20
Gorilla	44	Onion	16
Horse	64	Pine	24
Horsefly	12	Rice	24
Horseworm	4	Tobacco	48

24

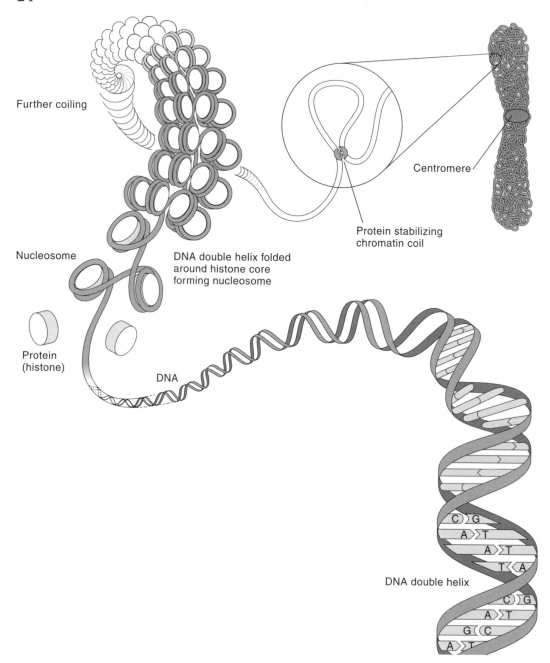

Figure 2.8
Different levels of structure in a eukaryotic chromosome. Each eukaryotic chromosome is a long DNA molecule complexed with many different types of proteins and characterized by several orders of folding and coiling, which greatly reduces its length. At the first level of folding, the DNA helix is wrapped around a histone protein core forming the nucleosome. The sequence of nucleosomes are then subjected to another level of folding. Higher-order folding also takes place. The centromere region of the chromosome is necessary for the partitioning of the chromosome during the mitotic and meiotic divisions.

of the entire set of chromosomes stained in a manner that distinguishes the different chromosomes is known as a **karyotype**. Two human karyotypes are shown in Fig. 2.9.

Genes determine the species. Horses differ from cats because the set of genes that specify the horse are different from those that specify a cat; that is, the horse genome and the cat genome are not identical. The relatedness of species is determined by the pool of genes that they share—that is, by the similarity of their genomes. Human beings are recognizable as such because every human being shares a common pool of genes with every other human being. The other primates, including chimpanzees, gorillas, and orangutans, are our closest relatives because we share more genes with them than with other species. Nevertheless, there are many genes that we share with other, more distantly related species simply because all living organisms share many genes. For example, all species have a gene that encodes the enzyme RNA polymerase. The function of this gene has remained the same throughout evolutionary history, and the nucleotide sequence of the RNA polymerase genes of bacteria and of mammals, while not identical, are very similar.

Species also differ in complexity. A multicellular organism is more complex than a unicellular one. Likewise, among multicellular organisms, some are clearly more complex than others. We might expect that the amount of DNA and, therefore, the number of genes would increase with the complexity of the species. More genes, for example, would be needed to specify a starfish than a protozoan, a vertebrate than an invertebrate, and so on. And, in general, this appears to be the case. We have reasonable estimates for the number of genes required to specify the simplest species, but we do not have accurate estimates for the number required to specify the more complex ones. The genome of the common bacterium *Escherichia coli* contains about 5000 genes. The genomes of most multicellular organisms contain much more DNA than is needed to specify the organism. Indeed, most of the DNA in these organisms does not code for proteins. For example, the human genome contains about 3 billion nucleotide pairs, enough to encode some 3 million genes. However, until recently most estimates suggested that humans are specified by 70,000 to 100,000 genes. The latest results from the Human Genome Project suggest that only 30,000 genes may be required. In other words, only 1 to 2 percent of the human genome may be informational. The precise function of the excess DNA is not known with certainty.

SUMMARY

All living organisms are composed of cells, the basic unit of life. There are two fundamentally different types of cells, prokaryotic and eukaryotic, the latter being defined by the presence of a nucleus. Archaea and Bacteria are the two extant groups of prokaryotic cells, whose appearance on the planet can be dated to over 3 billion years ago. Eukaryotic cells, appearing about 1.5 billion years ago, are the constituent cells of most other living organisms, including fungi, plants, and animals. Despite enormous differences in cell size, morphology, and function, the metabolic and genetic machinery of prokaryotic and eukaryotic cells are essentially the same. All cells synthesize and use the same four classes of molecules, carbohydrates, lipids, proteins, and nucleic acids. Hereditary information is stored in DNA in the form of nucleotide triplets, which specify the amino acids in a protein. A gene is a sequence of nucleotides that encodes a unique protein. Expression of a gene involves a two-step process, transcription and translation, that is fundamentally the same in all cells.

Species are defined by a unique set of genes. Closely related species have more genes in common than distantly related species. Genes are organized into chromosomes. In prokaryotic cells, the entire DNA content characteristic of a species, the ge-

A

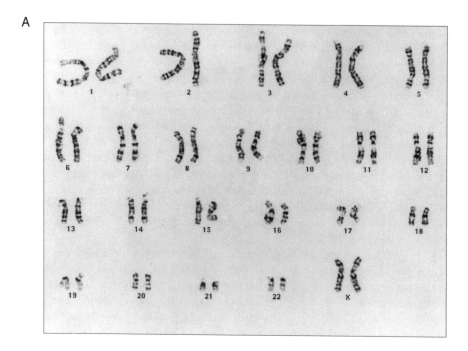

B

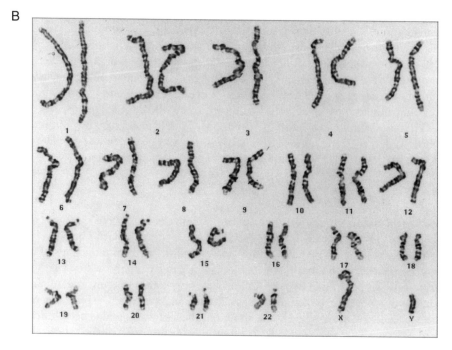

Figure 2.9
The human karyotype. (**A**) Normal female with 2 X chromosomes. (**B**) Normal male with an X and a Y chromosome. The chromosomes are numbered in the conventional manner. The alternating dark and light bands arise from differential binding of histological stains that have been applied to chromosome preparation. The banding pattern serves to distinguish the different chromosomes. Note the difference in size between the X and Y chromosomes. (Courtesy of Dr. O. W. Jones, UCSD School of Medicine.)

nome, consists of a single, circular chromosome, while in eukaryotic cells, the genome is partitioned into a number of linear chromosomes. Eukaryotic chromosomes are complex structures consisting of a DNA molecule complexed with different types of proteins. In general, genome size increases with species complexity. However, in most multicellular organisms, a very large fraction of the genome does not code for proteins. Human DNA, for example, consisting of about 3 billion nucleotide pairs, can encode about 3 million genes, but, current estimates indicate that the human genome carries only 30,000 to 70,000 genes.

SEX AND REPRODUCTION

In our everyday experience, sex and reproduction appear to be two sides of the same coin and we are apt to think of the two as inseparable, or to think that reproduction is impossible without sex. However, sex and reproduction are quite different processes. To appreciate the distinction, let us begin by considering the two fundamental modes of reproduction.

Asexual versus sexual reproduction

During the first 2 to 2.5 billion years of life on the Earth, living organisms reproduced **asexually.** Two types of asexual reproduction are commonly seen. One type is **binary fission,** in which a cell grows and divides in half, partitioning its contents equally (Fig. 2.10). This type of division is commonly seen in many bacteria. The second type, **budding,** is characterized by the "budding off" of one or more cells from the

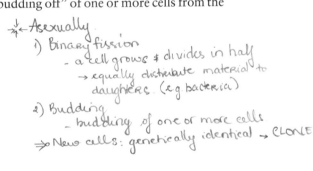

(handwritten margin notes)
‡← Asexually.
1) Binary fission
  – a cell grows & divides in half
    → equally distribute material to daughters. (e.g. bacteria)
2) Budding
  – budding of one or more cells
⇒ New cells: genetically identical → CLONE

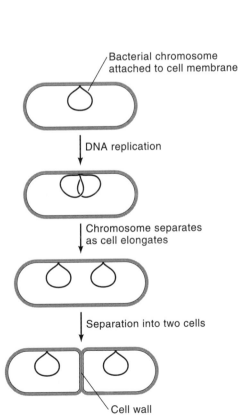

Bacterial chromosome attached to cell membrane

DNA replication

Chromosome separates as cell elongates

Separation into two cells

Cell wall

Figure 2.10
Asexual reproduction by fission in the common bacterium *Escherichia coli.* The circle represents the bacterial chromosome. After replication of the chromosome, the two daughter chromosomes are separated as the cell elongates. The formation of a new cell wall between the elongated cell marks the end of cell division.

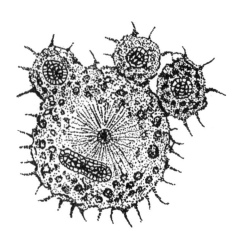

Figure 2.11
Asexual reproduction by budding in the unicellular microscopic species, *Acanthocystis*. Three buds are depicted. (Reprinted from G. W. Corner. 1947. *Hormones in Human Reproduction.* Princeton University Press, Princeton, NJ.)

parent organism (Fig. 2.11). In asexual reproduction, each new cell is genetically identical to the parent cell; hence, the progeny are **clones** of the parent cell. All prokaryotic cells reproduce asexually. Many eukaryotic unicellular and some primitive multicellular species can reproduce by budding.

An alternative, more complex, mode of reproduction, **sexual** reproduction, appeared about 1.5 billion years ago. Its defining feature is that each new organism is the product of the fusion of two cells rather than the result of the division of the parent cell. In sexually reproducing species, every individual has two parents (Fig. 2.12). The offspring are not clones of the parents, but are unique and different from the parents. The appearance of sexual reproduction did not lead to the disappearance of organisms that reproduce asexually. Instead, such organisms continued to proliferate and evolve, and we recognize these as members of the Archaea and Bacteria.

The earliest sexually reproducing species were unicellular eukaryotes. In such species each cell is the unit of organismic as well as cellular reproduction. With the emergence of more complex multicellular organisms (600 million years ago), reproduction of the organism had to be separated from the reproduction of the cells that made up the tissues and organs of the organism. Organismic reproduction became limited to a special subset of cells, the **germ line** cells. The cells of the tissues and organs that carried out the nonreproductive functions of the organism are referred to as **somatic cells.** Continuity of the species depended now on the germ cells. Each germ cell was the repository of information specifying and defining each species. The germ cell was the link between parent and progeny, between one generation and the next.

In time, two morphologically and functionally distinct forms of germ cells, or **gametes** (Greek, "marriage"), appeared, the **ovum** (egg cell) and the **spermatozoon** (sperm cell) in animals and pollen grain in plants (Figs. 2.13 and 2.14). The ovum is a

Figure 2.12
Simplified diagram of reproduction in most animal species. There are two sexes, each producing a specialized gamete, the egg cell by the female and the sperm cell by the male. Egg-sperm fusion generates the fertilized egg, or zygote. The zygote develops into either a male or a female, depending on the sex-determining system of the species.

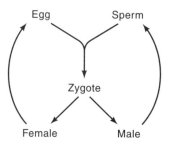

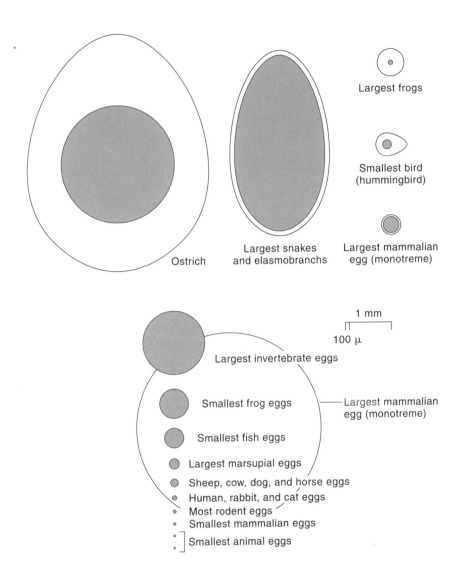

Figure 2.13
Comparison of eggs from different species. (**Top**) Relative sizes of the largest extant eggs. The largest known egg (9.5 inches long) is from an extinct flightless bird, *Aepyoris*. The most primitive mammalian species, the monotremes, produce the largest eggs. These, however, are small compared to avian eggs. (**Bottom**) Comparison of eggs from a variety of other species. Note the scale. The largest mammalian egg, from the monotremes, is indicated by the large circle. There are two other orders of mammals, the marsupials and eutherians (placental mammals), with the eutherians being the most recently evolved. Note that the largest marsupial eggs are larger than the largest placental mammalian eggs. The human egg is among the smallest of eggs. (Adapted from C. R. Austin. 1965. *Fertilization*. Prentice-Hall, Englewood Cliffs, NJ, with permission of Prentice-Hall.)

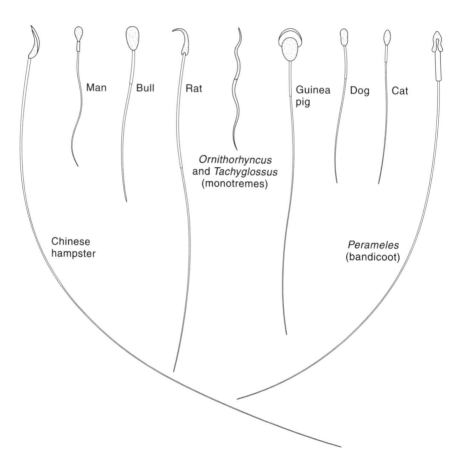

Figure 2.14
Diagrams of spermatozoa from mammalian and nonmammalian species. Again note that the human spermatozoon does not stand out in any particular way. (Adapted from C. R. Austin. 1965. *Fertilization.* Prentice-Hall, Englewood Cliffs, NJ, with permission of Prentice-Hall.)

reservoir of nutritional and informational molecules necessary for the development of the new organism. Egg size varies greatly, generally depending on the extent to which the development of the new organism is dependent on the mother (compare the egg sizes in mammals and birds). In mammals, the developing offspring is completely dependent on the mother, and receives its nutrients from her body through the placenta. In birds, on the other hand, the new organism is entirely independent of the mother and completely dependent on the nutritional content of the egg to support its development. Thus, avian eggs are larger than mammalian eggs. In animals, the sperm cell has to seek out the ovum and therefore has to be motile; that is, it has to have the ability to move on its own. Here, also, there are large differences in sperm size and form (Fig. 2.14).

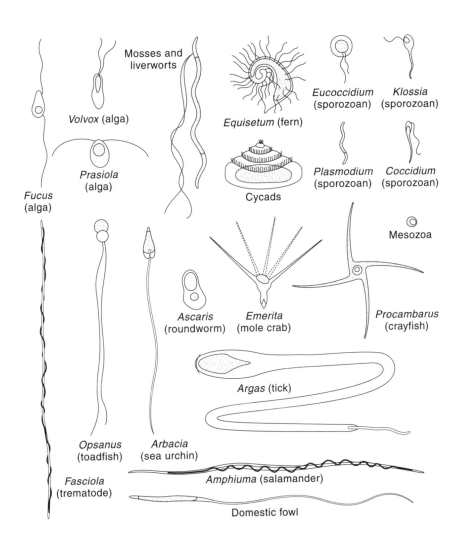

Mosses and liverworts

Volvox (alga)

Prasiola (alga)

Fucus (alga)

Equisetum (fern)

Cycads

Eucoccidium (sporozoan)

Klossia (sporozoan)

Plasmodium (sporozoan)

Coccidium (sporozoan)

Mesozoa

Ascaris (roundworm)

Emerita (mole crab)

Procambarus (crayfish)

Argas (tick)

Opsanus (toadfish)

Arbacia (sea urchin)

Fasciola (trematode)

Amphiuma (salamander)

Domestic fowl

Sex versus gender

Early sexually reproducing organisms did not exhibit the characteristic of "femaleness" or "maleness" that we find in animals. Such species had sex, but lacked *gender*. Gender, the distinction based on the type of gamete, egg or sperm, that an organism produces, emerged with the development of specialized tissues for the formation of the gametes. In animals, this tissue was the **gonad**. Eggs were produced in the **ovary**, while sperm were produced in the **testis**. In some animal species both types of gonads occur in the same individual. Such species are said to be **monosexual**, or **hermaphroditic** (from Hermes, the Greek god of war, and Aphrodite, the goddess of love). The

(margin note:) • Bisexual species
{ + ♀ : eggs
{ + ♂ : sperm

male-female distinction began to be made when animals carrying only one type of gonad appeared. Thus, in **bisexual** animal species, the eggs are produced by the female, and the sperm by the male. Perhaps because most familiar animals are bisexual, we accept the separation into two, fixed sexes, or genders, as a fundamental and unquestionable aspect of life. This circumstance is reinforced vividly—for example, in the book of Genesis recounting the creation of Adam and Eve, and in the story of Noah and the flood, and in all subsequent poetic imagination.

The ubiquity of the two-sex system in animals obscures the fact that other sexual systems exist. To appreciate the full complexity of sexuality we need to look at the world of the protists, whose sexual complexity is immense compared to that of plants or animals. The sexual practices of only relatively few of the thousands of extant species have been examined. A primitive form of multisexuality is found among some slime mold and fungal species. Individual organisms in these species are characterized by a genetically determined *mating type*, but are otherwise indistinguishable. Fusion (mating) of two organisms will take place only if they are of different mating type. The number of mating types is variable. In one species of slime mold, for example, 13 different mating types have been identified. Although such mating types are not sexes (genders) in the way we now understand that term, the essence of sexual reproduction is clearly present. Nevertheless, the multiple mating type possibility apparently never took hold and did not evolve, as far as we know, into multigender animal sexuality. What we can infer is that over long spans of evolutionary time, alternatives to the present ubiquitous two-sex (gender) system have appeared, but a full understanding of why such alternatives have not prevailed among animal species remains elusive.

(margin note:) Some sexually reproducing species:
→ sex is affected by temperature during embryotic development or individual's age.

In many animals, the two sexes (genders) are fixed: once a male (female), always a male (female). This need not be the case, however. Many examples of flexible or unstable sexes are seen today in some reptiles (alligators) and fishes in which the sex of the individual depends on the temperature at which embryonic development takes place. In other cases, individuals change from one sex to the other as they age. However, these cases represent only a small fraction of the total number of sexually reproducing species.

The chromosome cycle in sexual reproduction

Mitosis

Let us consider further the implications of sexual reproduction using the human life cycle as our example. Every individual begins life as a fertilized egg, the **zygote**, formed by the fusion of the egg from the female and the spermatozoon from the male. The zygote is a special type of cell because it has the ability, through a long series of divisions, to form all of the tissues of the body, the **soma**, including the heart, skeleton, and brain. A human zygote and the somatic cells that develop from it all contain 46 chromosomes (Fig. 2.15). Starting with the zygote, all the cell divisions that form the body preserve the chromosome number. Since the chromosomes carry the genes, every cell division must ensure that each daughter cell carries exactly the same number of chromosomes as the mother cell. **Mitosis** is the name given to the type of cell division that characterizes the development of the soma (Fig. 2.16). In mitosis each chromosome is replicated, and then one of each pair is partitioned to each of the two progeny cells.

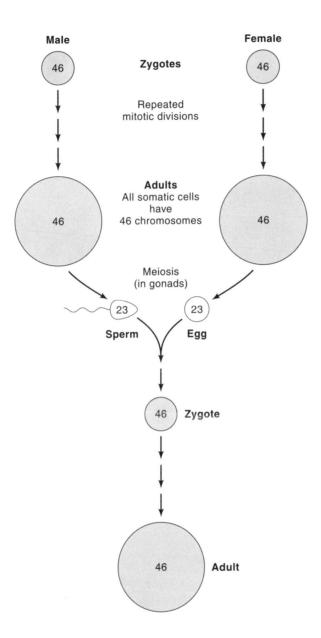

Figure 2.15
Chromosome cycle in humans. Alternation between the diploid state ($2n = 46$), the somatic number, and the haploid state ($1n = 23$), the gametic number.

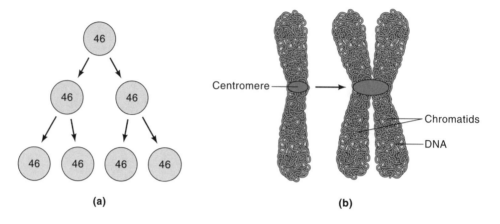

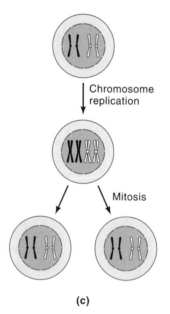

Figure 2.16
Simplified rendering of mitosis. (**a**) Mitosis preserves
the chromosome number: the two daughter cells have
the same number of chromosomes as the mother cell.
(**b**) Before each mitotic division, each chromosome is
replicated to form the characteristic "cross" structure,
consisting of two chromatids held together at the
centromere region. (**c**) In mitosis, the two chromatids
of each replicated chromosome separate in such
a way that each daughter cell receives one copy of
each chromatid. Each daughter cell will have the
same number and kinds of chromosomes as the
mother cell.

Meiosis

If mitosis were the only type of cell division, then the fusion of gametes to generate
a new individual would double the number of chromosomes at every fertilization. To
ensure that the human chromosome number is maintained at 46, there must be another
type of cell division that ensures that the chromosome number in the gametes is exactly
half that of somatic cells (Fig. 2.17). **Meiosis** is the name given to that division. Consider meiosis in humans. The 46 chromosomes of human somatic cells each exist as a
member of a pair. The 46 chromosomes are therefore 23 distinct pairs of chromosomes. The members of a given pair are said to be **homologous** because each carries the
same complement of genes. Somatic cells are said to have the **diploid** ($2n$) number of
chromosomes. Since every chromosome is paired, every gene is also present twice. In

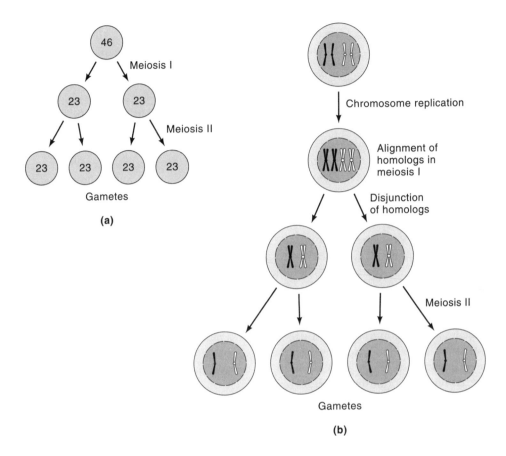

Figure 2.17
Meiosis. (a) Meiosis halves the somatic chromosome number so that each gamete receives the haploid number. Before meiosis begins, each chromosome replicates as in mitosis. (b) In the first meiotic division, homologous chromosomes exchange segments (not indicated in the diagram), and subsequently separate from each other (disjoin) forming two haploid cells. The second meiotic division is equivalent to a mitotic division in that the chromatids of the replicated chromosomes separate from each other. The consequence is that each of the four cells from the second meiotic division contains one member of each homologous pair of chromosomes. In general, there are four cells formed as a result of meiosis.

the formation of the gametes (**gametogenesis**), meiosis ensures that each gamete receives exactly one member of each homologous pair of chromosomes (Fig. 2.17). For example, each spermatozoon and each egg will contain one copy of chromosome 1, one copy of chromosome 2, and so on. Gametes, then, have one-half the diploid number of chromosomes—the **haploid** ($1n$) number. When spermatozoon and egg fuse to form the zygote, the pairs are restored. Therefore, one member of each homologous pair of chromosomes in each cell comes from the father (paternally derived), and the other from the mother (maternally derived).

Meiosis occurs in two stages, **meiosis I** and **meiosis II**. Two important events take place in meiosis. First, during the early phase, homologous chromosomes pair along their length and exchange DNA segments at random, a process known as **recombination**, or **crossing-over**. A consequence of recombination is that each homolog carries a unique mixture of paternally and maternally derived information. Second, after recombining, homologs **disjoin** from each other, and are partitioned into each of two daughter cells as meiosis I is completed. Meiosis I is referred to as the **reductional** stage of meiosis because the chromosome number is reduced from 46 to 23. In males, for example, each cell formed by meiosis I undergoes meiosis. The final result is four cells, each of which carries the gametic (haploid) number of chromosomes (Fig. 2.17). In females, however, only one of the daughter cells of meiosis is functional and will become the egg; the nonfunctional cell or cells are referred to as **polar bodies**. Polar bodies eventually degenerate. The reason for the sex difference in the meiotic divisions will become apparent in Chapter 5. Meiosis, therefore, has two fundamentally important consequences: first, it ensures that each gamete carries one member of each homologous pair of chromosomes; second, because of the recombination between homologs, each gamete carries a unique combination of paternally and maternally derived genes. Hence, for example, each gamete produced by the female carries a unique combination of genes from the female's mother and father. The male gamete will likewise carry a unique combination of genes from the male's mother and father.

In all multicellular organisms, meiosis occurs only in the germ line cells. In the female, the formation of eggs, or **oogenesis**, takes place in the ovaries. In the male, the formation of sperm, or **spermatogenesis**, takes place in the testes. When a spermatozoon fertilizes an egg to form a zygote, the diploid chromosome number is restored. Sexual reproduction, therefore, involves a cycle of chromosome doubling (fertilization) and chromosome reduction (gametogenesis). Meiosis contributes to the variability in the progeny that is characteristic of sexually reproducing species.

Meiosis contributes to the variability in the progeny → characteristic of sexually reproducing species

Sex chromosomes

Sex chromosome → sex determination (not all species)

Chromosomes that are found in one sex but not the other were first discovered around the end of the nineteenth century. These were later named sex chromosomes because they were considered to be involved in sex determination. However, it has turned out that not all kinds of organisms use a sex chromosome system to determine sex. For many species the sex-determining mechanism is unknown. Humans, in common with all other mammals, are characterized by an XX/XY sex chromosome system. The 23 pairs of human chromosomes are divided into 22 pairs of **autosomes** and a single pair of **sex chromosomes** (Fig. 2.9). Females carry two X chromosomes, while males carry one X chromosome and one Y chromosome. Genes carried on the autosomes are referred to as **autosomal genes**, while genes on the sex chromosomes are said to be **sex-linked**. Genes on the X chromosome are X-linked, while genes on the Y chromosome are Y-linked. In meiosis in the female, each egg receives 22 autosomes and one X chromosome. Males, on the other hand, produce two types of spermatozoa in equal proportions: one type carries 22 autosomes and one X chromosome, and the other type carries 22 autosomes and one Y chromosome.

Human $X \approx 3Y$ chromosome in length

X chromosome ~ 3000–4000 genes

Y chromosome ~ 1000–2000 genes

X & Y are not homologous

The human X and Y chromosomes are morphologically distinct. The human X chromosome is two to three times longer than the Y, and contains many more genes. Current estimates suggest that the X chromosome contains about 3000 to 4000 genes, but at present only about 10 percent of X-linked genes are known (a small sample of these is shown in Fig. 2.18). Based on the relative lengths of the X and the Y chromo-

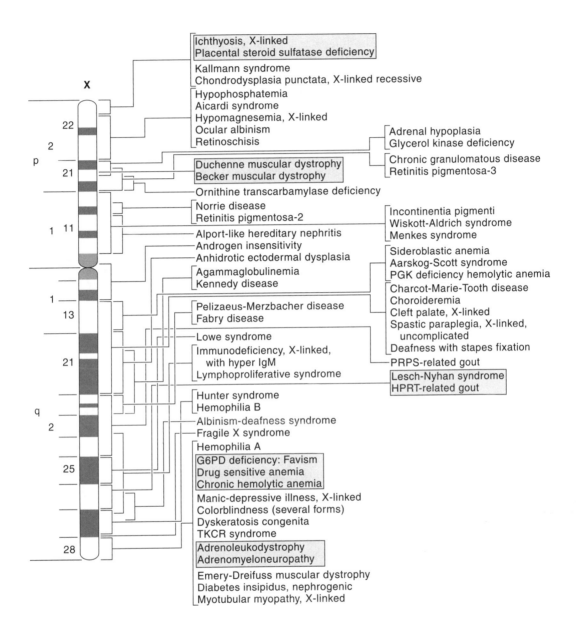

Figure 2.18
Schematic diagram of the X chromosome indicating the relative positions of a few X-linked genes. At present, over 300 genes have been localized to the X chromosome. Mutations in these genes lead to disease states or disorders of varying degrees of severity. (Adapted from V. McKusick. 1990. *Mendelian Inheritance in Man*, pp. cix. The Johns Hopkins University Press, Baltimore, MD.)

somes, we would expect the *Y* chromosome to carry at least 1000 to 2000 genes. However, the *Y* chromosome carries very few, perhaps fewer than 30 genes (Fig. 2.19). It appears that most of the DNA on the *Y* chromosome is nonfunctional and does not encode any proteins. The difference in size and gene content between the *X* and *Y* chromosomes means that they are not a homologous pair. Comparison of the *X* and *Y* chromosomes in the three classes of mammals—the **monotremes** (the most primitive), the **marsupials**, and the **eutherians** (placental)—suggests that the original mammalian *X* and *Y* were probably homologous. Since the origin of mammals, the *Y* chromosome has been losing genes. In Chapter 6 we will review the current hypothesis for the loss of informational DNA on the *Y* chromosome. The long-term future of the *Y* chromosome is an interesting subject for conjecture.

SRY gene → necessary for male development → form testis

The *Y* chromosome, despite its paltry gene content, is still very important, however, since it is essential for male development and male fertility. Maleness, in essence, means having testes for gonads. Working out the precise role of the *Y* chromosome in the development of the testes has turned out to be an extraordinarily difficult problem, and at present the most important clue we have is that only one gene on the *Y*—named *SRY*—is necessary for male development (Fig. 2.19). *SRY* is necessary for testes to form, but its precise function remains elusive. Some of the other genes on the *Y* chromosome are required for spermatogenesis and are therefore necessary for fertility in the male, but they are not necessary for any of the other attributes of being male. We shall consider the role of *SRY* in more detail in Chapter 4.

Why sex? or what use is sex?

We accept sex as part of the natural order of things. Since in our own lives and in the animal world around us we associate sex with reproduction, we may be surprised by the question itself, perhaps thinking that the answer to the question is self-evident. However, we can begin to appreciate the significance of the question by considering reproduction at the cellular level. At this level, sex and reproduction are direct opposites. Reproduction is the division of one cell to form two, while sex usually means the fusion of two cells to form one. Indeed, as we saw in the previous sections, sex is not necessary for reproduction. Bacteria reproduce very successfully without sex, and there are many multicellular organisms that can reproduce asexually as well. Sex may be necessary for animal reproduction, but animals represent only a small fraction of extant species. In a way, animals obscure the distinction between sex and reproduction. We can safely conclude, therefore, that in a broad biological sense, reproduction must not be the "function" of sex.

At cellular level
1) Reproduction: 1 cell → 2 cells
2) Sex: 2 cells fuse → 1 cell
→ Sex is not necessary for reproduction
→ reproduction (in broad biol sense): not the function of sex
indistinct

What, then, is the function or significance of sex? This is really two questions: "Why did sex come into being?" and "What maintains sex, i.e., why is it so prevalent?" We do not have a definitive answer to either question. Indeed, biologists have struggled with both questions for over a century. We cannot consider in this text the full complexity of the issues that these two questions raise, but perhaps we can point out why the answers to them remain elusive. A convenient starting point is to understand what "sex" and "function" mean in the biological sense. First, although there is no standard biological definition of sex, the possible definitions are all variations on a common theme. A useful one, for example, is to say that sex includes all processes whereby genetic material from different ancestors is brought together in a single descendant. This definition retains the common theme that sex is a "mixing" process, a device for bringing together different DNA molecules and fashioning a new set of molecules. This idea

Common def: sex is a mixing process to bring together ≠ DNA molecules & fashioning a new set of mol.

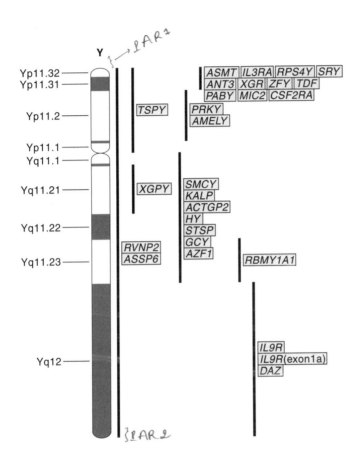

Figure 2.19
Schematic diagram of the Y chromosome indicating the distribution of its genes. The genes on the Y appear to be clustered in certain regions, rather than distributed uniformly along the length. The gene responsible for formation of the testis, *SRY*, is located near the tip of the short arm of the Y chromosome. Large regions of the Y chromosome do not carry any genetic information; that is, they do not encode proteins. (Adapted from GDB Mapview, http://www.gdb.org.)

PAR 1 & 2: homologus to X chromosome
↳ recombine w/ X chromosome,
cross over & exchange genetic material
↳ suggest that in the past Y & X chromosome might be very simmilar

is a very general one, and applies to all living organisms, even the asexually reproducing ones. Bacteria, for example, can engage in a primitive form of sex. They can take up fragments of DNA from other bacteria, from viruses, or even from their surroundings and incorporate them into their own chromosome. Many protists engulf other protists and incorporate some of the DNA of the engulfed cell into their own DNA. These examples once again show that sex and reproduction are separate processes.

Function: the advantage of something to contribute to the survival of individual organisms

The question of "function" is addressed in the way that evolutionary biology tries to deal with such questions. For example, the evolution of an organ is explained in terms of the advantage that it contributes to the survival of individual organisms. The liver, the heart, and the brain evolved because each conferred a survival advantage on the organism that has one. The placenta, or the pair-bonding between a female and a male, evolved because each contributes to the reproductive success of individuals and therefore, the species.

We can also ask the question *To whose advantage is sex?* Consider what happens in sexual reproduction. A female will only transmit 50 percent of her genes to each of her progeny. The other half comes from a male who, in most species, does not contribute resources to the offspring. A female who reproduced asexually would transmit 100 percent of her genes to her progeny. Sex, according to evolutionary biologists, introduces a "cost," or disadvantage, of 50 percent. Despite this great disadvantage, sex has been phenomenally successful. Since its appearance, sexually reproducing species have proliferated, so that now all animals and almost all species of plants reproduce sexually. The paradox of sex is that it does not appear to confer any advantage to the reproductive success of the individual. Yet the obvious success of sexual reproduction indicates that despite the clear disadvantages that it entails compared to asexual reproduction, it must possess some advantage. Sex in some way must be able to compensate for the 50 percent disadvantage.

Several hypotheses have been proposed to explain the paradox. One suggests that sex provides an advantage to the population as a whole, rather than the individual. By providing the advantage to a sexually reproducing group, sex overcomes the 50 percent disadvantage with respect to asexually reproducing species. The fact that there are relatively few asexual multicellular organisms has been used as an argument in support of this explanation.

sex → genetic variation → unique individual

Another hypothesis suggests that sex ensures genetic variation in any natural population. Since every gamete is unique, every fusion of gametes generates a genetically unique individual. According to this model, sex provides an efficient way for species to adapt to new environments and to change with time. However, many scholarly studies have shown that genetic variation can be found in abundance in asexual organisms, and alternatively, great uniformity can be observed in sexual organisms. It is not clear that asexual organisms are really inferior in terms of their survival rate or potential for evolutionary change.

The origin of sex is another difficult question. Sex may have arisen as a survival mechanism, perhaps as a protection against the harsh conditions of the planet when cellular life first arose. The ability to incorporate fragments of DNA from various sources, from other cells, or from the debris of dead cells in the environment may have been essential for survival in the earliest organisms. Perhaps later, sex provided protection against parasites. "Parasites" is used here in a very general sense; for example, viruses infecting a bacterial cell are parasites of that cell.

All such explanations are plausible, and a certain amount of evidence can be marshaled in their support. At the same time, however, these explanations are considered to be incomplete. It may be, as one theorist has suggested, that sex facilitates evolution only indirectly, by making extinction less likely, rather than by stimulating rapid adaptation. Evolutionary biologists cannot agree on the role of sex in evolutionary change, and its "purpose" remains as elusive and mysterious as ever.

SUMMARY

Sex and reproduction have become so inextricably interwoven in multicellular organisms that we tend to lose sight of the fact that they are distinct processes. Sex, in its broadest sense, is a mixing or fusion of DNA molecules from different organisms. Bacteria, which reproduce asexually, display a rudimentary form of sex in their ability to take up and incorporate DNA from other sources into their own genomes. Sex and reproduction began to be linked with the evolution of sexually reproducing organisms. In sexually reproducing species, each organism is the product of the fusion of two specialized cell types. The earliest sexually reproducing organisms were unicellular. Later, with the evolution of multicellularity, two fundamentally distinct types of cells appeared: somatic and germ cells. Somatic cells made up the tissues of the body, while in animals, germ cells gave rise to the gametes, the egg and the sperm. Specialization of egg or sperm production gave rise to the ubiquitous two-gender system familiar to us today.

The linkage of sex and reproduction is associated with the appearance of eukaryotic organisms and the cellular divisions known as mitosis and meiosis. Mitosis and meiosis are essentially mechanisms for partitioning the parent cell's chromosomes into the progeny cells. Mitosis preserves the chromosome number of the parent cell. In multicellular organisms, mitosis is the division by which all of the somatic tissues of the body are generated. The chromosome number in a somatic cell is referred to as the diploid number because chromosomes occur in pairs. In humans the diploid number is 46, consisting of 22 homologous pairs and a pair of sex chromosomes—XX in the female and XY in the male. Meiosis is the cell division that yields gametes. In meiosis, the diploid chromosome number of somatic cells is halved so that each gamete receives one member of each homologous chromosome pair. The female gamete receives in addition an X chromosome, while the male gamete receives either an X or a Y chromosome. Each gamete contains the haploid number of chromosomes. Meiosis takes place only in germ line cells, which are only found in specialized tissues. In animals, these are the gonads, the ovaries and testes. The ovaries produce the female gamete, the egg, while the testes produce the male gamete, the spermatozoon.

The origin and evolutionary significance of sex are profoundly interesting, unresolved questions. Sexual reproduction imposes an evolutionary "cost" due to the fact that the female transmits only 50 percent of her genes to her offspring rather than the 100 percent she would reproduce asexually. Evolutionary theorists have argued that this cost would imply that asexual reproduction would be favored over sexual reproduction. However, the opposite is true. A number of explanations for this circumstance have been proposed, but there is no general agreement regarding the function or significance of sex.

MUTATION

Nature of mutations

·Mutation: random
changes in the
nucleotide seq.
in DNA
↳ gene
↳ chromosome

Mutations are random changes in the nucleotide sequence in DNA. Many arise from single nucleotide substitutions, while others are additions or deletions of a small number of nucleotides. Many authors also include as mutations large-scale changes in a chromosome or the genome, such as changes in chromosome number, large-scale loss or gain of DNA, or reorganization of large segments of DNA within a chromosome. The small- and large-scale changes can be referred to as **gene** and **chromosome mutations**, respectively. The mechanisms that give rise to chromosome mutations are different from those that produce gene mutations, and we will consider some of those differences in Chapters 13 and 14. Our focus here is on gene mutations and the agents that produce them.

Importance of mutation
i) Mostly harmful
→ disease or disorder
⇒ state of health/
well–being.

2) Evolutionary
source → fraction
of beneficial effect

Mutations are important for two very different reasons. At the individual level, mutations, if they have any effect, are almost always harmful. Most changes in the nucleotide sequence within a gene will change the amino acid sequence in the protein encoded by the gene. The mutant protein, if it is produced at all, may be nonfunctional, or it may retain only some fraction of its normal functionality. The consequence is that some important cellular function will be perturbed. Some mutations have major effects and are recognized because they result in a significant disorder or disease state. Consider, for example, the disorders listed in Fig. 2.18 that are linked to mutations in *X*-linked genes. A mutation in the hemophilia *A* or *B* genes results in a blood-clotting disorder, while a mutation in the Duchenne muscular dystrophy gene results in a severe muscular degenerative disease. Other mutations may have effects that are not easily or distinctly recognized, but which may nevertheless affect overall health and well-being in subtle ways. Hence, mutations that have minor effects may also have a negative impact on the life of an individual, but one that may not be easily measurable.

On the other hand, mutations are indispensable to evolutionary change because they are the ultimate source of genetic variability. Only a very small fraction of mutations are likely to have a beneficial effect in an evolutionary sense, but without them, evolution would be impossible. Evolutionary change, then, occurs at the expense of individuals, since many mutations are deleterious to the individuals that carry them.

Types of mutations

Different mutations in a gene result in different forms of that gene. These alternative forms are known as **alleles**. The normal, or *wild-type*, form (allele) of a gene is the one that encodes a protein with normal function and is found in most members of the population. Depending on the site of the mutation, some mutant alleles have more severe effects than other mutant alleles. Since genes occur in pairs in sexually reproducing organisms, the effects of a mutant **allele** on one chromosome may be hidden by a wild-type allele on the other homolog. A mutation that leads to loss of function in one allele can generally be masked by the normal allele. Such a mutation is referred to as a **recessive mutation**, and will manifest itself only if the mutant allele is carried on both homologs. This will be true for autosomal genes, but not for sex-linked genes. Since males have only one *X* chromosome, mutations in *X*-linked genes will be expressed immediately in males. In other instances, a mutation may lead to the synthesis of a mutant protein that distorts or interferes with the function of the normal protein. Such a mutation will manifest itself in single copy, and is known as a **dominant mutation**. Dominant mutations tend to have major effects.

Mutations can be somatic or germinal. **Somatic mutations** arise in somatic cells and may be harmful to the individual in which they occur, for example, by leading to a severe disease. They are self-limiting in that the mutation is not transmitted to the offspring. **Germinal mutations**, on the other hand, occur in the germ cells, and hence, they can be transmitted through the gametes to the offspring. Germinal mutations are **heritable**, and their effects will be felt over many generations. It is for this reason that germinal mutations are of special concern, and why understanding how mutations are produced is particularly important to human welfare.

Mutagens and the spontaneous germinal mutation rate, μ

Mutations arise from damage to DNA, and agents that can damage DNA are known as **mutagens**. There are two broad categories of mutagens: physical agents that introduce breaks or other types of modifications in the DNA sequence, and chemical agents that modify DNA nucleotides by reacting chemically with them. Physical agents include **nonionizing radiation**, such as the ultraviolet (UV) radiation from the sun, and the more penetrating **ionizing** radiation. UV radiation is not energetic enough to penetrate through clothing and many layers of tissue to reach the gonads and, therefore, will not give rise to germinal mutations. UV radiation, however, can be quite harmful in a somatic sense. Short-term, unprotected exposure to the sun may result in a bad sunburn due to the massive destruction of skin cells by the UV radiation. Chronic unprotected exposure to the sun can also result in the production of mutations in skin cells. Exposure to UV radiation has been shown to be the most important risk factor in the development of skin cancers.

Ionizing radiation, being 1,000 times or more energetic than UV radiation, can penetrate into the gonads and, therefore, can produce germinal mutations. The number of mutations generated generally depends on (1) the dose of radiation absorbed by the gonads, and (2) whether the dose is acute (received over a short time) or chronic (received over an extended period of time). Acute doses are more harmful than chronic doses. The decay of radioactive elements, primarily radon gas, other radioactive elements in the soil and in the body, and cosmic rays are the major sources of ionizing radiation. Diagnostic X-rays, exposure to radiopharmaceuticals, and consumer products also contribute to the overall exposure (Table 2.3). These different types of radiation make up the background radiation to which every individual on the planet is continually exposed.

Chemical mutagens can be divided into two broad categories: (1) *exogenous agents*, including environmental toxins, such as pesticides, industrial chemicals, household products, cosmetics, compounds that we take in with our food, as well as tobacco smoke, and (2) the more recently identified *endogenous agents*, which are highly reactive compounds produced as by-products of normal metabolic reactions. These reactive compounds are known as **free radicals**.

Most mutations arise spontaneously from normal exposures to physical and chemical agents during the course of our lives, rather than from deliberate exposure to specific mutagenic agents. Hence, the important question is *What is the relative contribution of these different agents to the overall number of germinal mutations?* The standard parameter used to quantify germinal mutations is the *spontaneous mutation rate*,

Table 2.3
Annual Exposure of Human Beings in the United States to Ionizing Radiation

Source	Dose (in rems)
Natural radiation	
Radon gas	0.206
Cosmic rays	0.027
Natural radioisotopes in the body	0.039
Natural radioisotopes in the soil	0.028
Other radiation sources	
Diagnostic X-rays	0.039
Radiopharmaceuticals	0.014
Consumer products (TVs, clocks, building materials)	0.010
Fallout from weapon tests	<0.001
Nuclear power plants	<0.001
Total from all sources	0.363

Reference: National Research Council, Committee on the Biological Effects of Ionizing Radiations, *Health Effects of Exposure to Low Levels of Ionizing Radiation* (BEIR V), National Academy Press (1990).

denoted by the Greek letter μ, and which can be defined in two ways. First, μ can be defined as the number of mutations that occur spontaneously per gene per generation (a human generation being 30 years). We can represent this as $\mu(g)$. In practice, it is very difficult to measure $\mu(g)$ accurately in humans. Alternatively, μ can be defined as the number of mutations per nucleotide per generation, $\mu(n)$, which is easier to measure. From animal studies, current estimates suggest that $\mu(n)$ is between 1×10^{-8} and 2×10^{-8}. Given that the human genome contains 3×10^9 nucleotide pairs, we can calculate the expected number of new mutations, $\mu(n)$, in a human fertilized egg as follows. Consider first the lower estimate. Since the fertilized egg is the fusion of two gametes, the expected number of new mutations would be $2 \times 1 \times 10^{-8} \times 3 \times 10^9 = 60$. The higher estimate for $\mu(n)$ would lead to a value of 120. Therefore, between 60 and 120 new mutations would be expected to be present in each fertilized egg. Since only about 2 percent of human DNA is informational, we expect between $(.02)(60) = 1.2$ and $(.02)(120) = 2.4$ new mutations in genes per generation. This means that every individual born would be expected to carry 1 to 3 new mutations in genes that encode proteins. Is there any evidence to support this value? Two recent studies support this estimate. One study, which examined mutations in 46 genes, obtained a result of 4.2 mutations in genes per individual born, 1.6 of which would be deleterious and would be expected to result in a significant disorder. These mutations would be recessive, and their effects would be hidden by the normal allele on the other homolog. The second study, based on an exhaustive analysis of mutations in the hemophilia B gene, found 128 new mutations per zygote, and 1.3 deleterious mutations per human zygote. The estimates for the mutation rate from both studies agree substantially.

Two important questions are raised by these numbers. First, how can such high values for µ be sustained without leading to our extinction? From a theoretical point of view, such high mutations rates would, under normal conditions, have led to our extinction long ago. The paradox, however, is that we are not extinct. How do we reconcile the two? We do not have a clear answer as yet. Second, what are the agents responsible for this high mutation rate? The current view is that background radiation and environmental chemicals do not contribute significantly to the spontaneous mutation rate. In contrast, most germinal mutations are considered to arise from DNA damage due to endogenous compounds produced in normal metabolism. Let us summarize the evidence that supports this view.

The contribution of background radiation to µ

The question of how much background radiation contributes to the spontaneous mutation rate can be rephrased in the following way: *What dose of ionizing radiation will generate the same number of mutations as appear spontaneously?* This dose is known as the **doubling dose**. For example, if *n* mutations appear spontaneously, exposure to a doubling dose of a mutagen will also produce *n* mutations, making a total of 2*n* mutations. Consider, for example, the data shown in Table 2.3. Measured in the unit known as the *rem*, the estimate for the average annual exposure to background radiation in the United States is 0.363 rem. Over a generation (30 years), the average individual will receive a dose of $30 \times 0.363 = 10.89$ rems. The important question then is *How does this number compare to the doubling dose?*

Doubling dose?

The doubling dose is difficult to measure in humans. Ethical considerations do not permit experiments with humans that would give us estimates for the doubling dose. Initial estimates of the doubling dose came from animal experiments, but it was never clear how applicable these estimates were to humans. The most recent and most reliable estimates have come from the results of an unplanned human experiment: the atomic bombing of Nagasaki and Hiroshima that brought World War II to a close. After the end of the war, two international commissions were established to monitor the effects of radiation not only on the survivors of the explosions, but on their children as well. The most recent report summarized analyses of 7,986 children of survivors. The children, monitored for over 25 years since birth, have been screened using 8 different indicators of genetic disease. The surprising finding was that there was no evidence of increased genetic disease among the children of survivors compared to controls. This analysis also provided a new estimate for the doubling dose: 170–220 rems for acute doses, and 340–450 rems for chronic doses. Since the generational exposure to background radiation is 10.89 rems, much less than the doubling dose, we can conclude that background radiation plays only a minor role in the spontaneous mutation rate. It is important that this conclusion be understood correctly. It does not mean that we should not worry about exposures to ionizing radiation, for example, from diagnostic X-rays. Ionizing radiation is always mutagenic, and care should always be taken to limit exposure as much as possible. The conclusion indicates only that background radiation does not contribute significantly to the germinal mutation rate. Similar analysis indicates that background radiation is a minor contributor to the somatic mutation rate.

The contribution of chemical agents to µ

Most lines of evidence indicate that gene mutations are due to the effects of endogenous oxidizing compounds (free radicals) produced in cells in abundance during normal metabolic reactions. These compounds generate the primary DNA lesion (for example, a chemically modified DNA nucleotide). The higher the metabolic rate of the cell, the greater the number of free radicals formed, and the greater the number of lesions produced. Endogenous DNA damage is quite high, estimated to be about 10,000 lesions per cell per day in humans. Hence, normal metabolism can be thought of as generating a high background load of DNA lesions.

The cell has two protective mechanisms to limit the oxidative damage. First, the cell can make use of **antioxidants** to neutralize or eliminate free radicals. These antioxidants include endogenously produced compounds and compounds contributed by the diet, such as vitamins C, D, and E. Second, DNA lesions produced by oxidizing compounds are repaired by a very efficient DNA repair system that removes damaged DNA nucleotides. The repair system, however, is not 100 percent efficient, and some DNA lesions are not repaired. Unrepaired DNA lesions are converted to mutations when the cell divides. Hence, the number of mutations generated depends on the number of primary DNA lesions produced, the level of antioxidant activity, and the effectiveness of the DNA repair system.

Hundreds of environmental chemical agents with mutagenic potential have been identified. Many of these are industrial compounds used in manufacturing, agriculture, and cleaning industries; in principle these compounds could contribute to the high background of lesions produced by endogenous compounds. However, most experts consider that exposure of the population at large to these types of environmental agents is low, much lower than the dose levels that are generally used in animal studies to demonstrate a mutagenic effect. Moreover, to contribute to μ these agents must be able to get into the germ cells at a dosage high enough to contribute to the oxidative damage of DNA. Because gonadal exposures to most of these compounds are considered to be low, their contribution to μ is also considered to be small. Higher-level exposures to some of these agents may be job-related—for example, workers in the plants that manufacture industrial chemicals or farmworkers exposed to pesticides. At the present time, then, despite the fact that there is widespread apprehension about them, there is no compelling evidence that man-made environmental chemicals contribute significantly to the germinal mutation rate. This does not mean, however, that their contribution will remain small in the future.

Currently, smoking, an unbalanced diet, and chronic inflammation and infections are considered to be the most important agents contributing to somatic mutations. Smoke-induced DNA damage is clearly associated with the origin of many cancers, especially lung cancer, but also cancers of the mouth, esophagus, stomach, kidney, pancreas, and bladder. An unbalanced diet is thought to account for about one-third of cancers in the United States. Many epidemiological studies are beginning to show that the consumption of specific antioxidants and other compounds obtained from fruits and vegetables is associated with reduced risk of cancer, heart disease, and other degenerative disorders. This protective role is attributed to the effects of such compounds in minimizing oxidative DNA damage. Chronic inflammation and infections elicit the release of powerful oxidizing agents from cells of the immune system. These oxidizing agents are a normal and proper protective response of the body, but they also cause oxidative damage to DNA.

Do smoking, improper diet, and chronic inflammation contribute to μ? Diet and inflammatory disorders probably contribute to μ because of their potential to produce oxygen radicals, but we do not yet have enough information about their relative contributions. Tobacco smoke, on the other hand, is taken up readily by the gonads, and

Table 2.4
Qualitative Estimates of the Contribution of Mutagenic Agents to the Human Germinal Mutation Rate, μ

Class of agent	Contribution to μ
Physical—background radiation	Minor
Chemical	
Exogenous	
Environmental toxins	Minor
Tobacco smoke	Important
Endogenous	
Oxidative damage to DNA	Major

The contribution of background radiation is considered to be minor because the dose per generation is much less than the doubling dose. Reliable quantitative estimates for the contribution of environmental toxins, smoking, and oxidative damage is not yet available (see text for discussion).

there is much evidence now supporting the view that smoking, particularly paternal smoking, contributes importantly to genetic disorders and childhood cancers. We shall consider some of these effects in Chapter 13. The major contributions to μ probably come from natural chemicals found in our diet and normal metabolic processes (Table 2.4).

SUMMARY

Gene mutations are small-scale changes in the nucleotide sequence in DNA, while chromosome mutations involve changes in chromosome number, or large-scale re-arrangement of DNA within or between chromosomes. Such changes can have profound deleterious effects on cellular, tissue, or organismic function. Mutations in somatic tissues are not heritable, but will over time contribute to the development of cancers, heart disease, and many of the disorders associated with aging. Germinal mutations, produced in germ cells, are heritable, and their effects will be felt for many generations. Germinal mutations contribute to developmental abnormalities during pregnancy, birth defects, and many other genetic disorders that manifest themselves before and after birth. Despite the fact that mutations can be very harmful, mutations are still important because they are the ultimate source of genetic variability on which evolutionary change depends.

Radiation and chemical compounds are the major classes of mutagenic agents. Background radiation contributes only a small fraction of the total somatic and germinal mutation load. One exception to this general conclusion is the importance of UV radiation in the production of skin cancers. Current studies indicate that endogenously produced oxidizing compounds, which are by-products of normal metabolism, are responsible for most DNA damage. Despite elaborate cellular defense and repair mechanisms, some lesions to DNA nucleotides remain unrepaired, and these lesions give rise to mutations. The mutation rate, then, depends on the efficacy of the cellular antioxidant activity and on the DNA repair mechanisms.

Smoking, diet, and chronic inflammation and infections contribute to the mutation rate either because they contribute to the production of oxidizing compounds or because they counteract the normal cellular antioxidant defense activities. Strong epidemiological evidence indicates that reduction in smoking and improvement in diet, particularly ingestion of foods with antioxidant activity and reduced intake of foods that stimulate the productions of oxidizing compounds, will have beneficial effects in terms of reducing mutational damage. However, because we cannot affect the efficiency of the DNA repair system, the mutation rate will never be reduced to zero. In particular, the spontaneous germinal mutation rate, μ, is unlikely to decrease significantly beyond some minimum.

QUESTIONS

1. What evolutionary developments had to precede sexual reproduction?

2. What are the proposed roles for sexual reproduction in evolution?

3. Describe the nature of the chromosome cycle in humans.

4. What factors have played a role in the differentiation of the sex chromosomes?

5. Describe the utility of the concept of the "doubling dose."

6. In the context of the oxidative DNA damage model for the origin of the spontaneous mutation rate, what factors increase DNA damage, and which ones reduce DNA damage?

7. Library project: What *genetic* evidence supports the oxidative damage model?

SUPPLEMENTAL READING

Doolittle, W. F. 2000. Uprooting the tree of life. *Scientific American* 282(2), 90–95.

Margulis, L, and D. Sagan. 1997. *What Is Sex?* Simon and Schuster Editions, New York, NY.

Michod, R. E. 1995. *Eros and Evolution: A Natural Philosophy of Sex.* Addison-Wesley Publishing Co., Reading, MA.

Ridley, M. 1993. *The Red Queen: Sex and the Evolution of Human Nature.* Macmillan Publishing Co., New York, NY.

The Evolution of Sex. Nobel Conference XXIII. 1988. R. Bellig and G. Stevens, Eds. Harper and Row, Publishers, San Francisco, CA.

ADVANCED TOPICS

Ames, B. N., L. S. Gold, and W. C. Willett. 1995. The causes and prevention of cancer. *Proceedings of the National Academy of Sciences (USA)* 92, 5258–5265.

Bell, G. 1982. *The Masterpiece of Nature: The Evolution and Genetics of Sexuality.* University of California Press, Berkeley, CA.

Crow, J. F. 1994. Advantages of sexual reproduction. *Developmental Genetics* 15(3), 205–213.

Crow, J. F. 1997. The high spontaneous mutation rate: Is it a health risk? *Proceedings of the National Academy of Sciences (USA)* 94(16), 8380–8386.

Eyre-Walker, A., and P. D. Keightley. 1999. High genomic deleterious mutation rates in hominids. *Nature* 397, 344–347.

Giannelli, F., *et al.* 1999. Mutation rates in humans. II. Sporadic mutation-specific rates and rate of detrimental human mutations inferred from hemophilia B. *American Journal of Human Genetics* 65, 1580–1587.

Green, P. M., *et al.* 1999. Mutation rates in humans. I. Overall and sex-specific rates obtained from a population study of hemophilia B. *American Journal of Human Genetics* **65**, 1572–1579.

Neel, J. V., *et al.* 1990. The children of parents exposed to atomic bombs: Estimates of the genetic doubling dose of radiation for humans. *American Journal of Human Genetics* **46**, 1053–1072.

Williams, G. C. 1975. *Sex and Evolution.* Princeton University Press, Princeton, NJ.

Woodall, A. A., and B. N. Ames. 1997. Nutritional prevention of DNA damage to sperm and consequent risk reduction in birth defects and cancer in offspring. In *Preventive Nutrition*, A. Bendich and R. J. Deckelbaum, Eds. Human Press, Inc., Totowa, NJ.

Letters to the Editor

[*The Editor does not hold himself responsible for opinions expressed by his correspondents. Neither can he undertake to return, nor to correspond with the writers of, rejected manuscripts intended for this or any other part of* NATURE. *No notice is taken of anonymous communications.*]

Mass Excretion of Œstrogenic Hormone in the Urine of the Stallion

IN earlier investigations[1] it was shown that the largest quantities of œstrogenic hormone (folliculin—s. œstrin) are excreted in the urine of pregnant mares (100,000 mouse units per litre). I found this also to be the case in other equines (ass, zebra) during pregnancy, whereas, in the non-pregnant state, the excretion of hormone both in equines and in other mammals is very small, at most $0 \cdot 5$ per cent in comparison with that of the gravid animal. Curiously enough, as a result of further investigations, it appears that in the urine of the stallion also, very large quantities of œstrogenic hormone are eliminated.

Letter by Dutch scientist B. Zondek (1934) to the journal *Nature* describing the surprising finding that the urine of a stallion contains large quantities of what was then considered to be the female hormone. (B. Zondek. 1934. Mass excretion of oestrogenic hormone in the urine of the stallion. *Nature 133*, 209–210. © Macmillan Magazines Limited.)

CHAPTER 3

Reproductive Hormones

These chemical messengers . . . or "hormones" as we may call them, have to be carried from the organ where they are produced to the organ which they affect, by means of the blood stream, and the continually recurring physiological needs of the organism must determine their production and circulation through the body.

—E. H. Starling. 1905

The Croonian lectures on the chemical correlation of the functions of the body. *Lancet* ii, 339–341.

Just how each ductless gland produces its own special secretion, and why certain particular tissues, and these only, respond to a given hormone, are questions which must be solved by the physiologists and the chemists of the future.

—G. W. Corner. 1942

The Hormones in Human Reproduction.
Princeton University Press, Princeton, NJ, p. 28.

MULTICELLULARITY REQUIRES that the cells, tissues, and organs of the organism be able to communicate with each other to coordinate and regulate each other's activity. This communication can occur through cell-to-cell contact, or through chemical messengers, known as **hormones**. The existence of hormones (the term means "to arouse") was first recognized around the beginning of the twentieth century, but it was not until the 1930s that the first important clues about hormone function were obtained. The hormones first studied were produced by special ductless glands, such as the pituitary, thyroid, parathyroid, adrenal, and the pancreas, and secreted into the circulatory system to be transported to all other parts of the body (Table 3.1). These glands, referred to as the **endocrine glands**, stood in contrast to glands such as the salivary glands, sweat glands, kidneys, and liver, which are referred to as **exocrine glands** because they discharge their products through special ducts to the outside of the body or into other tissues or organs.

Table 3.1
Classical Endocrine Glands and Their Hormones

Gland	Hormone	Action
Thyroid	Thyroxine and triiodothyronine	Cell growth, bone growth, metabolic rate
Parathyroid	Parathyroid hormone (PTH)	Regulates calcium levels
Pancreas	Insulin	Lowers blood glucose
	Glucagon	Increases blood glucose
Adrenal cortex	Glucocorticoids	Convert stored fats to carbohydrates; antiinflammatory
	Mineralocorticoids	Sodium/potassium balance
Adrenal medulla	Epinephrine	Increases heart rate, oxygen consumption
	Norepinephrine	Increases blood pressure
Pituitary	Many hormones	Growth, reproduction, metabolism, lactation
Testes	Androgens	Male fertility; male sex characteristics
Ovaries	Estrogen	Female fertility; female sex characteristics
	Progesterone	Maintains pregnancy

The study of hormones gave birth to the field of **endocrinology.** Progress in understanding the functioning of the reproductive organs during the 1930s and 1940s showed clearly that a special set of hormones were involved in coordinating the activities of the pituitary and the gonads. We shall refer to the hormones involved in regulating reproductive function as *reproductive hormones*, although, as will become clear, the field of function for most of these hormones is not limited to reproduction (Table 3.2). An understanding of the important features of hormones, hormone action, and classes of reproductive hormones will be very helpful in understanding how the reproductive system is regulated. This chapter hopes to provide that understanding.

Table 3.2
Primary Reproductive Hormones

Hormone	Primary source
Steroid hormones	
Progestins	Ovary, placenta
Androgens	Testis
Estrogens	Ovary, placenta
Prostaglandins	
PGD_2	Uterus
PGE_2	Ovary, uterus
PGF_2	Corpus luteum, uterus
Protein hormones (polypeptides)	
Follicle-stimulating hormone (FSH)	Anterior pituitary
Luteinizing hormone (LH)	Anterior pituitary
Antiparamesonephric hormone (APH)	Fetal testis
Prolactin (PRL)	Anterior pituitary
Human chorionic gonadotropin (hCG)	Placenta
Human placental lactogen (hPL)	Placenta
Leptin	Adipose tissue
Peptide hormones (short amino acid chains)	
Gonadotropin-releasing hormone (GnRH)	Hypothalamus, placenta
Corticotrophic-releasing hormone (CRH)	Hypothalamus, placenta
Oxytocin	Posterior pituitary
β-endorphin	Hypothalamus, placenta
Adrenocorticotrophic hormone (ACTH)	Anterior pituitary
Amine hormones (neurotransmitters)	
Dopamine	Hypothalamus
Melatonin	Pineal gland
Serotonin	CNS

ESSENTIALS OF HORMONE ACTION

The field of endocrinology began to blossom in the 1960s with the development of techniques for purifying hormones and measuring the response of cells to their effects. These studies have shown that although hormones are structurally quite diverse, only a few principles govern their mode of action. This section will review the essential characteristics of hormone action.

Secretor and target cells

Hormones can be classified in different ways. Structurally, they fall into at least four classes: lipids, which include the steroids and eicosanoids; peptides and proteins; amino acids and amino acid derivatives; and gases. The reproductive hormones that

54

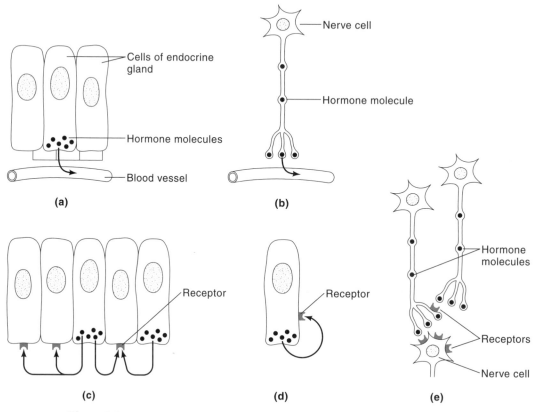

Figure 3.1
Modes of hormone action. (a) Endocrine. (b) Neuroendocrine. (c) Paracrine. (d) Autocrine.
(e) Neurotransmitter. (a) and (b) differ only in that the cell that releases the hormone in (b) is a
nerve cell. In both cases, the hormone is secreted into the circulatory system. (c) In paracrine
action, the hormone is released into the intercellular space and acts on neighboring cells, while
in autocrine action (d), a cell secretes and responds to the same hormone. (e) represents a
special type of paracrine action involving nerve cells.

we will consider in this text fall into the first three classes. Hormones can also be classified by their mode of action. Hormones were originally defined as substances produced
by specific tissues and secreted into the blood to activate tissues at some distance away.
This mode of action, known as *endocrine*, is global or organismic. Hormones, however, can act locally as well. Two major types of local action are recognized: **paracrine**
and **autocrine**. In paracrine action, the secretor and target cells are neighbors, and the
hormone is secreted into the intercellular spaces between the cells, not into the general
circulatory system. In autocrine action, the secretor and target cell are the same. In this
case the hormone is secreted to the exterior of the cell, after which it now exerts its action on the cell (Fig. 3.1). Many hormones have been found to act in all three ways.

Hormones not only can have multiple targets, but they also can have multiple
sources. For example, insulin and the sex steroid hormones, originally thought to be

produced by the pancreas and gonads, respectively, are now known to be produced by the brain as well. This promiscuity with respect to source, target, and mode of action has led to controversies in defining what a "proper" hormone is. Hormones, like languages, have resisted attempts to define them in what we would consider an orderly manner. It is perhaps best to think of them as agents that have been enlisted by organisms to solve different problems of communication between different tissues.

The response of a cell to a given hormone is generally multifaceted. In many, perhaps most, cases, the target cell responds by "turning on" a set of genes, resulting in the synthesis of a new set of proteins by the target cell. In other cases, the response results in the activation of a previously existing set of proteins. Both types of responses can occur in the same target cell. The common response of each cell of a given tissue is transformed into a characteristic, often dramatic, tissue or organ response. For example, under the stimulation of the steroid hormone *estrogen*, the prepubertal breasts in a young girl will be transformed into fully developed breasts. The growth spurt that takes place characteristically during the initial stages of puberty is probably stimulated by several different hormones. In the following chapters we will consider in more detail the functions of the different classes of reproductive hormones.

Hormone receptors

The essence of a hormone response is its specificity; that is, only certain cells respond to a given hormone. This implies that the target cell has to recognize the hormone to which it responds in some unique way. The basis for the uniqueness of the response began to be understood in the 1960s, and depends on recognition of the hormone by a specific molecule in the cell—the **receptor**. There are two major classes of hormone receptors—**nuclear** and **transmembrane**. Both classes of receptors are proteins that are synthesized by the target cell, and each receptor has the special property of binding to a specific hormone very tightly. In general, a given receptor will recognize only one hormone. A cell that responds to four different hormones synthesizes receptors specific for each hormone. If a hormone acts in an autocrine fashion, then the cell that synthesizes the hormone will also synthesize the receptor specific for that hormone. Binding of the given hormone to its receptor sets in motion a fixed set of reactions, that in turn result in the response typical for that particular cell-hormone combination. *A hormone-mediated action requires both the hormone and a functional receptor.*

Nuclear receptors

Nuclear receptors are members of a large family of related proteins that recognize the lipid class of hormones, the steroid and eicosanoid hormones. Both types of hormones diffuse easily through the cell membrane and, hence, are capable of entering any cell. Nuclear receptors reside in the nucleus of a cell. A steroid hormone that enters a cell that harbors its cognate nuclear receptor will bind to the receptor. The receptor-hormone complex binds to certain nucleotide sequences, known as **hormone recognition elements** (HREs), and triggers the transcription of the gene(s) that carries the HREs. This leads to the synthesis of a new set of proteins. The cellular response characteristic of that hormone depends on the functions carried out by the new set of proteins (Fig. 3.2). Response to steroid hormones is often referred to as a *genomic* response because it is invariably accompanied by the synthesis of new proteins (gene products).

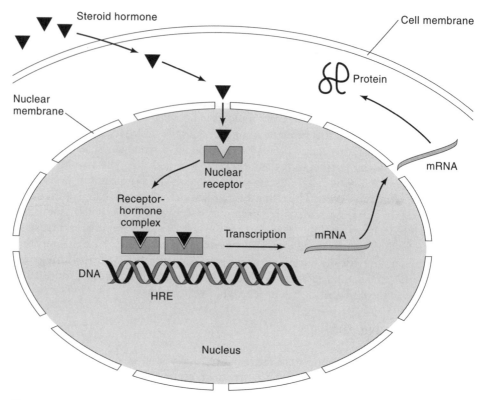

Figure 3.2
Schematic illustration of steroid hormone action. The steroid hormone diffuses into the nucleus of a cell where its receptor resides. Formation of the hormone-receptor complex results in the transcription of one or more genes determined by a characteristic hormone recognition element (HRE), and the synthesis of one or more new proteins (genomic action). The genes activated are characteristic of the tissue and hormone.

Transmembrane receptors

nongenomic action: activate previous existing pro
genomic action

- *recognize peptide/*
- *aa hormone*
- *on cell surface*
- *stimulate a*
- *cascade of rxn*
- *3 parts: exterior,*
- *transverse, interior*
- *(facing the cytoplasm)*

Peptide and amino acid hormones, because of their different structure, cannot diffuse into the cell as steroid hormones do. Instead, they are recognized at the cell surface by a receptor, also a protein, that is anchored to the cell membrane. The receptor has a part that faces the exterior of the cell, to which the hormone binds, a transmembrane segment that traverses the cell membrane, and finally a part that lies within the cell cytoplasm. Binding of the hormone to the membrane-anchored receptor generates an intracellular signal, known as the *second messenger*, which in turn triggers a cascade of reactions characteristic of that hormone and that target cell (Fig. 3.3). Cells can make use of different second messengers, each generated by a different *effector* system coupled to the receptor. Transmembrane-receptor responses typically involve *nongenomic* as well as genomic action. A nongenomic response is characterized by the activation of previously existing proteins.

Appropriate response
requires proper
fxn of # of factors.

Appropriate response to a hormone depends on proper functioning of a number of other factors. Since each factor itself is a protein, each the product of a different gene, response to a hormone is governed by many different genes. Mutation in any of these genes results in a defective hormone response.

cause
effec

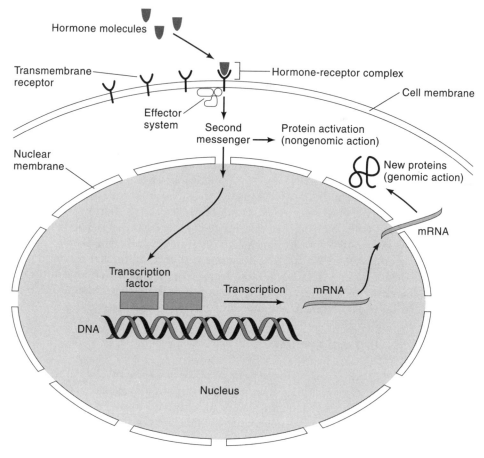

Figure 3.3

Schematic representation of a generalized transmembrane receptor–mediated hormone action. In this case, the hormone binds to a membrane-bound receptor. Formation of the hormone-receptor complex activates a signal transduction system, consisting of an effector system coupled to the receptor, which synthesizes a second messenger. The second messenger triggers the activation of a number of previously existing proteins—nongenomic action. In addition, transcription of certain genes is activated by specific transcription factors responding to the second messenger—genomic action. Several different effector systems and second messengers are used by cells.

Factors affecting hormone response

Under normal conditions, the response of a target cell to a given hormone **depends** on hormone and receptor concentrations. It is important to understand that most hormones are not produced and secreted continuously, but are generally released by the secretor cell in a periodic fashion, with the period varying from minutes to hours. The secretor cell itself may be responding to stimulation by other hormones. Consequently, hormone concentration fluctuates with time, and hence, target cells experience periodic changes in hormone concentration.

The response of the target cell also **depends** on the concentration of hormone receptors and whether the receptors are nuclear or transmembrane. Since a hormone response is elicited only if the hormone binds to its receptor, a low receptor concentration means a weak response irrespective of the hormone concentration. Different factors are **involved** in regulating hormone receptor concentration, one of which is the hormone itself. A hormone is said to **down-regulate** its receptor if, after continued exposure to the hormone, the concentration of the hormone receptor decreases. In other cases, the hormone can **up-regulate** its receptor, meaning that the receptor concentration increases after exposure to the hormone. Both of these responses are complex, and they serve to fine-tune the response of the target cell, rendering it more or less sensitive to the hormone. Down-regulation leads to the well-known phenomenon of *desensitization*, in which continuous exposure to a stimulus leads to a diminished response. For example, if a room air-conditioner were turned on as you are reading this paragraph, you would immediately be aware of the sound. However, after a short while, perhaps less than a minute, you would become desensitized to the noise; that is, you would not be consciously aware of the noise even though the sound is still being produced. The analogous phenomenon in the case of a hormone with its receptor is that continuous exposure to the hormone generally leads to a decrease in the concentration of hormone receptors, thereby reducing the response to the hormone. Important examples of both down-regulation and up-regulation will be encountered in later chapters.

Agonists and antagonists

The binding of any given hormone to its receptor is determined by the structural affinity between the two molecules. This means that molecules similar in structure to the naturally occurring hormone may also be able to bind to the receptor, with the strength or affinity of the binding depending on the extent of the structural "fit" between the receptor and the new molecule. Such molecules may themselves be naturally occurring or they may be synthetic analogs of the natural hormone—that is, they may be compounds synthesized in the laboratory with a structure similar to that of the natural hormone (Fig. 3.4). If the binding of such a compound to the receptor mimics the action of the natural hormone, the compound is referred to as an **agonist** at that particular receptor. On the other hand, if such a molecule elicits a response opposite to that

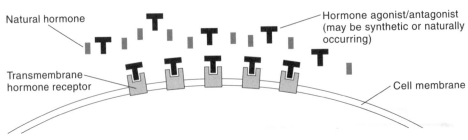

Figure 3.4
Mechanism of action of agonists/antagonists. Agonists/antagonists (synthetic or naturally occurring hormone analogs) at a given hormone receptor have a structure that permits them to bind to the receptor. If their binding affinity is high, they may displace the natural hormone from its binding site. Subtle structural details determine whether the hormone analog functions as an agonist or antagonist.

Table 3.3
Examples of Reproductive Hormone Agonists and Antagonists

Natural hormone	Agonist	Antagonist	Use
Progesterone	Levonorgestrel		Contraceptive
		RU-486	Abortifacient
Testosterone	Anabolic steroids		Testicular failure, increase muscle mass
		Flutamide	Prostate cancer
Estrogen	Mestranol		Contraceptive
		Raloxifene	Breast cancer
GnRH	Leuprolide		Precocious puberty
		Nal-Glu	Uterine fibroids
β-endorphin		Naloxone	Male erectile dysfunction

of the natural hormone, then the compound is referred to as an **antagonist**. The availability of synthetic agonists and antagonists at a number of hormone receptors has made possible a more precise analysis of the function of many hormones. A number of these have acquired extremely important therapeutic value (Table 3.3). We will consider in later chapters examples of the therapeutic use of some of these agonists and antagonists.

SUMMARY

Hormones are chemical messengers used by cells to communicate with each other. In multicellular organisms, hormones can have different modes of action. Some hormones are released into the circulatory system to activate target cells in many parts of the body (endocrine action). In other cases, hormones exert their effects in a localized manner, in neighboring cells (paracrine action), or the secretor cell may also be the target cell (autocrine action). Hormones can be grouped into four structural categories: lipids, peptides and polypeptides, amino acid derivatives, and gases.

The essence of hormone action lies in the fact that only certain cells respond to a given hormone. This specificity depends on the high affinity between a hormone and its receptor, a protein synthesized only in the target cells. The response of a cell or tissue to a hormone depends on the molecular events resulting from binding of a hormone to its receptor, and generally includes the expression of a specific set of genes or activation of a special set of proteins. For hormones of the lipid or gaseous group that are able to diffuse into a cell, the receptor is an intracellular protein, referred to as a nuclear receptor. Receptors recognizing peptide, polypeptide, or amine hormones are localized on the cell membrane and are known as transmembrane receptors. The intensity of the hormone response depends on hormone and hormone-receptor concentration, as well as on the dynamics of interaction between the hormone and its receptor. In some cases, a hormone can down-regulate its receptor, meaning that receptor concentration will decrease in response to its hormone. In other cases, a hormone can up-regulate its receptor, resulting in an increase in hormone-receptor concentration. Synthetic and non-synthetic analogs of the naturally occurring hormones that can bind to the respective hormone receptors can mimic the effect of the normal hormone or they can have an effect opposite to that of the natural hormone. The former are known as agonists, and the latter as antagonists. A number of synthetic agonists and antagonists have been shown to be of significant therapeutic importance.

MAJOR REPRODUCTIVE HORMONES

The classically important reproductive hormones include lipid and nonlipid hormones. The lipid class includes the **steroid** and **eicosanoid** hormones. The major group of eicosanoid hormones with a reproductive function is the **prostaglandins**. The nonlipid hormone class consists of the **polypeptide** hormones, which consist of long amino acid chains (greater than 40 amino acids); **peptide** hormones, which are short amino acid chains (less than 40 amino acids); and **amine** hormones, which are much simpler molecules, either amino acids or derivatives of amino acids (Table 3.2).

Steroid sex hormones

Steroids are ancient molecules whose existence can be traced to about 2 billion years ago. They appear to have functioned in primitive cells as intracellular regulators. In multicellular organisms they regulate the activity of diverse tissues. In mammals, the steroid hormone family includes five subfamilies: **progestins, androgens, estrogens, glucocorticoids**, and **mineralocorticoids**. All steroid hormones are related structurally (Fig. 3.5) and are synthesized from cholesterol, which is either supplied in the diet, or produced *de novo*. Each subfamily consists of several related molecules, all of which bind to the same receptor, but with different affinities. The progestins, androgens, and estrogens are known as the **steroid sex hormones** because they are produced primarily by the gonads. The glucocorticoids and mineralocorticoids, produced by the adrenal cortex, are extremely important in their own right. The glucocorticoid hormones, for example, corticosterone and cortisol, have essential functions in regulating carbohydrate metabolism, while the mineralocorticoid hormone aldosterone is necessary for regulating the potassium-sodium balance in the blood.

The synthesis of all steroid hormones from cholesterol follows a hierarchical scheme in which the progestin family is synthesized first, the progestins forming the substrate for the synthesis of androgens, and these in turn forming the substrates for the synthesis of the estrogens (Fig. 3.6). Transformation from one class to another along this synthetic pathway is carried out by a group of **steroidogenic enzymes**, special proteins synthesized only in the cells that produce steroid hormones. Hence, the adrenals produce the enzymes necessary to convert cholesterol to glucocorticoids, while the gonads produce the enzymes necessary to convert cholesterol to androgens and estrogens. This partitioning is not absolute, however, so that the adrenals carry out some synthesis of androgens and estrogens.

The effects of each subfamily of steroid sex hormones is mediated by a specific receptor—the progestin receptor (P-R), the androgen receptor (A-R), and the estrogen receptor (E-R). The members of each subfamily are distinguished from each other by their potency, a property that depends on the binding affinity of the hormone to its re-

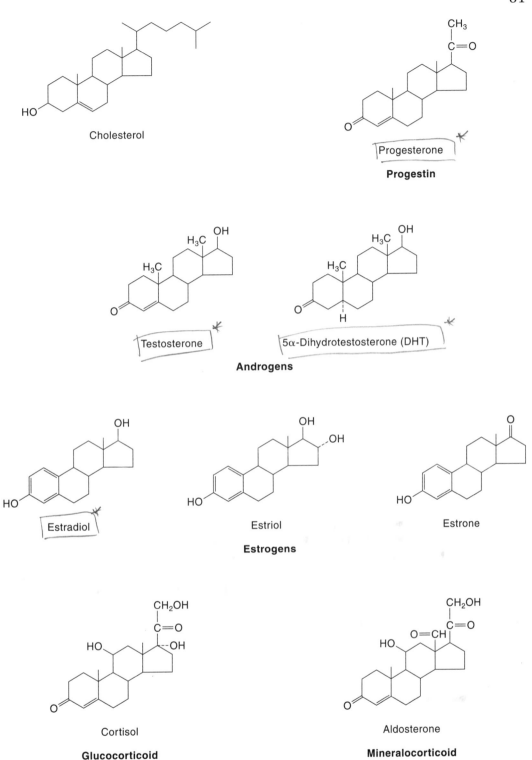

Cholesterol

Progesterone

Progestin

Testosterone

5α-Dihydrotestosterone (DHT)

Androgens

Estradiol

Estriol

Estrone

Estrogens

Cortisol

Glucocorticoid

Aldosterone

Mineralocorticoid

Figure 3.5
Structures of important members of the five families of steroid hormones. All steroid
hormones have as their structural core a four-ring structure derived from cholesterol.

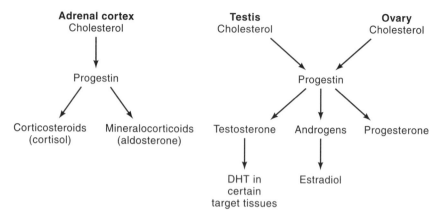

Figure 3.6

Highly simplified pathway for the synthesis of steroid hormones from cholesterol. The steroid sex hormones progesterone, estradiol, and testosterone are produced primarily by the gonads. The glucocorticoids and mineralocorticoids are produced by the adrenal glands. In these tissues, cholesterol may be dietary cholesterol or it may be synthesized *de novo*. Cholesterol is converted to members of the progestin family. In the adrenal cortex, progestins are converted to the corticosteroids and mineralocorticoids. In the ovaries of a nonpregnant female, the progestins are converted to androgens and then into the primary estrogen, estradiol. The ovaries are also the main source of progesterone. In a pregnant female, the placenta, rather than the ovary, is the primary source of steroid sex hormones (see Chapter 12). In the testis, progestins are converted to the primary androgen, testosterone. Testosterone is concerted to DHT in certain tissues (see Chapters 4 and 6).

ceptor. The higher the affinity, the more potent the response. Within each subfamily, the most important member in a biological sense is the one with the highest affinity for its receptor. **Progesterone**, for example, is the most important member of the progestin subfamily, **testosterone** and **dihydrotestosterone** (DHT) of the androgen subfamily, and **estradiol** for the estrogen subfamily (Table 3.4).

The steroid sex hormones have a number of **important functions**, but foremost are those associated directly with reproduction. Estrogens and androgens are essential for oogenesis and spermatogenesis and for the **development** of the secondary sex characteristics. Progestins prepare the uterus for implantation and maintain it during pregnancy (Table 3.5). In later chapters we will examine at length the specific functions of each of these hormones. Steroid hormone receptor **agonists** and **antagonists** have been very useful in studying the diverse functions of the three families of steroid sex hormones. An important class of recently synthesized agonists/antagonists are the **SERMs**, or *selective estrogen receptor modulators*. These compounds bind with different affinity to the two molecular forms of the estrogen receptor, and they display estrogen agonist or antagonist effects in a tissue-dependent manner (Fig. 3.7).

We infer that all three steroid sex hormones are involved in many different types of functions since progestin, androgen, and estrogen receptors are **distributed** widely throughout the tissues of the body, including the brain. The presence of such receptors is generally taken to mean that such cells respond to that hormone. In most cases, the **role** of these hormones is unknown. In a few cases we do have a sense of what they do

Table 3.4
Steroid Sex Hormone Families

Family	Members
Progestin	Progesterone Hydroxyprogesterone
Androgen	Dihydrotestosterone (DHT) Testosterone Dihydroepiandrosterone (DHEAS)
Estrogen	Estradiol Estriol Estrone

The members of each family are listed in order of potency. Testosterone is the main androgen in males, but DHT has greater binding affinity to the androgen receptor. Estradiol is the main estrogen in nonpregnant females, while estriol is the main estrogen in pregnant females. Estrone is the important estrogen in postmenopausal females.

Table 3.5
Principal Properties of the Steroid Sex Hormones

Hormone	Function
Progesterone	Prepares uterus to receive conceptus Maintains uterus during pregnancy Stimulates growth of mammary glands Regulates secretion of LH and FSH
Androgens	Required for spermatogenesis Induce and maintain secondary sex characteristics of males (growth of external genitalia, body and pubic hair, deep voice), and pubic and axillary hair of females Regulate secretion of LH and FSH Promote increase in muscle mass and tone
Estrogens	Necessary for oogenesis Stimulate growth and activity of mammary glands Stimulate growth of uterus Induce and maintain female body shape and fat distribution Regulate secretion of LH and FSH Required for skeletal maturation and bone mineralization in both males and females

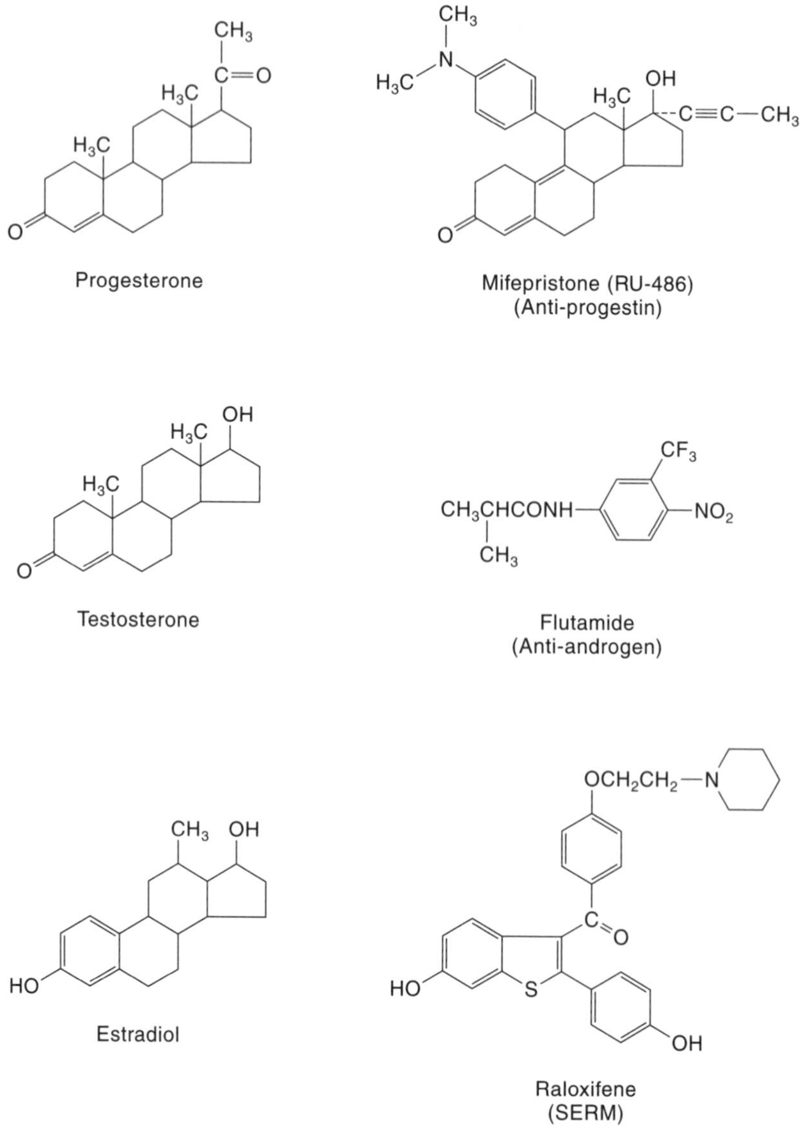

Figure 3.7
Structures of three therapeutically important steroid hormone receptor antagonists. Mifepristone (RU-486) is a progesterone receptor antagonist, and flutamide is an antagonist at the androgen receptor. Raloxifene is a SERM, which functions in some tissues as an agonist, and in others, as an antagonist.

(Table 3.5). The most abundant androgen, testosterone, for example, through its stimulation of protein synthesis, increases muscle mass. The more developed muscles in males is a reflection of the higher levels of testosterone in males. The so-called **anabolic steroids** taken illegally by many athletes, male and female, to increase their muscle mass and strength are synthetic testosterone agonists. The even more potent androgen, DHT, is unusual because it is synthesized only in a few tissues and it exerts its effects only in those tissues. We will consider the effects of DHT in later chapters.

Although in the past progestins and estrogens have been referred to as the "female" hormones and androgens as the "male" hormones, it was apparent even in the early days of endocrinology that such labels were not meaningful. For example, it was reported in 1934 that the urine of the stallion contained the highest levels of estrogen yet measured, much higher than found in the urine of mares. Today, we understand that because of the varied functions of steroid sex hormones, finding all of these hormones in both sexes is not surprising. A comparison of the levels of the three classes of hormones in males and females is shown in Table 3.6. In the female, because of the

Table 3.6
Relative Blood Levels of Cholesterol and
Steroid Sex Hormones in Males and Females

Hormone	Men	Women Prior to ovulation	After ovulation	Pregnant
Cholesterol	1.0	1.0	1.0	1.0
Progesterone	0.3	1.0	11.0	160.0
Testosterone	1.0	0.06	0.06	0.06
DHT	1.0	0.4	0.4	0.6
Estradiol	0.2	1.0	2.0	15.0
Estriol	—	—	—	100.0

These are relative values normalized to 1.0. After ovulation, progesterone levels increase about 11-fold compared to those before ovulation, and 160-fold in a pregnant female. Estriol levels are about 7-fold greater than estradiol levels in a pregnant female. (Adapted from M. H. Everitt and B. J. Johnson. 1995. *Essential Reproduction*. Blackwell Scientific Publications, Oxford, UK.)

complexity of her reproductive cycle, hormone levels fluctuate in accordance with the time of the menstrual cycle or with pregnancy. The large increase in **progesterone** levels during pregnancy indicates why progesterone is sometimes known as the hormone of pregnancy. In a nonpregnant, premenopausal female, estradiol is the predominant estrogen. Estriol is the most abundant estrogen in a pregnant female, while estrone is the main estrogen found in a postmenopausal female. Females have significant levels of androgens, especially DHT, the most potent of the androgens. Similarly, males have nonnegligible levels of estrogens.

Prostaglandins : *[handwritten: paracrine → in circulation : inactive when passing through lungs.]*

Prostaglandins (PGs) are a family of unusual fatty acidlike molecules produced probably in almost every cell in the body (Fig. 3.8). PGs act primarily in a paracrine

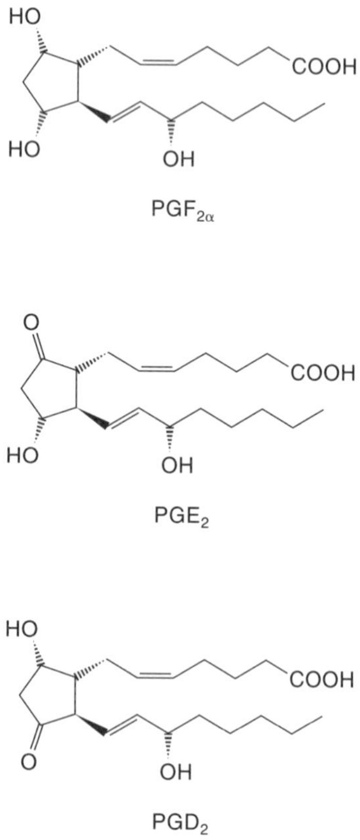

Figure 3.8
Structures of three prostaglandins
that have reproductive functions
(see Table 3.2).

[handwritten margin note: Fxn — ovulation — luteal regression — contraction of uterus → labor stimulation]

sense, and if they get into the circulatory system, they are inactivated during passage through the lungs. PGs participate in many different functions. In reproduction, PGs play a role in ovulation, luteal regression, preparation of the cervix for birth, and stimulation of contractions of the uterus, which are important in menstruation and labor. Three important PGs and their source tissues are given in Table 3.3.

Polypeptide hormones

Many polypeptide hormones are now known to play roles in reproduction, but for the purposes of this book, most of them would be considered of minor importance. We will consider here only a few of these, particularly the classically important ones.

Gonadotropins—LH and FSH *[handwritten: produced by anterior pituitary → affect the activity of gonad]*

Luteinizing hormone (LH) and **follicle stimulating hormone (FSH)** are produced by the **anterior lobe of the pituitary gland**, and their function is to regulate the activity

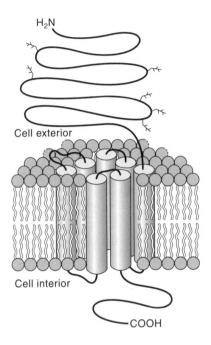

H₂N

Cell exterior

Cell interior

COOH

Figure 3.9
Schematic diagram of the LH receptor in the plasma membrane of a target cell. The top
portion represents the section of the polypeptide chain that faces the exterior of the cell and
to which LH binds. The seven cylinders are the sections that anchor the receptor to the
membrane; the short bottom segment faces the interior of the cell. Both LH and hCG are
recognized by the same receptor. The FSH receptor also has seven membrane-spanning
segments. (Adapted from K. E. McFarland, R. Sprengel, H. S. Phillips, *et al*. 1989. Lutropin-
choriogonadotropin receptor: an unusual member of the G protein–coupled receptor family.
Science **245**, 494–498. Copyright 1989 American Association for the Advancement of
Science.)

of the gonads. Because their primary target is the gonad, they are also referred to as
gonadotropins. Each consists of two peptide chains, a common alpha chain (89 amino
acids), and a beta chain (115 amino acids for FSH and 121 amino acids for LH). The
different beta chains confer specificity to each hormone. Each is recognized by a distinct membrane-bound receptor. A schematic illustration of the LH receptor is shown
in Fig. 3.9. The names of LH and FSH reflect the fact that both were first discovered
and characterized in females. The same hormones were later found to be present in
males and, as we shall see, to have complementary functions in the two sexes.

[handwritten margin notes: LH { 1α (89aa) / 1β (121 aa)   FSH { 1α (89aa) / 1β (115 aa) → distinct membrane-bound receptor]

hCG and hPL *[handwritten: → produced by the placenta]*

   Human chorionic gonadotropin (hCG) and **human placental lactogen** (hPL) are
produced by the placenta. Functionally, hCG is essentially identical to LH since it has
the same alpha chain and the beta chain is almost the same as the LH beta chain (Fig.
3.10). hCG is produced by the implanting embryo and is one of the earliest markers of

[handwritten margin notes: hCG: identical to LH (1α & 1β) → early implanting embryo → pregnancy test]

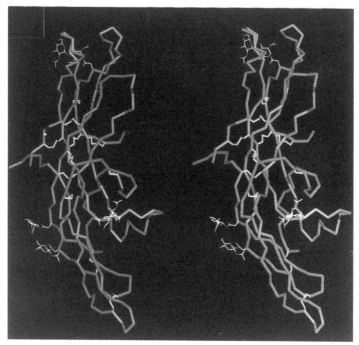

Figure 3.10
Molecular model of hCG. hCG, consisting of two polypeptide chains, is functionally
equivalent to LH. This model is based on x-ray crystallographic studies. (Reprinted from
A. J. Lapthorn, D. C. Harris, A. Littlejohn, *et al.* 1994. Crystal structure of human chorionic
gonadotropin. *Nature* 369, 455–461. Macmillan Magazines Limited, with permission.)

hPL → stimulation of fetal growth a clinically recognized pregnancy. For this reason, standard pregnancy tests measure
hCG levels. hPL has a number of functions, all related to the stimulation of fetal
growth (see Chapter 12).

Prolactin
produced by anterior pituitary
single peptide chain 199 (aa)
found in all vertebrate

Prolactin (PRL) is also produced by the anterior lobe of the pituitary. It consists
of a single peptide chain of 199 amino acids (Fig. 3.11). Prolactin, found in all vertebrates, is a highly versatile hormone, with about 85 different biological functions iden-

Figure 3.11
Schematic illustration of prolactin.
Prolactin is a single polypeptide chain
folded in a characteristic way by
interactions between sulfur-containing
amino acids.

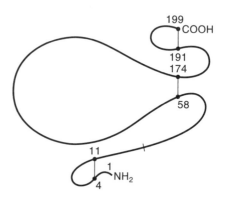

Table 3.7
Conditions Associated with Increased Prolactin Release

Condition	Feature
Lactation	Suckling
Sleep	Onset of sleep (declines before waking)
Eating	High-protein meal, especially at noon
Drinking alcohol	Chronic use
Stress	Physical and emotional stress
Ovulatory cycle	Late follicular and luteal phases
Pregnancy	Tenfold increase near term
Coitus	Orgasm
Postpartum	First three to four weeks
Neonate	Two to three weeks after birth

tified in different vertebrate species. In mammals, its best-known function is in maintaining lactation, and this function is the source of its name (Chapter 12). In humans, prolactin release occurs in many other instances, implying that it has many other functions (Table 3.7).

[handwritten: produced by testis in ♂, by ovary in ♀ (low level)]

APH
[handwritten: AMH; 2 peptide chains]

Antiparamesonephric hormone (APH), also known as Mullerian-inhibiting substance (MIS) or anti-Mullerian hormone (AMH), is a very large protein consisting of two peptide chains. In males, it is produced by the testes of the fetus and also during infancy and childhood, declining after puberty. In females, low levels of APH are produced by the fetal ovary during the latter part of gestation, and also during puberty, but its role in females remains unclear. We will consider the role of APH in more detail in Chapter 4.

[handwritten: produced by gonad (inhibit FSH) ; by placenta (unknown fxn)]

Inhibin
[handwritten: 1α (134aa) & 1β (116 aa)]

Inhibin is a protein consisting of two polypeptide chains, an alpha chain (134 amino acids), and a beta chain (116 amino acids) produced principally by the gonads. Gonadal inhibin plays an important role in feedback inhibitory control by the gonad on the pituitary secretion of FSH. Inhibin is also produced by the placenta, but the function of placental inhibin is unknown.

[handwritten: produced by fat-storing cell of adipose tissue; 1 peptide chain (167aa); regulate food intake/energy expenditure; signal puberty / regulate GnRH (?)]

Leptin

Leptin is a peptide containing 167 amino acids. It was discovered in 1994 during a study of a strain of genetically obese mice. Leptin is derived from the Greek word *leptos*, meaning thin. Since its discovery, leptin has been subjected to intense study. Leptin is produced by the fat-storing cells of adipose tissue. The great interest in leptin is that it appears to be involved in regulating food intake and energy expenditure. In our weight-conscious society, this guarantees that leptin will become one of the most popular and talked about hormones in the next few years. The function(s) of leptin in humans has not been clearly delineated. Very recent data indicate that it may be a necessary signal for the initiation of puberty and for regulating GnRH secretion after puberty (see Chapters 8 and 9).

Peptide hormones *c̄ neuropeptide, some fxn in reproduction*

Peptide hormones are short chains of amino acids, ranging in length from 3 to about 40 amino acids. They were first identified in brain tissue, especially the hypothalamus. Initially, they were known as **neuropeptides** because they were thought to be produced only by the brain. In fact, the field of neuroendocrinology began when the first neuropeptides were isolated and characterized in the early 1970s. Today many hormones of this type are known, but only a few are considered to have a role in reproduction. Recent studies indicate that some peptide hormones are also produced by the gonads and the placenta, which indicates that they must be involved in diverse functions.

GnRH

- produced by neurons in hypothalamus
- control synthesis of LH & FSH
- 10 aa peptide
→ easy to synthesize
agonist & antagonist

Gonadotropin-releasing hormone (GnRH), produced by neurons in different regions of the hypothalamus, controls the synthesis of LH and FSH by the pituitary. We will see in Chapter 7 why GnRH is sometimes known as the central regulator of reproduction. As we will see in later chapters, endogenous and exogenous factors that modulate reproductive activity exert their influence through their effects on GnRH secretion. GnRH is a 10-amino-acid peptide, very short compared to the polypeptide hormones described in the previous section (Fig. 3.12). Its relatively simple structure

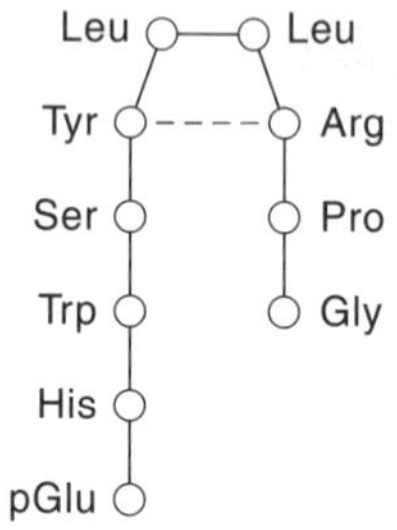

Leu — Leu
Tyr - - - - Arg
Ser   Pro
Trp   Gly
His
pGlu

Figure 3.12
Schematic illustration of GnRH, a 10-amino-acid peptide. Each circle represents an amino acid. The short peptide is folded by bonding between the indicated amino acids. (Adapted from M. Filicori. 1996. Ovulation induction: GnRH and its agonists. In *Reproductive Endocrinology, Surgery and Technology.* E. Y. Adashi, J. A. Rock, and Z. Rosenwaks, Eds. Lippincott-Raven Publishers, Philadelphia, PA.)

has facilitated the synthesis of a large number of synthetic GnRH analogs, some of which function as agonists and some as antagonists. Their therapeutic use is described in later chapters.

CRH [handwritten: stress hormone. → produced by neuron in hypothalamus → 41 aa → stimulate the synthesis of ACTH → adrenal cortex ⇒ glucocorticoid hormones]

Corticotropin-releasing hormone (CRH), also known as *corticotropin-releasing factor* (CRF), is a 41-amino-acid peptide produced by neurons in the hypothalamus. It stimulates the synthesis of **adrenocorticotrophic hormone** (ACTH) by the pituitary. ACTH in turn stimulates the adrenal glands to produce the glucocorticoid hormones. This pathway of activation is known as the **hypothalamic-pituitary-adrenal (H-P-A) axis.** Hypothalamic CRH is considered to be the stress hormone, and we will have occasion to refer to it in Chapter 9 when we discuss the effects of stress on reproduction. CRH is also produced by the placenta, and placental CRH is thought to regulate the timing of birth.

[handwritten: family of EOPs → well-being. (e.g. heroin, morphine) → produced by neuron in hypothalamus → 41 aa → role in reproduction]

β-endorphin

β-endorphin is one of the **endogenous opioid peptides** (EOPs). These peptides were first identified in the early 1970s as having strong effects on mood and behavior similar to those of the opiates, such as heroin and morphine. High levels of opioids are associated with a sense of well-being. The effects of the EOPs are mediated by a family of opioid receptors. Opiates function as nonpeptide, exogenous agonists at the opioid receptors, and this explains the potent actions of such compounds. The EOPs participate in many different functions, but only β-endorphin, a 27-amino-acid peptide secreted by neurons in the hypothalamus, appears to have a direct role in reproduction (Table 3.3). Its main function in this regard is to regulate GnRH secretion.

Amine hormones (neurotransmitters)

Reproductive functions are also regulated in different ways by several very simple compounds, derivatives of amino acids, that function as neurotransmitters—compounds that transmit impulses between neurons (Fig. 3.1). Neurons that secrete neurotransmitters are distributed throughout the CNS, and their effects are mediated by cell membrane–bound receptors. Neurotransmitters participate in many different functions in the body. A few—for example, **dopamine, melatonin,** and possibly **serotonin**—may also be involved in regulating GnRH secretion either independently or coordinately with β-endorphin (Fig. 3.13). Current models suggests that these hormones are mediators of environmental or internal stimuli that regulate different aspects of reproduction. The precise role of these hormones remains to be worked out.

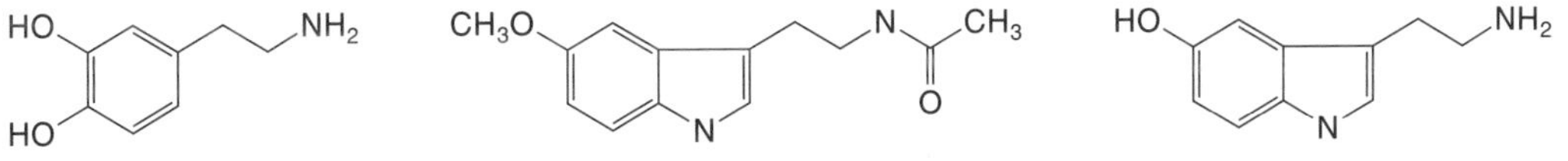

Dopamine   Melatonin   Serotonin

Figure 3.13
Diagrams showing the structures of the amine hormones dopamine, melatonin, and serotonin. All three are derived from amino acids.

SUMMARY

Many hormones function in reproduction. This text focuses on the classically important ones as well as on a few recently discovered ones that play major roles in regulating reproductive activity. These reproductive hormones fall into three structural classes: lipids, peptides and polypeptides, and amines. The sex steroids and prostaglandins are the most important of the lipid group. The sex steroids, produced primarily by the gonads, consist of the androgen, estrogen, and progestin subfamilies; these versatile hormones play essential reproductive roles in gametogenesis, the development of the secondary sex characteristics of males and females, maintenance of sexual drive, and pregnancy. They also have very important nonreproductive functions including bone mineralization, cholesterol metabolism, protein metabolism, and not-yet-well-understood regulatory effects in central nervous system function.

Some members of the peptide, polypeptide, and amine group of reproductive hormones are part of the hypothalamic-pituitary component of the H-P-G network of hormones. The major function of these hormones is to regulate gonadal activity in response to different types of endogenous and exogenous stimuli. Hormones of this group include GnRH, CRH, LH, FSH, leptin, β-endorphin, dopamine, melatonin, and serotonin. The development of the male internal genitalia depends on APH; lactation on prolactin, β-endorphin, and oxytocin; embryonic and fetal growth on hCG and hPL; and parturition on oxytocin.

QUESTIONS

1. Which hormones are primary gene products and which are derived from other compounds?

2. Despite their chemical and structural diversity, what is the one characteristic that all hormones have in common?

3. Does it make sense to talk about female and male hormones?

4. Some hormones have multiple functions; others have very restricted roles. List examples belonging to each category. Can you explain why some are multifunctional and others are not?

5. Library project: Examine the therapeutic applications of one or more of the hormone agonists or antagonists listed in Table 3.3.

SUPPLEMENTARY READING

Baulieu, E.-E., and P. A. Kelly. 1990. *Hormones: From Molecules to Disease*. Chapman & Hall, New York, NY.

Brown, R. E. 1994. *An Introduction to Neuroendocrinology*. Cambridge University Press, Cambridge, UK.

Corner, G. W. 1947. *The Hormones in Human Reproduction* (Revised edition). Princeton University Press, Princeton, NJ.

DeGroot, L. J., *et al.* (Eds.). 1995. *Endocrinology*. W. B. Saunders Co., Philadelphia, PA.

Felig, P., J. E. Baxter, and L. A. Frohman. 1995. *Endocrinology and Metabolism*, 3rd ed. McGraw-Hill, Inc., New York, NY.

Norman, A. W., and G. Litwack. 1997. *Hormones*. Academic Press, Inc., Orlando, FL.

Oudshoorn, N. 1994. *Beyond the Natural Body. An Archeology of Sex Hormones*. Routledge, London, UK.

Sandor, T., and A. Mehdi. 1979. Steroids and evolution. In *Hormones and Evolution*. E. J. W. Barrinton, Ed. Academic Press, New York, NY.

Henry Spencer Moore, *Two Forms*, 1936.
(The Philadelphia Museum of Art. Permission to reproduce from
The Philadelphia Museum of Art: Gift of Mrs. H. Gates Lloyd.)

PART

II

THE GONADS

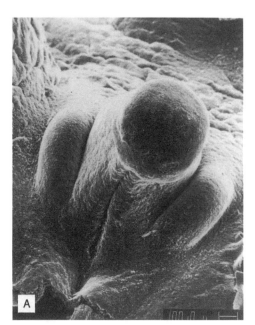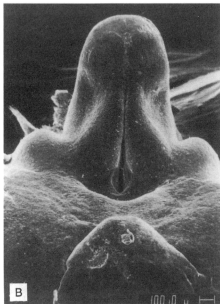

Scanning electron micrographs of the external genitalia of two human embryos before the sex of the embryo can be determined, 52–58 days after fertilization. (A) Genital tubercle, deep urethral groove, and urogenital folds. (B) Frontal view of the developing genital tubercle, urogenital folds, and the urethral groove with the external opening of the urogenital sinus. (J.E. Jirásek. 1983. *Atlas of Human Prenatal Morphogenesis*. Figs. 138 and 140. Martinus Nijhoff Publishers, with permission from Kluwer Academic Publishers.)

CHAPTER 4

Sex Determination and
Fetal Sexual Differentiation

*In my alternative scenario, the female is the ancestral sex,
and the male the derived sex . . . Every male must
contain evolutionary traces of femaleness.*

—D. Crews. 1994

Animal sexuality. *Scientific American* **270**(1), 108–114.

*Mammalian sex determination occurs
in the gonad of the developing embryo.*

—A. Swain and R. Lovell-Badge. 1997

A molecular approach to sex determination in mammals.
Acta Paediatrica Supplement **423**, 46–49.

BIRTH PRESENTS US with a *fait accompli* with respect to the sex of the infant. We are confronted with what appears to be a finished product, and we may not appreciate the complexity of the developmental process that led to that product. Indeed, it is only in the last 40 to 50 years that we have come to understand that the final sexual anatomy of every individual depends on two independent processes—the formation of the gonad and the formation of the internal and external genitalia. Quite different mechanisms are involved in these processes. It is commonly said that sex determination is controlled *genetically*, while sexual differentiation is determined *hormonally*. In this chapter, we will examine both processes. We will consider in some detail the genetics of sex determination by focusing on the consequences of errors of sex determination that occur in the human population. The consequences of errors of sexual differentiation will be considered in Chapter 14.

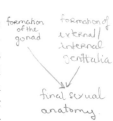

formation of the gonad

formation of external/internal genitalia

final sexual anatomy.

THE DEVELOPMENT OF SEX

The sexual anatomy of every individual develops in stages. At fertilization, the **chromosomal sex** is determined by the sex chromosome contribution of the sperm. The sex chromosome constitution of the zygote, *XX* or *XY*, in turn **determines** the **gonadal sex** of the embryo, that is, the nature of the gonads that will develop, ovaries or testes. The

sex chromosome of sperm determine the gonadal sex of the **77** *embryo*

*Phenotypic sex
→ internal & external genitalia

*♀ → don't require the present of ovaries
┌ internal:
│ • uterine tube
└ • uterus
┌ external:
│ • vagina
│ • clitoris
└ • labia

*♂ → require presence of testis
┌ internal
│ • deferens
│ • seminal vesicles
│ • prostate
└ • bulbourethral gland
┌ external
│ • penis
└ • scrotum
→ full sexual potent @ puberty

* sex determination
sex chromosome
↓
gonadal sex

* sexual differentiation
gonadal sex
↓
phenotypic sex

irreducible condition of maleness is the presence of functional testes; of femaleness, it is the presence of functional ovaries. **Phenotypic sex** is defined by the nature of the internal and external genitalia. The internal genitalia are the uterine tubes and the uterus in the female, and the epididymis, vas deferens, seminal vesicles, prostate, and bulbourethral glands in the male. The external genitalia are the vagina, clitoris, and labia in the female, and the penis and scrotum in the male. The external genitalia, although the most obvious manifestation of the sex of the individual, do not, as we will see below, necessarily define the sex of the individual. Male phenotypic sex requires the presence of testes, but female phenotypic sex does not require the presence of ovaries. These three stages of development take place during gestation and are completed by the time of birth. However, the full sexual potential of the individual is not realized until after **puberty**, during which the final growth and maturation of the internal and external genitalia, as well as the full expression of the behavioral differences between the two sexes, takes place. It is reasonable to suggest that puberty defines the fourth and final stage in the development of sex (Table 4.1).

The pathway from chromosomal sex to gonadal sex is referred to as **sex determination**, while the path from gonadal sex to phenotypic sex is referred to as **sexual differentiation**. Under normal conditions, chromosomal sex, gonadal sex, and phenotypic sex are concordant, meaning that an *XX* individual has ovaries for gonads and female internal and external genitalia, while an *XY* individual has testes for gonads and male internal and external genitalia. Infrequently, errors in sex determination result in a complex condition known as **sex reversal** or **true hermaphroditism**. Similarly, errors of sexual differentiation lead to conditions known as female or male **pseudohermaphroditism**.

Table 4.1
Stages of Sex

Stage	Female	Male
Chromosomal sex	XX	XY
Gonadal sex	Ovary	Testis
Phenotypic sex		
Internal genitalia	Uterus	Epididymis
	Uterine tubes	Vas deferens
		Prostate
		Seminal vesicles
		Bulbourethral glands
External genitalia	Vagina	Penis
	Labia	Scrotum
Puberty	Ovulation	Sperm production
	Menstruation	

The full sexual phenotype in humans develops sequentially. Chromosomal sex is set at fertilization. The gonadal sex period begins with the conversion of the indifferent gonad into an ovary or a testis. This is followed by the formation of the internal and external genitalia, which defines phenotypic sex. At puberty, the individual becomes fertile.

SEX DETERMINATION

In mammals, the crucial event in the development of the reproductive system is the formation of the gonad. The testis and ovary, although very different in their final form, develop early in embryogenesis from the same two progenitor tissues, the **genital ridges** and the **primordial germ cells** (PGCs). Both tissues are said to be *bipotential* because they are able to develop into either of two very different types of cells. The PGCs are the precursors of the gametes (ova and spermatozoa), and the lineage of cells from the PGCs to the gametes is known as the *germ line*. The genital ridges develop in the vicinity of the embryonic kidney, and consist of at least three different types of embryonic somatic cells that will eventually populate the mature gonad.

The indifferent gonad

The embryonic gonad begins to form at about the end of the third week after fertilization; it develops from interactions between the somatic cells of the genital ridge and the PGCs. A particularly interesting feature of the PGCs is that they do not originate in the genital ridge, but in fact migrate into the genital ridge from a distant (relatively speaking) site, the *yolk sac membrane* (Fig. 4.1). This phenomenon was first discovered in 1948 by E. Witschi in his classic study of early human embryogenesis. Witschi's project involved meticulous microscopic examination of thousands of thin sections of tissue taken from human spontaneous abortuses at various stages of embryogenesis. Witschi observed a group of large round cells along the yolk sac membrane, first detectable at the end of the third week of gestation. These cells could be distinguished

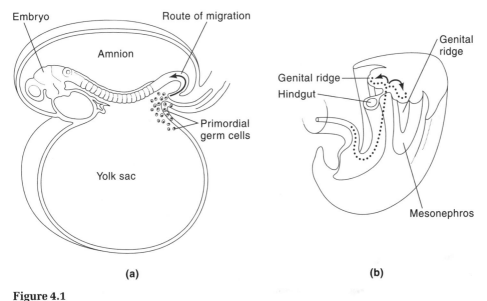

(a)   **(b)**

Figure 4.1
Schematic diagrams of a human embryo at the end of the third week of gestation.
(a) A longitudinal section showing the first appearance of the primordial germ cells (PGCs) along the yolk sac membrane. (b) Transverse view indicating the PGC route of migration along the hindgut and into the genital ridge.

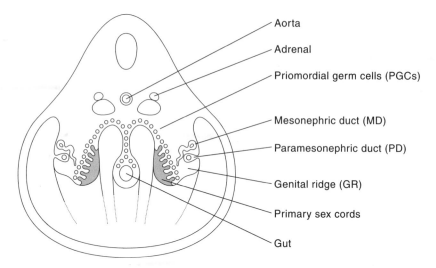

Figure 4.2
The indifferent gonad. Transverse view of the embryo at the sixth week of gestation showing the relative positions of the genital ridges and mesonephric (MD) and paramesonephric ducts (PD). The PGCs migrate along the hindgut and into the developing genital ridges. The primitive sex cords develop from the proliferation of the supporting cells along the cortex of the genital ridge. As the primitive sex cords grow into the interior (medulla) of the developing gonad, some of them envelope the incoming PGCs. The MD and PD begin to appear about the end of the fifth week. Up through the end of the sixth week, the gonad appears the same in *XX* and *XY* embryos and is known as the indifferent gonad.

from other cells of the yolk sac membrane by their large size and ability to take up a specific histological stain. By close examination of embryos at slightly later stages of gestation, Witschi inferred that these large round cells migrated from their **extraembryonic** position (outside of the embryo) during the next three weeks of development back into the embryo into the site at which the gonads would be developing. From their behavior, Witschi proposed that these migrating cells were the PGCs, an inference that was later shown to be correct. It is not absolutely clear why the PGCs originate extraembryonically. The extraembryonic origin of the PGCs is found in all mammals and in all vertebrates. One hypothesis is that the PGCs need to be protected from signals coming from the very early embryo that would convert them into somatic cells. Hence, to ensure that they remain PGCs, they are sequestered for a short period of time before they begin their migration back into the embryo proper.

Gonadogenesis, or the formation of the gonad, at least in a morphological sense, begins around the time the PGCs begin to reach the genital ridges. The genital ridge region also contains three other tissues which are important in the formation of the gonad and in the development of the internal and external genitalia—the *mesonephros* and the *mesonephric* and *paramesonephric* ducts (Fig. 4.2). At the time of their formation, the PGCs number perhaps about 500 cells. As the PGCs begin to move into the

[handwritten margin notes:]
Gonadogenesis begins when PGCs begin to Reach the genital ridges.
→ PGCs rapidly divide

genital Ridges ⎨→ mesonephros ⎬ important in
⎨→ mesonephric ⎬ formation of gonad
⎩→ paramesonephric ⎭ & development of internal/external genitalia

genital ridges, they undergo rapid cell divisions, increasing their numbers to about 50,000 in each developing gonad by the sixth week. The *primary sex cords* begin to develop along the periphery (the cortex) of the developing gonad through interactions of *supporting cells* of the genital ridge and cells from the mesonephros. As the primary sex cords grow, they penetrate into the interior (the medulla) region (Fig. 4.2). Some of the PGCs may be enveloped by the developing primary sex cords. Distributed between the primary sex cords are cells known as the *interstitial* cells or *steroidogenic* cells because they are the progenitors of cells that will be the main source of steroid hormones. Up through the end of the sixth week, no overt features mark the gonad as either a testis or an ovary, and hence, during this period the gonad is referred to as the **indifferent gonad.** Conversion of the indifferent gonad to a definitive testis or ovary begins to be discernible by changes in the internal organization of the developing gonad.

Formation of the testis

If the embryo is XY, the indifferent gonad begins to transform itself into a testis beginning at about the seventh week of gestation. A characteristic feature is the separation of the primary sex cords from the cortex and their elaboration and elongation into the medullary region becoming the medullary sex cords (Fig. 4.3). These will develop into the **testicular cords,** which in turn become the **seminiferous tubules** at puberty. The somatic cells of the cords differentiate into Sertoli cells, which become hormonally active and begin secreting antiparamesonephric hormone (APH) by about the end of the seventh week. APH causes the regression of the paramesonephric ducts. The interstitial cells differentiate into the fetal Leydig cells, which begin to secrete testosterone by about the eighth week. Testosterone stimulates the development of the mesonephric ducts. The PGCs, all of which are now enveloped by the Sertoli cells, differentiate into **spermatogonial cells.** At this point they cease dividing, and they enter and remain in a quiescent state until puberty when spermatogenesis is initiated.

Formation of the ovary

If the embryo is XX, the indifferent gonad develops into an ovary, but this transformation is delayed, at least morphologically, until about the tenth to twelfth week, with the definitive ovary traditionally being defined by the sixteenth week. In contrast to testicular development, major cellular proliferation takes place at the cortical region. The primary sex cords of the indifferent gonad degenerate and are replaced by cortical cords (Fig. 4.4). Over the next eight weeks the PGCs proliferate extensively, and then begin to differentiate into **oogonia,** and the oogonia themselves differentiate into **primary oocytes.** Meiosis is initiated in the primary oocytes, but is subsequently arrested early in the first division of meiosis. The stimulus for the arrest of meiosis appears to be enveloping of primary oocytes by the cortical cords. The cortical cords differentiate into **primordial follicles.** The primordial follicle consists of the primary oocyte surrounded by a monolayer of **follicular,** or **granulosa, cells.** There are about 7 million primordial follicles by about the twentieth week of gestation, at which time the formation of the ovary can be considered to be complete. The fetal ovary is also hormonally active and begins to produce estrogen quite early, even before the unmistakable morphological changes that define the gonad as an ovary begin to take place.

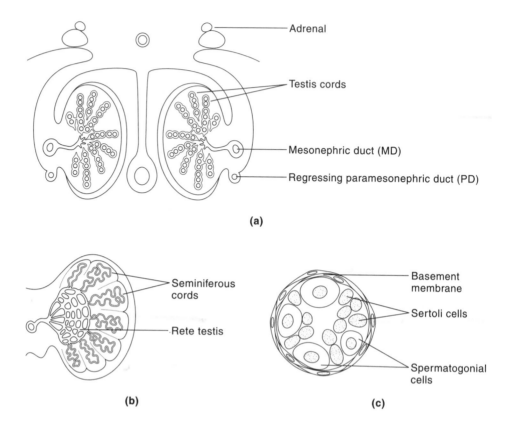

Figure 4.3

Formation of the testis. (**a**) Under the action of *SRY*, the Y-chromosome-linked testis-determining gene, the indifferent gonad begins to develop into a testis during the seventh week. A characteristic feature is the incorporation of the PGCs into the primary sex cords, the separation of the primary sex cords from the cortex, and their movement into the medullary region of the gonad. The PGCs differentiate into spermatogonial cells, but their subsequent development is arrested and they remain in a quiescent state until puberty. The primary sex cords develop into the testis cords, and cortical ends join the developing mesonephric duct. The supporting cells of the testis cords differentiate into Sertoli cells and begin to secrete APH, which causes the regression of the paramesonephric ducts. Cells of the interstitial region, the steroidogenic cells, develop into the Leydig cells and begin to secrete testosterone, which is necessary for the development of the mesonephric ducts. (**b**) By week 20 of gestation, the testis cords have matured into the seminiferous cords, which will become the seminiferous tubules at puberty. A dense network of cords, the rete testis, develops which connects the seminiferous cords with the developing mesonephric ducts. (**c**) Cross section of the seminiferous cords. Spermatogonial cells and Sertoli cells constitute the interior of the seminiferous cords, which is bounded by a basement membrane.

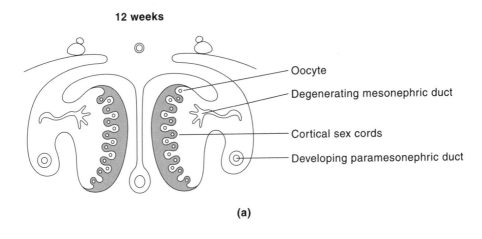

(a)

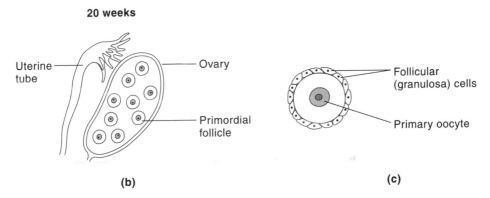

(b)   (c)

Figure 4.4
Formation of the ovary. (a) In an XX embryo (absence of SRY), conversion of the indifferent gonad into an ovary does not begin until about 10 to 12 weeks after fertilization. The PGCs proliferate, differentiate into oogonia, and these enter into the first meiotic division, becoming primary oocytes. The previously formed primary sex cords degenerate, and they are replaced by cortical cords that develop from the cortex of the gonad. The cortical cords envelope the primary oocytes, resulting in the arrest of meiosis. (b) By week 20, the cortical cords have differentiated into primordial follicles, consisting of the primary oocyte enclosed by a layer of follicular or granulosa cells. The follicular cells are derived from the supporting cells. Other cell types are present in the developing ovary. One of these is the steroidogenic cells, which later will become the theca cells, and will become part of the follicle as it matures. In the absence of testosterone, the mesonephric duct degenerates, while the paramesonephric ducts form the uterus and uterine tubes. (c) The primordial follicle consists of the arrested primary oocyte surrounded by a single layer of flattened follicular cells.

SUMMARY

The sex chromosome constitution of the embryo, set at fertilization, determines the nature of the gonad. However, the gonad does not begin to form until about the end of the third week after fertilization. The gonad forms from the interaction of two embryologically distinct tissues, the genital ridges (each forming next to the embryonic kidney), and the primordial germ cells (PGCs), which migrate from the yolk sac membrane, outside of the embryo. The PGCs are the progenitors of the germ cells and develop into oocytes or sperm cells in XX and XY individuals, respectively. Between the end of the third week of gestation when the PGCs begin to arrive at the genital ridges, and about the fifth or sixth week, the developing gonad is said to be indifferent because it does not yet have the morphological features that would identify it as an ovary or a testis. The formation of the definitive testis begins about the sixth week. A prominent feature of testicular development is the formation of the seminiferous cords, which contain Sertoli cells derived from the original cell population of the genital ridges and spermatogonial cells derived from the PGC population that colonized the genital ridges. The formation of the definitive ovary begins at about week 12 and is completed by about week 20. At that time, the prominent feature is the primordial follicles, consisting of the primary oocyte, itself derived from the PGCs, and surrounded by a layer of granulosa cells, derived from genital ridge tissue. Both the fetal testis and fetal ovary are hormonally active, the testis producing testosterone and antiparamesonephric hormone (APH), and the ovary producing estrogen.

FETAL SEXUAL DIFFERENTIATION

Internal genitalia

In the female, the internal genitalia are the **uterine tubes** and the **uterus**. The uterine tubes (also known as the oviducts or fallopian tubes) have two essential functions: the transport of sperm to the site of fertilization and transport of the fertilized egg to the uterus for implantation. The uterus provides the site of implantation and gestation, and also has an important sperm transport function. In the male, the internal genitalia, the **epididymis** and the **vas (or ductus) deferens** and urethra define the transport pathway that carries spermatozoa produced in the testis to the outside. The **seminal vesicle**, **prostate**, and **bulbourethral gland** produce different components of the seminal fluid in which the spermatozoa are bathed when they are ejaculated. This fluid also provides nutritional support for the spermatozoa after their ejaculation.

The internal genitalia develop from two progenitor tissues that appear during the indifferent gonadal stage, the **mesonephric ducts** and **paramesonephric ducts** (see Fig. 4.2). In an XX embryo, the mesonephric, or Wolffian, ducts degenerate, and the paramesonephric, or Mullerian, ducts develop into the uterine tubes and uterus, as well as the upper one-third of the vagina (Fig. 4.5). In the male, the internal genitalia develop from the mesonephric ducts, while the paramesonephric ducts degenerate. The factors governing the development of these two ductal systems are quite different. Development of the internal genitalia in the male is absolutely dependent on the presence of testosterone and APH produced by the fetal testis. Testosterone is required for the development of the mesonephric ducts, while APH is necessary to suppress development of the paramesonephric ducts. Hence, the degeneration of the paramesonephric ducts in an XY embryo is promoted actively by APH. In contrast, development of the inter-

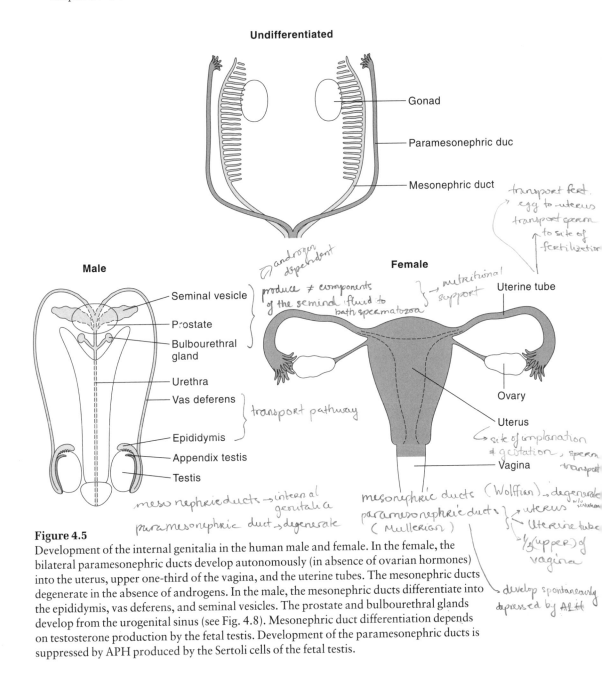

Figure 4.5

Development of the internal genitalia in the human male and female. In the female, the bilateral paramesonephric ducts develop autonomously (in absence of ovarian hormones) into the uterus, upper one-third of the vagina, and the uterine tubes. The mesonephric ducts degenerate in the absence of androgens. In the male, the mesonephric ducts differentiate into the epididymis, vas deferens, and seminal vesicles. The prostate and bulbourethral glands develop from the urogenital sinus (see Fig. 4.8). Mesonephric duct differentiation depends on testosterone production by the fetal testis. Development of the paramesonephric ducts is suppressed by APH produced by the Sertoli cells of the fetal testis.

nal genitalia in the female does not require ovarian products. The mesonephric ducts degenerate spontaneously in an XX embryo, while the paramesonephric ducts develop spontaneously. Development of the internal genitalia in the male takes place over a three- to four-week period beginning at about the ninth week. In the female, differentiation of the internal genitalia takes place over a longer period beginning about the end of the eighth week.

External genitalia

A similar distinction applies to the development of the external genitalia. The progenitor tissues here are the **genital tubercle, urogenital sinus,** and **urogenital folds** (Figs. 4.6 and 4.7). Development in the male depends on fetal testosterone and its

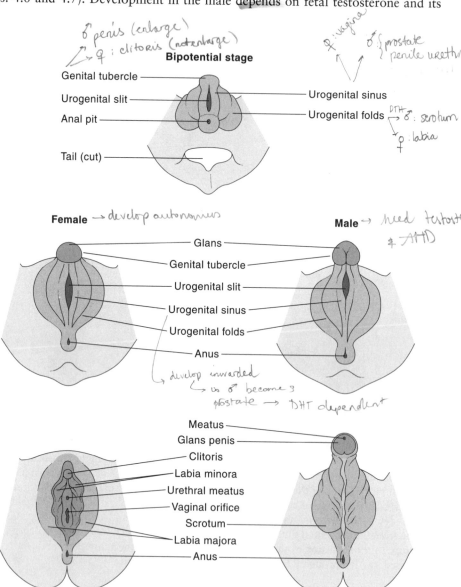

Figure 4.6

Differentiation of the external genitalia. In the female, the genital tubercle develops into the clitoris and the urogenital folds develop into the labia (major and minor). The urogenital sinus develops inwardly to form the lower two-thirds of the vagina. The development of these tissues does not require any ovarian hormones. In contrast, in the male, development of these tissues requires DHT. The genital tubercle develops into the penis, while the scrotum develops from the urogenital folds. The urogenital sinus develops inwardly to form the prostate, bulbourethral glands, and urethra.

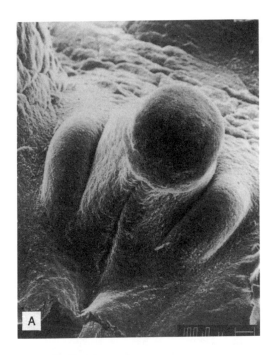

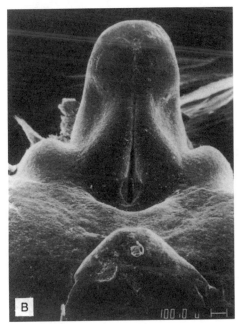

Figure 4.7
Two micrographs of external genitalia during the indifferent stage, 52–58 days after fertilization. (A) Genital tubercle, deep urethral groove, and urogenital folds. (B) Frontal view of the developing genital tubercle, urogenital folds, and the urethral groove with the external opening of the urogenital sinus. (Reprinted from J. E. Jirásek. *Atlas of Human Prenatal Morphogenesis*. 1983. Figs. 138 and 140. Martinus Nijhoff Publishers, with kind permission from Kluwer Academic Publishers.)

conversion to DHT. The testosterone enters the cells of these tissues and is converted to dihydrotestosterone (DHT). Under DHT stimulation, the genital tubercle and urogenital folds develop into the penis and scrotum, respectively, while the urogenital sinus develops into the prostate and the penile urethra (Fig. 4.8). In the female, the genital tubercle does not enlarge, but forms the clitoris, while the urogenital sinus and urogenital folds become the vagina and the labia (Figs. 4.6 and 4.8). Development of the external genitalia in the female does not require ovarian hormones. Differentiation of the external genitalia in the male begins during the ninth week and is completely by the end of the twelfth week. They continue to grow continuously throughout gestation. Differentiation of the external genitalia in the female, beginning also during the ninth week, is completed by the end of the twelfth week.

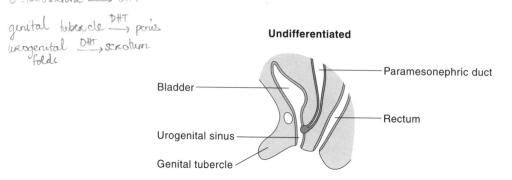

Undifferentiated

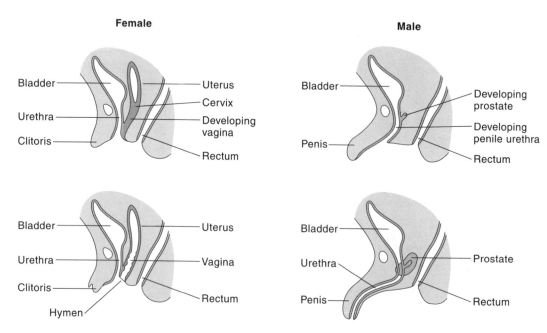

Figure 4.8
Differentiation of the urogenital sinus. In the female, the urogenital sinus develops into the lower two-thirds of the vagina. In the male, the urogenitalinus, under the action of DHT, is converted into the urethra, which connects to the vas deferens, the prostate, and bulbourethral glands.

Descent of the testes and ovaries

The testes and the ovaries both **descend** from their original high abdominal position. In both sexes, movement of the gonad depends on a ligamentous cord called the **gubernaculum**. In the male, the gubernaculum shortens and ultimately brings the testes into the scrotum (Fig. 4.9). Testicular descent appears to be dependent on testosterone and APH, and takes place primarily during the third trimester. Over 95 percent of male infants have the testes in the scrotum at birth. Failure of the testes to descend properly is known as **cryptorchidism** (hidden testes). When necessary, cryptorchid testes can be brought into the scrotum surgically. Untreated cryptorchidism leads to infertility due to failure of spermatogenesis and an increased risk of testicular cancer. In the female, the gubernaculum does not shorten, but nevertheless causes the ovaries to be moved into a peritoneal fold known as the broad ligament of the uterus.

[handwritten margin notes: gubernaculum — movement of gonad; ♂ gubernaculum shorten → testes to scrotum (depends on testosterone & APH); ♀ gubernaculum does not shorten → ovaries moved into a peritoneal fold]

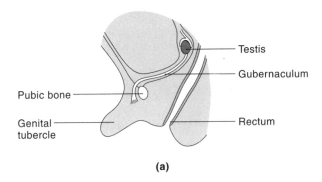

(a)

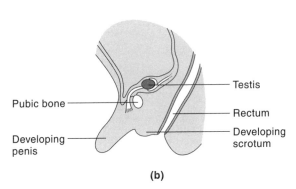

(b)

Figure 4.9
Schematic diagram of testicular descent. Migration of the testis is accomplished by shortening of the gubernaculum. (a) Early stage, testis is still in a high abdominal position. (b) Testis has moved into the inguinal position. (c) Testis reaches the scrotum. Testicular descent is androgen, and possibly APH, dependent.

[handwritten: Caused of non-descending, 1) genetic, gubernaculum deficiency 2)]

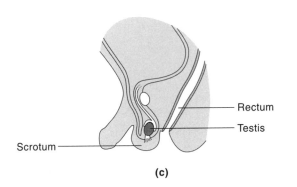

(c)

The induced versus the default sex

[handwritten margin notes: ♀: default pathway; ♂: induced pathway; XY human lacks testosterone → ♀ external genitalia; absence of ♂ internal genitalia]

Comparison of the mode of development of the internal and external genitalia reveals a profound difference between males and females. In the absence of any gonad at all, the internal and external genitalia take the female pattern of development, irrespective of the chromosomal sex of the embryo. In experimental animals, removal of the indifferent gonad results in internal and external genitalia that are completely femalelike, irrespective of the chromosomal sex. In XY human embryos, absence of testosterone action results in female external genitalia and absence of male internal genitalia (see Chapter 14). The female pattern of development is autonomous and is often referred to as the "default" pathway. In contrast, the male pattern is a deviation from the default and can take place only if a functional fetal testis is present. In this sense, male development can be said to be "induced." Only when this fundamental difference is understood does it make sense to think of the female as the "ancestral" sex, and male as the "derived" sex.

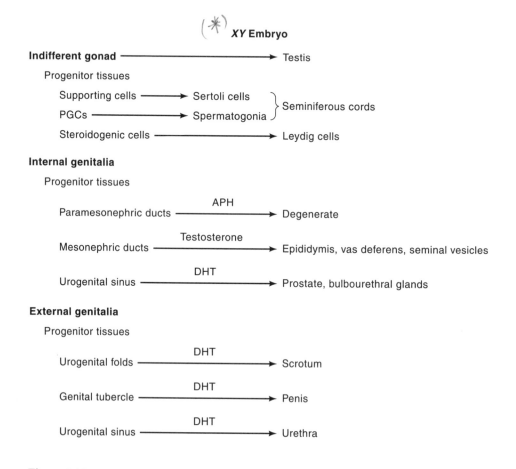

Figure 4.10
Summary of testis formation and fetal sexual differentiation in XY embryos. The gonad and internal and external genitalia develop from bipotential progenitor tissues. Three hormones—testosterone, DHT, and APH—determine the development of the internal and external genitalia in the male.

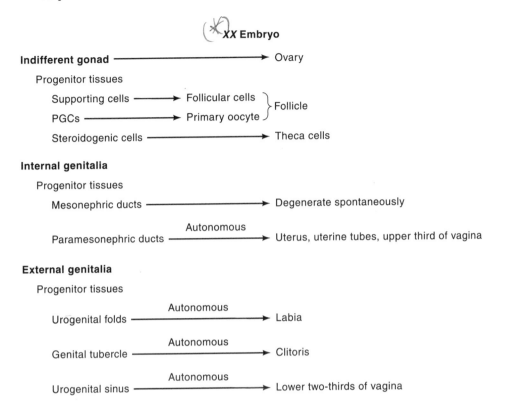

Figure 4.11
Summary of ovary formation and fetal sexual differentiation in XX embryos. Development of the internal and external genitalia does not require ovarian hormones.

The pathways of fetal sex differentiation are summarized in Figs. 4.10 and 4.11, while the timetable for the development of the different tissues is shown in Fig. 4.12. Note the differences in the timing of the initiation of testicular and ovarian development and the relatively long periods required for the development of the internal and external genitalia. As indicated above, fetal testicular testosterone production begins very early and continues throughout most of gestation, decreasing during the third trimester of pregnancy. Estrogen production by the fetal ovary follows a similar pattern. The precise function of fetal ovarian estrogen remains unclear.

SUMMARY

Both the internal and external genitalia in males and females are formed from progenitor tissues that are present by around the sixth or seventh week of gestation. The internal genitalia of the female (uterus and uterine tubes) are formed from the paramesonephric duct, while the internal genitalia of the male (epididymis, vas deferens, seminal vesicles, and prostate and bulbourethral glands) develop from the mesonephric duct and urogenital sinus. In both sexes, the external genitalia develop from the same progenitor tissues, the genital tubercle, urogenital sinus, and urogenital folds. There is a fundamental asymmetry in the nature of male versus female sexual differentiation. In

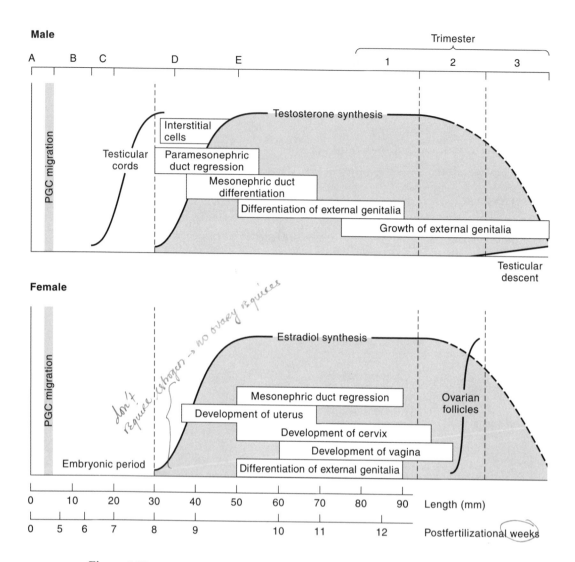

Figure 4.12

Timetable of major events in the human fetal sexual differentiation. Note the time difference between the initiation of testicular and ovarian function. Most sexual differentiation takes place during the first trimester (about the first 12 weeks). Testicular descent begins late in the second trimester and continues until the end of gestation. The mm scale indicates the length of the embryo at these different times in gestation. (Adapted from J. D. Wilson *et al.* 1981. The hormonal control of sexual reproduction. *Science* **211**, 1278–1284.)

males, the fetal testicular hormones testosterone and APH are required for the development of both the internal and external genitalia. In contrast, ovarian products are not required for development of the internal and external genitalia in the female. The male is said to be the induced sex, while the female can be considered to be the default sex. This difference indicates that sexual differentiation in the female is autonomous, while sexual differentiation in the male represents a deviation from the autonomous, or preprogrammed, pathway.

THE GENETICS OF SEX DETERMINATION

The SRY gene

The clear difference between fetal testicular and fetal ovarian functions with respect to the development of the internal and external genitalia was first reported by A. Jost in France in a series of studies carried out in the 1940s and 1950s. His main conclusion was that development of the full sexual phenotype was determined by the presence or absence of the fetal testis. If the fetal testis was present, male development took place, meaning male internal and external genitalia. Moreover, male development depended, as Jost showed, on two fetal testis products—testosterone, and another postulated (at that time) factor he referred to as *Mullerian-inhibiting substance*, later called *anti-Mullerian hormone*, or more recently, *antiparamesonephric hormone* (APH). If the testis was absent, which would be the case in an *XX* embryo, or if the indifferent gonad in an *XY* was removed before it became a testis, female development resulted. These observations indicated that the development of what we now call phenotypic sex took place in two stages. The first was the formation of the gonad, the critical step being the formation of the testis. Hence, "sex determination" really meant "testis" determination.

Jost proposed that sex determination depended on the presence or absence of a testis determination factor (TDF). If TDF was present, the indifferent gonad developed into a testis, and the development of phenotypic sex now depended on the two fetal testicular hormones, testosterone and APH. If TDF was absent, the indifferent gonad developed into an ovary, and the constitutive or default pathway of internal and external genitalia development followed. At the time of Jost's proposal, the nature of TDF was completely unknown.

An important clue about the nature of TDF came later when the *XX/XY* basis of sex determination in humans was discovered in 1959. The importance of the Y chromosome for male development was made apparent when individuals with an *XXY* genotype were discovered. In such individuals, despite the presence of two X chromosomes, the phenotypic sex of the individual was unambiguously male. The testes were smaller than in *XY* males and *XXY* individuals were generally infertile, but they were male in phenotype. Other cases of sex chromosome abnormalities were discovered, *XXXY* and even *XXXXY*. In all these cases, irrespective of the number of X chromosomes, the presence of the Y chromosome was always correlated with maleness—defined as having a testis for the gonad. The obvious conclusion was that TDF was in all likelihood a gene or genes on the Y chromosome.

Thus began a 30-year search for TDF, a period characterized by successively localizing TDF to smaller and smaller regions of the Y chromosome. Finally, in 1990, a gene named *SRY* and located close to the tip of the short arm of the Y chromosome was isolated and provisionally put forth as the candidate for TDF. Since then, many observations in humans and experimental work in mice are all consistent with *SRY* being the same as TDF. In a variety of experimental situations in mice, it has been shown that the presence of *SRY* in an otherwise *XX* background is sufficient to induce testis formation, showing that *SRY* is the only Y chromosome gene required for testis formation.

The next important question was *What does SRY do?* Unfortunately, we do not have a definitive answer to that question yet, but we are getting close. The most recent hypothesis about the nature of the *SRY* function has come from experiments with mice and analysis of errors of sex determination in humans. These errors result in two types of disorders, *sex reversal* and *true hermaphroditism*. Analysis of sex reversal in hu-

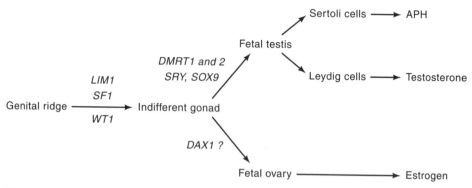

Figure 4.13

Genes involved in sex determination. This scheme has developed from studies of sex reversal in humans and experimental studies in mice. The genes *SF1*, *WT1*, and LIM1 appear to be required for the formation of the indifferent gonad itself. *SRY*, *SOX9*, DMRT1, and *DMRT2* are required for the conversion of the indifferent gonad into a testis, with *SRY* being the initiator of this conversion. *SRY* is a Y-linked gene; *SOX9*, *DMRT1*, and *DMRT2* are autosomal genes. Genes necessary to form the ovary have not yet been definitively identified, although *DAX1*, an X-linked gene, may belong to this group. *DAX1* appears to have a critical regulatory role in preventing the formation of testes in XX embryos (see Fig. 4.14).

SRY is part of a multistep bifurcated pathway for gonadogenesis mans and the counterparts in the mouse has indicated that *SRY* is part of a multistep, bifurcated pathway for gonadogenesis (Fig. 4.13).

The pathway begins with the formation of the indifferent gonad, which appears to require interactions between the cells of the genital ridge and the incoming primordial germ cells. Studies in the mouse suggest that at least three genes, *SF1, WT1,* and *LIM,* are required for this transition (Fig. 4.13). It is not clear whether these early steps take place in the same way in XX and XY embryos. If the embryo is XY, the effect of *SRY* is to initiate the conversion of the indifferent gonad into a testis. Once *SRY* acts, three other genes are required for testis formation: *SOX9, DMRT1,* and *DMRT2.* XY embryos that carry mutations in any of these genes develop ovaries instead of testes. Action of *SRY* and *SOX9* permits differentiation of the supporting cells into Sertoli cells and the steroidogenic cells into Leydig cells. The roles of *DMRT1* and *DMRT2* are less clear. Then under the action of *SF1*, the Leydig cells begin to secrete testosterone and the Sertoli cells, APH. Other genes are probably necessary as well, but at present we know very little about them. If the embryo is XX, another set of genes is presumably activated, in this case converting the indifferent gonad into an ovary. At present, no ovary-determining genes have been definitively identified, although some investigator suggest that *DAX1* may play a role. As the ovary is forming, the supporting cells differentiate into the follicular (granulosa) cells, and the steroidogenic cells into cells that will develop into the thecal cells at puberty. This schematic model is woefully incomplete and clearly much more work is necessary before more complete details of the sex determination pathway are available.

This short discussion has not indicated how this schematic model of sex determination was developed. The section below will review the observational and experimental evidence that has suggested this model. It is a good example of how a scientific model of a complex developmental process is constructed. Many experimental details have been omitted, and the arguments presented here are not necessarily the way or the order in which they appeared in the original publications. The focus is on using the observational data to develop a scheme for gonadogenesis.

A working model for sex determination

The focus of studies aimed at unraveling the pathway of sex determination has been to identify the genes involved in the process. Analysis of the consequences of naturally occurring mutations that disturb gonadal development in humans has been critically important in developing a model of sex determination. Let us consider how the different types of errors have contributed to the elaboration of the model.

Sex reversal and true hermaphroditism

Sex reversal occurs when XX individuals have testes rather than ovaries, or alternatively, when XY individuals have ovaries rather than testes. Sex-reversed individuals are infertile and in many cases have a number of other somatic abnormalities. Here we will be concerned primarily with their sexual phenotypes. In a general sense, five types of sex reversal can be distinguished. These can be indicated as follows:

1. $XX (SRY^+)$ males   2a. $XY (SRY^-)$ females

3. $XX (SRY^-)$ males   2b. $XY (SRY^m)$ females

4. $XY (SRY^+)$ females

where SRY^+ and SRY^- refer to the presence or absence of the SRY gene, respectively, while SRY^m indicates a mutated, nonfunctional form of the SRY gene. Hence, with respect to SRY function there are two classes of XX sex reversal—those with a functional SRY gene and those without a functional SRY gene. XY sex reversal can occur in the absence of SRY, with a nonfunctional SRY, or with a functional SRY gene.

True hermaphroditism is defined by the presence of both ovarian and testicular tissue in the same individual irrespective of the sex chromosomal constitution. This disorder is rarer than the sex reversal types discussed in the previous section. Several different combinations of gonadal tissue are possible, including one ovary and one testis, with the testis being found preferentially on the right side; one ovary and one ovotestis (a gonad that is both ovary and testis); one testis and one ovotestis; and two ovotestes. Although most hermaphrodites are infertile, there have been a few cases in which the individuals were considered to produce viable gametes, and there is even one report of a hermaphrodite who gave birth. Internal and external genitalia, as might be expected, are generally abnormal, with the particular pattern depending presumably on the "strength" of the testicular hormonal secretions, being more malelike, or more femalelike in different individuals. There may also be an ethnic difference in hermaphrodites. Among Caucasian/African populations, the majority of hermaphrodites are XX, while in the two Asian populations for which some information is available (Chinese and Japanese), the distribution is reversed, with most hermaphrodites being XY. There is no general model that accounts for this difference.

The autonomous pathway

Sex reversal and true hermaphroditism tell us, if nothing else, that sex determination cannot be a simple affair. Let us consider first how we might understand the different classes of sex reversal. Since an ovary forms in the absence of the Y chromosome, we can imagine that the transformation of the indifferent gonad into an ovary is an autonomous pathway. As a consequence, the formation of the testis must require that the ovarian pathway be suppressed and that the testicular pathway be turned on. Given the proposed role for *SRY*, sex reversal classes **1** and **2** can be explained in a reasonably straightforward fashion. For example, class **2b** is characterized by a mutation in *SRY* that renders the *SRY* protein nonfunctional, and therefore unable to transform the indifferent gonad into a testis. In the absence of testis formation, an ovary forms, and the individual is female in phenotype.

Class **1** sex reversal cases arise from occasions in which the *SRY* gene is translocated from the Y chromosome to the X chromosome. This occurs with a low frequency during meiosis in males and results in two types of spermatozoa: those carrying an X chromosome harboring the *SRY* gene and those carrying a Y chromosome missing *SRY*. A class **1** embryo is formed following fertilization by a spermatozoon carrying an X chromosome with the *SRY*. The formation of a testis in such an embryo is then consistent with the function of *SRY*. On the other hand, a class **2a** embryo arises from fertilization by a spermatozoon carrying a Y chromosome missing the *SRY* gene. In the absence of *SRY*, an ovary forms and a female phenotype follows. Class **1** and **2** sex reversal types, in fact, establish the role of *SRY* as the TDF in humans.

Other genes

Class **3** and **4** sex reversal seem paradoxical at first glance because they appear to be inconsistent with the role of *SRY* indicated above. The simplest way to resolve the paradox is to realize that these two types of sex reversal tell us that there must be other genes involved in sex determination. Consider class **4** first. Here, ovaries develop despite the fact that *SRY* is present and functional. We can imagine, for example, that *SRY* is the first of several genes whose action is required to convert the indifferent gonad into a testis. We can call this the testis pathway, and the genes the *testis-determining genes* (TDGs). A mutation in any one of the TDGs may result in failure to form a testis and failure to form a testis leads to the formation of an ovary. The validity of this picture has been confirmed by the identification of at least one gene, called *SOX9*, acting downstream of *SRY* and necessary for testis formation. Mutations in *SOX9* lead to sex reversal in XY embryos despite a functional *SRY* gene (Fig. 4.14).

Class **3** sex reversal, testis development in the absence of *SRY* function, presents a more involved challenge. We need to keep in mind that an explanation for class **3** sex reversal should be consistent with the model for sex determination that has been emerging in our discussion above. Since we have suggested that the TDGs are activated by *SRY*, class **3** sex reversal must indicate that under some circumstances the TDGs can be activated even in the absence of *SRY*. Moreover, since testis formation is the induced pathway, the TDGs must normally be suppressed in an XX embryo, while another set of genes, the *ovary-determining genes* (ODGs), must be turned on autonomously. We can surmise that there must be a gene or set of genes that function to suppress the TDGs in XX embryos. Therefore, we can imagine that class **3** sex reversal could arise if the putative suppressor(s) of the TDGs fail to work, for example, due to mutations in the genes that encode the suppressor(s) of the TDGs. Carrying the logic further, we can assume that in XY embryos, suppression of the TDGs must be overcome in order for testis formation to occur. The simplest possibility is that the function of *SRY* is to relieve TDG suppression, thereby permitting the TDG pathway to be activated. One possibility is that an interaction of the *SRY* protein with the putative suppressor pro-

[handwritten marginal notes:]
Class 4 (sex reversal) SOX9 (down stream of SRY) mutated

Class 3: — suppressor of TDGs fail to work.

SRY → relieve the inhibition of the expression of the TDG

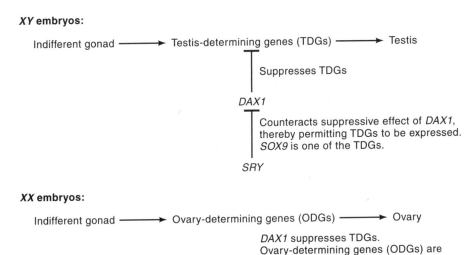

XY embryos:

Indifferent gonad ⟶ Testis-determining genes (TDGs) ⟶ Testis

Suppresses TDGs

DAX1

Counteracts suppressive effect of *DAX1*, thereby permitting TDGs to be expressed. *SOX9* is one of the TDGs.

SRY

XX embryos:

Indifferent gonad ⟶ Ovary-determining genes (ODGs) ⟶ Ovary

DAX1 suppresses TDGs.
Ovary-determining genes (ODGs) are expressed autonomously.

Figure 4.14
Model of sex determination indicating possible roles of *SRY* and *DAX1*. Experiments in mice indicate that *DAX1* can be considered to be an antitestis gene; that is, its normal role is to suppress the expression of the testis-determining genes (TDGs), such as *SOX9*. The function of *SRY* is to counteract the effects of *DAX1*, thus allowing the TDGs to be expressed and converting the indifferent gonad into a testis. This model is compatible with results obtained from the analysis of all types of sex reversal in humans (see text for discussion). (References: A. Swain and R. Lovell-Badge, 1997; K. McElreavey and M. Fellous, 1997; R. Jiménez and M. Burgos, 1998.)

tein inactivates the suppressor protein or otherwise relieves the inhibition of the expression of the TDGs.

Strong evidence for this model of *SRY* action has been provided by the discovery of a gene, *DAX1*, first discovered in humans, that appears to be the suppressor of the TDGs. The mouse Sry protein has been shown to interact with the mouse Dax1 protein. The consequence of that interaction depends on the ratio of the Sry to Dax1 protein. Ratios of four or five to one lead to expression of the mouse TDGs. In humans, *DAX1* is an X-linked gene, and this means that XX embryos have two copies, while XY embryos have one copy. The discovery of *DAX1* came about by the fortuitous identification of XY sex reversal in which the affected individual carried two copies of *DAX1*, the normal one on the X chromosome, and an extra copy on the Y chromosome. Two functional copies of *DAX1* are apparently sufficient to overcome the influence of one copy of *SRY*, and the consequence is that an ovary, rather than a testis, forms. Normally, XY embryos have one copy of *DAX1* and one copy of *SRY*, and under such conditions, *SRY* wins out, and the TDG pathway is activated. Hence, in humans the ratio of the SRY protein to DAX1 protein is also critically important in relieving the *DAX1*-mediated suppression of the TDGs.

These experiments and observations suggest a model in which *DAX1* has an "antitestis" function and in which *SRY* interferes with *DAX1* function (Fig. 4.14). In

[handwritten margin notes: X-linked; DAX1: suppressor of TDGs; 4 Sry : 1 Dax1 → express TDGs; Ratio of SRY:DAX1 is important in relieving the DAX1-mediated suppression of the TDGs]

an *XX* embryo, *DAX1* suppresses the TDGs and prevents testis formation. The ODGs are turned on by some as yet unknown gene. In contrast, in an *XY* embryo, *DAX1* suppression is relieved by *SRY*, thereby permitting the TDGs to be turned on. *SOX9* is one of the TDGs, and *SOX9* expression is necessary to drive the conversion of the indifferent gonad into a testis. Once the testis begins to form, testosterone and APH production begins, and these two hormones guide development of the male internal and external genitalia. Experiments in the mouse indicate that the formation of the indifferent gonad itself requires at least two genes, *SF1* and *WT1*, since mice with mutations in either gene fail to develop any gonad. The first case of an *SF1* mutation in humans was described recently. Gonads failed to form in the affected individual, indicating that *SF1* is also required to form the indifferent gonad in humans as well.

The scheme illustrated in Fig. 4.14 provides a framework for understanding the four classes of sex reversal. True hermaphroditism, however, remains a puzzle. A true hermaphrodite arises presumably when neither the testicular nor the ovarian pathway is completely suppressed. In an *XY* embryo, one possibility is that conversion of the indifferent gonad into a testis requires a certain threshold of SRY protein activity. Lower-than-normal SRY protein levels might permit both the testicular and ovarian pathways to be activated simultaneously. An alternative possibility, based on experiments in mice, suggests that the timing of *SRY* and *DAX1* action may also be important. Hence, one can propose that under normal conditions, *SRY* has to act within a certain "window of opportunity." A time mismatch between *DAX1* and *SRY* action, for example, could permit ovarian development to be initiated, but would not necessarily abort testicular development. Both suggestions, while plausible, are at this stage speculative. Moreover, the model will need to account for the unexpected and unexplained ethnic correlation of hermaphroditism as well. A more complete genetic dissection of the testicular and ovarian pathways is necessary before we will really understand how hermaphrodites arise.

SUMMARY

Analysis of sex reversal in humans has led to the discovery of a number of genes that are involved in sex determination. The current model, summarized in Fig. 4.14, embodies several elements. The critical gene is *SRY*, a Y-linked gene whose role is essential for testis formation in an *XY* embryo. Two sets of genes are proposed, testis-determining genes (TDGs) and ovary-determining genes (ODGs). Expression of the TDGs and ODGs transforms the indifferent gonad into a testis and an ovary, respectively. TDGs are programmed to be expressed earlier than ODGs. In an *XX* embryo, expression of the TDGs has to be suppressed while permitting expression of ODGs. Although no ODGs have yet been identified, one gene, *DAX1*, may function as the suppressor of the TDGs. In an *XY* embryo, expression of the ODGs must be suppressed while permitting expression of the TDGs. In order for the TDGs to be expressed, the inhibition by *DAX1* needs to be relieved. The model suggests that the role of *SRY* is to overcome the *DAX1*-mediated suppression of the TDGs, perhaps by interacting directly with the DAX1 protein. Suppression of the ODGs is maintained by a fetal testicular product, perhaps APH. At least one TDG, *SOX9*, has been identified. Although the current model represents a significant conceptual advance in our understanding of sex determination, particularly in a role for *SRY* and sex reversal cases in humans, much more work remains to be done in discovering other TDGs or ODGs.

QUESTIONS

1. Explain the results of the following two experiments:
   a. If an orchidectomized (testes removed) male embryo is given testosterone, both mesonephric and paramesonephric ducts will develop.
   b. If an intact male embryo is given an antiandrogen, no internal genitalia develop.

2. The development of phenotypic sex is said to be autonomous in the female, and induced in the male. Do these two terms apply in describing the development of gonadal sex? Why or why not?

3. In true hermaphrodites with ovotestes, the medullar region is the testislike part, while the cortex is ovarylike. Is this to be expected given the descriptions of the formation of the ovary and testis presented in this chapter?

4. An unusual type of *XY* sex reversal was observed recently. This individual carries a mutation in the *SF1* gene. What type of gonad and internal and external genitalia would you expect to see in this patient? Is this result compatible with the discussion in this chapter?

5. Why do you think it has been more difficult to identify genes required for the formation of the ovary than for the testis?

6. What do you think is the significance of the fact that testicular development takes place before ovarian development?

7. Library project: True hermaphrodites differ greatly in their anatomical and endocrinological features. How can this variability be explained?

SUPPLEMENTARY READING

Berkovitz, G. D., and J. Seeherunvon. 1998. Abnormalities of gonadal differentiation. *Bailliere's Clinical Endocrinology and Metabolism* 12, 133–142.

Capel, B. 1998. Sex in the 90s: *SRY* and the switch to male pathway. *Annual Reviews of Physiology* 60, 497–523.

Crews, D. 1994. Animal sexuality. *Scientific American* 270(1), 108–114.

Hiort, O., and P.-M. Holterhus. 2000. The molecular basis of male sexual differentiation. *European Journal of Endocrinology* 142, 101–110.

Moore, K. L., and T. V. N. Persaud. 1993. *The Developing Human*, 5th ed. W. B. Saunders and Co., Philadelphia, PA.

ADVANCED TOPICS

Achermann, J. C., *et al.* 1999. A mutation in the gene encoding steroidogenic factor 1 causes *XY* sex reversal and adrenal failure in humans. *Nature Genetics* 22, 125–126.

Jiménez, R., and M. Burgos. 1998. Mammalian sex determination: Joining pieces of the genetic puzzle. *BioEssays* 20(9), 696–699.

Jost, A. 1953. Problems of fetal endocrinology: The gonadal and hypophyseal hormones. *Recent Progress in Hormone Research* 8, 379–391.

McElreavey, K., and M. Fellous. 1997. Sex-determining genes. *Trends in Endocrinology and Metabolism* 8(9), 342–346.

Raymond, C. S., *et al.* 1999. A region of human chromosome 9p required for testis development contains two genes related to known sexual regulators. *Human Molecular Genetics* 8, 989–996.

Swain, A., and R. Lovell-Badge. 1997. A molecular approach to sex determination in mammals. *Acta Paediatrica Supplement* 423, 46–49.

Witschi, E. 1948. Migration of the germ cells of human embryos from the yolk sac to the primitive gonadal folds. *Contributions to Embryology* 32, 67–79.

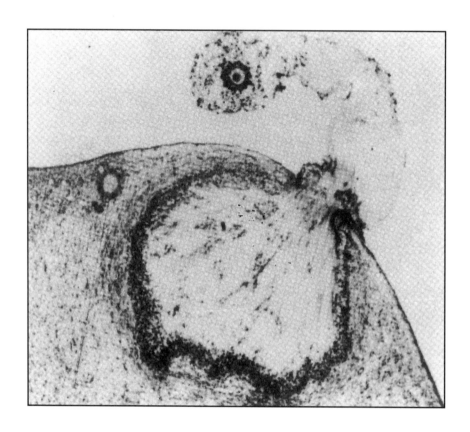

Ovulation. Photomicrograph showing the egg and its surrounding cloud of cumulus cells being expelled from the ovary. (R. J. Blandau. 1970. *Progress in Gynecology* 5, 58, with permission, W. B. Saunders Company, Philadelphia, PA.)

CHAPTER 5

The Ovary: The Ovulatory and Menstrual Cycles

*It has been completely wrong to regard the uterus as the
characteristic organ. . . . The womb, as part of the sexual canal,
of the whole apparatus of reproduction, is merely an organ of
secondary importance. Remove the ovary, and we shall have before us a
masculine woman, an ugly half-form with the coarse and harsh form, the
heavy bone formation, the mustache, the rough voice, the flat chest, the sour
and egoistic mentality and the distorted outlook . . . in short, all that admire
and respect in woman as womanly, is merely dependent on her ovaries.*

—R. Virchow (1817–1845), cited in V. C. Medvei. 1989

A History of Endocrinology. MTP Press, The Hague, p. 215.

*Concerning the influence of the ovary on menstruation,
it is probable that the ovary only regulates the condition.
The ovary is a protective, not a causative, factor in menstruation.
Symptoms of menstruation have been observed in the human subject
after double ovariotomy. In menstruation the whole organism is
involved and the ovary plays its role only as a part of the whole.*

—E. C. Dudley and H. M. Stowe. 1913

Gynecology. Practical Medicine Series, Vol. V.
The Year Book Publishers, Chicago, p. 181.

THE IMPORTANCE OF the ovary began to be appreciated by around the middle of the nineteenth century (see epigraphs above). Before then, the uterus occupied the most prominent place in female anatomy and physiology. Attention began to shift from the uterus to the ovary in accordance with increasing knowledge of the role of the ovary in most aspects of female physiology. An appreciation of the role of the ovary in orchestrating the changes in the uterus that are known as the **menstrual cycle** took longer, as the epigraph taken from a standard 1913 text on gynecology demonstrates. It also took some time to appreciate the changing activity and function of the ovary during the lifetime of the female.

We can discern four stages in the life of the typical ovary, each stage corresponding to the four developmental phases in the reproductive life of the female: the prenatal (fetal) ovary; the postnatal and prepubertal ovary; the postpubertal ovary; and finally, the postmenopausal ovary. We probably know most about the postpubertal and postmenopausal ovary. During the postpubertal period, ovarian activity dominates the entire physiology of the female, and more often than not, her sense of well-being is closely related to the status of the activity of her ovaries.

The postpubertal ovary has two important functions: (1) the production of the female gamete, the egg, or oocyte, and (2) the synthesis and secretion of steroid sex hormones, principally estrogens and progestins. As we shall see below, the hormone-producing activity of the ovary is inseparable from its egg-producing capability. Ovarian hormones, particularly estrogens, are involved in many functions. Estrogens determine the female physiognomy—body shape, distribution of fat, breast development. Moreover, the wide distribution of estrogen receptors throughout the central nervous system (CNS) implies important functions for estrogens in regulating CNS activity. The important point is that ovarian activity during the postpubertal period is not only essential for normal reproductive function, but is probably very important for many other nonreproductive functions as well. In this chapter we will focus primarily on the activity of the postpubertal ovary. We will consider the postmenopausal ovary in Chapter 10.

A SHORT HISTORY OF THE OVARY

A reasonably comprehensive and realistic understanding of the nature of ovarian function was achieved only in the latter half of the twentieth century. But we might well ask: *When was the ovary discovered, and why was it ignored for so long?* The most likely answer is that it is, after all, a small, rather inconspicuous, internal organ, certainly not as striking as the testis, and connecting it to reproductive functions is not at all obvious. No precise date for its discovery can be given; rather our understanding of ovarian function has grown slowly over the centuries. We do know, for example, that the spaying of female animals has been practiced for at least 2500 years. One of the earliest extant descriptions of this procedure comes from Aristotle (384–322 BCE), and it suggests an already long-standing practice.

> *The ovaries of sows are excised with the view of quenching in them sexual appetites and of stimulating growth in size and fatness. The sow has first to be kept two days without food, and, after being hung up by the hind legs, is operated on; they cut the lower belly about the place where the boars have their testicles, for it is there that the ovary grows, adhering to the two divisions (or horns) of the womb; they cut off a little piece and stitch up the incision. Female camels are mutilated when they are wanted for war purposes, and are mutilated to prevent their being got with young.*[1]

Although in the English translation the term "ovaries" is used, commentators have noted that it is not clear from the Greek text that the ovaries referred to are in fact the true ovaries. Nevertheless, it seems apparent that an empirically based practice for lim-

---

[1]Aristotle. *Historia Animalium*. In *The Works of Aristotle*, Vol. IV. 1910. Translated by D. W. Thompson, Clarendon Press, Oxford. Cited in Short, 1977.

iting the fertility of sows and female camels was well established by Aristotle's time. The origin of this practice remains unknown. Moreover, the behavioral aspects of spaying—"quenching in them sexual appetites"—were also recognized. What is surprising is that this empirical knowledge was not translated into a more significant understanding of their function. In fact, the recognition of the ovaries as distinct anatomical structures was not provided until about 400 years later by Soranus of Ephesus (ca. 80 CE), who referred to them as *didymi* (paired organs), and somewhat implicitly likened them to the testes:

> *Furthermore, the didymi are attached to the outside of the uterus, near its isthmus, one on each side. They are of loose texture, and like glands are covered by a particular membrane. Their shape is not longish as in the males; rather they are slightly flattened, rounded and a little broadened at the base. The seminal duct runs from the uterus through each didymus and, extending along the side of the uterus as far as the bladder, is implanted in its neck. Therefore, the female seed seems not to be drawn upon in generation since it is excreted externally.*[2]

The function of these structures remained unknown, and it was not until about 1500 years later that the ovary became the subject of new and more systematic anatomical investigation. Vesalius (1514–1564) in his detailed anatomical drawings provided one of the first description of the ovaries and some of the ovarian contents (Fig. 5.1). He provided perhaps the first recorded description of the pigmented **corpus luteum**, whose function we will discuss below. Note, however, that he referred to them as "testes."

> *The testes of women contain, besides blood vessels, some sinuses full of a thin watery fluid which, if the testis has not been previously damaged, but is squeezed and makes a noise like an inflated bladder, will spurt out like a fountain to a great height during the dissection. As this fluid is white and like a milky serum in healthy women, so I have found it to be a wonderful saffron yellow colour and a little thicker in two well-bred girls who were troubled before death with strangulation of the womb; the testis of one of the girls, or at any rate one of the sinuses in it, protruded like a rather large pea full of a yellow fluid, colouring the adjacent tissues just as in man the colon is coloured yellow where it passes beneath the liver, by the gall bladder. This colour rarely occurs in the tissues or the fluids; it also smelt very bad and had something poisonous and foul in it.*[3]

The first coherent suggestions of the ovary as the "female testis" came in the seventeenth century when several investigators proposed that the mammalian ovary was the counterpart of the ovary of the hen, and that it produced eggs. The Dutch anatomist Regnier de Graaf, along with most of his contemporaries, considered that the highly visible structure within the ovary—the **follicle**—was the egg, and took the extraordi-

[handwritten marginal note: ovaries ≃ ♀ testis (17th century)]

---

[2]Temkin, O. 1956. *Soranus Gynecology.* Johns Hopkins Press, Baltimore, MD. Cited in Short, 1977.

[3]Vesalius, A. 1543. *De Humani Corporis Fabrica Libri Septem.* (*The Illustrations from the Works of Andreas Vesalius of Brussels with Annotations and Translations.* 1950. J. B. de C. M. Saunders and C. D. O'Malley. The World Publishing Co., Cleveland, OH.)

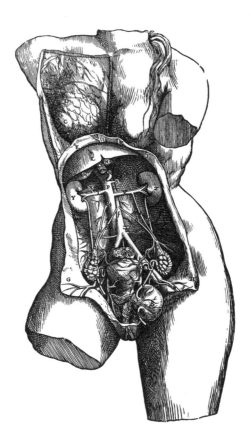

Figure 5.1
Andreas Vesalius' illustration of the female reproductive organs, appearing in *De Humani Corporis Fabrica Libri Septem*, 1543. The follicular nature of the ovaries is evident. (Reprinted from *The Illustrations from the Works of Andreas Vesalius of Brussels with Annotations and Translations*. 1950. J. B. de C. M. Saunders and C. D. O'Malley. The World Publishing Co., Cleveland, OH, with permission from the New York Academy of Medicine.)

nary step of removing the ovaries from a cadaver, cooking them, and showing that they tasted the same as a hen's egg (Fig. 5.2):

> *That albumen is actually contained in the ova of women will be beautifully demonstrated if they are boiled, for the liquor contained in the ova of the testicles acquires upon cooking the same colour, the same taste and consistency as the albumen contained in the eggs of birds.*[4]

---

[4]Cited in Jocelyn, H. D., and B. P. Setchell. 1972. Regnier de Graaf on the human reproductive organs. *Journal of Reproduction and Fertility* **5, Suppl. 17.**

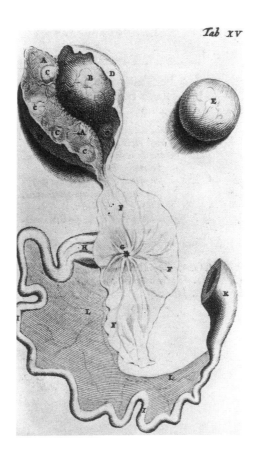

Figure 5.2
Regnier de Graaf's illustration of the ovary, uterine tube, and fimbriae of the cow. Small
follicles are labeled **C**, while **B** represents a large follicle. **E** is a large (Graafian) follicle
dissected out of the ovary. **F** is the fimbriae. de Graaf considered that the follicles were
the eggs themselves. The lower figure is a drawing of the uterine tube. (From R. de Graaf.
De Organis Generationem Inservientibus. Leiden, 1672.)

Hence, by the end of the seventeenth century the role of the ovary as the producer of
the egg was somewhat tenuously established, but the egg had not yet been identified
or isolated. Moreover, the nature of fertilization—the union of the egg and sperm—
although first proposed by Anton van Leeuwenhoek, the discoverer of sperm, was not
clearly understood. It was not until 1827 that K. E. von Baer was able to show that the
egg was a component of the follicle (Fig. 5.3). His report reveals clearly the excitement
he felt when he understood the significance of his discovery.

> *. . . Led by curiosity rather than by any thought that I had seen the ovules in*
> *the ovaries through all the layers of the graafian follicle, I opened one of the*

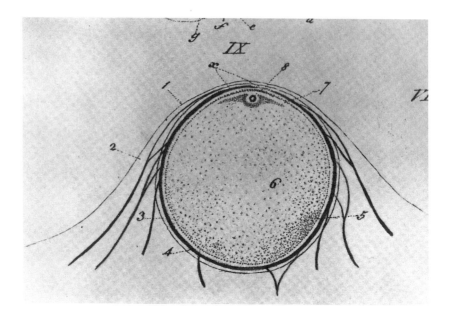

Figure 5.3
K. E. von Baer's drawing of a mammalian egg (oocyte) within the Graafian follicle.
1. Ovarian epithelium; 2. ovarian stroma; 3 and 4. theca externa and theca interna, respectively; 5. granulosa layer; 6. follicular fluid; 7. cumulus layer; 8. oocyte. (From K. E. von Baer. *De Ovi Mammalium et Hominis Genesi*. Lipsiae, 1827. Cited in Short, 1977.)

> *follicles and took up the minute object on the point of my knife, finding that I could see it very distinctly and that it was surrounded by mucus. When I placed it under the microscope I was utterly astonished, for I saw an ovule just as I had already seen them in the tubes, and so clearly that a blind man could hardly deny it. It is truly wonderful and surprising to be able to demonstrate to the eye, by so simple a procedure, a thing which has been sought so persistently, and discussed* ad nauseam *in every text-book of physiology, as insoluble!*[5]

By the end of the nineteenth century, the egg-producing capability of the ovary was established beyond doubt. The beginning of our understanding of its hormone-producing ability came about 20 years later, although here as well the antecedents were laid earlier. An interesting observation regarding ovarian function appeared quite early, but as often happens in science, it went unnoticed and its significance unappreci-

---

[5]von Baer, K. E. 1828. *De ovi mammalium et hominisi genesi*. Leipzig. Cited in Corner, G. W. 1933. The Discovery of the Mammalian Ovum. In *Lectures on the History of Medicine*, Mayo Foundation Lecture Series, 1926–1932, W. B. Saunders, Co., Philadelphia, PA. Translation from the German by G. W. Corner.

ated. This appeared in the description of the results of an operation carried out by the English surgeon W. Pott (1775):

Case XXIV An Ovarian Hernia

A healthy young woman about 23, was taken into St. Bartholomew's hospital on account of two small swellings, one in each groin, which for some months had been so painful, that she could not do her work as a servant.

The woman was in full health, large breasted, stout, and menstruated regularly, had no obstruction to the discharge per anum, nor any complaint but what arose from the uneasiness these tumours gave her, when she stooped or moved so as to press them.

Pott removed the herniated ovaries, and commented about the woman's recovery:

She has enjoyed good health ever since, but is become thinner and more apparently muscular; her breasts, which were large, are gone; nor has she ever menstruated since the operation, which is now some years.[6]

The significance of this observation was not followed up until more than a century later, when it was shown that the ovary was a "gland of internal secretion" (that is, that the ovary produced active substances that were circulated through the blood), a possibility implicit in Aristotle's description of the spaying of the sow. During the first three decades of the twentieth century, a large series of experiments by a number of investigators provided the evidence that these internal secretions were estrogens and progestins. Progesterone was isolated and characterized in 1934, and in 1936, in an epic study, estradiol was isolated from four tons of sows' ovaries.

By the end of the 1930s the twin functions of the ovary—producer of the egg and of steroid hormones—were clearly recognized. What remained was to understand how it accomplished these functions. In the following sections we will try to summarize what we know about ovarian function.

STRUCTURE OF THE FEMALE REPRODUCTIVE SYSTEM

The anatomical relationships of the postpubertal female internal genitalia—the uterus, uterine tubes, and ovaries—are illustrated in Fig. 5.4. The adult ovary is 2.5–5.0 cm along its long axis, 1.5–3.0 cm in width, 0.6–1.5 cm in thickness, and is connected to the pelvic wall by ligaments. For our purposes, it is useful to think of the ovary as consisting of two compartments: the **stroma**, which consists of a complex network of connective tissue and a variety of cell types, and the **follicle**, the functional unit of the ovary (Fig. 5.5). The stroma is derived from a progenitor population of cells that first appear in the fetal ovary. The follicle is a dynamic structure and has a developmental history that is visible within each adult ovary. In its preovulation stages, the follicle consists of the **oocyte**, which will become the egg, and a layer of cells recruited from the stroma. The smallest and simplest follicle is the **primordial follicle**, which includes the oocyte and a surrounding single layer of flattened cells known as **granulosa**

---

[6]Pott, W. 1775. *Chirurgical Observations*. London, UK. Cited in Short, 1977.

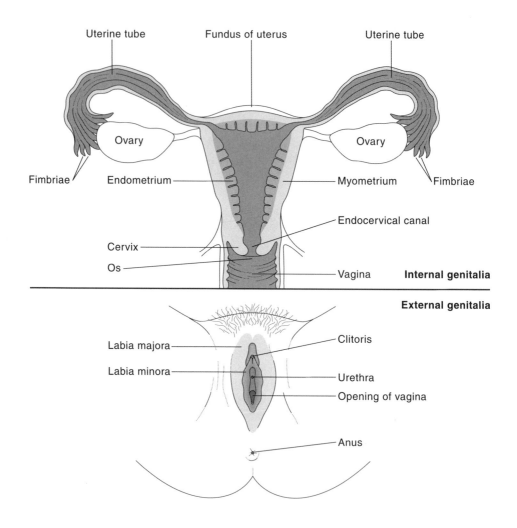

Figure 5.4
Schematic anterior view of the reproductive tissues of the female. Recall that the uterus,
uterine tubes, and upper part of the vagina are derived from the paramesonephric ducts. The
remaining part of the vagina and the external genitalia are derived from the urogenital sinus
and genital tubercle. The ovaries are held in place by ligaments that attach to the abdominal
wall. The innermost layer of the uterus is the endometrium, which is built up and sloughed
off every menstrual cycle. The underlying portion of the uterine wall is the myometrium.

↑ hormone during
gestation
↓ hormone during
infant /childhood
↑ hormone during
after puberty.

cells. You will recall that the primordial follicle was formed during the latter stages of
gestation. At the time of birth, about 95 percent of the oocytes in the ovary are present
as primordial follicles. During gestation, the ovary is active in producing estrogen. After birth, the ovary enters a relatively quiescent stage characterized by low hormonal
output. This stage is maintained all through infancy and childhood. The follicular developmental program is triggered at puberty. A section through a postpubertal ovary
captures, as it were, follicles at all different stages of development, from their birth—

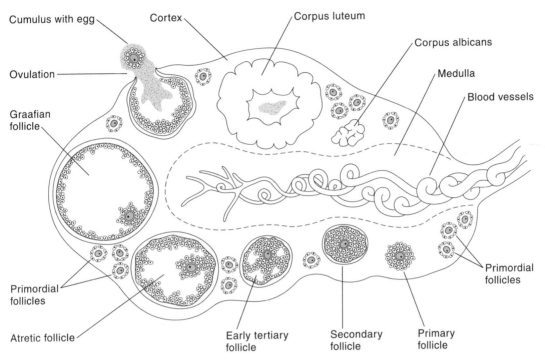

Figure 5.5
Schematic illustration of the internal architecture of the human ovary during the female's reproductive years. At any given time, follicles at different stages of development can be observed. Most of the follicular activity is found along the periphery (cortex), while arteries and nerves are found in the inner portion (medulla) of the ovary. (Adapted from Erickson, 1995. Courtesy of G. F. Erickson, UCSD School of Medicine.)

the primordial follicle—to their demise—the **corpus albicans** (Fig. 5.5). The stages of *corpus albicans* follicular development are discussed below.

ESSENTIALS OF OVARIAN FUNCTION

Ovarian function in a postpubertal and premenopausal female involves the periodic development of the primordial follicle, formed during fetal life, to its mature form and its final demise. These complex events reflect cyclic changes in ovarian and pituitary hormones required to produce the egg and to prepare for its fertilization. Let us consider the main features of these events.

Cyclicity

Gamete production in the female differs profoundly from gamete production in *♂ gamete produce* the male. Whereas in the male spermatozoa are produced in a continuous fashion, with *everyday (≈ 10⁵)* millions produced every day, in the female, at ovulation, one egg (very infrequently, more than one) is produced roughly every month. Indeed, this pattern of gamete production in the female is simply a reflection of the cyclic nature of postpubertal ovarian function. The most obvious manifestation of this inherent cyclicity is the monthly menstrual bleeding from the uterus that marks the beginning of puberty in the female and continues until the time of menopause.

As we shall see, the menstrual cycle is a consequence of a cyclic pattern of sex hormone secretion by the ovary. This cyclicity is necessitated by the more complex and demanding role that the female has in reproduction. The female not only has to produce the gametes, but must also be able to carry and nurture the newly formed embryo, the conceptus. Hence, in every ovulatory cycle, the first half of the cycle, known as the **follicular phase**, is dedicated to producing the egg, which is released from the follicle at the time of ovulation. After ovulation, during the **luteal phase**, the uterus is prepared for implantation of a fertilized egg. The normative cycle is generally taken as 28 days, although the length can vary from about 24 to 35 days, not only among different women, but also in the same woman. Except for extreme variations in length, it is very difficult to tell when the length of a cycle is "abnormal" and a source of concern. The luteal phase is relatively constant at 13 ± 1 days. If the length of the cycle becomes shorter or longer than the normative length, it is generally the follicular phase that changes. Much has also been written about the relationship between the ovulatory cycle and the lunar month. Although the effect of the moon on the ebb and flow of the tides has been rigorously established, there is absolutely no evidence that the moon has any effect on the timing of the ovulatory cycles. Hence, the similarity in the average length of the ovulatory cycle and the lunar month is simply a coincidence.

The hormonal environment during the follicular and luteal phases is quite different. Figure 5.6 illustrates the secretion pattern for estradiol and progesterone, and the gonadotropin hormones, LH and FSH, during a normal cycle. Conventionally, the beginning of menstruation—the initiation of **menses**—is taken as the beginning of the follicular phase. The follicular phase is characterized by rising estrogen and low progesterone levels, decreasing FSH and rapidly rising LH levels at the end. The luteal phase is characterized by high levels of both progesterone and estrogen, low LH and FSH throughout most of its length, and rising FSH toward the end. Although there is no precise point or event dividing the follicular from the luteal phase, two landmarks can be used in an operational sense. One is the LH peak, known as the **LH surge**, and the other is **ovulation**, which occurs about 36 hours later. The interval between the two defines the transition from the follicular to the luteal phase. The LH surge is of special importance because it is absolutely required for ovulation to take place. The end of the luteal phase is marked by the dramatic decrease of progesterone and estrogen levels and rising FSH. It is important at this stage to understand that the pattern of change of each hormone has a cause and purpose. In fact, it may not be an exaggeration to say that understanding the functional significance of the hormone profiles has been one of the success stories of ovarian endocrinology.

Oogenesis

The development and formation of the female gamete—the egg, or oocyte—is known as **oogenesis**. In humans, oogenesis is a prolonged affair, beginning during embryonic life (see Chapter 4) and continuing throughout the reproductive life of the female. Important stages in oogenesis in humans are summarized in Table 5.1. The PGC population in the embryonic ovary proliferates by successive mitotic divisions and dif-

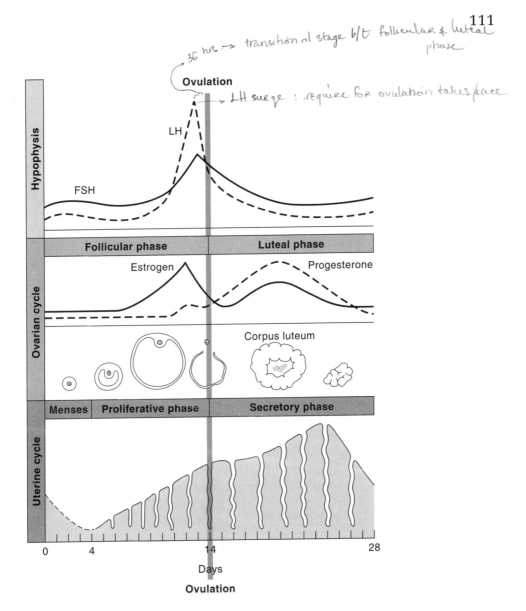

36 hrs → transitional stage b/t follicular & luteal phase

LH surge : require for ovulation takes place

Figure 5.6

Schematic representation of the estradiol, progesterone, LH, and FSH profiles during the normative 28-day ovulatory cycle. The initiation of menstruation conventionally marks the beginning of the cycle. The follicular phase is characterized by the selection and development of the dominant follicle, the Graafian stage follicle, rising estrogen levels culminating in the estrogen surge, and low progesterone levels. The time of ovulation is marked by the vertical line. The estrogen surge takes place before the gonadotropin peaks. The LH and FSH surges precede ovulation by about 36 hours. The luteal phase is characterized by the formation, maturation, and demise of the corpus luteum, and hormonally by high levels of estrogen and progesterone. FSH levels, which begin to rise toward the end of the luteal phase, are considered to initiate a new round of follicular development that will take place during the following follicular phase. The lower part of the figure indicates the changes taking place in the endometrium. During the follicular phase, the endometrium is built up. During the luteal phase, it undergoes a number of changes, and finally, the built-up layer is destroyed. (Adapted from G. F. Erickson, 1995. Courtesy of G. F. Erickson, UCSD School of Medicine.)

Table 5.1
Stages of Oogenesis in Humans

Stage	Gestation time	Approximate number (both ovaries)
PGC	Week 4–5	500–1000
Oogonium	Week 6–7	100,000
Oocyte (primordial follicle)	Week 8–9	600,000; meiosis begins
Quiescent oocyte (primordial follicle)	4–5 months	6,800,000; arrest of meiosis
	Birth	1,000,000
	7 years	300,000
Primordial to mature follicle	Puberty on	40,000
Ovulated secondary oocyte	Puberty on	One per month Secondary oocyte completes meiosis only if fertilized

ferentiates into **oogonia** during weeks 5, 6, and 7 after fertilization. During weeks 8 and 9, oogonial cells begin to differentiate into **primary oocytes** from which they enter the first stage of meiosis. The primary oocytes do not complete the first meiotic division, but arrest at an early stage of meiosis beginning perhaps by week 12. Proliferation of the oogonia and their differentiation into primary oocytes continues until about the fifth month of gestation, by which time the ovary contains 6–7 million primary oocytes. Oogonia are still found in the ovary at this stage, but most of these will not differentiate into primary oocytes and will be lost. The cessation of oogonial proliferation means that the oocytes form only during one stage of a female's life, and her supply of oocytes is nonrenewable.

The arrest of meiosis in the primary oocyte appears to be coincident with the formation of the primordial follicle, which consists of the primary oocyte surrounded by a single layer of granulosa cells. Integrity of the follicle is important in maintaining the arrest of meiosis, since if the oocyte is removed from the follicle, it resumes meiosis. From the reservoir of primordial follicles present at birth, some will commence growth in a random but continual fashion beginning in early infancy. Hence, during infancy and childhood, the ovary contains a large pool of nongrowing primordial follicles and a smaller pool of follicles that have initiated their growth trajectory. However, the primary oocytes in all of these follicles remain arrested in the first stage of meiosis. Reentry into meiosis begins at puberty, but in only one or two follicles just before ovulation. This pattern of development continues throughout the reproductive life of the female. This means that the earliest ovulated oocytes have been arrested for 12–13 years (from birth to puberty), while those that are ovulated at 40 have been arrested for 40 years.

Oogenesis is an asymmetric process, yielding only one functional gamete at the completion of meiosis. In contrast, the equivalent process in males, spermatogenesis, is symmetric and yields four functional gametes (Fig. 5.7). In oogenesis, completion of the first meiotic division yields the **secondary oocyte** and the **first polar body**, which is extruded into the space between the oocyte membrane and the zona pellucida (ZP).

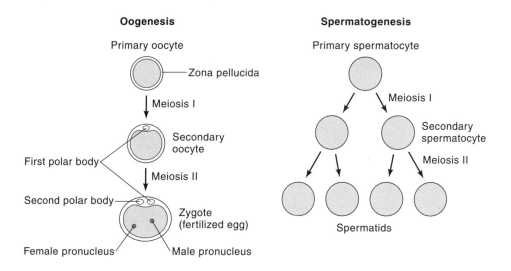

Figure 5.7
Illustration of the asymmetry in the first and second meiotic divisions in oogenesis. The primary oocyte, arrested since fetal life, is triggered to complete the first meiotic division, yielding the secondary ooctye and the first polar body, both of which are retained within the zona pellucida. The secondary oocyte is ovulated, but it only completes the second meiotic division if it is fertilized. This division is also highly asymmetric, yielding the second polar body. The two nuclei within the fertilized egg are the maternal and paternal pronuclei.

Essentially the entire cytoplasmic contents of the primary oocyte are now contained within the secondary oocyte, and the first polar body is degraded and resorbed after some time. The secondary oocyte enters the second meiotic division, but arrests very quickly afterward. Some 36 hours after being formed, the secondary oocyte is ovulated; that is, it is expelled from the ovary into the uterine tube (Fig. 5.8), there to await fertilization. In humans, the secondary oocyte resumes and completes the second meiotic division only if it is fertilized. Entry of the spermatozoa into the secondary oocyte triggers resumption of meiosis, and the result, again, is a highly asymmetric cytoplasmic division, and extrusion of the **second polar body**. The fertilized secondary oocyte now becomes the **zygote**, containing one set of paternal and one set of maternal chromosomes, and this is the beginning of embryogenesis (Fig. 5.7). If the secondary oocyte is not fertilized, the oocyte degenerates and is resorbed by the uterine tube tissue.

In humans and the other higher primates, the hormonal environment that regulates follicular and oocyte development generally ensures that only one oocyte is ovulated at each cycle. Single births are the norm, and multiple births are the exception. This is in contrast to many other mammalian species that normally ovulate several eggs at the same time, and therefore have large litters. The reason for this difference most likely is that the human uterus will carry optimally only one fetus at a time, and the human ovulatory dynamics has been selected for ovulating one egg at a time.

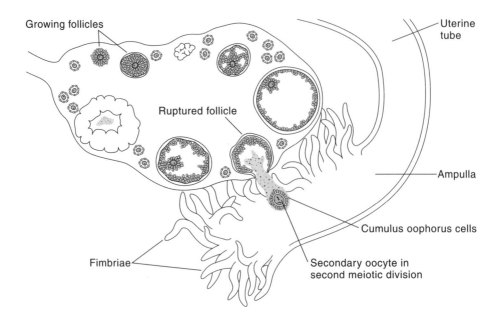

Figure 5.8
Diagram depicting the expulsion of the secondary oocyte (ovulation) into the uterine tube. During ovulation, the fimbriae of the uterine tube are thought to "sweep" over the ovary at the site of rupture perhaps to ensure that the oocyte will land in the uterine tube.

Atresia : the loss of oocytes

Beginning at about 20 weeks of gestation another remarkable process begins to take place in the ovary: this is the loss of oocytes, which is known as **atresia**, and was first recognized early in the twentieth century. After the peak at about 5 months gestation, there is a precipitous loss of oocytes during the remaining part of gestation (Table 5.1 and Fig. 5.9). At birth, about 1,000,000 oocytes remain in the ovary. The loss continues after birth, and in fact all through the lifetime of the female, although at a slower rate, until the ovary is empty of oocytes. This marks the time of **menopause**. The rate of atresia is responsible for the timing of menopause. Atresia can remove a follicle at any stage of its development right up to the time it will be ovulated. Despite atresia and the fact that oogenesis is a nonrenewable process, the female has far more oocytes than she will ever use. During the lifetime of the female very few of the oocytes initially produced develop to any extent, and even fewer are ovulated. If we assume that the first ovulation takes place at 13 years and the last one at 50 years, and that ovulation occurs once each month, then we can estimate that about $37 \times 12 = 444$ oocytes will be ovulated during the female's lifetime. Hence, out of the 7 million oocytes produced, only about 1 of every 20,000 oocytes are ovulated. It is difficult to understand such profligacy. Indeed, we can say that the fate of almost every follicle is to be destroyed, and that ovulation enables a few follicles to be rescued. We do not really understand the precise reason for generating such a large reservoir of oocytes when only a few will be utilized. We do know, however, from studies in the mouse, that the atretic process is a good ex-

Only a few follicles are ovulated

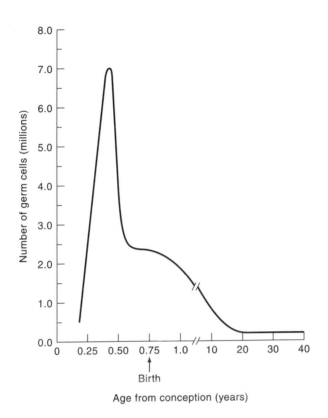

Figure 5.9
The change in number of primary oocytes in the ovary with the age of the female. After
their arrival at the genital ridges, the PGCs undergo a period of rapid proliferation and
differentiation into oogonia beginning at 25–30 days after fertilization. Primary oocytes
begin to appear at 50–60 days, and reach a maximum at about 5 months (20 weeks) gestation.
Thereafter, atretic loss depletes the original store of primary oocytes. (Adapted from T. G.
Baker and W.S.O., 1976.)

ample of a type of cell degeneration known as **apoptosis**, in which cell death occurs in a *Death of oocytes*
closely controlled manner. These studies also suggest that the atretic signal is generated *are signaled w/ in*
within the ovary. The atretic process in humans is much less well understood. *the ovary.*

Follicular stages

The central activity of the ovary is follicular development. A complete understanding of the factors that regulate the growth and maturation of the follicle is not yet
possible. Nevertheless, it is fair to say that we have a good grasp of the fundamentals.
Although the process is continuous, we can recognize a few distinct stages in the life of *6 stages in a life of*
a follicle. These can denoted as the *preantral, antral, preovulatory, ovulatory, corpus* *a follicle*
luteum, and *luteolytic* stages. Morphological as well as endocrinological differences
distinguish the different stages. Let us consider these stages briefly.

Preantral stage : *during fetal & postnatal period*

change in shape & size of granulosa

single flat layer
↓
cuboidal shape
↓
↑ follicular diameter

primordial follicle → 1° follicle

Three phases of the preantral follicle can be recognized based on morphology and size. The primordial follicle consists of the primary oocyte surrounded by a single layer of flattened granulosa cells (Fig. 5.10). The change of the granulosa cells into cuboidal shapes and the increase in follicular diameter from about 40 to about 100 micrometers marks the conversion of the primordial follicle into the primary follicle. The primary

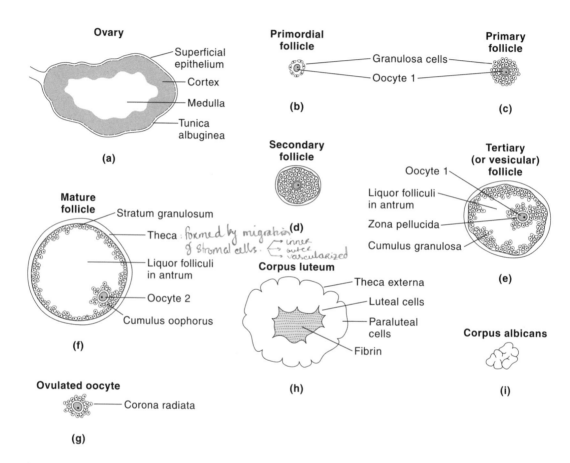

Theca : formed by migration of stromal cells. → inner → outer vascularized

Figure 5.10
Schematic depiction of the developmental stages of a follicle.(a) Primordial follicles form along the cortex of the ovary. (b, c, d) Preantral (primordial, primary, and secondary, respectively) stages of the follicle. Preantral development is driven primarily by intraovarian factors. (e and f) Antral (tertiary and Graafian) stages. The follicle at the tertiary stage is highly vascularized and, in intimate contact with the circulatory system. At the preovulatory Graafian follicle stage, the oocyte assumes a polar position and remains surrounded by several layers of granulosa cells, the cumulus layer. (g) Ovulated oocyte with its cloud of cumulus cells (corona radiata). (h) Corpus luteum, derived from the remnants of the ovulated follicle. The residual granulosa and theca layers differentiate rapidly into lutein cells. The antral cavity of the ruptured follicle is filled with the protein fibrin. (i) Eventually, the corpus luteum degenerates to form the corpus albicans. (Adapted from R. O'Rahilly and F. Müller. 1996. *Human Embryology and Teratology*. Wiley-Liss, New York, NY.)

follicle is converted to a secondary follicle by: (1) the proliferation of the granulosa cells to form a region several cell layers thick, (2) the secretion of the **zona pellucida** (ZP), a proteinaceous layer between the oocyte and the inner granulosa cell layer, and (3) formation of a follicular membrane. In a later chapter we will consider the important function of the ZP. The secondary follicle is 200 micrometers in diameter, still too small to be seen by the naked eye. Preantral growth occurs continually during the fetal and postnatal period, but significant growth beyond the secondary stage is infrequent. Development during the preantral stage is thought to be stimulated by as yet poorly defined locally produced ovarian factors.

Antral stage

The characteristic feature of the antral period is the formation of a fluid-filled cavity—the *antrum* (Fig. 5.10). The follicle increases from about 120 micrometers to about 2 millimeters in diameter and is now visible to the naked eye. The large increase is due to a large increase in the number of granulosa cells and to the accumulation of follicular fluid in the antrum. The early antral stage is sometimes referred to as a *tertiary follicle*, and is characterized in part by the beginning of the separation of the cells surrounding the oocyte into the inner layer—the cumulus oophorus cells—and multiple outer layers. In addition, the *theca* layer, formed by migration of stromal cells to the outer surface of the follicle, as well as its differentiation into an inner and outer theca layer, becomes apparent. Moreover, the theca layer becomes highly vascularized, meaning that its penetration by blood capillaries brings the follicle in intimate contact with the general circulatory system. This is important not only because the follicle now becomes sensitive to hormones in the circulatory system, particularly the gonadotropins, but also because the hormones secreted by the follicle itself will now be distributed rapidly to many other tissues in the body. It has been estimated that it takes about 300 days for a primordial follicle to grow to the early tertiary stage.

The final phase of the antral follicle is known as the *Graafian stage*, named after the Dutch embryologist who first described it (Fig. 5.10). At this stage the follicle is about 20 to 30 millimeters in diameter, large enough to bulge out of the ovary. The follicular diameter has increased an impressive thousand-fold, or about one billion-fold in volume. In the Graafian follicle the oocyte has separated from the mass of granulosa cells, has moved to one pole of the follicle, and remains in communication with the other cells of the follicle by a thin stem of cumulus granulosa cells.

Preovulatory stage

The event defining the preovulatory stage is the resumption of the first division of meiosis, which occurs about 17 hours before the LH surge, and is thought to be triggered in part by the rapidly rising levels of LH. The first meiotic division is completed, the first polar body is formed and extruded between the oocyte membrane and the zona pellucida. The secondary oocyte now arrests at the beginning of the second meiotic division. The factors involved in the resumption of the first meiotic division or in the subsequent arrest at meiosis II, although not yet well defined, are thought to be produced by the granulosa cells.

Ovulatory stage

The ovulatory stage, which involves the expulsion of the secondary oocyte from the follicle and ovary into the uterine tube, is the shortest, but also the most dramatic,

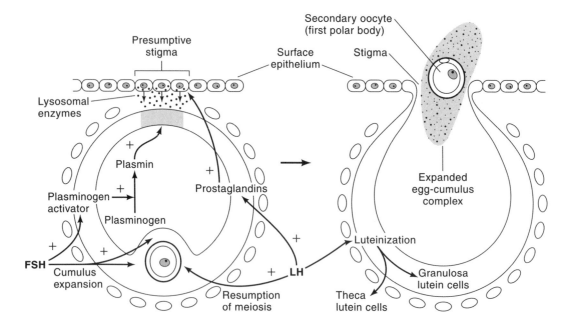

Figure 5.11
Schematic diagram of the preovulatory and ovulatory follicle. The stigma defines the site on the ovarian epithelium through which the oocyte will be expelled. Stigma formation begins in a specialized population of epithelial cells which will release hydrolytic enzymes initiating the progressive destruction of the cellular layers surrounding the oocyte. LH is believed to be the primary trigger for ovulation. The rapidly rising LH levels that culminate in the LH surge activate progesterone production by the granulosa cells. The LH surge also stimulates the resumption of meiosis and conversion of the primary into the secondary oocyte. LH and FSH in concert are responsible for the release of prostaglandins PGE and PGF, and the protease plasmin, which in turn promote stigma formation and expulsion of the oocyte from the ovary. LH is also responsible for the conversion of the granulosa and theca cells into lutein cells. (Adapted from Erickson, 1995. Courtesy of G. F. Erickson, UCSD School of Medicine.)

36 hrs after
LH surge

LH → cascade of
hormonally controlled
events → localized
rupturing of ovarian
wall → expulsion of
oocyte

No duct connecting
ovary & uterine
tube
→ fimbriae catches
exited oocyte

of the follicular stages. Although the LH surge is absolutely required for ovulation, ovulation does not take place until about 36 hours after the LH surge, at a time when LH levels are falling. What appears to happen is that the LH surge triggers a cascade of hormonally controlled events that eventually lead to the localized rupturing of the ovarian wall and expulsion of the oocyte (Fig. 5.11). A release of prostaglandin triggers contractions of the ovary that may help expel the oocyte, and this sometimes causes a mild abdominal ache. The release of the oocyte from the ovary appears precarious because there is no duct connecting the ovary and the uterine tube. Instead, a successful ovulation requires that the expelled oocyte be caught by the fimbriae of the uterine tube (see Figure 5.8).

Corpus luteum

After expulsion of the secondary oocyte, the residual cells of the follicle collapse and differentiate to form the cells of the *corpus luteum* (CL), which means "yellow

follicle collapse
↓
corpus luteum
→ accumulation
of fatty pigment

body" (Fig. 5.10). The color associated with the CL is due to the accumulation of a *CL ≠ hue in ≠ species* fatty pigment—*lutein*—which has a different hue in different mammalian species. Much of the early work on the CL was done on the cow, which has a yellow CL, which explains the origin of the name. In the sow, sheep, rat, and rabbit, the CL is white, while the human CL has an orange hue. The CL, dependent particularly on (decreasing) LH, is the dominant structure in the ovary during the luteal phase of the ovulatory cycle, and its two main hormonal products, progesterone and estrogen, are secreted by the lutein cells.

Luteolysis

If fertilization has not occurred by about 8 days after ovulation, the CL begins to *[fertilization → CL hormonal output ↓]* involute. Its hormonal output declines, and by about 14 days, the decay process is complete, resulting in a shrunken structure known as the *corpus albicans*. In humans, the primary trigger for luteolysis, the involution and decay of the CL, remains elusive, but it appears that the CL has an internal clock. It may be that the relative constancy of the *[CL internal clock]* luteal phase is due to the internal CL clock. If fertilization and implantation take place, one of the first products of the implanting conceptus is human chorionic gonadotropin *[fertilization & implantation → hCG → CL hormonally active]* (hCG). The hCG keeps the CL hormonally active, particularly in producing progesterone, for the first two months of gestation, until the placenta can take over the progesterone-producing function (more on this aspect later).

SUMMARY

In the postpubertal female, ovarian activity is cyclic and is divided into two stages, follicular and luteal. The cyclicity is necessary because of the twin functions of the female in reproduction, i.e., production of the egg and development and nurturing of the fertilized egg. The follicular phase, which is dedicated to the production of the egg, culminates at ovulation when the egg is expelled from the ovary into the uterine tube. During the luteal phase, the uterus is prepared to receive the newly formed conceptus if fertilization occurs. The production of an egg at ovulation is a prolonged process that begins during fetal life with the primordial germ cells (PGCs). During the formation of the ovary, the PGCs differentiate progressively into oogonial cells and primary oocytes, in which the first meiotic division is initiated. A characteristic feature of oogenesis is that meiosis is arrested soon after it begins. The primary oocyte remains in the arrested condition for many years. At puberty, as a consequence of a complex maturation process, generally one primary oocyte will be triggered to complete the first meiotic division. The asymmetric first division yields one secondary oocyte, which is released at the time of ovulation, and one polar body, which degenerates. Hence, when we speak of the egg, we are referring to the secondary oocyte released midway through the ovulatory cycle. If fertilization does not take place, the secondary oocyte degenerates and is absorbed by the uterine tube tissue. The second meiotic division takes place only if the secondary oocyte is fertilized.

Each germ cell is contained in a larger structure called the follicle, which also contains many cells derived from the stroma of the ovary. The simplest follicle, the primordial follicle, is formed during gestation, and a total of about 7,000,000 such structures are formed by about five months gestation. From then on, follicles are continually lost by a degenerative process known as atresia, leading eventually to the final depletion of follicles from the ovary when the female is about 50 years old. At puberty, groups of follicles are triggered at intervals to enter a growth trajectory that will convert the primordial follicle into successively larger and more complex structures. A follicle that completes this growth phase successfully becomes a Graafian stage follicle, which becomes the dominant structure during the follicular phase. At ovulation, the secondary oocyte is expelled from the Graafian follicle. After ovulation, the cellular remnants of the Graafian follicle reorganize to form the corpus luteum, which becomes the dominant structure during the luteal phase of the ovulatory cycle. The corpus luteum has a limited lifetime, and begins to self-destruct about eight days after it is formed unless fertilization occurs.

OVARIAN ENDOCRINOLOGY

It is probably not an exaggeration to say that the day-to-day well-being of a female depends to some extent on her ovaries. The female's role in reproduction is very complex. An elaborate system of checks and balances has evolved to deal successfully with the multiple demands that are placed on her system. Significant disturbances in normal ovarian function can have many consequences, ranging from loss of fertility to physical discomfort or pain to alterations in mood and behavior. The focus of this section will be on ovarian endocrinology during the ovulatory cycle. We will use the hormone profile shown in Fig. 5.6 as our blueprint, since that profile forms the basis of any understanding of ovarian endocrinology in the human. But keep in mind that the profile is like the tip of an iceberg, revealing only a small part of the marvelously complex interactions that ultimately determine ovarian function. To look beyond the tip we will approach ovarian function in a series of approximations, first by considering how the sex hormone secretion pattern is established, and secondly, by relating that pattern to follicular dynamics.

The principal hormones involved in ovarian function are the sex steroids, estrogen and progesterone, and the gonadotropins, LH and FSH. It is important to keep in mind that progression through the ovulatory cycle involves an elaborate feedback circuit between the ovary and the pituitary, and the ovary and hypothalamus. At this stage we will consider only the ovarian-pituitary axis. We also need to recognize that in addition to the four primary hormones, there are many other ovarian factors that play important roles in fine-tuning follicular development and in ensuring that pituitary stimulation is well coordinated with the follicular stage. We will not consider the additional factors here, but will concentrate on the function of the four main players in this drama. We consider first the production of steroid sex hormones by the ovary.

Estrogen and progesterone production

Steroid hormone production during the ovulatory cycle depends principally on the largest tertiary/Graafian follicle, the **dominant follicle**, during the follicular phase, and on its derivative, the CL, during the luteal phase. During the follicular phase, steroid sex hormone biosynthesis requires a special type collaboration between the granulosa and theca cells of the follicle, a collaboration generally referred to as the *two-cell model of estrogen synthesis* (Fig. 5.12). Theca cells convert cholesterol to androgens under LH stimulation from the pituitary. However, theca cells cannot convert the androgens to estrogens because they lack the necessary enzyme, **aromatase**. The androgens produced by the theca cells diffuse into the granulosa cells, which convert the androgens to estrogens, primarily estradiol, under FSH stimulation. Hence, the

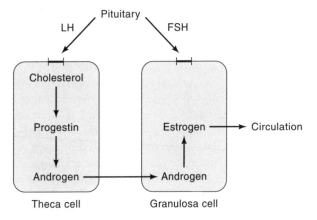

Early follicular phase

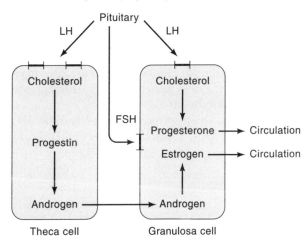

Late follicular phase (beginning of luteinization)

Figure 5.12
The two-cell model of estrogen synthesis. During the early follicular period, the theca cells, stimulated by LH, synthesize androgens, which then diffuse into the granulosa cells. The granulosa cells, stimulated by FSH, convert androgens to estrogens. In the latter stages of the follicular phase, rising levels of estrogen stimulate the synthesis of LH receptors by the granulosa cells, changing the endocrinology of the granulosa cells so that they begin to secrete progesterone. This switch is necessary for ovulation and for the transformation of the follicle into the corpus luteum after ovulation.

ability of the follicle to produce estrogens depends first on the androgen-producing ability of theca cells, and secondly, on the aromatase activity of granulosa cells.

The follicular phase

During the early part of the follicular stage, estrogen production begins to increase under the combined action of LH on theca cells and FSH on granulosa cells (Fig. 5.12). FSH and perhaps estrogen (estrogen acting in an autocrine fashion) in turn stimulate granulosa cell proliferation. Hence, as the number of granulosa cells in the follicle in-

[margin notes:] ⇒ estrogen surge ↓ endocrinological switch ↑estrogen → ↑ LH /effect. - resumption of meiosis - ovulation - estrogen → progesterone

creases, estrogen output increases. Estrogen levels increase rapidly, reaching a peak known as the **estrogen surge** (Fig. 5.6). The estrogen surge triggers a remarkable "endocrinological switch," in which the high levels of estrogen exert a positive-feedback effect on LH secretion by the pituitary. Positive feedback means that rising estrogen levels stimulate LH secretion by the pituitary. As a consequence, LH levels, which have been relatively low up to this point, begin to increase rapidly, culminating in the LH surge about a day after the estrogen peak (Fig. 5.6). The LH surge is perhaps not only the most distinctive landmark during the ovulatory cycle, but it is also of absolute importance, since resumption of the meiosis in the arrested primary oocyte, ovulation, and the shift in estrogen to progesterone synthesis in the granulosa cell population all depend on the precipitous rise in LH levels. That estrogen plays a crucial role in generating the LH surge has been established in experiments that show that disturbance of either estrogen secretion or estrogen action abolishes the LH surge. Despite its importance, the precise mechanism by which the rising levels of estrogen bring about the LH surge is poorly understood, but it is known that action of estrogen at both the pituitary and the hypothalamus is required.

[margin notes:] fall increase / estrogen shifts to progesterone / FSH estrogen / 2ⁿᵈ endocrinological switch / ↑ synthesis of LH receptor

FSH and estrogen action on the granulosa cells also bring about a second endocrinological switch: induction of the synthesis of LH receptors on the granulosa cells by the combined action of FSH and estrogen leads to a change from estrogen to progesterone synthesis and secretion (Fig. 5.12). As a consequence, estrogen levels begin to fall and progesterone levels begin to increase (Fig. 5.6). Although progesterone levels are quite low at this point, they turn out to be necessary for generating the LH surge. Hence, the two sex hormones—high levels of estrogen and low levels of progesterone—are required for the LH surge. These changes set the stage for the beginning of the luteal phase.

[margin notes:] high estrogen low progesterone → LH surge → luteal phase begins

The luteal phase *require low level of LH & FSH*

[margin notes:] high level of progesterone / CL (two type of luteal cells) (LH/FSH) / estrogen progesterone / ⊖ feedback

After ovulation, progesterone synthesis and secretion continues at a high level with the reorganization and reprogramming of the residue of the ovulated follicle. Moreover, after a delay, estrogen secretion by the CL is resumed. The CL is thought to have two different types of luteal cells, one specialized for producing progesterone, and the other for estrogen. The relationship between the two types of cells is unclear, but both appear to require LH and FSH for their activity even though both LH and FSH levels are quite low for most of the luteal phase. The low LH and FSH levels are due to negative feedback effects of estrogen and progesterone, that is, the rising sex hormone levels inhibit LH and FSH secretion by the pituitary. Note also that estrogen levels during the luteal phase are quite high, certainly as high as they are during most of the follicular phase, and high enough in principle to generate another LH surge. An LH surge is probably blocked by the strong negative-feedback effects of progesterone on the hypothalamus. Through the combined actions of progesterone on the hypothalamus and estrogen on the pituitary, secretion of LH and FSH by the pituitary is maintained at low levels.

[margin notes:] (estrogen) high enough for another LH surge but blocked by ⊖ feedback effects of progesterone on hypothalamus

[margin notes:] No implantation signal (hCG) / CL demise / ↓ estrogen → ↓ negative feedback / ↓ progesterone / ↑FSH at the end

Involution of the CL begins about eight days after the LH surge if no signal that implantation has occurred is received. The signal is hCG, a hormone produced by the implanting conceptus, but functionally equivalent to LH. hCG maintains progesterone at the level necessary to prevent the implanted conceptus from being expelled from the uterus. If no implantation occurs, with the demise of the CL, estrogen and progesterone levels decline. With the drop in estrogen, the negative-feedback effects of estrogen on the pituitary decrease, allowing FSH levels to begin to rise at the end of the luteal phase.

[margin note:] hCG → maintain progesterone level → prevent the implanted conceptus from being expelled

It is important to keep in mind that one structure, the dominant follicle during the follicular phase, and then the corpus luteum during the luteal phase of the ovulatory cycle, is responsible for essentially the entire steroid sex hormone output in the female. Moreover, this output is absolutely dependent on the pituitary hormones LH and FSH. We now look at how the dominant follicle is selected.

one structure is responsible for the entire steroid sex hormone output

The dominant follicle

follicle → (ovulation) dominant follicle

The dominant follicle is the follicle that undergoes ovulation, and hormonally, is the most important structure in the ovary. What may not be as clear is that it is the product of an ongoing developmental process that begins long before ovulation. Follicular development from the primordial to late preantral and occasionally the early antral stages does take place in the prepubertal female, but growth beyond these stages is quite sporadic and abortive. This suggests that the factors necessary to sustain further development are not present in the concentrations required. The transition from childhood to puberty is accompanied by significantly increased LH and FSH levels, and these may permit changes in the intraovarian environment that promote follicular development. The time required for the development from the primordial to the late Graafian stages is not known in any precise way, but estimates suggest periods of at least nine months. In other words, the ovulating follicle in any given cycle began its growth trajectory at least nine months earlier (Table 5.2 and Fig. 5.13). It must also be kept in mind that atretic loss of follicles continues unabated during this time.

childhood ↓ ↑LH, ↑FSH puberty ↓ change in intraovarian ↓ ↑follicular development

ovulating follicle growth trajectory at least 9 month earlier

Table 5.2
Growth Trajectory of the Ovulated Follicle

Phase	Follicular stage (diameter)	Period
Nongrowing pool	Primordial (0.03 mm)	Generated during the fetal period
Initiation (dependent on intraovarian factors)	Primordial to primary (0.12 mm) to secondary (0.20 mm)	May take more than 1 year. Occurs sporadically during childhood. Begins on a regular basis at puberty
Recruitment (accelerated growth; gonadotropin-dependent)	Secondary to tertiary (5.0 mm)	70 to 14 days before ovulation
Selection	Tertiary (5.0 mm)	14 days before ovulation
	Dominance (16.0 mm)	10 days before ovulation
Preovulation	Preovulatory (20 mm)	1 day before ovulation
Ovulation	Ovulatory (30 mm)	—

Adapted from Gougeon, 1996.

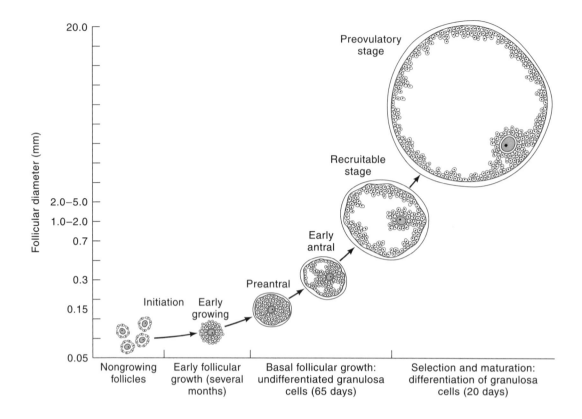

Figure 5.13

The temporal pattern of human follicular growth. Folliculogenesis is initiated from a nongrowing pool of primordial follicles formed originally during fetal life. At the onset of puberty, primordial follicles from this reservoir are triggered to begin developing. Over a period of several months, the stimulated primordial follicles progress through primary and secondary stages. Approximately 85 days before ovulation during the current cycle, a small cohort of preantral follicles enters the gonadotropin-dependent (recruitment) stage. The follicle that will be ovulated (the dominant follicle) is selected from a pool of surviving late-tertiary-stage follicles at the end of the luteal phase, just prior to the beginning of the cycle in which the dominant follicle will be ovulated. (Adapted from A. Gougeon. 1986. *Human Reproduction* 1, 81–91.)

[handwritten notes in margin:]
initiation
recruitment → formation of
selection   dominant follicle

The formation of the dominant follicle can be roughly divided into three phases—**initiation**, **recruitment**, and **selection**, although these terms are not used consistently by all authors. Follicular growth is initiated from the pool of resting or nongrowing primordial follicles, which represents what is known as the *ovarian reserve*. It appears that once initiation takes place, only two outcomes are possible: the follicle either undergoes ovulation or it is lost by atresia. Initiation transforms the primordial follicle into the primary and secondary follicle, a process that takes about nine months to a year (Table 5.2). During most of this time, growth is not dependent on gonadotropins, but is most likely driven by intraovarian factors. The recruitment step may begin as

[handwritten notes across bottom:]
primordial follicle — initiation → ovulation → 1° follicle → 2° follicle (intraovarian factors) — recruitment → 90% lost / 10% → 3° follicle — selection → 1 survivor (most sensitive)
atresia / lost
dominant follicle — synthesize/secret estrogen / suppress other follicle

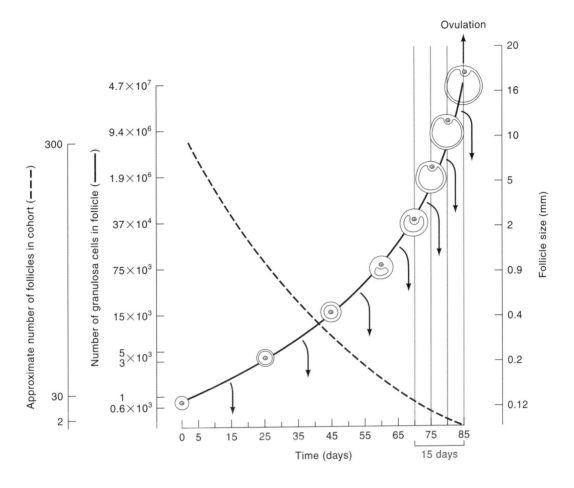

Figure 5.14
Characteristics of the recruitment, growth, and selection of the dominant follicle. A cohort of about 300 follicles initiate their growth trajectory about 3 months before the current cycle. Two competing processes take place: atresia and development into late-tertiary follicles. About 10 percent of the original cohort participate in the final selection phase that begins about 14 days before ovulation. The number of granulosa cells per follicle increases exponentially from 600 in the primary follicle to 47 million in the preovulatory follicle. Follicles are lost by atresia at all stages of their development (indicated by the arrows). (Adapted from A. Gougeon. 1986. *Human Reproduction* **1**, 81–91.)

early as about three months (85 days) prior to the day of ovulation of the current cycle, and appears to involve recruitment of cohorts of about 300 secondary-stage follicles. During the next two months about 90 percent of the follicles in this cohort become atretic and are lost (Fig. 5.14). The survivors are tertiary-stage follicles. Recruitment requires low levels of FSH and LH, as well as other, not-well-characterized ovarian factors. The conditions that determine which follicles survive remain obscure. Initiation and recruitment take place in the background and independent of the main events during each cycle (Fig. 5.13).

Roughly 14–15 days before the next LH surge, that is, at the end of the previous cycle, the remaining 30 or so tertiary follicles enter the selection stage (Figs. 5.13 and 5.14). The rising FSH levels at the end of the luteal phase play the critical role in determining which of the remaining follicles survive. At this stage, continued survival depends on sensitivity to FSH, which translates into the ability of the follicle to synthesize and secrete estrogen. The follicle with the greatest response to FSH will probably become the dominant follicle. It is not really clear what prevents some of the other follicles from achieving dominance. Some studies suggest that the follicle that achieves dominance suppresses the development of the other follicles in the cohort, perhaps because of its enormous estrogen-producing ability. One suggestion is that estrogen and inhibin act by negative feedback on the pituitary to reduce FSH levels, thereby depriving most of the follicles of the FSH support they need to keep growing. The dominant follicle is probably selected by about 5 days after the end of the previous luteal phase, at which point it can be said to enter the Graafian stage. The preovulatory stage occurs about a day before ovulation (Table 5.2).

The enormous capacity of the dominant follicle to produce estrogen can be appreciated simply by considering that the number of granulosa cells in the follicle increases from 600 in the secondary-stage follicle to about 47,000,000 in the Graafian follicle (Fig. 5.14). Although we have focused on FSH and estrogen in the selection of the dominant follicle, many other intraovarian factors and hormones are likely to participate in the recruitment and selection process, but at present, we know relatively little about the identity of such factors or their roles. Since there are two ovaries, these events involve follicles in both ovaries. Eventually the dominant follicle emerges in one ovary. Because of the vascular connections of the dominant follicle, the estrogen produced by the dominant follicle in one ovary quickly registers in the other ovary to suppress follicular development in that ovary.

In every mammalian species there is normally a match between the number of oocytes ovulated and the number of offspring that can be supported by the uterus. In the higher primates, this number is generally one. Hence, the hormone profile and follicular dynamics in the human female is one that ensures conditions for one ovulation per cycle. We know, of course, that infrequently two, and even more infrequently, three or more oocytes are ovulated. In the first case, two dominant follicles must have been selected, leading to fraternal twins. If three follicles survived the selection process, then fraternal triplets would result. Given our understanding of the role of FSH in the recruitment and selection process, we would conclude that fraternal twins and higher-order fraternal births are attributable to higher-than-normal FSH levels at critical times in the selection process. That this conclusion is correct has been demonstrated in many standard *in vitro* fertilization protocols in which multiple ovulations are stimulated by administration of FSH.

SUMMARY

The complexity of the ovulatory cycle can be kept in perspective by recalling the two functions of the postpubertal ovary: first, generating the gamete, and second, preparing for implantation of a fertilized egg. The generation of the gamete takes place over a long period of time. Most of follicular development takes place more or less hidden from view, off-stage as it were. Intraovarian factors and gonadotropins transform the primordial follicle into tertiary-stage follicles. Most follicles that begin this process perish at different points along the way. The final stage in this developmental program is the selection of the dominant follicle. We are most aware of this stage because of the enormous changes in follicular diameter and hormone output that characterize it. This stage begins at the end of the luteal phase of the ovulatory cycle, when, under the influence of FSH, a small number of tertiary-stage follicles enter the selection phase of their growth trajectory. Generally only one follicle—the dominant follicle—survives this phase, and this is the follicle that undergoes ovulation. The dominant follicle is also responsible for most of the estrogen produced by the ovary during the follicular phase. After ovulation and expulsion of the secondary oocyte, the remnants of the dominant follicle regroup and convert themselves into the corpus luteum. The CL becomes the most important structure during the luteal phase and is responsible for the production of estrogen and progesterone, which together prepare the uterus for implantation in case the oocyte is fertilized. Despite all we have learned about follicular dynamics, follicular selection, dominance, and atresia remain profoundly mysterious processes.

THE MENSTRUAL CYCLE

→ periodic building up + subsequent destruction of the endometrium (outermost layer of uterus)

The menstrual cycle consists of the periodic building up and subsequent destruction of a specialized tissue in the uterus, the **endometrium**. The endometrium is the outermost layer of the uterine lining (Figs. 5.4 and 5.6), and it undergoes a series of developmental changes that are regulated by the ovarian hormones estradiol and progesterone. The menstrual cycle is orchestrated by the ovulatory cycle, and hence, its periodicity is determined by the periodicity of the ovulatory cycle. Menstruation has probably been one of the most puzzling and mysterious of phenomena for most of the history of humankind. The loss of blood from any tissue has always been seen as evidence of injury or sickness. Hence, how was the monthly shedding of blood that every female experiences to be understood? What was its purpose? Why did a young girl not menstruate, and what was the relationship between reproductive maturity and menstruation?

Regulated by estradiol + progesterone

An understanding of the nature of the relationship between the ovary and menstruation is very recent. The earliest histological descriptions of the changes in the endometrium during the cycle were carried out around the turn of the century, but it was not until the remarkable experiments of J. E. Markee (1940) that menstruation as a continuous process was followed. Markee removed small pieces of the endometrium from a female Rhesus monkey and transplanted small bits of the tissue into the anterior chamber of the same animal's eye, placing the tiny grafts just behind the cornea. Because this area is well-supplied with blood capillaries from the iris, the grafts are assured an ample blood supply. The grafts not only survive, but because of their connection to the circulatory system, they develop and destruct just as if they were still part of the uterus. If the ovaries of the female are removed, the grafts immediately atrophy, demonstrating that the endometrial tissue depends on ovarian hormones. Menstruation in the graft occurs in synchrony with that of the uterus and ceases at the same time. The small amount of blood released from the hemorrhaging graft is released into the aqueous humor of the eye but after a few days it is cleared away. Markee was able to observe the changes in the small bit of endometrial tissue with a microscope. His observations of the changes taking place immediately prior to the initiation of bleeding have been particularly valuable.

Menstrual versus estrus cycles

It must have been apparent to our ancestors that the animals with which they had the most contact did not menstruate. Why was menstruation confined to human fe-

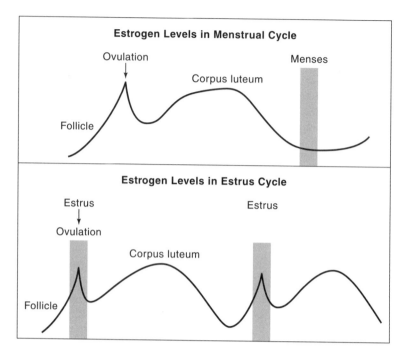

Figure 5.15
Schematic comparison of the estrus cycle of most mammalian species with the menstrual cycle of the higher primates. In estrus species, the endometrium develops only if fertilization occurs. In menstruating species, endometrial development is independent of fertilization.

[handwritten margin notes: Estrus: current period during which P is receptive too. ovulation occurs ↳promote mating ↳fertilization]

males? This is not a trivial question, for it forces us to try to understand a very significant aspect of human reproductive biology. Surprisingly, the full significance of menstruation remains elusive. To appreciate this question, it is instructive to contrast the menstrual cycle with the reproductive cycle of nonmenstruating mammalian species. Such species are characterized by an *estrus* cycle (Fig. 5.15). Estrus is the term, coined in 1901, for the recurrent periods during which the female is receptive to the male, commonly called "heat." It is during this period that ovulation occurs, and the female signals her receptivity to the male in a variety of species-specific ways. Hence, mating is promoted at the opportune time for fertilization to occur. In humans, the development and destruction of the endometrium is autonomous in that it does not depend on the sexual activity of the female or whether fertilization takes place. In humans and the higher primates, the female is continuously receptive to the male. In contrast to estrus species, the human female's sexual receptiveness is disassociated from ovulation. Moreover, there is no outward or public manifestation by the female to indicate that she is approaching ovulation.

[handwritten margin notes: Human: menstruation ↳ sexual receptiveness disassociated from ovulation]

    Many wild animals have estrus cycles tuned to the seasons, timed so that the birth of the young occurs when food is plentiful. In domesticated species, a variety of estrus cycles are found. The shortest cycle is that of the hen, which can lay an egg once a day. Cows, mares, and sows have 21-day estrus periods during most of the year, while sheep have several estrus cycles during the summer, but during the rest of the year they are

[handwritten margin notes: Estrus cycle → season]

anestrus, that is, sexual activity ceases completely. Dogs and cats generally have two or three estrus periods per year, while many carnivores are **monestrus**, having only one estrus period per year.

Endocrinology of the menstrual cycle

≈ Endometrial → proliferative ≈ follicular → secretory ≈ luteal

The menstrual cycle is conventionally divided into two phases—the **proliferative (estrogenic) phase** corresponding to the follicular phase, and the **secretory (progestagenic) phase** corresponding to the luteal phase of the ovulatory cycle, respectively. In reality, it is the **endometrial cycle**, since it is the endometrium that is built up and destroyed. The relationship between the endometrial and ovulatory stages, as marked by the development of the dominant follicle in the first half of the cycle, and the CL during the second half is shown in Figs. 5.6 and 5.16. The endometrium consists of an interior stroma (stromal cells), covered by a thin layer of epithelial cells. Secretory glands are formed by invaginations of the epithelial cells into the stromal matrix (Fig. 5.16). The

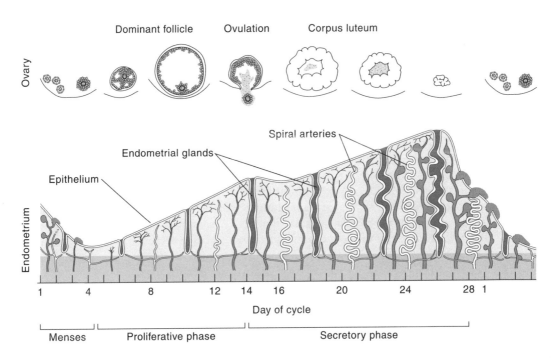

Figure 5.16
Relationship of the ovulatory and endometrial cycles. The first 4–5 days of the follicular phase are characterized by the destruction of the endometrial layer built up during the previous cycle (menses). Beginning at day 5–6, estrogen from the dominant follicle stimulates the endometrium to proliferate. After ovulation, progesterone and estrogen from the corpus luteum promote extensive differentiation of the endometrium to prepare for implantation in case fertilization takes place. If fertilization does not occur, the corpus luteum begins to involute, and without the continued progesterone and estrogen support, the endometrial layer begins to shrink (ischemic phase). The menstrual phase begins when the regressing endometrial layer is removed and expelled. The ovulatory cycle is controlled by LH and FSH, and ultimately by the hypothalamic hormone, GnRH.

rising estrogen levels during the follicular phase stimulate the proliferation of the stromal cells, leading to the thickening of the stromal matrix. The epithelial cells also respond to the increasing levels of estrogen with increases in the number and size of the glandular invaginations and fluid secretions. These are conditions that favor the reception and transport of spermatozoa from the cervix to the uterine tube. Hence, during the first half of the endometrial cycle, two complementary functions take place: first, the growth and proliferation of the endometrial tissue provides the substrate for the second half of the cycle; second, the facilitation of sperm transport maximizes the opportunity for fertilization.

The second crucially important function of estrogen during the first phase of the cycle is to induce the synthesis of progesterone receptors in the endometrial cells. This occurs principally before estrogen levels peak, and by the time of ovulation, the endometrium is primed to bind the progesterone that is now beginning to be produced by the CL. This is the beginning of the progestagenic phase of the menstrual cycle, which prepares the uterus to receive the conceptus from the uterine tube and nourish it. This is accomplished by a progesterone-induced differentiation of the estrogen-primed endometrium. Progesterone has several effects. It stimulates the secretion of a fluid rich in sugars, amino acids, and proteins by the endometrial glands. The stromal cells enlarge and become filled with fluid (edematous), and the spiral arteries grow and develop fully. Some of these changes are potentiated or facilitated by estrogens produced by the CL during the luteal phase (Fig. 5.16).

If fertilization does not take place, the hormonal support provided by the CL begins to diminish once the CL begins to involute. One of the first signs of progesterone withdrawal is the restriction of blood to the endometrium. This occurs, as Markee was able to see in his endometrial-eye grafts, by the contraction of the spiral arteries, and eventual rupturing of the ends of the arteries. The cut-off of the blood supply leads very quickly to the death and rupturing of the endometrial cells. The dead tissue together with the blood released from the ruptured spiral arteries is shed via the cervix and the vagina. Expulsion may be aided by the secretion of prostaglandins, which induce small contractions of the uterine wall, beginning at about day 25 to 26 of the cycle. The pain or ache felt by some women at the beginning of menstruation is due to uterine contractions induced by the release of prostaglandins. After removal of the built-up endometrial tissue, a repair process sets in to restore the tissue so that another cycle can be initiated.

Why do women menstruate?

Menstruation, being unique to humans and higher primates, is a fairly late arrival in evolution. Modern explanations for the origin of menstruation have generally sought to propose an adaptive significance to it. In other words, menstruation provided some type of advantage to the organism who had it over those who did not. These types of explanations suggest that menstruation has a protective function for the female. One version of this hypothesis argues that menstruation evolved as a way for the body to expel bacteria and other potential pathogens introduced into the female during copulation. In menstruating species such as humans in which the female is continually receptive to the male, the possibility of transferring pathogenic organisms would be greater than in an estrus species, and the female would require more protection. Menstruation, by its periodic restructuring of the uterus, provides an essential way for the body to remove pathogenic organisms.

Another version of the protective hypothesis suggests that menstruation is a way of removing defective embryos before a pregnancy has proceeded very far. A substantial amount of evidence does suggest that a significant fraction of human embryos that are formed are defective (see Chapter 14). If such defective embryos do not send the appropriate signals to the maternal system, hormonal support for the endometrium from the corpus luteum is withdrawn, initiating menstruation and eliminating the defective embryo before it can develop any further. A third version suggests that menstruation conserves energy. If it did not occur, energy would have to be expended in maintaining the endometrium in the highly differentiated state that it reaches about midway through the menstrual cycle. These explanations, each appealing in its own way, have been criticized because they fail to take into account the relationship of menstruation to the other functions of the uterus.

[handwritten margin notes: 2) remove defective embryos before a pregnancy proceed very far 3) conserves energy However, these fail to describe relationship b/t menstruation & fxn of uterus]

The most recent model links the occurrence of menstruation to the nature of the implantation process itself and to the way in which the endometrium is triggered to grow and develop. In menstruating species, the endometrium is stimulated to develop and differentiate solely by ovarian hormones, whether fertilization occurs or not. In nonmenstruating species, endometrial development and differentiation, although still dependent on ovarian hormones, occurs only if the fertilized egg makes contact with the endometrium. In the case of the mouse, for example, stromal differentiation, which occurs as the fertilized egg initiates implantation, can be mimicked by placing a drop of oil in the uterus. This suggests that endometrial development and differentiation in the human is controlled in a fundamentally different way than in the mouse. In the human, endometrial development is autonomous and constitutive, while in the mouse it is induced by attachment of the fertilized egg to the endometrium. Moreover, in primates, the implantation process is much more invasive than it is in nonprimates, and endometrial differentiation is much more extensive, presumably to prepare for and accommodate the primate fertilized egg. As a consequence of these two features, the extensively built-up endometrium needs to be removed at the end of each cycle if fertilization has not occurred. According to this view, menstruation is simply a consequence of the nature of the implantation process in primates, and it is not necessary to postulate another function independent of that.

SUMMARY

Menstruation occurs only in humans and a few higher primates. In these species, the menstrual cycle is defined by the periodic development and destruction of the uterine endometrium. The menstrual cycle is orchestrated by the ovarian hormones estrogen and progesterone. Estrogen, during the follicular cycle, stimulates the proliferation of endometrial cells, while progesterone, during the luteal phase, stimulates differentiation of the built-up endometrium to provide a suitable environment for the implantation of the conceptus. If fertilization does not take place, progesterone and estrogen support of the endometrium is withdrawn when the corpus luteum begins to involute. As a consequence, the built-up endometrial layer and the maternal vascular network that provided the blood supply undergo massive cell destruction, initiating the menstrual blood flow.

The function of menstruation has been debated for at least 2500 years. The most likely explanation for the origin of menstruation is that it is a consequence of evolutionary changes that took place in the uterus to permit the type of gestation typical of humans and other higher primates. In these species, in contrast to nonmenstruating, or estrus, species, development of the endometrium takes place independently of fertilization. Moreover, the early embryo is highly invasive, and much more extensive development of the endometrial stroma is required to accommodate implantation.

QUESTIONS

1. The synthetic compound mifepristone (RU-486) is an antagonist at the progesterone receptor. Administration of RU-486 during the late follicular phase will inhibit ovulation. What conclusion do you draw from this observation?

2. What is the functional significance of the rise in FSH at the end of the luteal phase?

3. What is the role of the two endocrinological switches during the follicular phase?

4. Describe the factors that play a role in the selection of the dominant follicle.

5. Compare the roles of estrogen and progesterone in the menstrual cycle.

6. What is the critical difference between estrus and menstrual species?

7. Which explanations for the significance of menstruation do you find the most compelling, and why?

8. Library project:
    a. Describe the latest method(s) for following the development of the dominant follicle. Under what circumstances is this necessary or useful?
    b. Involution of the corpus luteum is the precipitating trigger for menstruation. What factors play a role in the demise of the CL?
    c. The LH surge is a key event during the ovulatory cycle. What is known about the mechanism that brings it about?

SUPPLEMENTARY READING

Adashi, E. Y. 1996. The ovarian follicular apparatus. In *Reproductive Endocrinology, Surgery, and Technology, Vol. 1.* E. Y. Adashi, J. A. Rock, and Z. Rosenwaks, Eds. Lippincott-Raven Publishers, Philadelphia, PA, pp. 17–40.

Corner, G. W. 1933. The discovery of the mammalian ovum. In *Lectures on the History of Medicine*, Mayo Foundation Lecture Series. 1926–1932. W. B. Saunders, Co., Philadelphia, PA.

Corner, G. W. 1942. *Hormones in Human Reproduction.* Princeton University Press, Princeton, NJ.

Erickson, G. F. 1995. The ovary: Basic principles and concepts. In *Endocrinology and Metabolism*, 3rd ed., pp. 973–1015.

Johnson, M., and B. Everitt. 1996. *Essential Reproduction*, 4th ed. Blackwell Scientific Publications, Oxford, UK.

Markee, J. E. 1940. Menstruation in intraocular endometrial transplants in the Rhesus monkey. *Contributions to Embryology* **28**, 219–308.

Oudshoorn, N. 1994. *Beyond the Natural Body. An Archeology of the Sex Hormones.* Routledge, London, UK.

Short, R. V. 1977. The discovery of the ovaries. In *The Ovary*, 2nd ed., Vol.1. Lord Zuckerman and B. J. Weir, Eds. Academic Press, New York, NY.

Yen, S. S. C. 1999. The human menstrual cycle. In *Reproductive Endocrinology: Physiology, Pathophysiology, and Clinical Management*, 4th ed. S. S. C. Yen, R. B. Jaffe, R. L. Barbieri, Eds. W. B. Saunders Co., Philadelphia, PA.

ADVANCED TOPICS

Baker, T. G., and W.S.O. 1976. Development of the ovary and oogenesis. *Clinics in Obstetrics and Gynecology* **3**, 3–26.

Clarke, J. 1994. The meaning of menstruation in the elimination of abnormal embryos. *Human Reproduction* **9**, 1204–1206.

Finn, C. A. 1994. The adaptive significance of menstruation. *Human Reproduction* **9**(7), 1202–1207.

Finn, C. A. 1998. Menstruation: A nonadaptive consequence of uterine evolution. *Quarterly Review of Biology* **73**, 163–173.

Gougeon, A. 1996. Regulation of ovarian follicular development in primates: Facts and hypotheses. *Endocrine Reviews* **17**(2), 121–155.

Profet, M. 1993. Menstruation as a defense against pathogens transported by sperm. *Quarterly Review of Biology* **68**, 335–386.

Strassmann, B. I. 1996. The evolution of endometrial cycles and menstruation. *Quarterly Review of Biology* **71**, 181–220.

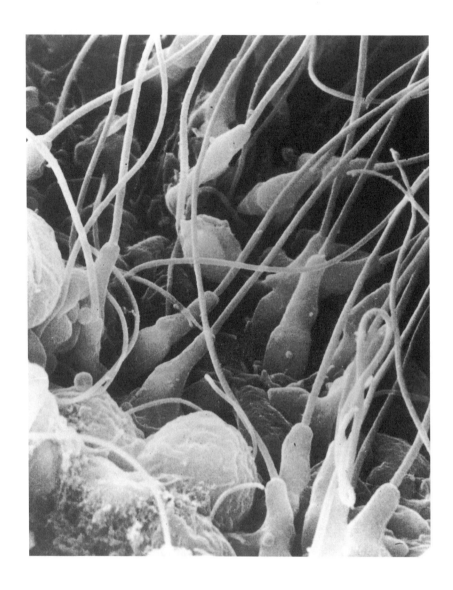

Scanning electron micrograph of human sperm about to be released from the Sertoli cells. (A. H. Holstein and E. C. Roosen-Runge. 1981. *Atlas of Human Spermatogenesis*. Grosse Verlag, Berlin, Germany. Courtesy of A. H. Holstein.)

The Testis and Testicular Function

The rooster wears a comb, which is, so to speak, the flag that is hoisted to announce the presence of his testes to the hens. On removal of the testes the flag is lowered: the comb atrophies and the rooster has become a capon.

—M. Tausk. 1978

Organon: De geschiedenis van een bijzondere Nederlands Onderneming.
Dekker en Van de Vegt, Nijmegen, the Netherlands. Cited in N. Oudshoorn. 1994.
Beyond the Natural Body: An Archeology of Sex Hormones. Routledge, London, UK, p. 49.

One thing can be predicted with certainty—more functions in the male that have traditionally been ascribed to androgen action will ultimately prove to be either regulated or modulated by estrogens.

—R. M. Sharpe. 1998

The roles of estrogen in the male.
Trends in Endocrinology and Metabolism 9(9), 371–377.

THE TESTES, unlike the ovaries, never had to be discovered. They have always been there, palpable, and the subject of experimentation and speculation throughout most of human civilization. The testis, as we saw in Chapter 4, is formed early in embryogenesis and goes through at least three developmental stages—the fetal, the postnatal and prepubertal, and the postpubertal. Since males do not undergo the same relatively sudden loss of gonadal function as females do in menopause, there is probably no testicular equivalent to the postmenopausal ovary. Testicular aging does occur, but whether it represents a significant qualitative difference in testicular function remains to be seen. Our focus in this chapter is on the adult or postpubertal testicular function, but we will also review briefly fetal and prepubertal testicular function.

A SHORT HISTORY OF THE TESTIS

The relationship of the word itself, "testis" (Latin for "witness" or "spectator"), and also of the words "testify" and "testament," to the organ is not at all clear. Possibly because the testes were evidence of virility, and only adult males could be witnesses in Roman law (boys, eunuchs, and women could not be witnesses), the act of testifying be-

135

came associated with the organs themselves. Indeed, in some societies the witness was required to hold his or someone else's testes in the act of testifying. A good example comes from the Book of Genesis in the Old Testament where the "thigh" was a euphemism for testes:

> And Abraham said unto his eldest servant of his house, that ruled over all that he had. Put, I pray thee, thy hand under my thigh: And I will make thee swear by the Lord, the God of heaven, and the God of earth. (Genesis 24:2–3)

The Greeks had at least two words for testes; one was *orcheis*, from which words like *orchitis* (inflammation of the testes) and **cryptorchid** (hidden testes, meaning undescended testes) are derived. Apparently this word was considered to be too vulgar, and the more acceptable alternative *didymis* (meaning "twins"), from which the word epididymis is derived, was used more commonly.

The fact that removal of the testes, **orchidectomy** or **castration**, does not ordinarily lead to death was probably recognized very early in our history. Evidence of the castration of animals dates at least to about 7000 BCE, around the period when animals were first domesticated. The major effects of testes removal on reproductive function in both animals and men must have been recognized very early. In the historical record, castration was the punishment for adultery in the Babylonian Code of Hamurabi (ca. 2000 BCE) and in Egyptian law of the 20th Dynasty (1200–1085 BCE). In Assyria (ca. 1500 BCE) castration was prescribed for sexual offenses. This suggests that the effect of castration on the male sexual drive and erectile function was already known. Indeed, castration for social purposes, especially of prepubertal boys, to produce eunuchs to guard the women in harems, is an ancient practice. It is thought to have originated in the Middle East, but was adopted by most of the high cultures of China, India, and Europe. In Europe, although the practice was condemned by the Roman Church in 325 CE, the castration of young boys to maintain a supply of male soprano singers was continued until the late nineteenth century. Many primitive societies also practiced castration, in many cases for ritualistic purposes. The cannibalistic Caribs of Brazil were reported to castrate their prisoners of war in order to increase their weight and the tenderness of their flesh before they were killed for a ritual feast.

The effects of castration were described clearly by Aristotle, who differentiated between the effects of prepubertal and postpubertal castration. In a passage describing the effects of castration on roosters he continues:

> . . . after they are full grown, the comb turns yellow and they cease to crow and no longer desire sexual intercourse. If they are not full grown, these parts never reach perfection. The same is the case with human subjects, for if a boy is castrated, the hair that is produced after birth never appears, nor does his voice change; but if a full grown man is castrated, all the hair produced after birth falls off except that on the pubes, this becomes weaker, but still remains. In the eunuch, the hair present at birth does not fall off, for the eunuch never becomes bald.[1]

Despite Aristotle's fairly accurate observations of castration effects, the importance of the testes for fertility remained in doubt. This was probably because of the ob-

---

[1]Aristotle. *Historia Animalium*. In *The Works of Aristotle*, Vol. IV. 1910. Translated by D. W. Thompson, Clarendon Press, Oxford. Cited in Bremner, 1981.

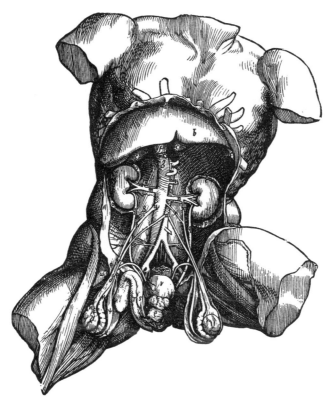

Figure 6.1
Anatomical drawings of the male reproductive organs made by Vesalius (1538). (Reprinted
from *The Illustrations from the Works of Andreas Vesalius of Brussels with Annotations and
Translations*. 1950. J. B. de C. M. Saunders and C. D. O'Malley. The World Publishing Co.,
Cleveland, OH, with permission from the New York Academy of Medicine.)

servation that a bull was still able to impregnate a cow during the first ejaculations after being castrated. It took another 2000 years before this question was answered
definitively. Vesalius provided some of the first anatomically accurate drawings of the
male reproductive organs (Fig. 6.1). Van Leuwenhoek (1667) discovered what he
called "animalcules" in seminal fluid. This observation led to what was called the
"spermist" school of reproduction, the notion that a human being was contained in
miniature in one of the animalcules (see Fig. 1.1). De Graaf (1668) showed definitively
that the fertilizing potential of the seminal fluid was produced by the testes, and not by
the other tissues of the male internal genitalia. Finally, van Koelliker (1840) demonstrated that van Leuwenhoek's animalcules were in fact cells, and not miniature animals, and that they developed in the seminiferous tubules by processes analogous to
the development of other cells.

138

Figure 6.2
The effect of castration on the comb and the wattles of the rooster. Normal rooster on the left and castrated one on the right. (Reprinted from P. C. Clegg and A. G. Clegg. 1969. *Hormones, Cells, and Organisms*. Heinemann Educational Books Ltd., London, UK, with permission.)

In 1849, Berthold performed a classic experiment by analyzing the effect of removal and transplantation of the testes of birds. He showed that removal of the testes resulted in a reduction of comb size and loss of crowing (Fig. 6.2). The loss of these male traits could be prevented if the testes were transplanted to an abnormal site, even when all nerve connections of the testes were severed. He concluded, correctly, that the effects of the testes must be exerted by some substance or substances that are transmitted through the bloodstream. Berthold had provided the first clear evidence for the chemical messengers that, some 50 years later, were to be called hormones. Finally, the active components of testicular secretions, androgens, were isolated and purified in the 1920s and 1930s from animal urine. Research on human testicular hormones was complicated by the fact that no readily available source of human testes was available. This was not the case for ovaries, since the removal of ovaries from women as therapy for a number of "female disorders" was a standard clinical practice. No comparable clinical practice for removal of testes existed, and consequently, human testes were in short supply. Endocrinologists turned to human urine as a possible source of testicular androgens. The first human androgen was isolated from 25,000 liters of men's urine collected in the police barracks in Berlin. In a landmark success for the field of synthetic chemistry, testosterone was synthesized in 1931. "Dynamite, gentlemen, pure dynamite," was how one of the discoverers of testosterone referred to its effects.

One of the more curious features of the fascination with the testes has been the association of the testes and its secretions with rejuvenation and vigor, perhaps first stated clearly by Aretaeus of Cappadocia (150 CE):

> *For it is the semen, when possessed of vitality which makes us to be men, hot, well-braced in limbs, well-voiced, spirited, strong to think and act.*[2]

---

[2]Rolleston, J. 1963. *The Endocrine Glands in Health and Disease*. Oxford University Press, London, UK. Cited in Bremner, 1981.

The association of vitality with potency was probably even by Aretaeus' time already an ancient one, but it was not until the nineteenth century that this notion was put to experimental test. The first reports came in 1889 by Brown-Séquard (then 72 years old), a well-regarded French physiologist, who injected himself with extracts of dog testes and claimed to have recovered his potency. In his famous lecture, he mentioned (with an obvious meaning to all the members of the audience) that he had "paid a visit" to his wife that morning. The reported effects were eventually shown to be invalid, and Brown-Séquard was forced to accept the likelihood of a very strong placebo effect. In the early twentieth century, two other procedures for potency were introduced. One was the Steinach operation, which involved occlusion of the vas deferens, essentially equivalent to a vasectomy. This procedure was promoted as a way to improve blood flow to the testis, and therefore testicular function. The Voronoff procedure, involving transplanting testes from monkeys or other males (executed criminals), was more drastic. The effects in both cases, despite the promotion, were short-lived or nonexistent. Nevertheless, these practices continued, and private clinics in Europe, promising a return of youthfulness and an increase of sexual activity, prospered during the 1920s and 1930s, and even after World War II. Indeed, there were reports in the 1950s that Pope Pius XII underwent such treatments. After the practice disappeared in Europe, it continued in other countries. As late as the 1970s, a clinic in Tijuana, Mexico, across the border from San Diego, California, was still attracting patients.

STRUCTURE OF THE MALE REPRODUCTIVE SYSTEM

The complete male reproductive system includes the testes, the internal genitalia (the epididymis, vas deferens, prostate, seminal vesicles, and bulbourethral glands), and the two external structures, the penis and the scrotum (Fig. 6.3). The testes produce the spermatozoa, while the prostate, seminal vesicles, and bulbourethral produce the bulk of the seminal fluid. The postpubertal testis has two essential functions, the production of the male gametes (spermatogenesis), and the synthesis and secretion of androgens, principally testosterone. These two functions are separated anatomically into two compartments within the testis: the seminiferous tubules (ST), which contain the germ cells and Sertoli cells, and an interstitial region between the seminiferous tubules (Figs. 6.3 and 6.4). The interstitial regions contain a variety of cell types, including the most important androgen-producing cell type, the Leydig cell, as well as blood vessels and nerves. The testis is covered by a capsule containing arteries and veins that establishes a connection to the general circulatory system. The ends of the ST empty into a network of ducts (the rete testis), which leads eventually into the epididymis, a single highly convoluted tubule that connects to the vas deferens (also known as the ductus deferens) (Fig. 6.3). The vas deferens together with its connection to the urethra forms the transport system for the sperm to the exterior.

The scrotum, the external sac that houses the testes, is found only in certain orders of mammals (for example, primates, ruminants, and marsupials). However, in probably all mammalian species, the testes migrate from their site of origin to their final anatomical position. The extent of migration is quite variable so that the final position of the testes varies considerably between different orders of mammals (Fig. 6.5). In scrotal species such as humans, migration of the testes from an abdominal position near the kidneys into the scrotum (referred to as testicular descent) occurs during the latter stages of fetal life or early in the postnatal period. Over 90 percent of male infants at the time of birth have descended testes. In the remainder, descent generally takes place

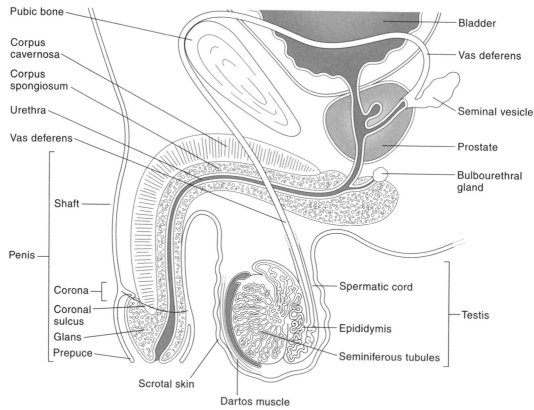

Figure 6.3

The male reproductive tissues. Recall that the epididymis, vas deferens, seminal vesicles, and bulbourethral glands developed from the mesonephric ducts, while the prostate developed from the urogenital sinus. The prostate, seminal vesicle, and bulbourethral glands provide the components of seminal fluid that is mixed with the spermatozoa at the time of ejaculation. The vas deferens connects to the urethra and together the two form the conduit for sperm ejaculation. The testes are housed in the scrotum. Spermatogenesis takes place in the seminiferous tubules. The total length of the seminiferous tubules is over 100 meters. The spermatozoa produced are moved down the lumen of the tubules into the efferent ductules, and into the epididymis and then into the vas deferens. Movement of the sperm through these tissues is androgen dependent.

within two or three months after birth. Failure of the testes to descend is known as cryptorchidism. It has long been known that men who are bilaterally cryptorchid (both testes fail to descend) are always infertile, while men who are unilaterally cryptorchid (only one testis fails to descend) are generally fertile. Hence, if the testes have not descended by themselves after a few months after birth, they are brought into the scrotum surgically.

The reason for the infertility in cryptorchidism was not clear until 1924, when it was discovered that spermatogenesis in species with scrotal testes is very sensitive to temperature and does not take place at normal body temperature. Placing the testis of a guinea pig together with its blood supply back into the abdominal cavity resulted in

140

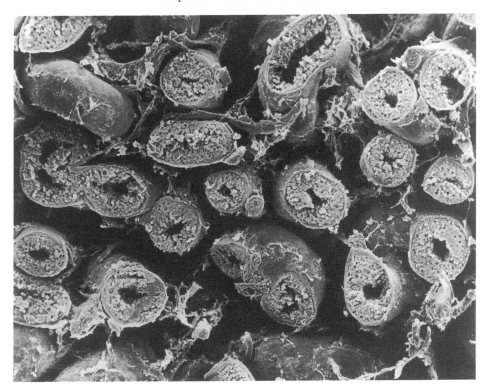

Figure 6.4
Scanning electron micrograph of a cut through a testis from a 51-year-old man. The gross features of the seminiferous tubules are seen: basement membrane surrounding each tube; although not very visible, the male germ cells in various stages of spermatogenesis; and the hollow center of the tube, the lumen. In the region between the tubes are seen some Leydig cells and other cell types. (Reference: A. H. Holstein and E. C. Roosen-Runge, 1981. Courtesy of A. H. Holstein.)

a complete disorganization of the ST within one week and a rapid cessation of spermatogenesis. If the testis was left too long in the abdominal cavity, the damage to the ST was irreversible, and spermatogenesis could not be restored when the testis was brought back into the scrotum. Demonstration of the sensitivity to temperature did not require an operation. Wrapping the scrotum of a fertile ram, for example, with woolen socks so that the temperature of the testes is the same as that of the body also results in a cessation of spermatogenesis. In humans the scrotum is 2–3°C lower than body temperature, a very small temperature difference but apparently enough to permit spermatogenesis to proceed. Actually, the connection between fertility and testes temperature has been known for a long time. An early form of birth control was the Roman custom of taking very hot baths, a practice which is still followed, for the same reasons, in some parts of Japan. Fifteen-minute exposures to 100–104°F in a hot spa every day will result in over an 80 percent decrease in sperm production in a two-week period. Although this may seem to be an impressive decrease, it is not nearly enough to consider hot baths an effective birth control method.

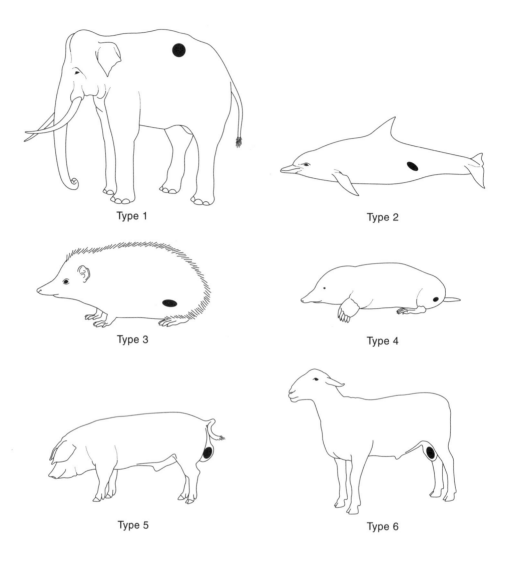

Type 1

Type 2

Type 3

Type 4

Type 5

Type 6

Figure 6.5

The approximate anatomical positions of the testes in different mammalian species. *Type 1:* Testes caudal to the kidneys, found in the elephant family, ungulate mammals, many species of insectivorous mammals, and monotremes. *Type 2:* Testes near the bladder, close to the posterior abdominal wall, found in marine mammals (dolphins), many species of toothless mammals, and two large families of insectivores. *Type 3:* Testes near the anterior abdominal wall, found in the spiny-coated mammals (hedgehog), and many species of insectivores, the seal, and bat families. *Type 4:* Testes in a sac near the base of the tail without producing any external swelling, found in species of the shrew and nocturnal burrowing mammals (moles). *Type 5:* Testes in a nonpendulous scrotum, found in nonruminants (swine), rodents, other gnawing mammals, carnivores, and many species of tree shrews. *Type 6:* Testes in a pendulous scrotum with a distinct neck, found in ruminants (rams), most marsupials, and primates. (Adapted from B. P. Setchell. 1978. *The Mammalian Testis.* Cornell University Press, Ithaca, NY.)

SUMMARY

The testis is a composite structure, consisting of the seminiferous tubules in which spermatogenesis takes place, and the interstitial tissue containing the Leydig cells, which are the primary source of testosterone. All of the anatomical elements of the male reproductive system, the internal and external genitalia, depend on testicular androgens for their formation and functional maintenance. The internal genitalia in turn not only provide a transport path for sperm produced in the seminiferous tubules, but their secretions are important in ensuring fertility. The human testis is somewhat unusual in that in common with a few other mammalian species, but in contrast with most mammals, it is found outside the body in an external sac, the scrotum. Spermatogenesis is particularly sensitive to temperature, and will not take place at normal body temperature. The scrotum, about 3 degrees cooler than the interior of the body, provides the proper temperature environment for spermatogenesis to take place successfully.

SPERMATOGENESIS *continually renewable process*

Spermatogenesis begins at puberty in response to signals ultimately originating, as we shall see in Chapter 8, in the hypothalamus. The spermatogonial population of cells produced during fetal life remains quiescent during infancy and childhood, and at puberty the cells are triggered to initiate a process that, once started, appears to continue until late in the individual's life. Hence, in contrast to oogenesis, spermatogenesis is a continually renewable process.

Stages of spermatogenesis *3 stages*

It is convenient to divide spermatogenesis into three major stages—*proliferation and regeneration*, *meiosis*, and *spermiogenesis* (Fig. 6.6).

Stage 1: Proliferation and regeneration

The continuous production of spermatozoa requires a continuous supply of cells (*spermatogonial stem cells*) able to embark on the spermatogenic pathway. This is accomplished by having two pathways for spermatogonial differentiation. One pathway provides cells that can undergo meiosis, the transitions designated A *(dark)* to A *(pale)* to B in Fig. 6.6 and secondly, the other pathway regenerates the stem cell population, A *(dark)*. The **dark-pale** distinction is based on the staining characteristics of the two cell types. The precise distinctions between the different spermatogonial cell types and the mechanism that directs these differentiative events are not well understood and remain the subject of much study.

All the different types of spermatogonial cells are embedded in Sertoli cells along the basement membrane of the ST (Figs. 6.7 and 6.8). The Sertoli cells extend from the basement membrane inwardly into the lumen of the seminiferous tubule (ST). The Sertoli cells have a complex morphology determined by the cell types at all stages of spermatogenesis that are embedded in the Sertoli cell membrane. At the periphery of the ST, adjacent Sertoli cells form very specialized junctions, referred to as *tight junctions*, which are responsible for what is known as the **Sertoli cell barrier** (Fig. 6.8). The Sertoli cell barrier isolates the ST from most molecules that are present in the interstitial region of the testis. Effectively, the tight junctions regulate the uptake of many com-

144

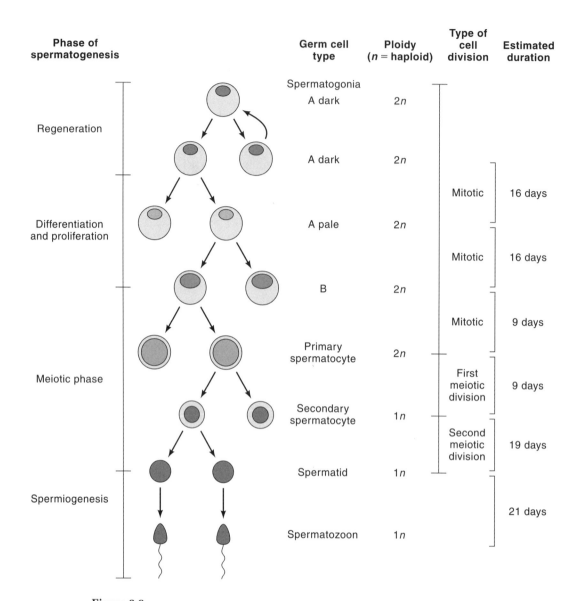

Figure 6.6

The stages of spermatogenesis from the spermatogonial cell to the spermatozoon. A quiescent pool of spermatogonial cells formed during fetal life is present in the prepubertal testis. At the onset of puberty, the spermatogonia begin to proliferate and differentiate. The A (dark) spermatogonia form the regenerative population on which spermatogenesis depends. They also give rise sequentially to the type A (pale) and B spermatogonia. The B spermatogonia in turn give rise to the primary spermatocytes which signal the entry into the meiotic divisions. The products of the first meiotic division are the secondary spermatocytes, and these undergo the second meiotic division, forming the spermatids. Finally, during spermiogenesis, the spermatids are transformed into spermatozoa. Spermatogenesis is estimated to take about 80–90 days.

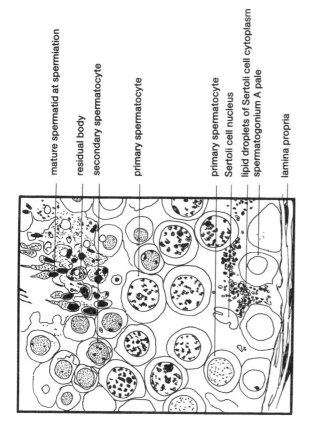

mature spermatid at spermiation

residual body

secondary spermatocyte

primary spermatocyte

primary spermatocyte

Sertoli cell nucleus

lipid droplets of Sertoli cell cytoplasm

spermatogonium A pale

lamina propria

Figure 6.7

Arrangement of cell types in the seminiferous tubule (ST) during spermatogenesis. (Right) Light microscopic view of a partial cross-section of an ST showing several stages in spermatogenesis. (Above) Drawing of the same photograph identifying the cell stages. Type A (pale) spermatogonia are found along the basement membrane (lamina propria). Two phases of the primary spermatocyte and a secondary spermatocyte are also evident. The mature spermatid at the lumen of the ST is well on its way to becoming a spermatozoon. Parts of the cytoplasm removed from the spermatid to form the spermatozoon (the residual bodies) are visible. They will be eventually resorbed by the Sertoli cell. (Reprinted from A. H. Holstein and E. C. Roosen-Runge, 1981. Courtesy of A. H. Holstein.)

146

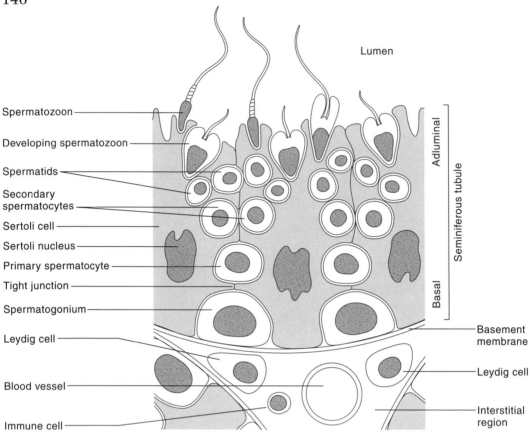

Spermatozoon

Developing spermatozoon

Spermatids

Secondary spermatocytes

Sertoli cell

Sertoli nucleus

Primary spermatocyte

Tight junction

Spermatogonium

Leydig cell

Blood vessel

Immune cell

Lumen

Adluminal

Basal

Seminiferous tubule

Basement membrane

Leydig cell

Interstitial region

Figure 6.8
The organization of the Sertoli and germ cells within the seminiferous tubule. The Sertoli cells envelope the developing germ cells and form tight junctions between themselves (the Sertoli cell barrier). The basal compartment contains the spermatogonial population, while the adluminal compartment contains the spermatocytes, spermatids, and spermatozoa. Progression through the stages of spermatogenesis means movement from the basal to the adluminal compartment. The mature spermatozoa are released into the lumen of the seminiferous tubule. The interstitial region, containing the androgen-producing Leydig cells, cells of the immune system, blood vessels, and other cell types, is separated from the Sertoli cells by the basement membrane which encases the seminiferous tubules.

Sertoli cell barrier
↳ tightly controlled
environment necessary
for spermatogenesis
to occur (↗ environment
for higher development
germ cells)

pounds into the interior of the ST. The precise reason for the Sertoli cell barrier is not known, but it seems reasonable to infer that it provides the tightly controlled environment that is necessary for spermatogenesis to occur. The Sertoli cells also form distinct types of junctions with germ cells at all stages of germ cell development; in each case the junctions are specialized to provide the proper environment for the further development of the germ cells. Integrity of the Sertoli cell is an absolute requirement for successful spermatogenesis. All stages of spermatogenesis are regulated and directed by the Sertoli cells.

Stage 2: Meiosis

Type B spermatogonia differentiate into **primary spermatocytes,** which begin meiosis. The primary spermatocytes undergo the first meiotic division, forming **secondary**

Type B spermatogonia
↓
1° spermatocyte
↓ 1st meiotic division
2° spermatocy —2nd meiotic division→ spermatids

spermatocytes, which in turn quickly undergo the second meiotic division, forming the **spermatids** (Fig. 6.6). As spermatogenesis proceeds, the cells move from the basement membrane in a radial direction toward the lumen of the ST (Figs. 6.7 and 6.8).

Stage 3: Spermiogenesis   spermatid ⟶ spermatozoon

The last stage of germ cell development, **spermiogenesis,** is perhaps the most remarkable in a morphological sense, involving a dramatic change from the spermatid to form the spermatozoon (Figs. 6.7, 6.9, 6.10, and 6.11). Two particularly important

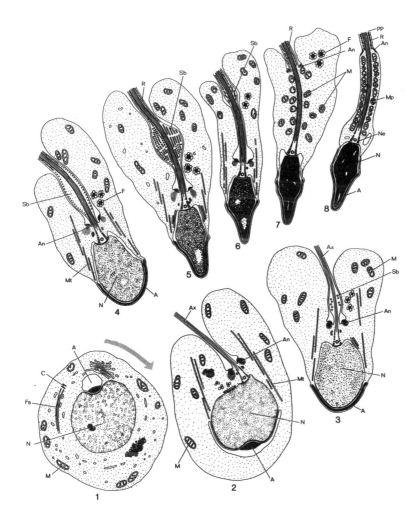

Figure 6.9
Spermiogenesis—the transformation of the spermatid into a spermatozoon. Amid the complexity of changes that take place are three of major importance: first, the condensation of the nucleus forming the compact nucleus seen in step 8; second, the formation of the acrosome which is wrapped around the sperm head (steps 3–8); and third, the formation of the flagellum, which provides the motive force for the spermatozoon (steps 2–8). **A,** acrosome; **M,** mitochondria; **M$_p$,** middle piece; **Ne,** neck; **N,** nucleus; **PP,** principal piece. (Reprinted from A. H. Holstein and E. C. Roosen-Runge, 1981. Courtesy of A. H. Holstein.)

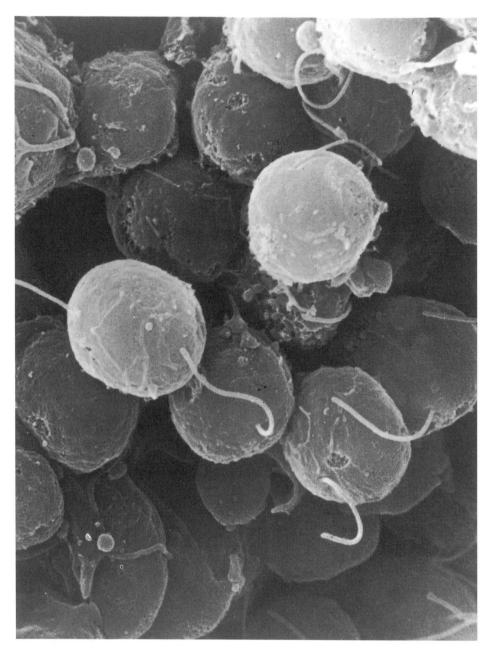

Figure 6.10
Scanning electron micrograph of spermatids at the beginning of spermiogenesis. Initial stages
of flagellar formation are evident. Magnification, 4000. (Reprinted from A. H. Holstein and
E. C. Roosen-Runge, 1981. Courtesy of A. H. Holstein.)

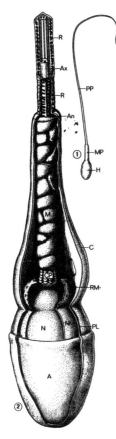

Figure 6.11
Schematic diagram of a human spermatozoon.
(1) H, head; Mp, middle piece; PP, principal
piece; EP, end piece. (2) A, acrosome; N, nucleus;
M, mitochondrion; CP, connecting piece; Ax,
axoneme. (Reprinted from A. H. Holstein and
E. C. Roosen-Runge, 1981. Courtesy of A. H.
Holstein.)

changes are the formation of the **acrosome** and the **flagellum**. During these changes,
most of the cytoplasm of the spermatid is removed and resorbed by the surrounding Sertoli cell. Each step in spermiogenesis is tightly regulated by the specialized connections between the Sertoli cells and germ cells. Once spermiogenesis is complete,
the spermatozoon is released into the lumen of the ST and begins its journey to the
epididymis.

Spermatogenesis in time

In humans, about 20 Type B spermatogonia per second enter Stage 2 of spermatogenesis. Since a maximum of 64 spermatozoa can be produced for each Type B spermatogonium, about 80,000 spermatozoa per minute, or about 110 million spermatozoa per day can be produced. This is an average figure, with actual values ranging from
20 million to 270 million per day being reported and considered spanning the normal
range. The process from Type B spermatogonium to spermatozoa takes about 60 days,
with spermiogenesis itself requiring about 21 days (Fig. 6.6).

In all the species studied to date, spermatogenesis is very inefficient. Cell loss can
occur at each of the three stages of spermatogenesis. In rats, for example, most of the
loss occurs in Stage 1; in humans about 50 percent of the loss takes place during meiosis, especially during the second meiotic division, while very little loss occurs during

Table 6.1
Daily Sperm Production (DSP) in Different Mammalian Species

Species	Paired testes weight (gram)	DSP (10^6/gram)	Total DSP (10^6)
Rabbit	6.4	25.	160.
Rat	3.7	24.	90.
Rhesus monkey	49.	23.	1,100.
Boar	720.	23.	16,200.
Ram	500.	21.	9,500.
Bull	775.	13.	8,900.
Human	40.	4.	160.
Gorilla	25.	1.6	40.
Chimpanzee	70.	5.8	400.

These are average figures. In humans, DSP may range from 20 million to 270 million per day. (References: Sharpe, 1994; and Jansen, 1995.)

spermiogenesis. For reasons that remain unknown, the efficiency of spermatogenesis in humans is significantly lower than in other well-studied mammalian species (Table 6.1).

The mechanisms of degeneration and wastage in a normal testis are a matter of debate, although most studies indicate that small disturbances in Sertoli cell function or loss of Sertoli cells are to blame. Most cases of **oligospermia** (low sperm count) in human males are **idiopathic**, i. e., of unknown cause. It seems likely that many such cases involve increased germ cell wastage due to perturbations in Sertoli cell function. The reduction in daily sperm production with increasing age appears to be associated with loss of Sertoli cells.

Hormonal control of spermatogenesis

Both spermatogenesis and steroidogenesis (the synthesis and secretion of steroid hormones) depend primarily on stimulation by the pituitary gonadotropic hormones LH and FSH, with a possible secondary role played by another pituitary hormone, prolactin. These relationships were first recognized in the 1930s from analysis of experiments involving the removal of the pituitary (**hypophysectomy**) in mice. Removal of the pituitary leads quickly to testicular atrophy and loss of both steroidogenesis and spermatogenesis. Examination of the intratesticular tissues shows that within a day of pituitary removal, the Leydig cells begin to involute. This is followed by morphological changes in the ST characterized by a disorganization of the normal cellular arrangements of the different germ cell stages.

In one experiment, testosterone was administered exogenously immediately following hypophysectomy. The Leydig cells still involuted, but normal spermatogenesis could be maintained, at least for a time. Two important conclusions could be drawn from these observations. First, testosterone is required for spermatogenesis. Second, the Leydig cells must be the normal site of testosterone synthesis, and they must respond to some pituitary hormone. By testing pituitary extracts for their ability to re-

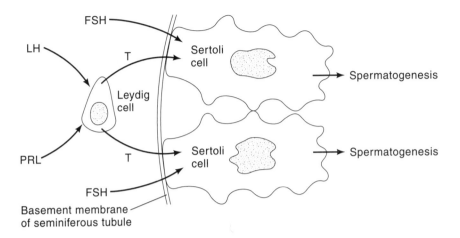

Figure 6.12
Gonadotropic control of spermatogenesis. Testosterone (T) production by the Leydig cells is regulated by LH and prolactin (PRL). T and FSH together are required for spermatogenesis.

store Leydig cell function, a hormone called **interstitial cell simulating hormone** (ICSH) was identified and isolated. ICSH was shortly thereafter shown to be identical to the hormone LH, previously identified in females.

In a second experiment, the investigators waited a few days until the ST began to show signs of disorganization. At that point, they found that administration of testosterone by itself was no longer sufficient to restore spermatogenesis. Clearly some other component from the pituitary was needed. By testing different types of pituitary extracts, they isolated a hormone that together with testosterone was able to restore spermatogenesis. This hormone turned out to be identical to FSH, again already known from the studies on ovarian function.

These experiments showed that LH and FSH have two qualitatively different roles in the testis. LH is required for spermatogenesis indirectly, through its ability to stimulate the Leydig cells to produce testosterone. LH, then, is necessary for the steroidogenic function of the testis. FSH, on the other hand, must have a direct role in spermatogenesis itself. Additional insight into the contrasting roles of LH, FSH, and testosterone was obtained when it became clear that the germ cells do not have receptors for any of the three hormones. The Leydig cells are the only cells in the testis that have receptors for LH, while the Sertoli cells have both androgen and FSH receptors. This suggests that FSH and testosterone regulate spermatogenesis by modulating Sertoli cell function (Fig. 6.12).

One interpretation of the animal experiments is that FSH has an *initiation* function with respect to spermatogenesis, while LH (and therefore testosterone) may have a *maintenance* function. The precise distinction between these two presumed roles is not really clear, but attempts to treat gonadal failure in humans are consistent with a significant difference in the LH and FSH roles. For example, hCG (equivalent to LH) therapy has been successful in treating cases of postpubertal gonadal failure. On the other hand, hCG therapy is not sufficient to restore fertility in cases of prepubertal gonadal failure. In those cases, FSH and LH are required to restore some fertility. One major difference between these two types of failure is that in the former case, spermatogenesis was initiated normally at puberty, while in the latter case, spermatogenesis had not yet been initiated. The requirement for FSH in prepubertal failure implies that FSH has an essential role in initiating spermatogenesis. However, in both cases, both LH and FSH are required to maintain quantitatively normal levels of spermatogenesis. The stages of spermatogenesis at which these hormones act are far from clear, especially in humans. The Sertoli cell has multiple crucial roles in spermatogenesis: regulation of germ cell development, maintaining ST structure and compartmentalization, releasing mature spermatozoa, and secreting the components of ST fluid. All of these functions depend on FSH and testosterone.

Prolactin also appears to have a role in spermatogenesis. The evidence comes from studies on mice. For example, Leydig cells have prolactin receptors, suggesting that prolactin is important in Leydig cell function. Consistent with this interpretation, prolactin has been shown to potentiate the LH stimulation of testosterone production of Leydig cells in culture (Fig. 6.12). Moreover, mice carrying mutations in two genes, *dw* and *df*, are **aspermatogenic** (do not produce sperm), but spermatogenesis can be restored by administration of prolactin. In humans, however, a definitive role for prolactin in spermatogenesis has not been demonstrated. However, it has been observed that high levels of prolactin suppress spermatogenesis. Elevated levels of prolactin, due, for example, to tumors in the prolactin-producing cells of the pituitary, lead to a condition known as **hyperprolactinemia**. Hyperprolactinemic males are characterized by sterility and/or erectile dysfunction. These two conditions appear to be independent of testosterone levels, since they are found in individuals with high testosterone levels. That the effect is due to high prolactin has been demonstrated by the success of *bromocriptine* therapy. Administration of bromocriptine, a dopamine receptor antagonist, inhibits prolactin secretion and has been found to restore fertility and erectile function.

SUMMARY

Spermatogenesis is initiated at puberty and, in contrast to oogenesis, is a continually renewable process, with millions of sperm produced daily. The different stages of spermatogenesis take place within the tightly controlled environment of the seminiferous tubules in which Sertoli cells play the primary role. Beginning with the spermatogonial population, successive populations of germ cells, primary and secondary spermatocytes, spermatids, and eventually mature spermatozoa, move from the periphery of the seminiferous tubule, embedded between adjacent Sertoli cells, into the lumen. Two pituitary hormones, LH and FSH, play the primary role in regulating spermatogenesis, although the precise role of each has not yet been defined. LH stimulates the Leydig cells to secrete testosterone, which then passes into the Sertoli cells. FSH and testosterone control spermatogenesis. Animal studies, and some case studies in humans, suggest that FSH may play an important role in the spermatogonial stages of spermatogenesis, while testosterone may be important during the meiotic divisions and during spermiogenesis. In rodents, prolactin is also important, but its role in humans has not been clarified.

TESTICULAR ENDOCRINOLOGY

Testicular function depends on feedback interactions between testicular hormones and the pituitary (the testis-pituitary axis) and the hypothalamus (the hypothalamus-pituitary-testis axis). We consider the first axis in this section; the second is considered in Chapter 7.

The testis-pituitary axis

The pituitary and the testes are part of a short feedback circuit, analogous to that of the pituitary-ovarian axis. The pituitary and testis regulate each other's activity so as to maintain stable rates of spermatogenesis and androgen production by the testis. Removal of the testes in animals, and hence, removal of the main source of androgens, results in a marked increase in the LH and FSH levels. Two testicular hormones are primarily involved in the feedback regulation of LH and FSH. One is inhibin, a Sertoli cell hormone that suppresses FSH secretion by the pituitary. The second is testosterone. LH and FSH levels can be brought back down to normal levels by exogenous administration of testosterone. In human males, exogenous testosterone suppresses LH and FSH secretion sufficiently so that spermatogenesis is also suppressed. The testosterone-induced suppression of spermatogenesis, in fact, has formed the basis of experimental contraceptive regimens for males. In several pilot studies reported recently, sperm production has been lowered to very low levels. However, to be effective in suppressing spermatogenesis effectively, the testosterone had to be administered by weekly injection, which limited the efficacy and acceptability of this protocol.

Androgen-dependent negative feedback effects have been assumed, until very recently, to be due to testosterone itself. However, the discoveries of estrogen deficiency and estrogen receptor abnormality in two human males (see below) have shown that it is estrogen, rather than testosterone, that is responsible for the negative feedback. Testosterone is converted into estrogen, and it is estrogen working through estrogen receptors that regulates LH and FSH levels in the male. Estrogens and progestins are, in fact, powerful inhibitors of pituitary function in males. Indeed, a synthetic progestin analog, **depoprovera**, a common component of birth control pills for women, has been in use in some countries in Europe, and has been approved for use in California, for the "physiological castration" of rapists and child molesters. The progestin analog, by suppressing pituitary secretion of LH and FSH, severely inhibits testicular function, suppressing, in particular, androgen production by the testes. This in turn leads to loss of libido and potency. This type of therapy may make the individual unable to rape someone, but it is not clear that the decrease in androgen production will always lessen the urge to abuse children.

Androgens

The postpubertal testis is a prodigious testosterone factory. Over 95 percent of the testosterone in the male is produced by the Leydig cells. A large fraction of this testosterone remains within the testis so that the testosterone levels within the testis are more than 50-fold greater than they are in the circulatory system. The reason for such high testosterone requirements within the testis or how testosterone exerts its effects at such high concentrations is unknown.

Testosterone (T) is not the only important androgen. Dihydrotestosterone (DHT) is the other major functional androgen. Both T and DHT exert their effects through the

Table 6.2
Contrasting Functions of Testosterone and Dihydrotestosterone (DHT)

Stage	Testosterone	DHT
Fetal	Stimulates development of mesonephric duct derivatives to form the vas deferens, epididymis, and seminal vesicles	Stimulates development of the prostate, bulbourethral glands, and external genitalia
Childhood	Stimulates musculoskeletal development	Stimulates growth of external genitalia
Puberty and later	Required for spermatogenesis	Stimulates maturation of external genitalia and appearance of body, axillary, and facial hair
	Stimulates muscle mass increase, pigmentation of penis and scrotum, vocal cord enlargement, and skeletal growth	Head hair regression
	Regulates LDL/HDL cholesterol ratio	Prostate enlargement
	Regulates synthesis of liver enzymes	Acne
	May play a role in maintenance of hemoglobin levels	

same androgen receptor. DHT has a binding affinity to the androgen receptor about 5–10 times greater than that of testosterone, and hence, it is more potent than testosterone. However, DHT represents about 5 percent of the total androgens in the male (about 20 percent within the testis). The functions of T and DHT are divided in a very interesting, but perplexing, way (Table 6.2). Within the testis, most of the DHT is produced by the Sertoli cells, whereas outside of the testis, most DHT is produced by its target tissues. The target tissues are those that contain the enzyme **5-alpha-reductase**, which converts T to DHT upon the entry of T into the cell. This enzyme is found in great abundance in the penis, scrotum, and prostate, and is required for the development and growth of these tissues during puberty, as well as their maintenance afterwards. 5-alpha reductase is also found widely distributed in the skin, and appears to be responsible for the distribution of body hair that is typical for males (Table 6.2).

In addition to their purely sexual functions, T and DHT also have a number of important somatic functions. Perhaps the best studied of these is the effect of T on muscle mass. Testosterone, by stimulating protein synthesis, increases lean body mass and body weight. Generally, the effect of T is to increase the size of muscle cells rather than their number. This is why many athletes have routinely used synthetic androgens, known as *anabolic steroids*, to increase muscle mass and perhaps to improve their performance.

Androgens have a number of other functions. They regulate the synthesis of a number of liver enzymes, including several blood-clotting factors. They regulate the ratio of high-density lipoprotein (HDL) to low-density lipoprotein (LDL), which is generally lower in men than in women. This lower ratio may be an important factor in the higher incidence of coronary heart disease in men. Certain types of anemia are associated with low T levels, indicating that T is an important factor in the mainte-nance of hemoglobin levels.

DHT is probably the active androgen in the *sebaceous* glands of the skin. *Sebum* production by these cells increases significantly during the onset of puberty, resulting in the well-known problem of acne. Severe acne is not associated with higher T levels, but instead with an increased conversion of T to DHT in certain cells of the skin. Body hair distribution seen in males is also determined by androgens. Axillary (underarm) hair and hair in the low pubic area is produced by relatively low androgen levels, and hence, is found in both males and females. On the other hand, higher androgen, and in particular, DHT, levels are required to produce hair on the face, chest, and in the upper pubic area, and hence, this type of hair is found normally in postpubertal males.

Estrogens

We saw in Chapter 2 that although estrogens are commonly thought of as the fe-male hormones and androgens as the male hormones, the hormonal differences be-tween males and females are quantitative rather than qualitative. Recall that andro-gens are intermediates in the synthesis of estrogens, with the conversion of androgens to estrogens being carried out by the enzyme aromatase. The testes are not the primary source of estrogens in the male. Sertoli cells are thought to account for 20–25 percent, and many other tissues contribute to the total estrogen levels. Both androgen and es-trogen receptors have also been detected in many nonreproductive tissues. Both estro-gen receptors and aromatase appear to be present in the same tissues (Table 6.3). These findings have suggested that estrogen must carry out physiologically important func-tions in the male.

Table 6.3
Distribution of Androgen and Estrogen Receptors in Males

	Androgen receptors	Estrogen receptors
Reproductive tissues		
Testes	Yes	Yes
Prostate	Yes	Yes
Seminal vesicles	Yes	?
Epididymis	Yes	Yes
Penis	Yes	?
Vas deferens	Yes	Yes
Nonreproductive tissues		
Muscle	Yes	Yes
Sebaceous glands	Yes	?
Adipose tissue	?	Yes
Bone	Yes	Yes
Spinal cord	Yes	?
Kidney	Yes	Yes
CNS	Yes	Yes
Cardiovascular system	?	Yes

References: Sharpe, 1998; and others.

[Handwritten margin notes: low [T]; ↳ anemia → T maintainance of hemoglobin levels. DHT active in sebaceous → acne; ↑T → DHT ⇒ severe acne; androgen level ↳ hair distribution. androgen —aromatase→ estrogen; Sertoli cell (produce 20-25%); estrogen → in both reproductive/nonreproductive tissues.]

Estrogen receptor deficiency

Until quite recently, it was difficult to assign a specific function to estrogen in males. Indeed, until 1994 it was believed that a functional estrogen receptor was essential for viability. Complete or partial deficiencies, for example, in the androgen receptor function (see discussion in Chapter 14) had been described in humans and other species, and shown to be quite compatible with viability. However, no deficiencies in estrogen receptor function had ever been observed, that is, until 1994, when estrogen receptor mutations were engineered in the mouse. Unexpectedly, mice without functional estrogen receptors were viable. However, both males and females were sterile. This was expected for the female, but the sterility of the male mice indicated that estrogen is also required for fertility in males. Both sexes also had demineralized bones, showing for the first time that estrogen was required for bone stability in both males and females.

Does estrogen have equivalent roles in human males? The answer came with the serendipitous discoveries in 1994 and 1995 of two new human syndromes, one resulting from a mutation in the estrogen receptor gene, and the second from a mutation in the aromatase gene. The first syndrome was seen in a 28-year-old man who went to an orthopedic surgeon seeking correction of pronounced knock-knees. The man had gone through normal puberty, was fully masculinized, and was tall (6 ft. 8 in.). He had highly demineralized bones resulting in osteoporosis, and a bone age of 15 years, the latter meaning that his long bones (legs and arms) were still growing. A follow-up report indicates that he continues to grow at a rate of 1 cm per year. The important clue about the cause of his condition came with the finding that although testosterone levels were normal, estrogen and gonadotropin levels were significantly elevated. Such high estrogen levels would be expected to result in pronounced feminization of the body (for example, breast development), but none was seen. The high estrogen levels indicated some form of insensitivity to estrogen.

This phenotype was similar to that observed in the mice with mutant, nonfunctional estrogen receptors. With the mouse estrogen-receptor gene in hand, it was possible to show very quickly that the young man was carrying a mutation in each of his *own* estrogen-receptor genes, and as a consequence, his estrogen receptors were nonfunctional. Analysis of his family provided a probable explanation for his situation. His parents were third cousins, and he had inherited the mutation from both of his parents. To date he remains the only example of estrogen-receptor deficiency in humans.

Fertility in the young man had not been tested as of this writing. His sperm count is in the low normal range, but with evidence of higher-than-normal fraction of sperm with abnormal morphology. Although it is still too early to tell, it is possible that in humans, as in mice, estrogen may play a role in spermatogenesis.

Aromatase deficiency

Another syndrome that produces the same phenotype occurred in male and female siblings carrying a mutant gene for the enzyme aromatase. Both individuals are characterized by tall stature, very low bone mineralization resulting in osteoporosis in two individuals in their early twenties, elevated androgens and gonadotropins, but very low estrogen levels. Hence, the same phenotype is obtained by two different mechanisms: in the first case, estrogen deprivation occurs because of abnormal estrogen receptors; in the second, it occurs because of an abnormal aromatase enzyme.

Aromatase deficiency, in contrast to estrogen-receptor deficiency, is correctable by administering estrogen. *Would estrogen replacement be effective in restoring bone mass in the male?* A recent study shows that it is. The male sib agreed to be treated with estrogens. The estrogen doses were adjusted to minimize any feminization effects, and the patient reported no change in libido or other side effects even after three years of therapy. The effects on growth and bone mass were quite dramatic. Linear growth ceased immediately and stabilized at 204 cm. Bone mass increased continually, and by three years, averaged about 15 percent when measured in the lumber spine, femoral neck, and wrist. This study demonstrates clearly that estrogen is essential for the establishment of peak bone mass in growing boys and for the maintenance of bone mass in adult men.

These two cases show that life without estrogen is possible, but at a high price. Estrogen is essential for normal skeletal maturation and skeletal proportions and the maintenance of normal bone mineral density in both human males and females. The requirement is a very specific one, since in both clinical cases described above, high androgen levels could not substitute for the absence of estrogen action. Recent experiments in mice indicate that consistent with the wide tissue distribution of estrogen receptors and aromatase, estrogen is very likely involved in functions that previously had been attributed to androgens (Table 6.3).

Estrogen excess

In males, estrogen excess, or probably more correctly, an increase in the estrogen/androgen ratio, is manifested by the development of breast tissue, a condition known as **gynecomastia**. Most cases of gynecomastia are nonpathological, and are frequently seen during three developmental stages in the male's life: neonatal, pubertal, and senescent. In the first two cases, the gynecomastia is transient. For example, gynecomastia seen in 50 percent or more of male infants is attributed to the high estrogen levels present during gestation. It normally disappears after a few months. Breast development, lasting for up to two years, is observed in about 50 percent of normal pubertal boys. Recent studies using careful examination techniques have reported that 36 percent of military recruits had palpable breast tissue, suggesting that gynecomastia may be more prevalent than previously thought. Breast enlargement is also common in older men. Some estimates indicate that 40 percent or more of males over 50 years of age experience gynecomastia. Several factors contribute to this effect. Although not understood, conversion of androgens to estrogens in peripheral tissues increases as the male ages. In some cases the accumulation of adipose tissue with age is an important contributing factor. Adipose tissue contributes a substantial fraction of the circulating estrogens in males, and gynecomastia is seen more frequently in overweight or obese males.

Infrequently, gynecomastia can also be produced by disturbances in steroid hormone synthesis, various disease states, tumors, and other conditions that lead to an increase in the estrogen/androgen ratio. In rare cases, breast cancer can develop. A few rather bizarre cases have also been reported. One interesting case concerned a mortician who began developing breasts. It turned out that the source of the estrogen was the cream that he applied to the face of the bodies to make them "look better." Such creams contain estrogens, and in the case of the mortician, when applied with the ungloved hand, estrogen in the cream was absorbed through the skin. The increased estrogen levels were responsible for the development of breasts in the mortician. Gynecomastia underscores the essential similarity between male and female breast tissue. Indeed, with proper hormonal stimulation the male breast can be made to produce milk.

SUMMARY

The pituitary and testis define a short feedback loop, connected by the gonadotropins LH and FSH, testicular androgens, and the polypeptide hormone inhibin, a product of the Sertoli cells. In both cases, the feedback regulation is negative, with androgens suppressing LH and FSH release from the pituitary; inhibin also regulates FSH secretion. Recent studies in mice and in humans indicate that the negative feedback effects that had previously been attributed to testosterone are in fact due to estrogen. Hence, testosterone is converted to estrogen in the pituitary gonadotrophs, and its effects are mediated by estrogen receptors, rather than androgen receptors.

Most of the testosterone produced by the testis remains in the testis, presumably because high intratesticular testosterone concentrations are required for spermatogenesis. A small fraction of the testosterone produced is distributed by the circulatory system to the different androgen-dependent tissues. In some tissues, testosterone is first converted into DHT by the enzyme 5-alpha reductase, and it is DHT that is responsible for the androgen-dependent effects. The recent discovery and analysis of estrogen-receptor deficiency and aromatase deficiency in humans has shown unexpectedly that estrogen in males has an essential role in bone mineralization and skeletal maturation. In addition, estrogen, not testosterone, is responsible for the negative feedback at the pituitary. Both functions were previously attributed to androgens. The recent understanding that estrogen receptors and aromatase activity are found in the same tissues indicates that estrogen has previously unsuspected roles in males. Current views suggest that the estrogen/androgen ratio may be a critical parameter in many tissues. Perturbation of this ratio, for example, has observable clinical consequences, as occurs in gynecomastia.

QUESTIONS

1. Distinguish between the roles of LH and FSH in spermatogenesis.

2. Explain the roles of testosterone and DHT in different tissues.

3. What evidence demonstrates that in males the feedback effects of testosterone in the H-P-G axis are mediated by estrogen?

4. What are the consequences of estrogen excess and deficiency in males?

5. Indicate ways in which the powerful negative-feedback effects of steroid hormones in the male can be used therapeutically.

6. Library project:
    a. What are some of the factors that affect the rate of spermatogenesis?
    b. What are some of the health risks of taking anabolic steroids?

SUPPLEMENTARY READING

Bremner, W. J. 1981. Historical aspects of the study of the testis. In *The Testis.* H. Burger and D. de Kretser, Eds. Raven Press, New York, N.Y.

Grudzinskas, J. G., and J. L. Yovich. 1995. *Gametes: The Spermatozoon.* Cambridge University Press, Cambridge, UK.

Holstein, A. H., and E. C. Roosen-Runge. 1981. *Atlas of Human Spermatogenesis.* Grosse Verlag, Berlin, Germany.

Matsumoto, A. M. 1996. Spermatogenesis. In *Reproductive Endocrinology, Surgery, and Technology.* E. Y. Adashi, J. A. Rock, and Z. Rosenwaks, Eds. Lippincott-Raven Publishers, Philadelphia, PA.

Oudshoorn, N. 1994. *Beyond the Natural Body: An Archeology of Sex Hormones*. Routledge, London, UK.

Sharpe, R. M. 1998. The roles of estrogen in the male. *Trends in Endocrinology and Metabolism* **9**(9), 371–377.

van Basten, J. P., *et al.* 1996. Fantasies and facts of the testes. *British Journal of Urology* **78**, 756–762.

ADVANCED TOPICS

Bilezikian, J. P., *et al.* 1998. Increased bone mass as a result of estrogen therapy in a man with aromatase deficiency. *The New England Journal of Medicine* **339**(9), 599–603.

Glass, A. R. 1994. Gynecomastia. *Endocrinology and Metabolism Clinics of North America* **23**(4), 825–839.

Jansen, R. P. S. 1995. Bioethics and the spermatozoon: Reproductive destiny. In *Gametes: The Spermatozoon*. J. G. Grudzinskas and J. L. Yovich, Eds. Cambridge University Press, Cambridge, UK.

Morishima, A., *et al.* 1995. Aromatase deficiency in male and female siblings caused by a novel mutation and the physiological role of estrogens. *Journal of Clinical Endocrinology and Metabolism* **80**(12), 3689–3698.

Sharpe, R. M. 1994. Regulation of spermatogenesis. In *Physiology of Reproduction*, 2nd ed. E. Knobil and J. D. Neills, Eds. Raven Press, New York, NY.

Smith, E. P., and K. S. Korach. 1996. Estrogen receptor deficiency: Consequences for growth. *Acta Paediatrica Supplement* **417**, 39–43.

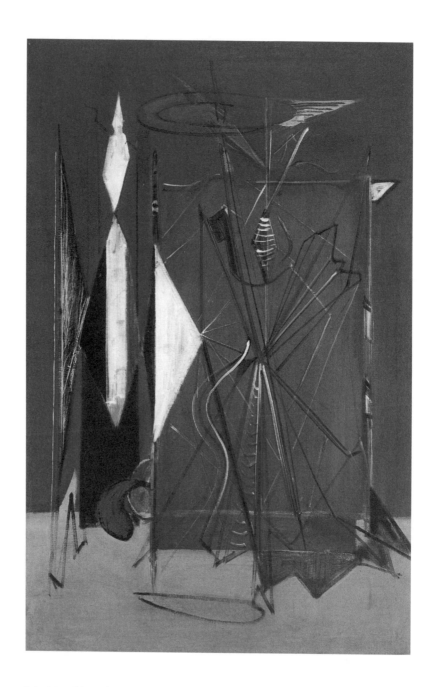

Mark Rothko, *Phalanx of the Mind*, c. 1945. (Gift of The Mark
Rothko Foundation. © Board of Trustees, National Gallery of Art,
Washington, D.C.)

THE H-P-G AXIS

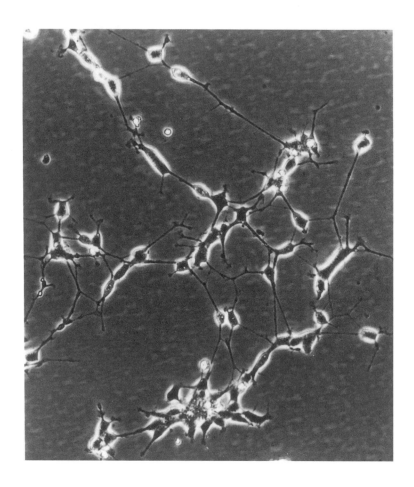

Micrograph of GnRH-secreting cell line (GT1-3) obtained from the mouse hypothalamus growing in culture. GnRH is the primary regulator of reproductive function in mammals. (Reprinted from P. L. Mellon *et al.* 1990. Immortalization of hypothalamic GnRH neurons by genetically targeted tumorigenesis. *Neuron 5*, 1–10, with permission from Excerpta Medica Inc.)

CHAPTER 7

The Brain and Reproduction

In the aggregate of physiologic events involved in the reproductive process, the gonad plays only a responsive role, prepared to adapt according to the cues it may receive.

—E. C. Amoroso. 1969

Physiological mechanisms in reproduction.
Journal of Reproduction & Fertility 6, 5–18.

The results were literally breathtaking. We observed strikingly large, rhythmic oscillations in plasma LH concentrations with a period of approximately 1 hour, leading us to coin the term "circhoral" (about 1 hour) to describe this phenomenon . . .

—E. Knobil. 1992

Remembrance: The discovery of the hypothalamic gonadotropin-releasing hormone pulse generator and of its physiological significance. *Endocrinology* 131(3), 1005–1006.

Anyone who has contemplated reproductive and sexual activity will conclude that there is more to reproduction than the gonads. The behavioral effects of castration of male animals, for example, in making domestic animals less aggressive and more manageable, have been known since prehistoric times. The variations in sexual activity in birds and animals triggered by seasonal changes in temperature and the length of the days have also long been recognized. In humans, severe emotional disturbances and many different kinds of substances have long been known to affect reproductive function. These simple examples suggest that tissues other than the gonads must play a role in reproduction. The brain was a logical suspect, and this was first suggested by the Greek physician Hippocrates (400 BCE). However, a full elaboration of that idea had to wait until the twentieth century. The first steps in this effort were taken toward the end of the nineteenth century, and working out the anatomy, physiology, and endocrinology of the brain is still ongoing. We now understand that the brain is the final arbiter and determinant of reproductive activity in animals.

This chapter has three objectives. First, we will examine the upper tier of the H-P-G axis, in particular, the properties of the GnRH pulse generator, and its regulation, via the pituitary, of gonadal activity in both males and females. Gonadal hormones, in turn, exert feedback control on both pituitary and hypothalamic activity. Second, we will review the essential features of modulation of the H-P-G axis by gonadal hormones and by nongonadal, central nervous system (CNS) hormones. The CNS mediates internal and external stimuli that regulate reproductive activity. Third, we will consider three conditions that illustrate the neuroendocrine modulation of the H-P-G axis: pregnancy, lactation, and PMS (premenstrual syndrome). In Chapter 9 we will review in more detail the way in which metabolic and psychogenic status affect reproductive function.

THE HYPOTHALAMUS-PITUITARY CONNECTION

Our current understanding of the brain-reproduction connection rests on work beginning at the turn of the twentieth century and culminating in the 1970s with the recognition that gonadal activity is controlled sequentially by hormones produced by the pituitary and the hypothalamus, forming what is called the hypothalamus-pituitary-gonadal (H-P-G) axis. This shorthand notation describes a complex regulatory network that includes control of gonadal function by the hypothalamus and the pituitary, feedback regulation of the hypothalamus and pituitary by gonadal hormones, and modulation by internal and external (environmental) stimuli. The distinction between internal and external stimuli is not very precise or exact, but it is nevertheless a useful one for the purposes of our discussion. In both cases, the central nervous system (CNS) plays a critical role in modulating the H-P-G axis. Different elements of the CNS convert the stimuli into neuroendocrine signals whose primary target is the hypothalamus. For the most part we remain woefully ignorant of the way in which environmental signals are sensed or perceived. Moreover, even though some of the CNS hormones that are part of the conversion process have been identified, we have only a superficial understanding of the mechanisms that link specific stimuli to well-defined effects on reproduction.

Hypothalamic-releasing hormones

The hypothalamus is a small area of the basal part of the brain found in all vertebrates (Fig. 7.1), and is an element of the most ancient part of the brain. Its small size belies its importance, however. It is involved in the regulation of the autonomic nervous system, blood pressure, body temperature, water and electrolyte balance, sugar metabolism, and complex behavioral activities, such as eating, drinking, sleep, and reproductive activity. Even emotions—anger, fear, joy, love, and hate—are under hypothalamic control. Essentially all of its activities are unconscious; we are not aware of them, nor can we control them in a conscious way.

The different functions of the hypothalamus are carried out by groups of neurons, referred to as **hypothalamic nuclei**, localized in different regions of the hypothalamus. Each hypothalamic nucleus produces a distinct hypothalamic hormone. Through its many connections to other parts of the brain, the hypothalamus acts as a receiver of visual, olfactory, nutritional, and environmental signals transmitted to it by different regions of the CNS. We can think of the hypothalamus as functioning as a sort of "clearinghouse" to integrate the different inputs that come to it from the different regions of the CNS.

The pituitary gland is a very small ovoid tissue lying just below the hypothalamus (Fig. 7.1). It weighs about 500 mg (about 0.02 oz.) in the adult human. During the

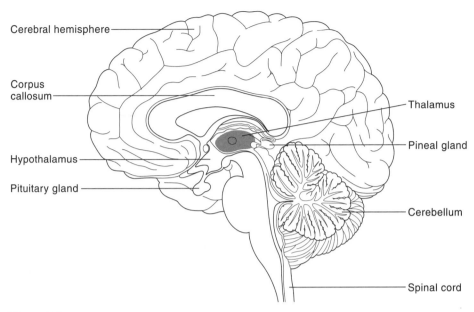

Figure 7.1
Diagram indicating the relative positions of the hypothalamus, pituitary, and pineal gland at the base of the brain.

early part of the twentieth century the pituitary was referred to as the "master gland," since it was the source of such important hormones as LH, FSH, ACTH, GH, and prolactin. Only later did it become clear that the pituitary is not an autonomous organ, but that it functions in conjunction with the hypothalamus, forming the hypothalamic-pituitary axis (H-P axis).

Evidence for the importance of the pituitary gland in reproduction came initially from experiments conducted in the 1930s and 1940s, which demonstrated that removal of the pituitary in rats and mice resulted in loss of gonadal function and gonadal atrophy. Transplantation of the pituitary to other sites, even other sites in the brain, did not restore gonadal function. These observations indicated not only that the pituitary controlled gonadal activity, but that its ability to do so depended on its anatomical position and its connections to regions farther up in the brain. Lesions deliberately introduced into the hypothalamus, or severing the connections between the hypothalamus and pituitary, led to cessation of pituitary function. The functional relationship between the hypothalamus and the pituitary was demonstrated by experiments carried out in the 1950s using explants of pituitary and hypothalamic tissue. For example, pituitary tissue extracted from experimental animals could be induced to produce a variety of pituitary hormones if it was co-incubated with hypothalamic fragments. Experiments of this type led directly to the hypothesis that "releasing factors" produced by the hypothalamus regulated the secretion of pituitary hormones. Then, in what must surely rank as one of the epic searches in biology, 5 million sheep brains were ground up in an attempt to isolate the putative hypothalamic-releasing factors. Eventually five hundred tons of brain tissue (50 tons of hypothalamic fragments) were used in this search. In 1969, the first successful product of this search was reported: 1 mg of the hypothalamic hormone now known as thyrotropin-releasing hormone (TRH) was purified from 300,000 hypothalamic fragments. TRH stimulates the pituitary to produce thyrotropin, or thyroid-stimulating hormone (TSH). This was followed in 1970 by the isolation of the hormone now known as gonadotropin-releasing hormone (GnRH), which regulates pituitary production of LH and FSH.

The pituitary

The pituitary has two major functionally important regions, the **anterior** and **posterior lobes,** that differ in their embryological origin and in their connections to the hypothalamus (Fig. 7.2). The posterior lobe, or *neurohypophysis*, is connected to the hypothalamus by a large tract of nerve fibers originating from two nuclei in the hypothalamus. The cell bodies in these two nuclei synthesize two hormones, **vasopressin,**

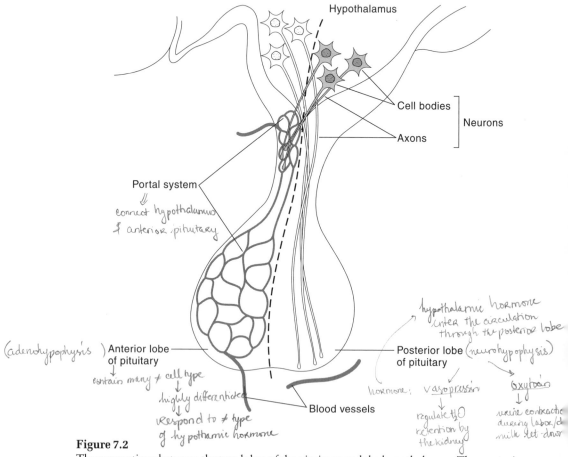

Handwritten annotations:

Portal system — connect hypothalamus & anterior pituitary

(adenohypophysis) Anterior lobe of pituitary → contain many ≠ cell type ↓ highly differentiated ↓ respond to ≠ type of hypothalamic hormone

hypothalamic hormone enter the circulation through the posterior lobe

Posterior lobe (neurohypophysis)

hormone: vasopressin ↓ regulate H₂O retention by the kidney

oxytocin ↓ uterine contraction during labor/d milk let-down

Figure 7.2

The connections between the two lobes of the pituitary and the hypothalamus. The oxytocin and vasopressin neurons of the posterior lobe originate in the hypothalamus, but secrete their respective hormones into the bloodstream at the posterior lobe. Secretion of anterior lobe hormones is controlled by releasing hormones produced by hypothalamic neurons terminating at the vessels of the portal system. The hypothalamic-releasing hormones are transported to the anterior lobe and act on cell types bearing appropriate receptors for these hormones.

whose main function is to regulate water retention by the kidney, and **oxytocin**, which stimulates uterine contractions during labor and childbirth and also milk let-down by the mammary glands. These hormones move down their respective axons, which make contact with capillaries of the posterior lobe; the hormones enter the capillaries and pass into the general circulation. These two hormones, first characterized in 1952, were the first neuropeptide hormones identified. Although they are still referred to as pituitary hormones, they are really hypothalamic hormones that enter the circulatory system through the posterior lobe.

The anterior lobe, or *adenohypophysis*, is a more complex tissue than the posterior lobe. It consists of several types of cells, each type secreting a particular hormone in response to stimulation by a particular hypothalamic-releasing hormone (Table 7.1). The connection between the anterior lobe and the hypothalamus is vascular rather than neural; communication occurs through a restricted circulatory system, the *portal* system. Neurons originating from different hypothalamic nuclei terminate at the portal capillaries, where they release their hormones. The hormones then travel into the anterior lobe and stimulate the appropriate cell type (Fig. 7.2).

The cell population in the anterior lobe is highly differentiated. For example, cells known as **thyrotrophs** respond to TRH to produce thyrotropin (TSH). The **corticotrophs** secrete ACTH in response to CRH. The **gonadotrophs** secrete the two gonadotropins LH and FSH in response to GnRH (Table 7.1). The secretion of each anterior lobe hormone is more complicated than this simple picture might indicate. In general, several factors influence the secretion pattern of each pituitary hormone, but the synthesis of each depends on a specific hypothalamic-releasing hormone. LH and FSH secretion, for example, not only depends on the GnRH signal, but are also subject to control by steroid hormones and other peptide hormones. Some aspects of this regulatory network will be examined in more detail in later chapters.

Table 7.1
Some Hypothalamic-Releasing Hormones and Their Anterior Pituitary Hormones

Hypothalamic-releasing hormone	Pituitary cell target	Pituitary hormone
Tyrotropin-releasing hormone (TRH)	Thyrotrophs	Thyroid-stimulating hormone (TSH)
Corticotropin-releasing hormone (CRH)	Corticotrophs	Adrenocorticotropic hormone (ACTH)
Growth hormone-releasing hormone (GHRH)	Somatotrophs	Growth hormone (GH)
Prolactin-releasing hormone (PRH)	Mammotrophs	Prolactin (PRL)
Gonadotropin-releasing hormone (GnRH)	Gonadotrophs	Luteinizing hormone (LH) Follicle-stimulating hormone (FSH)

The GnRH pulse generator

One of the most important and far-reaching discoveries in reproductive biology in the last 25 years was the finding that GnRH secretion by the hypothalamus is *pulsatile.*^{vibrate} In turn, LH and FSH release from the pituitary also occurs in a pulsatile fashion. This discovery was the consequence of studies begun in the 1960s that were aimed at analyzing the dynamics of LH secretion. Technological improvements in the design of catheters to permit frequent sampling of blood led to the unexpected discovery that LH was secreted in a pulsatile fashion. Sampling blood from the hypothalamic portal system (a procedure that required a considerable amount of technical skill) showed that the immediate source of LH pulsatility was GnRH pulsatility. The results of one such experiment in the ewe are shown in Fig. 7.3. LH pulses appear with a period of about one pulse every two hours, with each LH pulse immediately preceded by a GnRH pulse. Similar experiments in the Rhesus monkey showed that the characteristic LH pulse frequency in this species was about one pulse per hour. Pulsatile LH secretion was soon found in a large number of mammalian species, although the pulse frequency varied with the species. Hence, the source of LH pulsatility lies in the ability of

[margin handwritten note: GnRH secretion & LH/FSH secretion are pulsatile]

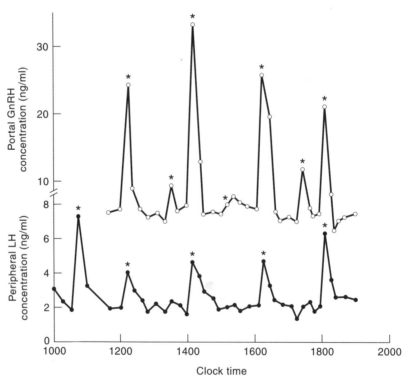

Figure 7.3
Pulsatility in GnRH and LH release in the ewe. GnRH and LH concentrations were measured over a 10-hour period in the portal system and jugular veins, respectively. The asterisks indicate secretory pulses of GnRH and LH. Note that strong GnRH pulses elicit a corresponding LH pulse; weak ones do not. (Adapted from I. J. Clarke and J. T. Cummins. 1982. *Endocrinology* **111**, 1737–1739. © The Endocrine Society.)

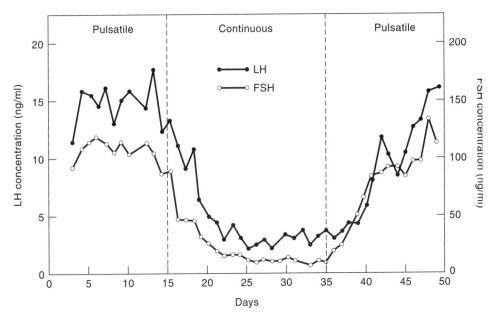

Figure 7.4

Demonstration of the consequences of continuous versus pulsatile GnRH on LH and FSH levels. Endogenous GnRH secretion was abolished by introducing a lesion into the GnRH-secreting nuclei in a Rhesus monkey. Normal gonadotropin levels were maintained by exogenous pulsatile infusion of GnRH (0 to 15 days). Switching from pulsatile to continuous infusion results in a dramatic drop in LH and FSH levels, and a return to pulsatile infusion restores LH and FSH levels. (Adapted from P. E. Belchetz *et al.* 1978. *Science* **202**, 631–633. Copyright [1978] American Association for the Advancement of Science.)

the GnRH neurons to secrete GnRH in a synchronized fashion. The set of neurons that coordinate and synchronize their GnRH discharges is referred to as the **GnRH pulse generator**.

Severing the neural connections between the hypothalamus and the CNS did not abolish LH pulsatility, demonstrating that the source of the pulsatility was in the hypothalamus itself, and not from other regions of the CNS. GnRH pulsatility, therefore, is an intrinsic property of the GnRH-secreting neurons. The functional significance of GnRH pulsatility remained obscure for about 8 years after its discovery. The first clues about that significance came in experiments in Rhesus monkeys in which endogenous GnRH secretion had been abolished by producing lesions in the hypothalamus. These surgical procedures required great technical skill, and were made possible by an extensive knowledge of Rhesus monkey anatomy and physiology built up during the previous 80 years. Hypothalamic lesions that abolished GnRH secretion resulted in a dramatic decrease in LH and FSH levels (Fig. 7.4). LH and FSH levels could be restored, however, by exogenous pulsatile infusion of GnRH, administered by means of a pumping system. This experimental system permitted varying the pulsatile frequency at will. The prelesion LH and FSH levels were restored only if the pumping frequency was the same as the natural GnRH pulse frequency. Surprisingly, continuous infusion of GnRH did not restore LH and FSH secretion by the pituitary. Indeed, under contin-

Table 7.2
Some Therapeutic Uses of GnRH and Its Analogs

GnRH
    Delayed puberty
    Functional hypothalamic amenorrhea
    Hypogonadotropic hypogonadism

GnRH agonists
    Precocious puberty
    Endometriosis
    Breast cancer
    Prostate cancer
    Premenstrual syndrome
    Polycystic ovary syndrome
    Ovulation inhibition
    Suppression of spermatogenesis

GnRH antagonists
    Benign uterine fibroid tumors
    Gonadotropin suppression (in clinical trials)

uous infusion of GnRH, LH and FSH levels fell to essentially undetectable levels. Changing the pulse frequency to two, three, and five pulses per hour also resulted in a decrease in LH and FSH levels.

These landmark experiments demonstrated in a dramatic and convincing fashion that the GnRH pulse frequency has to be controlled relatively precisely in order to maintain normal LH and FSH levels. There is little doubt, therefore, that the GnRH pulse generator is the central regulator of reproduction and that understanding the mechanisms that control its activity is the key to understanding the factors and conditions that control reproductive activity. In fact, we have come to recognize that modulation of reproductive function in most instances is accomplished by modulation of the GnRH pulse generator. A new field of reproductive biology was born.

The decrease of LH and FSH in the presence of continuous GnRH stimulation is an excellent example of the phenomenon of hormonal desensitization (described in Chapter 3). Technically, what happens is that continuous secretion of GnRH down-regulates the GnRH receptors in the gonadotrophs in the anterior pituitary. In effect, continuous occupancy of the GnRH receptors leads to their loss from the gonadotroph cell membrane, and they are restored only when normal GnRH pulsatility itself is restored. The understanding that came from these studies has had far-reaching effects. Within a few years it formed the basis of a number of therapeutic programs that made use of GnRH or its analogs (Table 7.2). Some of these programs, for example, treatment of infertility due to hypothalamic dysfunction and inhibition of pituitary function in certain types of precocious puberty, will be discussed in later chapters.

SUMMARY

The central regulator of reproductive activity is the GnRH pulse generator, a group of neurons in the hypothalamus that secrete the peptide hormone GnRH in a synchronized fashion. GnRH is one of several hypothalamic-releasing hormones secreted by different groups of hypothalamic neurons (known as hypothalamic nuclei), whose targets are corresponding cells in the anterior pituitary. The target cells of GnRH in the anterior pituitary are the gonadotrophs, which respond to GnRH by secreting LH and FSH. GnRH is secreted by the GnRH-secreting neurons in a pulsatile fashion with a pulse frequency and amplitude that is unique for each species. GnRH pulsatility results in LH and FSH pulsatility. In humans, the pulse frequency is about one pulse every 60 minutes. Variations in GnRH pulse amplitude or frequency result in significant changes in LH and FSH production by the anterior pituitary. GnRH pulse frequency is an intrinsic property of the GnRH pulse generator in that the pulsatility does not depend on input from other regions of the CNS.

GONADAL REGULATION OF
THE GnRH PULSE GENERATOR

GnRH pulse generator activity can be modulated by many different factors. Particularly important are the regulatory effects exerted by steroid sex hormones. This feedback regulation is a normal part of the dynamic character of the H-P-G axis. Let us consider these cases here.

In the male

In both males and females the feedback effects of steroid sex hormones can be seen most easily in rather extreme circumstances, for example, as a consequence of a **gonadectomy** (removal of the gonads), or in the absence of gonadal function, as in the postmenopausal female. In humans, GnRH cannot be measured directly, but the activity of the GnRH pulse generator can be inferred by measuring LH levels and LH pulsatility. Since LH pulse frequency depends only on the GnRH pulse frequency, changes in LH pulse frequency reflect modulation of the GnRH pulse generator. On the other hand, changes in LH levels can be due either to changes in pituitary sensitivity or to changes in GnRH pulse amplitude. In males, removal of the testes results in a dramatic increase in LH levels and also in pulse frequency (Fig. 7.5). In this experiment, LH pulsatility in a male red deer was compared before and after removal of the testes. The increase in LH pulse frequency, in particular, indicates that some testicular product or products normally inhibit the activity of the GnRH pulse generator. The effect of castration indicates that under normal conditions, the H-P-T axis is maintained in a dynamic equilibrium, with testicular products exerting negative feedback effects on both gonadotropin and GnRH secretion.

Until recently, testosterone was considered the most important testicular product exerting a negative feedback effect on the pituitary and hypothalamus. For example, administration of exogenous testosterone is very effective in suppressing GnRH pulse generator activity, reducing LH and FSH levels, and in turn suppressing spermatogenesis quite significantly. Indeed, this effect has been the basis for attempts to develop a testosterone-based contraceptive regime for males. But as we learned in the previous chapter, it is estrogen that is the direct effector. Estrogen exerts its effects on both the pituitary and hypothalamus. Progesterone is also a powerful inhibitor of the GnRH pulse generator. In males, progesterone is not normally an important modulator of GnRH secretion. However, the potent inhibitory effects of progesterone are utilized in the "physiological castration" of child molesters, as mentioned in the previous chapter.

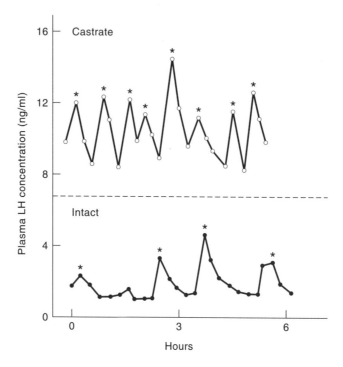

Figure 7.5
Comparison of LH levels in an intact and castrate male red deer. LH amplitude and pulse frequency increase in the castrate as a consequence of the loss of negative-feedback inhibition by the testis. The increase in LH pulse frequency indicates that the site of feedback is at the hypothalamus. (Adapted from M. H. Johnson and B. J. Everitt, 1995.)

In all cases, the feedback effects of steroid sex hormones at the pituitary and hypothalamus are always negative. Other nonsteroidal testicular products are also known to exert feedback effects. For example, **inhibin**, a peptide hormone secreted by the Sertoli cells, is important specifically in feedback control of FSH release by the gonadotrophs. A summary of these feedback effects is shown in Fig. 7.6.

In the female

postmenopause ♀
↑LH↓, ↑FSH

    The situation in the female is more complicated than in the male. In the postmenopausal female, the absence of ovarian activity results in a large increase in LH and FSH levels, as well as in their pulse frequency (Fig. 7.7). Indeed, the increase is so large that LH and FSH are secreted in the urine. When extracted from the urine, the combination is known as **human menopausal gonadotropin** (HMG). HMG has been used as a convenient and relatively inexpensive source of LH and FSH in many experiments. The increase in gonadotropin levels reflects the release from the negative feedback effects of ovarian steroids at both the pituitary and the hypothalamus, but it is difficult to separate the two contributions precisely. On the other hand, the increase in pulse frequency can clearly be interpreted as the release of the GnRH pulse generator from the inhibitory effects of sex steroids.

    In the cycling female the pattern of LH secretion varies significantly during the

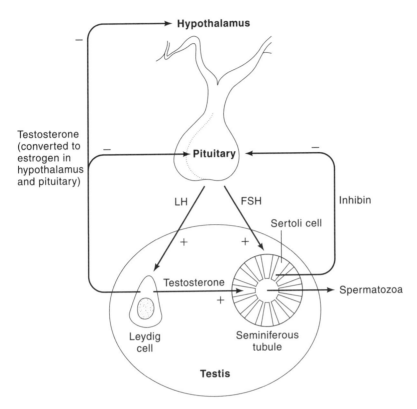

Figure 7.6
Feedback circuits involving the testis, pituitary, and hypothalamus. Testosterone and the peptide hormone inhibin, secreted by the Sertoli cells, account for the major regulatory effects. As discussed in Chapter 6, the effects of testosterone are indirect and depend on the conversion of testosterone into estrogen in the pituitary and hypothalamus. The effects of testosterone are ultimately mediated by the estrogen receptor.

ovulatory cycle (Fig. 7.8). During the follicular phase, LH pulse frequency does not change appreciably, but LH levels increase dramatically as the LH surge approaches. This dramatic increase is due to the rapid rise in estrogen during the latter part of the follicular phase. The feedback effects of estrogen switch from negative to positive, and it appears to exert its effect on both the pituitary and the hypothalamus, although the mechanism remains poorly understood. The positive feedback effect of estrogen is unique to females, since an LH surge cannot normally be generated in males.

After the LH surge, there is a marked reduction in the LH pulse frequency and amplitude. The reduction in the LH pulse frequency during the luteal phase reflects the suppression of the GnRH pulse generator, primarily by the increasing levels of progesterone during the luteal phase. What is the significance of this suppression? It can be argued that the inhibition of GnRH release prevents another LH surge that could in principle be generated since estrogen levels are high enough to do so. Perhaps more importantly, the decrease in GnRH pulse frequency during the luteal phase may be essential for the maintenance of cyclicity. There is some evidence that reduction in the GnRH pulse frequency prepares the pituitary so that when progesterone and estrogen levels decrease toward the end of the luteal phase it can release the FSH that has been

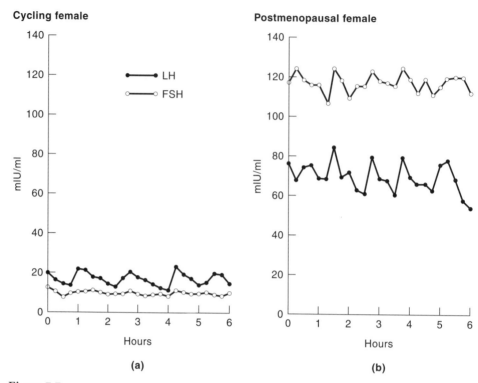

Figure 7.7
Comparison of the LH and FSH levels in a **(a)** cycling (follicular phase) and **(b)** postmenopausal female. The increase in level and pulse frequency in the postmenopausal female is attributed to release from inhibition by ovarian hormones. (Adapted from S. S. C. Yen and R. B. Jaffe, 1978.)

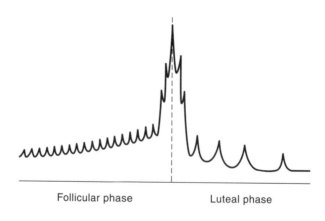

Figure 7.8
Schematic illustration of the pattern of LH pulsatility during the ovulatory cycle. During the follicular phase, LH pulse frequency remains constant and the pulse amplitude increases gradually. Shortly before the LH surge, LH pulse amplitude increases dramatically. After the LH surge, the most dramatic change is the significant decrease in LH pulse frequency, while maintaining significant pulse heights. The decrease in LH pulse frequency is due to the inhibitory effect of progesterone on the GnRH pulse generator. (Adapted from S. S. C. Yen and R. B. Jaffe, 1978.)

synthesized during the luteal phase. You may recall that the increase in FSH at the end of the luteal phase is necessary for the selection phase of the dominant follicle for the next cycle.

SUMMARY

Feedback regulation of the GnRH pulse generator by gonadal steroid hormones is a central feature of the H-P-G axis. In males, steroid sex hormones exert a negative feedback on the pituitary and the GnRH pulse generator, and this effect is manifested by a decrease in LH amplitude and LH pulse frequency. Feedback effects by testosterone are mediated not by testosterone, but by estrogen, after testosterone has been converted to estradiol. In the female, estrogen can have both negative and positive feedback effects at the level of the hypothalamus and the pituitary, while progesterone exhibits primarily a negative effect. The LH surge represents a change in LH amplitude rather than LH pulse frequency, and depends on the switch of estrogen from negative to positive feedback. High levels of progesterone inhibit GnRH pulse generator activity during the luteal phase, an effect that appears necessary for maintaining normal cyclicity. The feedback effects of ovarian steroids on the pituitary and the hypothalamus can be summarized as indicated in Figs. 7.9 and 7.10.

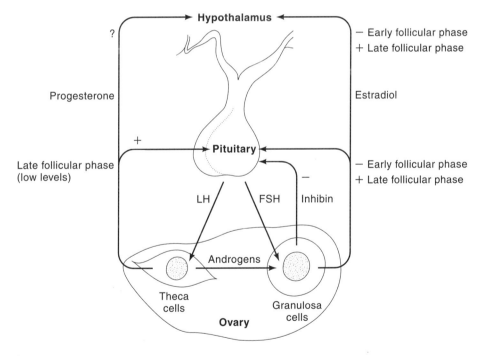

Figure 7.9
Feedback circuits between the ovary and the pituitary and hypothalamus during the follicular phase. Estrogen has a negative feedback effect during the early follicular phase. As estrogen levels rise past a certain threshold by the midfollicular phase, its effect changes dramatically from negative to positive feedback. This endocrinological "switch" is responsible for the LH surge. Low levels of progesterone in the late follicular phase also have a positive effect and are required for the LH surge as well.

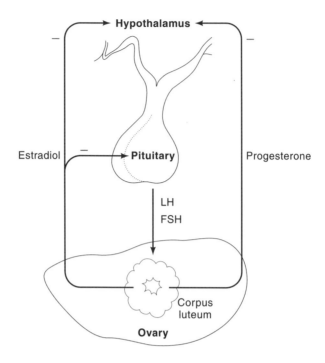

Figure 7.10
H-P-O interactions during the luteal phase of the cycle. During this period both estrogen and progesterone exert a negative feedback effect. Estrogen exerts its negative effects primarily at the pituitary, while progesterone acts at the hypothalamus.

CNS MODULATION OF REPRODUCTIVE ACTIVITY

A progressive understanding of the etiology of many reproductive disorders has led us to recognize the variety of factors or conditions that influence reproductive function. In many cases, these disorders can be traced to disruption or alteration of neuroendocrine hormones that control the activity of the GnRH pulse generator. Many studies have shown that a number of CNS neuroendocrine systems can have profound effects on reproductive activity and behavior. These systems are identified as distinct groups of neurons, distinguishable by the type of hormone each secretes, and each responsive to other hormonal signals. These CNS elements function as mediators of internal and external stimuli. Normal reproductive function, therefore, depends on the integrity of these regulatory neuroendocrine networks. We will consider here the better characterized of these neuroendocrine systems.

Endogenous opioids

Evidence linking **opiates**, compounds such as morphine and heroin derived from plants belonging to the poppy family, and reproduction came initially from many observations during the early part of the twentieth century of heroin-addicted individuals. Ovarian dysfunction and **amenorrhea** (cessation of menstruation) were typical consequences in female addicts, while subfertility, loss of libido, and erectile dysfunction were commonly seen in male addicts. Injections of morphine were shown to block ovulation in rats. The mechanism by which these compounds exerted their effects remained unknown for many years. A significant conceptual advance came with the discovery in 1971 of opiate (opioid) receptors in the brain, since this demonstrated that opiates exerted their effects through specific receptors. Another advance came in 1975 with the discovery of naturally occurring **endogenous opioid peptides** (EOPs), which served as the endogenous hormones for the opiate (opioid) receptors. The EOP family of peptides, consisting of three classes of related peptides, enkephalins, endorphins, and dynorphins, is involved in many different functions. One of the most important is the modulation of mood and behavior. The important opioid in reproduction appears to be β-endorphin, a member of the endorphin family.

Many experiments have demonstrated the inhibitory effects of opiates on reproduction. One example is the classic experiment showing the effects of morphine on the GnRH pulse generator (Fig. 7.11). Electrodes were implanted into the *mediobasal hypothalamus* of a Rhesus monkey; this is the region containing the GnRH-secreting neurons. The implanted electrodes made it possible to measure the activity of the GnRH-secreting nuclei. Every pulse of GnRH released from the GnRH nuclei is accompanied by an electrical current, referred to as multiunit activity (MUA). In the ex-

[handwritten margin notes: heroin-addicted ♀ → ovarian dysfxn → amenorrhea; ♂ → erectile dysfxn; EOPs family → many ✓ fxn → modulation of mood & behavior; β-endorphin → important opioid in reproduction; Opiates → inhibit reproduction]

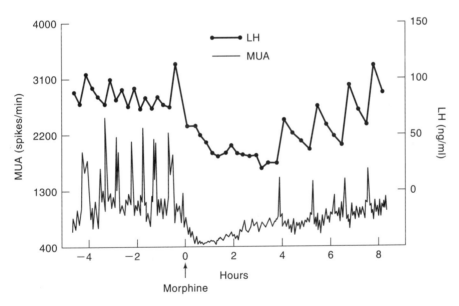

Figure 7.11
Effect of morphine on the GnRH pulse generator (MUA, lower trace) and LH pulsatility (upper trace). A two-minute infusion of morphine was begun at time 0 in an ovariectomized Rhesus monkey bearing electrodes implanted into the mediobasal hypothalamus. The MUA measures the firing activity of the pulse generator resulting in the release of GnRH. GnRH pulse generator activity and LH levels ceased about one minute after the morphine injection and resumed at a slower frequency after about four hours. (Adapted from J. S. Kessner *et al.* 1986. *Neuroendocrinology* **43**, 686–688. S. Karger AG, Basel.)

periment, MUA volleys and the resulting LH pulses were measured before and after a single intravenous injection of morphine. The effect of morphine was immediate and dramatic: within 1 to 2 minutes after the morphine injection, both MUA activity and LH levels dropped precipitously and stayed low for about 4 hours. At 4 hours, recovery began, but at a significantly lower pulse frequency. Normal activity returned after several more hours. This experiment showed clearly that morphine has a potent inhibitory effect on the GnRH pulse generator, and demonstrates that H-P-G axis impairment seen in opiate addicts is due to suppression of GnRH secretion.

From this and many other experiments it appears that the normal role of β-endorphin with respect to the GnRH pulse generator is inhibitory. Moreover, since anatomical and physiological analysis has shown that the β-endorphin-secreting neurons and the GnRH-secreting neurons are anatomically and functionally connected to one another, the effect exerted by β-endorphin is probably a direct one. It appears likely that the hypothalamic β-endorphin system functions as an intermediary for other hormonal stimuli that ultimately regulate the activity of the GnRH pulse generator. Indeed, there is good evidence that the steroid sex hormone modulation of the GnRH pulse generator described above is mediated by the β-endorphin system. Steroid sex hormones activate the β-endorphin system, and in turn, β-endorphins suppress the GnRH pulse generator (Fig. 7.12). In human females, the opioid tone (that is, the β-

[margin note: β-endorphin affects directly]

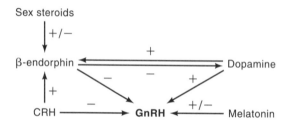

Figure 7.12
Some neuroendocrine systems that regulate GnRH release. The + / − symbols indicate stimulation and inhibition of GnRH release, respectively. The β-endorphin and dopamine neurons are anatomically and functionally connected to the GnRH neurons, and they are considered to mediate the effects of the steroid sex hormones and CRH. The functional relationships are complex, however. Melatonin may exert a more direct effect on GnRH release. Other neurotransmitter systems quite likely participate in the control of GnRH release, but the details remain unclear.

endorphin concentration) fluctuates during the cycle, being highest in the luteal phase and lowest in the follicular phase. Hence, the decrease in the LH pulse frequency observed during the luteal phase, due to the high progesterone levels, is really the consequence of β-endorphin inhibition of the GnRH pulse generator.

Corticotropin-releasing hormone (CRH) *→ stress hormone*

CRH is known as the stress hormone because CRH release is associated with the physiological and behavioral changes that are concomitants of the stress response (discussed in Chapter 9). In rodents, the stress response can be induced simply by injecting

CRH into the animal. An important consequence of stress is the suppression of GnRH secretion. We will see the rationale for this effect in Chapter 9 when we examine the effects of different types of stress on reproduction. The anatomical and functional coupling relationships between CRH-secreting and GnRH-secreting neurons and other cell types in the hypothalamus are extremely complex. Experiments in both rodents and monkeys have shown that in some circumstances CRH can inhibit GnRH directly, while in others, the inhibitory effects are mediated by the β-endorphin system (Fig. 7.12).

(handwritten margin note: CRH → ↓GnRH)

Neurotransmitters (amine hormones)

GnRH release is also influenced by other less-well-characterized neuroendocrine systems, including melatonin and dopamine (Fig. 7.12). Melatonin is the main hormonal product of the pineal gland (Fig. 7.1). The pineal gland is absolutely essential for regulating the reproductive cycles of seasonally breeding animals. In such mammals melatonin is secreted only during the hours of darkness and hence provides reliable measure of the seasonal changes in the length of the day and night. As days shorten to signal the approach of autumn and winter, the duration of the melatonin signal increases; after the winter solstice, as the day begins to lengthen, the melatonin signal decreases. These changes in melatonin levels result in significant modulation of GnRH secretion, but the mechanism by which such modulation is accomplished has not yet been answered definitively.

(handwritten margin notes: Pineal gland | secret melatonin ↓ essential for regulating the reproductive cycles of seasonally breeding animals. ↙ secreted during hrs of darkness.)

In nonseasonal breeding species, such as humans and other primates, melatonin secretion also follows a day-night rhythm, but it is not at all clear that such fluctuations have reproductive significance. On the other hand, in certain types of pathological conditions, excess levels of melatonin in humans result in suppression of GnRH secretion. Higher-than-normal levels of melatonin have been found in some women with stress-induced and exercise-induced amenorrhea, and the amenorrhea has been shown to be due to inhibition of GnRH release. Higher-than-normal levels of melatonin have also been found in some infertile men. The suppressive effect of high doses of melatonin suggested that it might be possible to develop a melatonin-based contraceptive for females, and exploratory trials with an oral melatonin-progesterone combination were held a few years ago.

(handwritten margin notes: In human high melatonin level ↓ ↓GnRH secretion ↗ ↖ infertile ♂ | stress, exercise ↑)

Dopamine has a well-established and important role in regulating prolactin secretion. The control is inhibitory: high levels inhibit and low levels permit prolactin secretion from the pituitary lactotrophs. The role of dopamine in reproductive function is less clear cut. Dopamine-secreting neurons appear to be coupled anatomically and physiologically to both GnRH- and β-endorphin-secreting neurons. In experimental animals, dopamine has been shown to stimulate GnRH secretion directly. In other circumstances dopamine exerts an inhibitory effect mediated by the β-endorphin system (Fig. 7.12). The precise role of the dopamine and other neurotransmitter systems in regulating the human H-P-G axis remains unknown.

(handwritten margin notes: Dopamine ↓⊖ prolactin secretion | ↑GnRH secretion directly ↑β-endorphin ↓ ↓GnRH.)

SUMMARY

Modulation of the GnRH pulse generator by different CNS elements is an essential aspect of the normal reproductive system. The β-endorphin, dopamine, and CRH systems are the best characterized, but a full understanding of their precise roles remains elusive. All three systems respond to other stimuli. The β-endorphin system, for example, mediates the effects of steroid sex hormones. CRH (as we will see in Chapter 9) mediates response to stress. Both CRH and β-endorphin exert primarily an inhibitory effect on GnRH release. Dopamine may have both an inhibitory and stimulatory effect on GnRH release, and much less is known about the stimuli to which the dopamine system responds. The role of melatonin in human reproductive activity is not understood, although it is known that excess levels have an inhibitory effect on GnRH release. In animal studies, other neurotransmitter systems have been shown to modulate GnRH levels, but little is known about their effects in humans.

THE H-P-G AXIS IN PREGNANCY AND LACTATION

Pregnancy

The pregnant female is in a **hypogonadotropic** state, that is, her ovaries have ceased functioning, and she has returned to a noncycling condition. Steroid sex hormone levels are very high, but these levels are maintained by the placenta rather than the ovaries. The hypogonadotropic condition is a consequence of the suppression of the GnRH pulse generator. During the first half or more of gestation, the GnRH pulse generator is suppressed by the high levels of progesterone produced by the placenta. The inhibitory effects of progesterone are mediated primarily by β-endorphin, which is released at high levels during the first half of gestation. However, during the latter part of gestation, β-endorphin levels fall, but the suppression continues. The mechanism at work is unknown, but it appears that during the late stages of pregnancy the hypothalamic status of the pregnant female returns to a prepubertal state.

The hypogonadotropic state of pregnancy is reversed soon after delivery. The recovery from pregnancy with respect to LH and FSH secretion in a nonbreast-feeding woman is illustrated in Fig. 7.13. Intermittent LH pulses, reflecting intermittent GnRH stimulation of the pituitary, appear some 19 days after delivery, with increased amplitude pulses observed during the early morning hours. Full restoration of FSH and LH levels sufficient to support follicular development and ovulation occurs at about 45 days postpartum, with the first menstruation occurring some 14 days later. Generally, the first and second cycles are characterized by low corpus luteum function, and a normal cycling pattern is established by the third menstrual cycle postpartum.

Lactation

Breast-feeding has sometimes been called Nature's contraceptive because of the state of infertility that it produces. This contraceptive effect is considered to be the main factor in the control of human populations in traditional societies without access to modern birth control methods, and certainly must have been at work during most of humankind's history. In developing countries today, breast-feeding probably prevents more births than modern contraceptive methods. The infertile period consists of two phases, an initial period characterized by amenorrhea, followed by a period in which menstrual cyclicity returns but in which the cycles may be anovulatory, or if ovulatory, are characterized by a condition known as **luteal suppression**, in which corpus luteum function is inhibited to some extent. Although the amenorrheic period in a fully breast-feeding woman can last up to 10 to 12 months, or even longer in certain societies or in individual cases, under most circumstances full protection (98 percent) is normally considered to be effective for only the first six months postpartum. Since menstruation associated with an anovulatory or inadequate luteal phase cycle generally precedes the

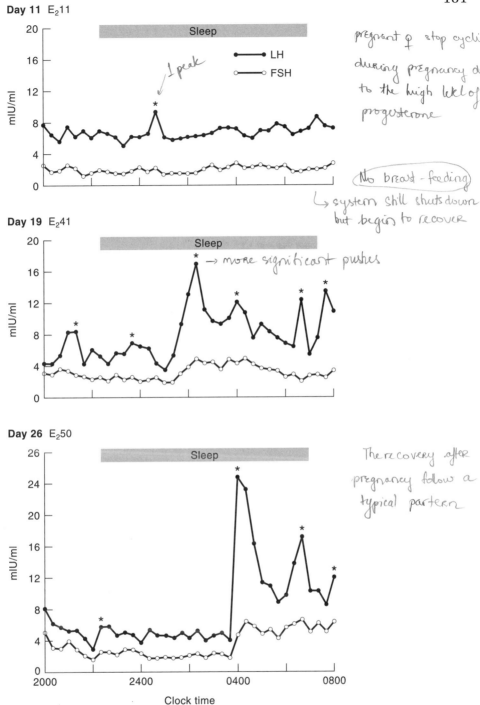

Figure 7.13
LH and FSH profiles during a 12-hour period in a nonlactating woman on days 11, 19, and 26 postpartum. Reactivation of the GnRH pulse generator begins during the sleep period in a manner similar to that during puberty. The asterisks indicate significant LH pulses. (Adapted from J. H. Liu and K. L. Park. 1988. *Journal of Clinical Endocrinology and Metabolism* 66, 639–642, © The Endocrine Society.)

first potentially fertile ovulation, resumption of menstruation signals the time at which breast-feeding is no longer sufficiently effective for contraceptive purposes. At this time, other forms of contraception are required to prevent a pregnancy.

The initial amenorrheic stage is characterized by suppression of GnRH pulse generator activity. We know this is the case because exogenous pulsatile infusion of GnRH in breast-feeding women readily induces ovulation, and removal of the exogenous GnRH results in a return of the women to a state of ovarian inactivity. The agent responsible for the suppression of GnRH release has until recently been considered to be prolactin. Prolactin levels in breast-feeding women are maintained at high levels, a situation that could be characterized as physiological hyperprolactinemia. Because of the well-known suppressive effects on ovarian function by high prolactin levels in other contexts, a tempting inference has been that prolactin *per se* was responsible for lactational amenorrhea, and indeed, this suggestion has appeared in most textbooks.

An alternative hypothesis derived from a number of recent studies is that it is the intensity and duration of suckling that is more important in inhibiting GnRH release. Intense suckling (frequent and long episodes of suckling) has been shown to delay the resumption of ovarian function for periods of 1 to 3 years, even after prolactin levels have decreased significantly. The mechanism of GnRH suppression appears to depend on β-endorphin release. After birth the nipple becomes very sensitive to tactile stimulation. The basis for the increased sensitivity remains unclear. Suckling generates neural inputs from the nipple via the spinal cord to the hypothalamus that result in the release of β-endorphin in the hypothalamus, which in turns suppresses GnRH pulse generator activity. The greater the suckling frequency the stronger the suppression. Hence, the generally intense suckling activity characteristic of the first few months of breast-feeding is responsible for the amenorrhea seen during this period.

Despite the clear relationship between suckling intensity and infertility, it has proven extremely difficult to define a suckling parameter that can be used to insure a certain period of lactational amenorrhea in any given female. A reduction in suckling frequency or suckling duration very quickly leads to resumption of normal ovarian activity. Reduction in suckling intensity typically begins when supplementary feeding is introduced. The consequences of supplement introduction in the breast-feeding schedule is illustrated in Fig. 7.14. In this experiment, changes in the LH and prolactin profiles were measured during a 24-hour period in a breast-feeding woman at 4 weeks and at 16 weeks postpartum. Supplementary feeding was begun at 16 weeks. Activation of the GnRH pulse generator, inferred from the LH pulses, is coincident with the initiation of supplementary feeding, when the suckling intensity decreased. In this particular instance, the woman remained amenorrheic until 22 weeks. Reduction in the intensity of suckling is correlated with entry into the second phase of lactational infertility, when cycling returns, but the pattern of GnRH pulse generator activity has not yet returned to the prepregnancy condition.

The prolactin profile in Fig. 7.14 also shows that prolactin is not secreted in a continuous fashion. Each suckling episode results in the release of prolactin, with the amount of prolactin released proportional to the suckling intensity. The intensity of the suckling pattern determines the degree of hyperprolactinemia. Hence, the release of prolactin is a reliable indicator of hypothalamic sensitivity to the suckling stimulus: high release when the hypothalamus is particularly sensitive to suckling, and low release as the hypothalamus becomes less sensitive. Average prolactin levels decline as suckling intensity declines. Prolactin release is secondary to the suckling stimulus; it is simply a monitor of hypothalamic sensitivity to the suckling stimulus and is not the primary cause of the suppression of GnRH release.

Breast feeding

Four weeks of lactation

Introduction of supplements → *reducing suckling episode.*

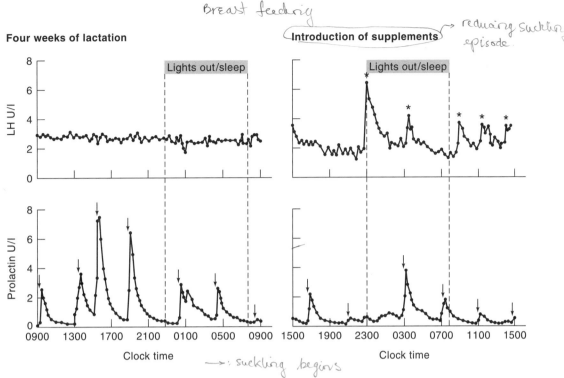

→: *suckling begins*

Figure 7.14
LH and prolactin profiles during a 24-hour period in a lactating woman at 4 weeks postpartum and at 16 weeks postpartum after the introduction of supplements. The arrows indicate suckling episodes. The intensity of the suckling episodes is greater at 4 weeks than at 16 weeks. Note that prolactin is released with each suckling episode. Reactivation of the GnRH pulse generator after the introduction of supplements begins during the sleep period, as is seen in the nonlactating female. (A. S. McNeilly. 1993. *Endocrine and Metabolism Clinics of North America* 22[1], 59–73, with permission, W. B. Saunders Company.)

suckling
↳ neuron signal
+ hormone signal

SUMMARY

Pregnancy is characterized by the complete suppression of the maternal H-P-G axis due to the high progesterone secretion by the placenta. After birth in a nonlactating female, H-P-G axis activity returns slowly and is characterized by sleep-entrained reactivation of the GnRH pulse generator. In a breast-feeding female, the maternal H-P-G axis is suppressed for a period of several months. Hence, lactation not only provides the primary food source for the newborn, but it contributes to the postpartum infertility of the breast-feeding woman in traditional societies. Lactational infertility depends on the intensity of suckling by the infant, and its duration varies considerably among women and among societies. Suppression of the GnRH pulse generator in a lactating female appears to depend on the release of β-endorphin stimulated by the infant suckling at the nipple. As the suckling intensity decreases and supplementary feeding is introduced, the inhibition of GnRH release is reduced, and GnRH pulsatility and normal fertility return.

PREMENSTRUAL SYNDROME

Premenstrual syndrome (PMS), known clinically as *premenstrual dysphoria*, refers to a set of physical, psychological, and behavioral alterations that are linked to the menstrual cycle. Although over 50 physical and psychological symptoms have been reported, the most common kinds of complaints in approximate order of occurrence are listed in Table 7.3. Until relatively recently this disorder was viewed primarily as hav-

Table 7.3
Common Symptoms of PMS

Physiological (somatic)	Behavioral (affective)
Breast tenderness	Depression
Abdominal bloating	Irritability; angry outbursts
Headaches	Anxiety; confusion
Swelling (legs, ankles)	Social withdrawal

ing a psychiatric, rather than a physiological origin. The variability of the symptoms among women, with some experiencing severe and others minimal disturbances, and difficulties in agreeing on suitable diagnostic criteria prevented the full acceptance of PMS symptoms as a clinically defined disorder. Probably the first serious attempt to define the premenstrual syndromes dates to 1931, when premenstrual syndrome was defined as consisting of "*the cyclic recurrence in the luteal phase of psychological, and/or behavioral changes of sufficient severity to result in the deterioration of interpersonal relationships and/or interference with normal activities.*" Since then the diagnostic criteria for the premenstrual syndromes have been progressively refined, and the most recent compilation, which attempts to distinguish between the mild and severe forms of PMS, appears in the fourth edition of the *Diagnostic and Statistical Manual of Mental Disorders* (DSM-IV). The 1931 definition, however, still embodies the essential features of PMS, in particular, the recurrence of the symptoms during the luteal phase of the ovulatory cycle, and disappearance during the follicular phase.

Epidemiology

Epidemiological surveys suggest that perhaps as many as 50 percent of women experience some symptoms of PMS. In most such women, the symptoms are mild and do not require medical attention, and a variety of effective self-management techniques have been developed to help in the relief of symptoms. The gamut of treatments or therapies is wide, ranging from vitamin and trace element supplementation, exercise, yoga, herbal remedies, and even hypnosis. In 3–8 percent of women, however, PMS symptoms are much more severe and debilitating. Such women do not respond to the conventional self-management methods, and require medical or psychiatric intervention. This circumstance has led to the delineation of **premenstrual dysphoric disorder** (PMDD) as a separate disorder within the PMS categories. The PMDD criteria defined in DSM-IV are much stricter than for simple PMS, requiring in particular a minimum number of symptoms and severe functional impairment. It is fair to say that the diagnosis of PMDD indicates a pathological condition, while ordinary PMS does not. Despite some skeptics, the existence of PMS, and especially PMDD, is an accepted clinical condition. In the popular culture an indication of the full acceptance of premenstrual syndrome is clearly indicated by the use of PMS as a verb—for example, in variations of the expression "She is PMSing."

In the criminal literature, an association between crime and PMS has also been noted. Crimes by women apparently are committed most frequently during the premenstrual period, while fewer crimes are committed during the ovulatory or preovulatory stages. A number of studies have also shown that the incidence of child battering during the premenstrual period may be relatively high. In most such cases, the women involved may be perfectly normal and maternal during other stages of the ovulatory cycle. Since the first half of the nineteenth century, courts in France have taken the menstrual cycle into account when dealing with female offenders. For example, acquittals for shoplifting because of what we now would call PMS took place on a more or less regular basis. Indeed, in France, PMS-induced changes have been recognized legally, so that women who commit a crime during the few days before menses can claim temporary impairment of sanity. In England, in some circumstances PMS has been considered a mitigating factor in crimes committed by women. This defense has not had much success in the United States.

Etiology and therapy

Treatment for the relief of PMS symptoms has generally been based on some hypothesis about the origin of PMS. Although the PMS symptoms manifest themselves during the luteal phase, inducing menstruation shortly after ovulation (that is, shortening the luteal phase) does not abolish the PMS symptoms. This indicates that PMS is probably due to a series of events that are set in motion during the follicular phase. In addition, some ovarian factor(s) produced in a cyclic manner must be the precipitating factor in PMS. For example, menopause abolishes PMS. Before the 1960s, the only effective therapy for the most debilitating cases of PMS was the removal of the ovaries. Treatments that inhibit ovarian function have been shown to be very effective in severe cases. A number of hypotheses have been proposed to explain the origin of PMS symptoms, particularly for the more severe forms; these hypotheses include progesterone deficiency, β-endorphin withdrawal, vitamin B complex deficiency, thyroid dysfunction, and prolactin excess. However, despite much effort, it has not been possible to identify unambiguously any specific ovarian product that can be associated with PMS.

An alternative way of looking at the origin of PMS has been emerging in the last few years. This hypothesis suggests that PMS originates not from an abnormal production of some putative ovarian factor, but rather, that it may reflect a CNS response to normal ovarian cycling. Cyclical mood changes may occur without necessarily considering them pathological, and most likely the mild forms of PMS are of this type. PMDD, on the other hand, may reflect cyclically triggered *abnormal* responses in the CNS that result in abnormal changes in neurotransmitter levels. One current hypothesis suggests that alterations in serotonin levels may be responsible for PMDD. PMDD shares many of the features of many mood and anxiety disorders, disorders that are considered to be characterized by disturbances in serotonin levels. In some studies, PMDD women have been shown to have a lower level of platelet serotonin. A number of recent studies have reported significant success in treating PMDD women with a class of compounds known as **selective serotonin reuptake inhibitors** (SSRIs), the best known of which is fluoxetine (Prozac). SSRIs have also been shown to be efficacious in the treatment of depression and other mood disorders. However, it is still premature to consider that all cases of PMDD arise from a disturbance in serotonin metabolism.

First, not all PMDD women respond to SSRIs (perhaps only 50–60 percent do). Second, the relief experienced diminishes after the second or third month of treatment. These findings suggest disruptions in other neuroendocrine hormones may also be important. The partial therapeutic success with SSRIs is welcome news for some PMDD sufferers, and indicates that the focus on an abnormal neuroendocrine disturbance to ovarian cycling may be the best approach for understanding the nature of premenstrual dysphoria. There is good reason to be optimistic about the future.

SUMMARY

Premenstrual dysphoria, or PMS, refers to a set of somatic, psychological, and behavioral disorders of varying degrees of severity that manifest themselves during the latter part of the luteal phase of the ovulatory cycle. The origin of PMS remains unclear, but the current view suggests that the symptoms associated with PMS arise from normal responses of the CNS to the periodic variation in ovarian hormones during the cycle. Particularly severe PMS symptoms may arise from abnormal CNS responses. This view is supported by some success in relieving severe PMS symptoms with selective serotonin reuptake inhibitors (SSRIs). Since relief appears to be transient and not all severe PMS cases respond to SSRIs, it is likely that disorders in the metabolism of neurotransmitters other than serotonin are involved in the etiology of PMS.

QUESTIONS

1. What is the significance of the fact that pulsatile GnRH is an intrinsic property of GnRH-secreting neurons?

2. Describe the different ways in which gonadotropin release is controlled.

3. Despite the importance of the hypothalamus and pituitary, why is it still reasonable to consider their role in the maintenance of cyclicity as permissive, rather than determinative?

4. Measuring β-endorphin levels in the circulatory system is not very informative because β-endorphin functions primarily in a paracrine manner. Can you think of an effective way to determine the effects of β-endorphin on a specific process, for example, whether β-endorphin plays a role during the ovulatory cycle?

5. What is the importance of CNS control over GnRH release?

6. How would you demonstrate that lactational infertility is due to inhibition of the GnRH pulse generator? What is the evolutionary significance of lactational infertility?

7. The administration of long-acting GnRH agonists has been found to be effective in relieving severe PMS symptoms in many instances. Why would you expect this to be the case? Is this observation compatible with the recent hypothesis about the origin of PMS?

8. Library project:
   a. β-endorphin-mediated suppression of GnRH secretion is one model for explaining lactational infertility. What other model(s) have been proposed?
   b. What is the latest view about the role of melatonin in reproduction?
   c. What is known about the mechanism by which GnRH-secreting neurons synchronize their release of GnRH?

SUPPLEMENTARY READING

Gold, J. H. Premenstrual dysphoric disorder: What's that? 1999. (Editorial.) *The Journal of the American Medical Association* **278**(12), 1024–1026.

Guillemin, R., and R. Burgus. 1972. The hormones of the hypothalamus. *Scientific American* **227**(5), 24–33.

Hotchkiss, T., and E. Knobil. 1996. The hypothalamic pulse generator: The reproductive core. In *Reproductive Endocrinology, Surgery, and Technology.* E. Y. Adashi, J. A. Rock, and Z. Rosenwaks, Eds. Lippincott-Raven Publishers, Philadelphia, pp. 123–162.

Johnson, M. H., and B. J. Everitt. 1995. *Essential Reproduction.* Blackwell Science, Oxford, UK.

Short, R. V. 1993. Lactational infertility in family planning. *Annals of Medicine* **25**(2), 175–180.

Vanezis, P. 1991. Women, violent crime, and the menstrual cycle: A review. *Medicine, Science, and the Law* **31**(1), 11–14.

Worthman, C. M. 1995. Hormones, sex, and gender. *Annual Review of Anthropology* **24**, 593–617.

Yen, S. S. C., and R. B. Jaffe. 1978. *Reproductive Endocrinology,* 1st ed. W. B. Saunders & Co., Philadelphia, PA.

ADVANCED TOPICS

Cagnacci, A. 1996. Melatonin in relation to physiology in adult humans. *Journal of Pineal Research* **21**, 200–213.

McNeilly, A. S. 1993. Lactational amenorrhea. *Endocrine and Metabolic Clinics of North America* **22**(1), 59–73.

Rasmussen, D. 1991. The interaction between mediobasohypothalamic dopaminergic and endorphinergic neuronal systems as a key regulator of reproduction: An hypothesis. *Journal of Endocrinological Investigation* **14**, 323–352.

Steiner, M. 1997. Premenstrual syndromes. *Annual Review of Medicine* **48**, 447–455.

Yen, S. S. C. 1999. Chronic anovulation due to CNS-hypothalamic-pituitary dysfunction. In *Reproductive Endocrinology,* 4th ed. S. S. C. Yen, R. B. Jaffe, and R. L. Barbieri, Eds. W. B. Saunders & Co., Philadelphia, PA.

Yonkers, K. A., U. Halbreich, E. Freeman *et al.* 1997. Symptomatic improvement of premenstrual dysphoric disorder with sertraline treatment. *Journal of the American Medical Association* **278**(12), 983–988.

William H. Johnson, *Jitterbugs (II)*, c. 1941. (Smithsonian American Art Museum, Washington, D.C./Art Resource, New York.)

CHAPTER 8

Puberty

*Foremost among the unsolved mysteries of modern biology
are the precise neurophysiological mechanisms that drive the
onset of adolescence, and direct the individual's progress through
normal puberty into the transitional state of young adulthood.*

—J. D. Veldhuis. 1996

Neuroendocrine mechanisms mediating awakening of the
human gonadotropic axis in puberty. *Pediatric Nephrology* **10**, 304–317.

Puberty is the period between the end of childhood and the attainment of sexual maturity. For all societies these changes have been a source of fascination. In most traditional societies, these changes have been celebrated in many, often unusual, ways. In the female, the onset of puberty is most clearly signaled by the first menstruation, the **menarche**, while in the male, perhaps the ability to ejaculate or experiencing spontaneous nocturnal ejaculations (so-called "wet dreams") are the indicators of the onset of puberty. But many other changes are involved as well, including the growth and maturation of the gonads, with concomitant endocrinological changes resulting in the elaboration of the secondary sexual characteristics that will mark the obvious differences between the female and the male. Less obvious are many complex psychological changes, such as those affecting mood, cognitive behavior, and social adaptation, that also take place. Our objective in this chapter is to review briefly the changes associated with puberty, and to provide a perspective on the mechanisms and factors that are considered important in initiating the onset of puberty.

SOMATIC AND REPRODUCTIVE CHANGES WITH PUBERTY

Many changes take place during the pubertal transition. Some involve the reproductive tissues, and some involve somatic tissues and result in development of the secondary sexual characteristics. Conventionally, the morphological changes in breast development, appearance and growth of pubic hair in females, and in external genitalia in males have been standardized in what are known as the "Tanner stages," with Stage 1 representing the prepubertal state, and Stage 5 the completion of pubertal changes. These are used diagnostically to measure the progression of the pubertal changes in girls (Figs. 8.1 and 8.2, and Table 8.1) and boys (Fig. 8.3 and Table 8.2). It is important

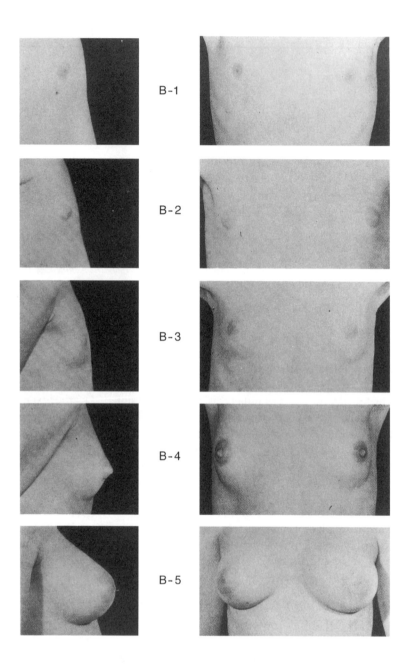

Figure 8.1
Tanner stages in breast development. See Table 8.1 for a description. (From J. C. van
Wieringen *et al.*, 1971, with permission.)

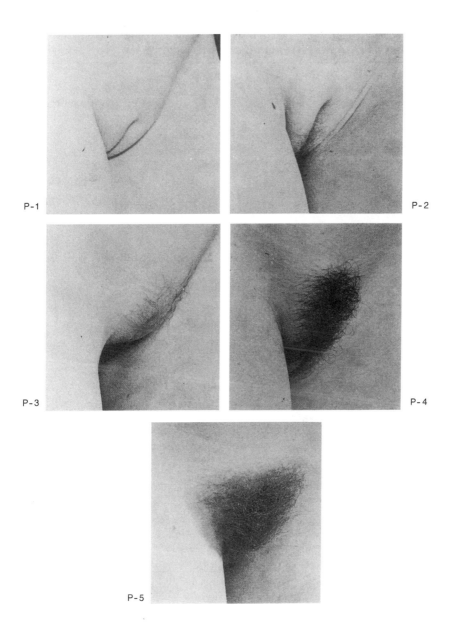

Figure 8.2
Pubic hair development in girls. See Table 8.1 for description. (From J. C. van Wieringen *et al.*, 1971, with permission.)

Table 8.1
Tanner Stages of Breast and Pubic Hair Development

	Breast		
	Stage	Mean age	Range (years)
I.	Prepubertal; slight elevation of papilla		
II.	Elevation of breast and papilla; areola diameter increases	11.2	9.0–11.3
III.	Enlargement of breast tissue; no separation of breast and areola	12.2	10.0–14.3
IV.	Areola and papilla form a secondary mound above the level of the breast	13.1	10.8–15.3
V.	Mature stage; erect papilla projecting above areola	15.3	11.9–18.8
	Pubic hair		
	Stage	Mean age	Range (years)
I.	Prepubertal; no pubic hair		
II.	Sparse, curly, pigmented hair appearing along the lower labia	11.7	9.3–14.1
III.	Spread of darker, coarser hair across the lower pubis	12.4	10.2–14.6
IV.	Abundant adult type hair, but limited to labia area	13.0	10.8–15.1
V.	Spread of pubic hair to form an inverted triangle; spread of hair along the upper inner thigh	14.4	12.2–16.7

References: Tanner, 1962; Marshall and Tanner, 1969.

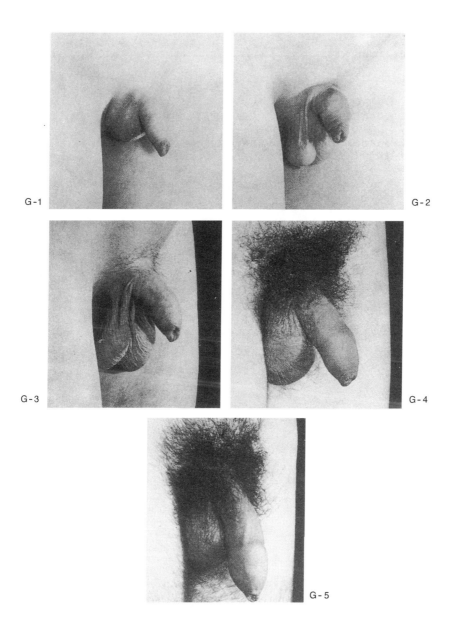

Figure 8.3
Tanner stages in the development of the external genitalia and pubic hair in boys. See Table 8.2 for a description of the changes. (From J. C. van Wieringen *et al.*, 1971, with permission.)

Table 8.2
Tanner Stages of External Genitalia and Pubic Hair in Males

External genitalia			
	Stage	Mean age	Range (years)
I.	Prepubertal		
II.	Enlargement of scrotum and testes; scrotum becomes pigmented	11.6	9.0–14.7
III.	Growth of penis; continued growth of scrotum and testes	12.9	10.3–15.5
IV.	Increase in length and breadth of penis; growth of glans penis	13.8	11.2–16.3
V.	Adult size and shape	14.9	12.2–17.7

Pubic hair			
	Stage	Mean age	Range (years)
I.	Prepubertal; no pubic hair		
II.	Sparse growth of slightly curled hair along the base of the penis	13.4	10.8–16.0
III.	Spread of darker, coarser hair above penis	13.9	11.4–16.5
IV.	Abundant adult type of hair, but limited to genitalia	14.4	11.7–17.1
V.	Adult hair in type and quantity; spread of hair along the inner thigh and above the penis	15.2	12.5–17.9

References: Tanner, 1962; Marshall and Tanner, 1970.

to keep in mind that these stages measure progression along a continuum. Growth of the uterus and ovaries in the female (Fig. 8.4) and the testes in the male (Fig. 8.5), for example, takes place in a gradual fashion. Nonetheless, the Tanner stages are generally correlated with identifiable endocrinological milestones. In the female, these milestones are adrenarche (pubarche), thelarche, and menarche, while in the male the corresponding endocrinological landmarks are known as adrenarche and gonadarche.

Adrenarche, the activation of adrenal gland steroid hormone synthesis, particularly androgen synthesis and secretion, seems to be unique to human beings and chimpanzees. A consequence of adrenarche is the first appearance of pubic hair in both males and females. The earliest hormonal changes detected in the peripubertal period are those associated with adrenarche, in particular, the increased secretion of two adrenal androgens, DHEA and DHEAS. This secretion is unusual, since other adrenal hor-

[handwritten margin notes: stages measure progression along a continuum; tanner stage → identifiable endocrinological milestone; Adrenarche activation of adrenal gland → (pubic hair); ↑DHEA ↑DHEAS]

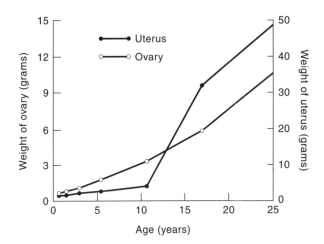

Figure 8.4
Growth of the uterus and ovary with age. (Adapted from J. M. Tanner. 1962. *Growth at Adolescence*. Blackwell Science, Oxford, UK.)

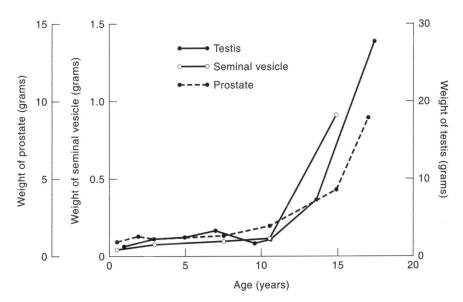

Figure 8.5
Growth of a single testis, seminal vesicle, and prostate gland with age. (Adapted from J. M. Tanner. 1962. *Growth at Adolescence*. Blackwell Science, Oxford, UK.)

mones are not secreted at this time. The significance of this secretion remains obscure. The first appearance of pubic and axillary (underarm) hair growth is generally attributed to these two weak androgens, but it is not clear what other function these hormones have. **Thelarche**, the onset of breast development and usually the first sign of puberty in females, may also depend on adrenarche (in this case, adrenal androgens are converted to estrogens). The relationship between adrenarche and the definitive signs of puberty—menarche and **gonadarche** (testicular enlargement)—remains obscure, since they can occur independently of each other. Menarche is the last obvious marker of sexual maturation in the female, while gonadarche is generally the first sign of puberty in the male.

Secondary sexual characteristics involve changes that include those of nonreproductive tissues (Table 8.3). The major immediate causes of these changes are dramatic increases in the production of gonadal steroid sex hormones. The target tissues for these hormones are the genitourinary ducts, hair follicles, apocrine glands, fat cells, muscle cells, bones, and even the brain. The pubertal "growth spurt," due to the combined effects of steroid hormones and increased secretion of GH (growth hormone), is initiated earlier in females than in males, as easily verified by comparing the heights of

Table 8.3
Secondary Sexual Characteristics in Females and Males Associated with Puberty

Change	Mean age
Females	
Initiation of growth spurt and deposition of fat	10
Widening of the pelvis	11
Growth and maturation of the internal genitalia and vagina	12
Axillary hair begins to appear	13
Skeletal growth decreases; sweat and sebaceous gland development, sometimes accompanied with acne; first ovulation	14
Voice deepens slightly	15
Adult height reached	16
Males	
Initiation of androgen production after childhood quiescence	10
Fat deposition begins	11
Skeletal growth begins; spontaneous erections; growth of seminal vesicles and prostate gland	12
Spontaneous nocturnal ejaculations begin	13
Growth of vocal cords and deepening of voice; appearance of axillary hair and hair on upper lip	14
First fertile ejaculation	15
Appearance of chest, body, and facial hair; sweat and sebaceous glands develop, often with acne; loss of body fat	16
Muscle growth and increase in muscle strength; broadening of shoulders	17
Adult height may be reached, although often growth may continue into the early 20s.	18

Individuals can vary greatly in the timing of these changes without being abnormal. (References: Marshall and Tanner, 1969; Marshall and Tanner, 1970; Reynolds and Wines, 1948; Reynolds and Wines, 1951.)

students in junior high school (Table 8.3). Perhaps in part because of differences in estrogen levels between males and females, skeletal maturation occurs earlier in females than in males, so that females generally reach their final height at around 16 years of age. In males, the growth spurt starts later, but also continues for a longer time, so that males do not attain their final height until late in their teens or in their 20s. The growth spurt ends with cessation of long bone growth, known as *epiphyseal plate fusion*. The growth spurt does not occur in other mammalian species, even other primates, and seems to be unique to humans.

Body composition changes differentially in females and males. In females, fat deposition begins earlier, and the total gain of fat is much greater than in males. In males, there is an early accumulation of fat, but slightly later, an absolute loss of fat takes place (Table 8.3). Muscle growth is similar in boys and girls up to about the age of 12, but thereafter, boys accumulate more muscle tissue so that at the end of puberty, males have on the average about 56 percent more muscle tissue than females. Androgens have a role in both sexes. Small increases in androgens, from the ovaries as well as the adrenals, lead to the appearance of pubic and axillary hair, maturation of the apocrine and sweat glands that results in the first appearance of a characteristic body odor, and the bane of many adolescents, *acne*, due to increased conversion of testosterone to DHT by skin cells. In males, testosterone and its conversion to DHT in specific target tissues lead to the marked development of facial, axillary, and pubic hair, as well as the darkening of most other body hairs, and finally, the appearance of acne. Androgens also stimulate the growth of the vocal cords, which leads to the deepening of the voice, much more pronounced in males than females.

Attainment of full reproductive capacity is gradual. In females, the establishment of the cyclic pattern of gonadotropin secretion and the coordination of hypothalamic-pituitary-ovarian activity sufficient to support ovulation take place over a period of several months. For example, ovulatory cycles are not established at the time of menarche, but develop only after perhaps a year. Similarly, the ability to ejaculate in males occurs before spermatogenesis has really been established, so that in the early pubertal stages, the male ejaculate is azoospermic.

Psychological and behavioral changes, some subtle and some not, also accompany the physical ones. Intellectual ability, preference for abstraction, increased capacity for making judgments, and other cognitive changes are associated universally with pubertal changes. Behavioral changes that often lead to conflicts within the family, with authorities, and with other adolescents are also characteristic. Interest in sexual activity is awakened, and sexual orientation and gender identity may also be established definitively during this period.

TIMING OF PUBERTY

The time interval between birth and puberty varies considerably among species. This interval must reflect both the average life span of the species and the length of time necessary to reach sexual maturity. For example, mice reach puberty 6 weeks after birth, while elephants require 15 years. In the cod and haddock, 4 years is typical; 10–15 years in the sturgeon; 22 years in the European eel, probably the longest period recorded; 2–4 years in amphibians such as frogs and toads; and 1–4 years in birds.

In humans, the interval between birth and the onset of puberty has generally been attributed to the comparatively long period that is required for the human child to become autonomous. Human societies, even the most primitive by the standards of modern industrialized societies, are extremely complex. It has been argued that it makes

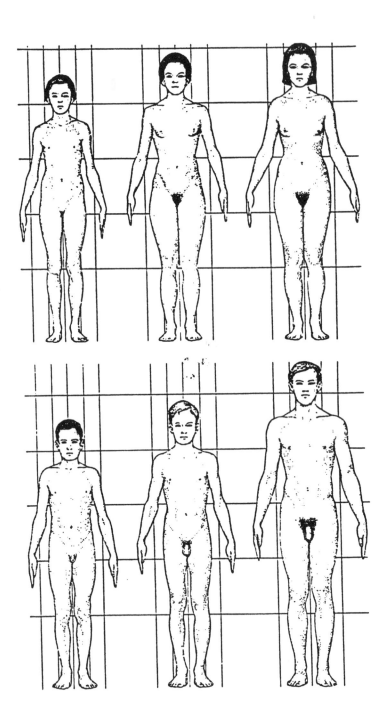

Figure 8.6
Differences in the extent of sexual maturation in individuals of the same age. The males are 14.75 years old, and the females are 12.75 years old. (From J. M. Tanner. 1962. *Growth at Adolescence*. Blackwell Science, Oxford, UK.)

sense that sexual maturity be postponed until all individuals have attained the physiological and psychological maturity necessary to participate in the society to which they are born. Social, linguistic, and intellectual skills have to be developed. Females have to attain a size and weight that will permit them to sustain a pregnancy. Males, presumably, have to have acquired the skills and size to participate in the protection of the offspring. Both also will be needed to take care of the infant and to share in the raising and training of the young. These conditions and constraints probably applied throughout most of humankind's history. At present, however, in the industrialized countries, puberty occurs at least 10 years before the intellectual and psychological maturity necessary for full participation in a technologically complex society is attained. But, as we will see below, puberty takes place when attainment of a certain metabolic and/or nutritional status is achieved, regardless of cognitive or intellectual maturity.

The timing of puberty is strikingly variable. Hence, although the mean age of menarche in the United States is 12.3 years, the normal range is from 9.7 to 16.7 years. Similarly, the normal range for thelarche is 7.9 to 13.9 years, while gonadarche can occur normally between 8.5 and 15.9 years. This variability results in significant differences in pubertal changes at the same chronological age (Fig. 8.6). The timing of puberty is considered to be abnormal only if it takes place at ages outside those limits. Despite this normal variation, it is clear that over the last century the mean age of puberty onset has been decreasing steadily (the *secular trend* in puberty) in Western Europe and the United States. This phenomenon is most clearly seen in the decrease in the mean age of menarche (Fig. 8.7), a trend that apparently stabilized in the 1990s at a little over 12

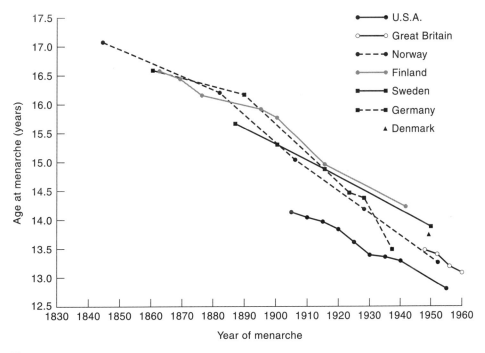

Figure 8.7
The "secular trend" in the mean age of menarche in the girls from Western Europe and the United States. (Adapted from J. M. Tanner. 1962. *Growth at Adolescence*. Blackwell Science, Oxford, UK.)

years. Changes of this type have suggested that environmental factors play an important role in determining the triggering of puberty.

SUMMARY

Puberty is marked by dramatic changes in growth and maturation of many tissues in the body. All of these changes are the consequence of the development of the gonads and substantial increases in their steroid hormone ouput. Increased estrogen and DHT ouput in the female is responsible for the female physiognomy, while increased androgen output is responsible for the male physiognomy. Puberty ends when the individual attains reproductive competence, which means establishment of regular ovulatory and menstrual cycles in the female and high levels of spermatogenesis and erectile function in the male. Pubertal changes begin earlier in the female than in the male, but the reason for this difference is not really known. The age of puberty onset is highly variable among individuals. In females, the mean age of menarche (the first menstruation) is about 12.3 years, but the normal range is from 9.7 to 16.7 years. The mean age of menarche has declined from 16 to 17 years to the present value over the last century in the industrialized countries. This secular trend in the age of puberty onset is most likely due to significant improvement in the nutritional status of the population.

THE ONSET OF PUBERTY

Early clues about the origin of pubertal changes were obtained from observations and experiments that began early in the twentieth century. A type of delayed puberty known as Frohlich's syndrome associated with lesions in the brain was described in 1904. Experiments reported in the 1930s showed that ovaries removed from an immature animal were able to ovulate when transplanted to adult animals. These results demonstrated that the onset of puberty was not initiated by the gonads. The demonstration in the 1950s of the essential role of the hypothalamus in regulating gonadotropin secretion, and the experimental induction of premature puberty by producing lesions in certain regions of the hypothalamus left no doubt that the hypothalamus was the brain center responsible for the initiation of puberty.

The "reawakening" of the GnRH pulse generator

There is general agreement that the immediate cause of puberty is the reactivation of the GnRH pulse generator, which is in a quiescent state during infancy and childhood. It is known that the GnRH pulse generator is active during fetal life. It is operative by about the 12th week of gestation and reaches a peak of activity around the beginning of the third trimester of pregnancy, after which its activity declines to a very low level by the end of gestation. Although GnRH levels in humans cannot be measured directly, the activity of the GnRH pulse generator can be inferred by monitoring the LH and FSH levels.

In the male, the developmental history of the GnRH pulse generator activity produces the testosterone profile shown in Fig. 8.8. We have seen in Chapter 6 that the fetal testis is active in producing testosterone. Testosterone levels reach a peak around the middle of the second trimester of pregnancy, and then decrease markedly to the end of gestation. There is an abrupt rise just after birth, a change known as the **neonatal**

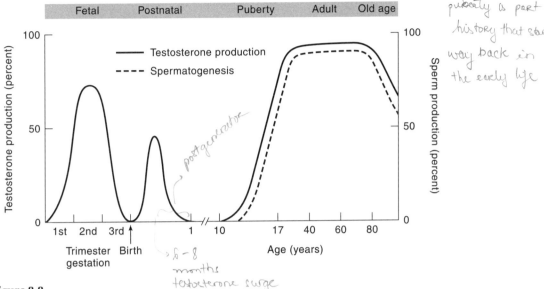

[Handwritten margin note: puberty is part of history that starts way back in the early life]

Figure 8.8
The testosterone profile in males as a function of age. A special feature is the postnatal testosterone surge whose function remains unclear. Note also that although the fetal testis produces large amounts of testosterone, no spermatogenesis takes place. Spermatogenesis begins only at puberty. (Adapted from J. S. O. Winter *et al.*, 1976; and J. E. Griffin and J. D. Wilson, 1980.)

testosterone surge. The significance of this surge is unclear, but some observations suggest that it is physiologically important. For example, inhibition of the neonatal surge in rodents results in disturbances in adult male copulatory behavior. In humans, studies in individuals who suffer from Kallman's syndrome suggest that the neonatal testosterone surge may be important in ensuring that spermatogenesis begins normally at puberty. Four to six months after birth, testosterone levels decline and remain extremely low during the rest of infancy and throughout the early and middle childhood period. During this long interval, the GnRH pulse generator remains relatively quiescent. Testosterone levels begin to increase again with the onset of puberty, a reflection of the reawakening of the GnRH pulse generator and its consequent stimulation of the pituitary and gonads. Testosterone levels remain high during the long postpubertal period, and may decline somewhat in old age.

In the female, the developmental history of the GnRH pulse generator is reflected by the LH and FSH profiles shown in Fig. 8.9. As in the male, LH and FSH levels reach a peak during the second trimester and then decline to the end of gestation. A sharp increase in LH and FSH levels takes place after birth, again reflecting a transient activation of the GnRH pulse generator, after which both decrease to very low, but not zero, levels by 1 to 2 years. The LH and FSH patterns are not completely coincident, probably reflecting differences in the effects of other factors that modulate the release of LH and FSH. After the long childhood slumber, the GnRH pulse generator is reactivated, and LH and FSH levels begin to rise, and establishment of gonadal function generates the cyclical release of LH and FSH. During the postpubertal period gonadal steroids

[Handwritten margin notes: neonatal testosterone surge ↓ spermatogenesis?; ↓ testosterone level after birth; ↑ testosterone at puberty → remain high; ↓ w/ age; ♀ GnRH → LH/FSH; Sharp increase in LH/FSH after birth; ↓ ↓ very low]

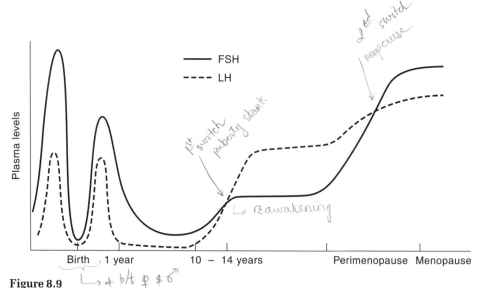

Figure 8.9
LH and FSH profiles in females as a function of age. Note the inversion of LH and FSH levels
in going from the fetal to childhood to postpubertal stages. These changes are considered to be
due to ovarian hormones. (Adapted from J. S. O. Winter *et al.*, 1976.)

are responsible for the relatively low, but regulated LH and FSH levels. There is a
marked shift in the relative levels of LH and FSH, an effect attributed to the activation
of ovarian function. FSH may be particularly sensitive to negative-feedback inhibition
by inhibin and possibly other nonsteroidal gonadal factors. At menopause, gonadotropin levels increase significantly, and this rise is due to the cessation of ovarian function and consequent decrease in gonadal hormones. Due to the difference in sensitivity
to gonadal hormones between LH and FSH, FSH secretion by the pituitary is greater
than that of LH in the postmenopausal woman.

Sensitive methods for measuring gonadotropin levels have shown that although
GnRH pulse generator activity during infancy and childhood is very low, it is never really zero. During the long childhood period, the GnRH pulse generator is strongly inhibited, although it is not clear what causes the inhibition. At some point, however, the
restraint begins to be relaxed, and pulse generator activity begins to manifest itself by
the increase in gonadotropin pulsatile secretions. Particularly interesting is the fact
that during the initial pubertal stages the reactivation of the pulse generator is **sleep-
entrained**. This is manifested by the first appearance of LH and FSH pulses during the
sleep period. A good example of this phenomenon is that shown by the experiment
summarized in Fig. 8.10. In this study, the LH and FSH profiles in Tanner stage 1

Figure 8.10 *(facing page)*
Temporal reactivation of the GnRH pulse generator. LH and FSH profiles during a 24-hour
period at different stages of puberty in females. Note the increase in LH and FSH that takes
place during the sleep period (gray bars) during the early pubertal stages. Asterisks mark true
peaks in LH and FSH levels. (Adapted from D. Apter *et al.* 1993. *Journal of Clinical
Endocrinology and Metabolism* **76**, 940–949. © The Endocrine Society.)

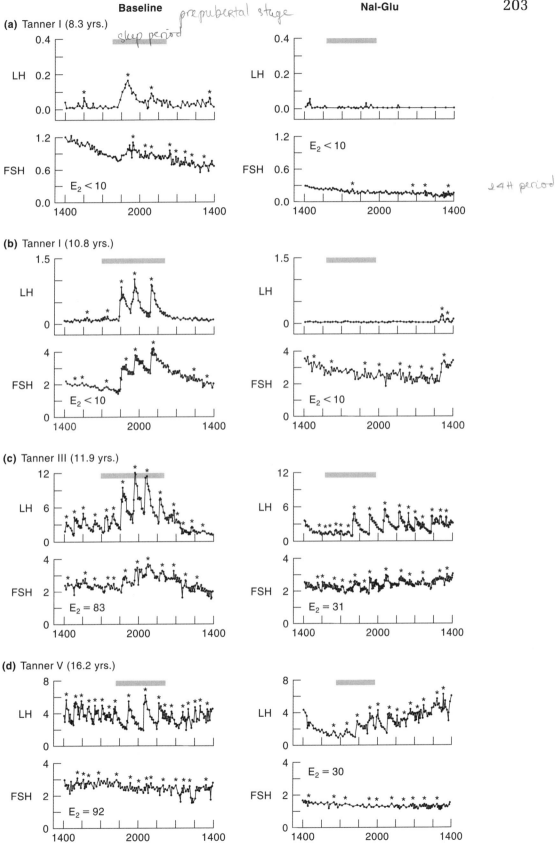

Baseline — prepubertal stage · sleep period

Nal-Glu

(a) Tanner I (8.3 yrs.)

LH / FSH — $E_2 < 10$ (Baseline); $E_2 < 10$ (Nal-Glu)

↓4H period

(b) Tanner I (10.8 yrs.)

LH / FSH — $E_2 < 10$ (Baseline); $E_2 < 10$ (Nal-Glu)

(c) Tanner III (11.9 yrs.)

LH / FSH — $E_2 = 83$ (Baseline); $E_2 = 31$ (Nal-Glu)

(d) Tanner V (16.2 yrs.)

LH / FSH — $E_2 = 92$ (Baseline); $E_2 = 30$ (Nal-Glu)

Clock time

big activation of LH/FSH
is very sensitive

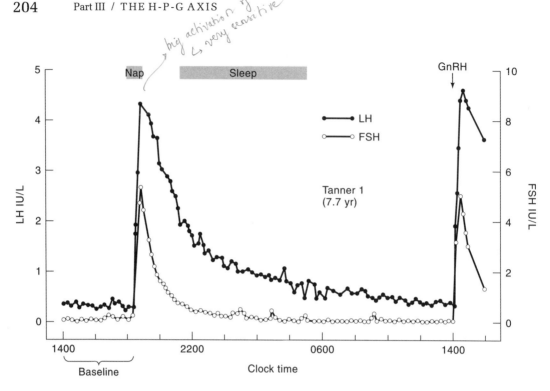

Figure 8.11

Acute sensitivity of GnRH pulse generator activation to sleep during the prepubertal period. LH (open circles) and FSH (closed circles) levels during a 24-hour period in a Tanner 1 stage prepubertal girl (7.7 years old). The girl fell asleep at 1800 (6:00 P.M.) for 1 hour. There was an almost immediate release of LH and FSH during this time, certainly comparable to that elicited by an infusion of GnRH (see arrow). This study demonstrates the acute sensitivity of the GnRH pulse generator to sleep during the prepubertal stage. (Adapted from D. Apter *et al.* 1993. *Journal of Clinical Endocrinology and Metabolism* **76**, 940–949. © The Endocrine Society.)

through stage 5 young girls were monitored using a very sensitive method for measuring gonadotropin levels. During the early Tanner stages, reactivation of the GnRH pulse generator occurs only during the sleep period. In fact, GnRH secretion is particularly sensitive to the effects of sleep. This is demonstrated by the profile shown in Fig. 8.11. In this instance, one of the subjects of the study, a young girl, 7.7 years old (Tanner stage 1), fell asleep at about 6 o'clock in the evening, perhaps because she was tired. This resulted in an almost immediate activation of the GnRH pulse generator, reflected by the sharp increase in LH and FSH. Interestingly, after her short nap, she later went to sleep at the normal time, but at this later time GnRH pulse generator activation did not take place. Although we do not understand why the first sleep period interferes with or inhibits the subsequent activation of the pulse generator, it is clear that sleep at this stage is a potent activator of the GnRH pulse generator. The mechanism and the significance of the sleep-dependent activation of the GnRH pulse generator during the early pubertal phase remains a fascinating question.

sleep in early stage of Tanner is a potent activator of the GnRH pulse generator → cause ??

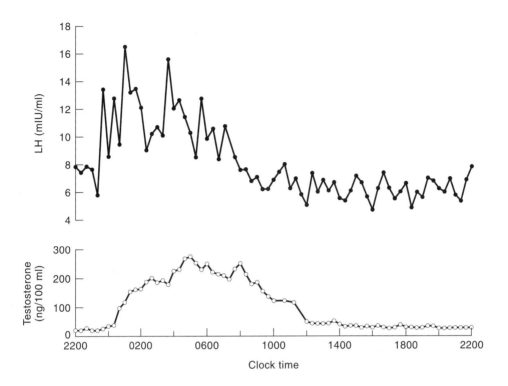

Figure 8.12
Nocturnal testosterone release in a pubertal male. Plasma LH and testosterone levels in a 12-year-old boy. The rise in testosterone is coincident with the rise in LH. (Adapted from R. M. Boyar *et al.* 1974. *The Journal of Clinical Investigation* 54, 609–618, American Society for Clinical Investigation).

The middle-to-late pubertal period is characterized by significant increases in LH pulse amplitude, but without any marked sleep-entrainment effect (Fig. 8.10). The adult pattern exhibits somewhat lower LH pulse frequency at night but with some evidence that the amplitude of the LH pulses is higher at night. The increased gonadotropin pulses result in concomitant increases in gonadal steroids. This effect is particularly apparent in males, in which the high LH pulses result in pulsatile release of testosterone from the testes (Fig. 8.12). During the sleep-entrained period, the LH pulses and high levels of testosterone are thought to be responsible for the spontaneous nocturnal emissions (wet dreams) that are experienced by most adolescent males.

Attainment of the postpubertal pattern of gonadotropin secretion is not immediate, but develops over a period of 3 to 4 years. Some decrease in the pulse frequency at night has been noted. Some investigators have suggested that the slowing down of the pulse generator at night may be due to melatonin secretion, since the day-night rhythm of melatonin secretion appears to be established about this time. However, the significance of the decrease in GnRH pulse frequency at night remains unknown.

The puberty clock and metabolic cues

Experiments in rodents and rhesus monkeys have shown that infusion of exogenous GnRH into a prepubertal individual is sufficient to induce all of the changes that would occur during a normal puberty. Similar findings have been obtained in humans, not as a consequence of experiments, but of attempts to treat delayed puberty. Individuals who experienced delayed puberty, but who retrospectively were shown to have a constitutional delay of growth and puberty, were administered pulsatile GnRH to induce puberty. The therapy was successful in that the exogenous pulsatile GnRH initiated the onset of puberty, but when GnRH infusion was stopped after a few months, the individuals returned to their prepubertal stage. They then underwent normal puberty spontaneously at a later age. These observations demonstrate that although exogenous GnRH can induce the initiation of puberty, it does not alter the individual's own "puberty clock." Moreover, in studies of gonadectomized animals and human patients with **gonadal dysgenesis** (failure of the gonads to develop properly), reawakening of the GnRH pulse generator occurs at the same time as in the gonadally intact individuals. Hence, the onset of puberty is a gonad-independent event. These examples indicate that the reactivation of the GnRH pulse generator in the prepubertal individual must be controlled by a biological clock that is not only independent of gonadal hormones but also independent of the pulse generator itself. Exogenous GnRH can activate the H-P-G axis, but cannot by itself reset the clock.

Given that puberty is initiated by the reactivation of the pulse generator, what do we know about the factors that control the timing of its reactivation? A significant clue has come from trying to understand the decline in the mean age of menarche (depicted in Fig. 8.7). Significant improvement in living conditions and the standards of health care, apparent especially in the industrialized countries, have generally been considered to have played a major role in the timing of puberty. The importance of adequate food intake and its effects on reproduction has been noted in experimental studies with animals and in observations in humans. The Lapps, for example, a nomadic people who live in the Arctic regions of Scandinavia and who did not experience the general improvement in living conditions and nutrition as the other people of Scandinavia in the years between 1870 and 1930, did not exhibit any decrease in the mean age of menarche. In a 1942 study, Japanese-American girls experienced menarche more than a year and a half earlier than their counterparts in Japan. In the rural areas of Bangladesh, where malnutrition is widespread, menarche occurs 4 to 5 years later than in the more affluent areas, and growth persists for several years. The historical record reveals that puberty has occurred at a relatively early age during periods characterized by affluence and prosperity at least for the upper classes. In the fifth and sixth century in Rome, females and males reportedly went through puberty by 12 and 15, respectively. Similarly, wealthy females in seventeenth-century Austria underwent puberty between 12 and 13, while their poorer counterparts from the rural areas did so at 17.

In studies with rodents, reducing food intake to maintain body weight about 80 percent below normal results in cessation of estrus cycles. Restoration of a high protein diet immediately restores fertility. Sheep farmers have long exploited a practice known as "flushing," that is, allowing sheep to graze in rich pastureland, which leads to earlier onset of estrus and lambing.

These observations have indicated that some aspect of nutritional status must be a critical parameter in the timing of puberty. Circumstantial evidence for this possibility has come from the finding that the body weight at the time of menarche in popula-

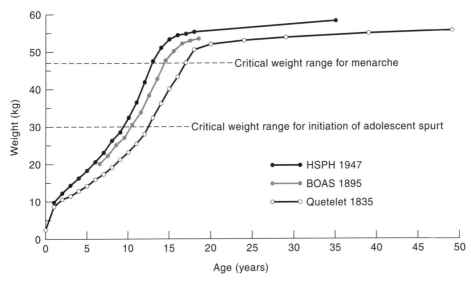

Figure 8.13
Weight at menarche remains constant. Body weight as a function of age in three populations of girls in 1835 (Belgium), 1895 (America), and 1947 (America). Note the constant weights at the beginning of the growth spurt (30kg) and at menarche (47kg). (Adapted from R. E. Frisch. 1972. *Pediatrics* 50, 445–450.)

tions which have experienced an earlier menarche has remained remarkably constant, at about 47 kg (106 lb.) (Fig. 8.13). The weight at which the onset of the growth spurt occurs has also remained quite constant. The decrease in the age of menarche is then a reflection of the improved nutritional status, so that individuals in the industrialized countries reach the necessary body weight at an earlier age. In males, the corresponding weight is about 55 kg. These correlations have led to the "critical weight" hypothesis for the onset of puberty and continued reproductive function, according to which the necessary signal for initiating puberty depends on a critical metabolic parameter in some way related to weight.

It seems reasonable to suppose that puberty would not occur unless the body (and here we mean the female body) has reached a size sufficient for the demands that pregnancy would place upon it. Malnutrition, for example, is generally associated with delayed menarche, while moderately obese girls tend to have an earlier menarche than thin girls. Disturbance of menarche or menstrual function is typically common in individuals with eating disorders, such as anorexia nervosa. Typically, puberty is delayed in prepubertal anorectic females, but in a refeeding program the onset of puberty is correlated with reaching the critical weight of around 47 kg.

The critical-weight hypothesis in its original form does not account properly for all aspects in the triggering of puberty. A significant addition to the hypothesis has been to include the percentage of body fat as another important parameter. A number of

[handwritten margin notes: "∴ puberty occurs when body size sufficient for the demands of pregnancy"]

[handwritten margin notes: "% body fat → important parameter that triggers puberty"]

fat body ratio < 20%
inhibition of GnRH

studies suggest that the critical body fat ratio, that is, the percentage of body weight due to fat, is about 22 percent. At values greater than 22 percent no inhibition of the GnRH pulse generator takes place, while at values less than 20 percent, inhibition of the pulse generator begins to take place. Weight and fat content probably reflect some global aspect of metabolic and nutritional status that serves as an important signal for the initiation of puberty.

strenuous physical activity → delay puberty (but not clear)

Strenuous physical activity may also play a role in delaying puberty. In rodents, puberty can be experimentally delayed by physical activity, and the suppression of the H-P-G axis has been shown to be exerted at the level of the hypothalamus, since pulsatile infusion of GnRH restores normal puberty. In humans, there is considerable evidence that delay of puberty is correlated with strenuous exercise, particularly in females, such as runners, ballet dancers, and athletes. Nevertheless, the correlation is not yet considered definitive (see further discussion in Chapter 9). Separating the effects of physical activity and nutrition is difficult, since generally food intake tends to be more restricted in individuals who engage in strenuous physical activity.

A possible metabolic signal

The critical-weight and/or critical-fat hypothesis is an attractive one, but it is not yet clear how weight and/or fat content, or any other measure of metabolic readiness, is converted into a neuroendocrine signal that can initiate GnRH pulse generator activity. Some recent studies suggest that a polypeptide hormone, leptin, may be part of the regulatory circuitry that monitors nutritional and metabolic status. Leptin is produced by adipose tissue. Studies in mice have demonstrated that leptin is involved in the control of body weight and energy expenditure. It appears to have a role in reproduction as well. Mutations in the *Ob* mouse gene that result in leptin deficiency lead to increased food intake, obesity, infertility, hypothermia and cold tolerance, and other endocrine and metabolic abnormalities, all of which can be reversed by leptin administration. Leptin, moreover, appears to act at the level of the hypothalamus. In addition, leptin treatment of prepubertal female mice can accelerate the onset of puberty. These findings suggest that leptin might be the long-sought signal that transmits information about metabolic status to the hypothalamus.

leptin → hypothalamus → accelerate the onset of puberty
long-sought signal transmit info about metabolic → hypothalamus

The roles of leptin in humans, both with respect to food intake and weight control and in initiating puberty, have not yet been worked out. Several studies published recently have measured leptin levels before and during puberty. Prepubertal leptin levels in boys and girls appear to be similar. They rise abruptly and reach a peak during Tanner stage 2. In boys, however, leptin levels decline to prepubertal levels afterwards. In girls, leptin levels remain constant through midpuberty, and then increase through late puberty. These studies, although suggestive, cannot be considered definitive.

Until very recently no mutation equivalent to that in *Ob* mice had been found in humans. Obesity in humans has not been found to be associated with leptin deficiency. In fact, leptin levels generally increase as weight increases, in contrast to what is seen in mice. Nevertheless, in a recent survey in a Turkish population, three members of a large family were found to carry a mutation in their leptin gene. All three individuals (2 females and 1 male) were obese. The male (34 years old) was hypogonadic (underdeveloped gonads) and had never entered puberty. One of the females was 6 years old, while the other (22 years) had never gone through menarche. The phenotypes of the two adults in this family suggest that leptin may be a *necessary* signal for the initiation of puberty in humans. However, since leptin appears to affect multiple neuroendocrine circuits, it is still too early to know if it is the primary signal.

Summary

Puberty represents a particularly dramatic stage in the developmental history of the GnRH pulse generator. Its activity begins by the 12th week of gestation, reaches a peak about halfway through gestation, and then declines significantly just prior to birth. After a brief period of activation for a few months after birth, GnRH pulse generator activity enters a quiescent period that extends all through infancy and childhood. The onset of puberty is defined by the reactivation of the GnRH pulse generator; its reactivation is sleep entrained (release of GnRH during sleep). The postpubertal pattern of daily GnRH secretion is established after one to two years.

There is broad agreement that none of the three elements of the H-P-G axis—the gonads, the pituitary, or the GnRH pulse generator—is the limiting factor for the onset of puberty. Although increases in GnRH pulses are both necessary and sufficient in initiating puberty, the timing of puberty depends on neuroendocrine elements distinct from those of the H-P-G axis. It has generally been considered that during infancy and childhood the GnRH pulse generator is under an inhibitory neuroendocrine restraint whose nature is not fully understood. The onset of puberty may come about by a relaxing of this restraint, or alternatively from stimulation by excitatory neuroendocrine signals that overcome the restraint. The nature of these inhibitory or excitatory signals remains unknown. There is ample evidence that the metabolic/nutritional status reflected by weight and fat content of the individual is critical in determining when puberty begins. However, it is not yet clear which parameters are the critical measure of the appropriate metabolic/nutritional status, or how this putative status is converted to the appropriate neuroendocrine signal or signals that eventually impinge on the GnRH pulse generator. Recent studies suggest that the hormone leptin, involved in energy and weight balance, is part of a neuroendocrine pathway that regulates the onset of puberty.

QUESTIONS

1. What evidence indicates that pulsatile GnRH release is both necessary and sufficient to initiate puberty? What is the role of the gonad in triggering puberty?

2. What might be the nature of the "biological clock" that determines the onset of puberty?

3. How might you account for the large variation in the age at which puberty begins?

4. What evidence indicates that metabolic parameters are the crucial determinants in controlling the onset of puberty? Where does leptin enter into this equation?

5. Library project: GnRH pulse generator activation is acutely sensitive to sleep during the early pubertal period. What might be the significance of this fact?

SUPPLEMENTARY READING

Apter, D. 1997. Development of the hypothalamic-pituitary-ovarian axis. *Annals of the New York Academy of Sciences* **816**, 9–21.

Grumbach, M. M., and D. M. Styne. 1992. Puberty: Ontogeny, neuroendocrinology, physiology, and disorders. In *Williams' Textbook of Endocrinology*, 8th ed. J. D. Wilson and D. W. Foster, Eds. W. B. Saunders Co., Philadelphia, PA.

Hall, J. G., *et al.* 1989. *Handbook of Normal Physical Measurements*. Oxford University Press, Oxford, UK.

Hofmann, A. D., and D. E. Greydanus. 1997. *Adolescent Medicine*, 3rd ed. Appleton and Lange, Stamford, CT.

Tanner, J. M. 1962. *Growth at Adolescence*. Blackwell Science, Oxford, UK.

Tanner, J. M. 1973. Growing up. *Scientific American* **229**(3), 34–43.

Veldhuis, J. D. 1996. Neuroendocrine mechanisms mediating awakening of the human gonadotropic axis in puberty. *Pediatric Nephrology* **10**, 304–317.

ADVANCED TOPICS

Apter, D., *et al.* 1993. GnRH pulse generator activity during pubertal transition in girls. *Journal of Clinical and Endocrinological Metabolism* **76**, 940–949.

Boyer, R. M., *et al.* 1974. Human puberty: Simultaneous augmented secretion of LH and testosterone during sleep. *Journal of Clinical Investigation* **54**, 609–618.

Frisch, R. E. 1972. Weight at menarche: Similarity for well-nourished and undernourished girls at differing ages and evidence for historical constancy. *Pediatrics* **50**, 445–450.

Griffin, J. E., and J. D. Wilson. 1980. The testis. In *Metabolic Control and Disease*, 8th ed. P. K. Bondy and L. E. Rosenberg, Eds. W. B. Saunders Co., Philadelphia, PA.

Kiess, W., W. F. Blum, and M. I. Auber. 1998. Leptin, puberty and reproductive functions: Lessons from animal studies and observations in humans. *European Journal of Endocrinology* **138**, 26–29.

Marshall, W. A., and J. M. Tanner. 1969. Variations in pattern of pubertal changes in girls. *Archives of Diseases in Childhood* **44**, 291–303.

Marshall, W. A., and J. M. Tanner. 1970. Variations in pattern of pubertal changes in boys. *Archives of Diseases in Childhood* **45**, 13–23.

Reynolds, E. L., and J. V. Wines. 1948. Individual differences in physical changes associated with adolescence in girls. *American Journal of Diseases of Children* **75**, 329–340.

Reynolds, E. L., and J. V. Wines. 1951. Physical changes associated with adolescence in boys. *American Journal of Diseases of Children* **82**, 529–542.

Strobel, A., T. Issad, L. Camoin, M. Ozata, and A. D. Strosberg. 1998. A leptin missense mutation associated with hypogonadism and morbid obesity. *Nature Genetics* **18**, 213–215.

van Wieringen, J. C., *et al.* 1971. *Growth Diagrams: 1965 Netherlands. Second National Survey on 0 to 24-Year-Olds.* Netherlands Institute for Preventive Medicine TNO Leiden, Wolters-Noordhoff, the Netherlands.

Winter, J. S. D., I. A. Hayes, F. I. Reyes, and C. Faiman. 1976. Pituitary-gonadal relations in infancy. 2. Patterns of serum gonadal steroid concentrations in man from birth to two years of age. *Journal of Clinical Endocrinology and Metabolism* **42**, 679–686.

Paul Klee, *Das Fräulein vom Sport* (*Sportive Young Woman*), 1938.
(© 2001 Artists Rights Society [ARS], New York/VG Bild-Kunst, Bonn.)

CHAPTER 9

Stress and Reproduction

*Mr Duke's daughter in S. Mary Axe, in the year 1684 and the
eighteenth year of her age, in the month of July fell into a total
suppression of her monthly courses from a multitude of cares and
passions of her mind but without any symptom of the green-sickness
following upon it . . . I do not remember that I did ever in all my practice
see one, that was conversant with the living so much wasted with the
greatest degree of a consumption, (like a skeleton clad only in skin)
yet there was no fever . . . only her appetite was diminished . . .
she was after three months taken with a fainting-fitt, and died.*

—R. Morton. 1696

Opera medica liber primus pathisologiae, donatum donati amstelodami.
Cited in J. C. Nehimiah. 1950. Anorexia nervosa, *Medicine* 29, p. 225.

D r. Morton's vivid description in 1696 of a condition that we now recognize as anorexia nervosa, observed most commonly in female adolescents and characterized by self-starvation and amenorrhea, is perhaps one of the earliest reports of the effects of nutritional deprivation on reproductive function. We saw in the previous chapter the importance of nutrition on the onset of puberty. Reduced food intake, however, is only one of a number of conditions that are known to perturb or alter reproductive function. A contemporary way of expressing this understanding is to say that the activity of the H-P-G axis is compromised by a variety of factors or conditions that can be referred to as **reproductive stressors** (Table 9.1). Stressors induce a *stress*

Table 9.1
Reproductive Stressors

Physical
    Strenuous exercise; competitive sports activity
Metabolic
    Loss of weight; altered body fat ratio; nutritional deficiency; eating disorders
Psychogenic
    Emotional distress or trauma

Table 9.2
Behavioral and Physiological Changes Associated with Stress

Behavioral
  Increased arousal and alertness
  Increased cognition, vigilance, and focused attention
  Suppression of feeding behavior
  Suppression of reproductive behavior

Physiological
  Oxygen and nutrients directed to the CNS
  Increased blood pressure and heart rate
  Increased respiratory rate
  Increased CRH and cortisol levels
  Suppression of reproduction

Adapted from Chroussos, 1998.

reaction, which in turn leads to marked changes in a number of behavioral and physiological parameters (Table 9.2). Response to stress is multifaceted, and the specific changes that take place depend on the stressor, its strength, and its duration. The concept of stress derives from the ancient Greek idea of health being a balance between harmonious and disharmonious forces. The Greek term *homeostasis*, meaning "steady-state," was coined in the early 1900s to describe this state of harmony or balance. Stressors are forces or conditions that disturb homeostasis, and they are countered in the body by activation of the stress response system, which tends to restore homeostasis. Our objective in this chapter is to review briefly the nature of the stress response and to examine the effects of reproductive stressors on the H-P-G axis.

THE STRESS RESPONSE SYSTEM

Our understanding of the concept of stress and the way the body responds to it rests conceptually on the work of Walter Cannon in the early 1900s and Hans Selye in the 1930s. Cannon described the role of the sympathetic nervous system, adrenal hormones, and emotional arousal in the often-cited "flight or fight" response. Selye developed and broadened the stress concept and provided the first descriptions of stress-response-system disorders. The current view is that the stress response system depends on the coordinated activity of different neurosensory elements of the peripheral and central nervous systems. The stress response system receives and integrates diverse neurosensory stimuli (visual, auditory, olfactory, gustatory, visceral, higher cortical), and hormonal signals that impinge on its different elements. The utility of the stress concept is that it has provided a comprehensive and unified way of understanding the multifaceted responses of the body to a large variety of physical, chemical, and emotional stimuli.

A central component of the stress response system is the corticotropin-releasing hormone—H-P-A (hypothalamic-pituitary-adrenal)—system (Fig. 9.1). CRH is referred to as the "stress hormone" because its release appears to be not only necessary, but sufficient as well, for the stress reaction. The sufficiency condition has been clearly

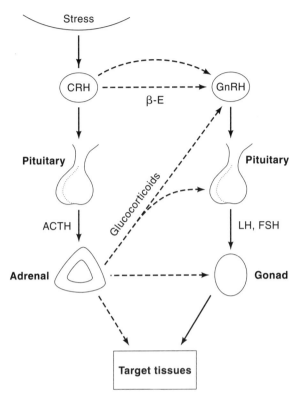

Figure 9.1
Stress inhibits the H-P-G axis. Stress in its various forms results in CRH release, and CRH in turn activates the H-P-A axis and suppresses GnRH release directly or indirectly via the β-endorphin system. Glucocorticoids, the products of H-P-A axis activation, may also contribute to suppression of reproductive function by working at the gonad, pituitary, or possibly hypothalamic levels. (Adapted from E. O. Johnson, T. C. Kamilaris, G. P. Chroussos, and P. W. Gold. 1992. *Neurosciences and Biobehavioral Reviews* **16**, 115–130.)

demonstrated by showing that the stress responses listed in Table 9.2 can be elicited in animals merely by intracerebral injection of CRH. In humans, CRH release is associated with a number of stress responses, including inhibition of reproductive function (Table 9.3). Stress-induced inhibition of reproductive function, induced by a rise in CRH levels, is characterized by low gonadotropin levels with reduced LH pulse frequency and by activation of the H-P-A axis, which is monitored by an increase in the adrenal hormone cortisol. These observations have raised the question of whether stress-induced suppression of the H-P-G axis is due to increased production of CRH directly, or to other components of the H-P-A axis, for example, increased cortisol levels. The current view is that H-P-G axis inhibition is primarily due to inhibition of the GnRH pulse generator directly by CRH, or indirectly, through the activation of the β-endorphin system (Fig. 9.1). In the Rhesus monkey, for example, β-endorphin-mediated CRH suppression of the GnRH pulse generator appears to be the major effect. Nevertheless, adrenal glucocorticoids working at the level of the hypothalamus or pituitary cannot be excluded and may be important in different circumstances.

Table 9.3
Actions of CRH in Stress

Activation of the H-P-A axis
Increase in oxygen consumption, cardiac output, blood pressure, and blood glucose
Suppression of GnRH release
Stimulation of respiration
Altered immune/inflammatory responses
Increased arousal
Anorexia
Decreased sexual activity

Reference: Chroussos, 1998.

Females appear to be much more sensitive to stress than males. This sex difference is generally attributed to the more complex and demanding role that the female plays in reproduction. Under certain conditions inhibition of reproduction can be considered to be beneficial for both the individual and the species. Particularly for the female, suppression of reproductive activity will protect her from the physical and psychological demands of pregnancy under nonoptimal conditions, and may serve as a natural form of population control. Viewed from this perspective, stress-induced alterations in reproductive functions can be beneficial.

It is convenient for our discussion to divide reproductive stressors into the categories listed in Table 9.1. Strenuous physical activity (**physical stress**) and restriction of food intake (**metabolic stress**), for example, have long been associated with menstrual dysfunction. Similarly, a correlation between menstrual dysfunction and emotional upheavals (**psychogenic stress**) has also been noted for many years. In the female, these stress conditions alone or in combination produce a continuum of disturbances in ovarian function (Table 9.4), the most severe disorders being complete or partial suppression of menstruation, amenorrhea or oligomenorrhea, respectively. In the male, disturbances of the H-P-T axis are not as obvious, but may include lowering of testosterone levels, loss of libido, and lowered sperm count (Table 9.4). More recently, some attention has focused on the effects of recreational drugs, such as alcohol, nicotine, and

Table 9.4
Stress-Related Reproductive Disorders

Females	Males
Follicular suppression	Decreased sperm count
Luteal suppression	Decreased sperm motility
Anovulation	Altered sperm morphology
Oligomenorrhea	Reduced testosterone levels
Amenorrhea	Loss of libido; erectile dysfunction

In terms of reproductive function, females are more sensitive to stress than males. Severe, chronic stress can lead to amenorrhea. Mild stress will often lead to luteal suppression. In males, the disorders listed are seen less frequently. (Reference: Chroussos, 1998.)

cocaine, on reproductive activity. We will consider the effects of drugs, both therapeutic and recreational, in Chapter 16.

The response to stress depends on many factors. Response to an *acute* versus *chronic* stress can differ considerably. Suppression of the H-P-G axis, for example, in response to an acute stressor, such as a very strict dietary program, will normally be transient, and normal reproductive activity will be restored after the stressor is removed. Chronic stress, on the other hand, places special demands on the stress response system. The continued activation of the stress response system may itself become the major stressor. Under such conditions, disturbance or impairment of the H-P-G axis may no longer be transient. The most important negative consequences of the inhibition of the H-P-G axis, particularly in females, are the loss of bone mineralization and abnormalities in coronary vasculature. Both of these effects are a consequence of reduced estrogen levels due to the suppression of ovarian function. Significant changes in the lifestyles of females in the industrialized countries, such as increased physical activity and dietary restriction, may have unintended negative consequences, despite the otherwise beneficial effects of physical activity and reduced food intake. The age at which the stress is experienced is beginning to be recognized as being extremely important with respect to the long-term consequences of stress, especially in females. In particular, severe or prolonged stress during the maturational phase of the H-P-G axis, roughly the 10 years following menarche, may result in ovarian or menstrual dysfunction that will plague the individual for many years.

Physical, metabolic, and psychogenic stress are not exclusive categories; there is often considerable overlap between them. For example, rigorous exercise is often associated with an alteration in dietary intake. Similarly, eating disorders may have a strong psychological component. We will see below that there are deep connections between these different types of stress, and that they probably all share the same underlying neuroendocrine mechanisms.

SUMMARY

The stress response system is the mechanism by which the body responds to physical, chemical, or emotional stimuli that affect homeostasis. One essential feature of the stress reaction is the release of the hypothalamic hormone CRH. Most, if not all, of the diverse consequences of stress can be attributed directly or indirectly to the effects of CRH. Strenuous physical exercise, persistent reduced food intake, and emotional upheavals have long been known to lead to a suppression of reproductive activity. It is now clear that the suppressive effects of all three conditions are a consequence of the activation of the stress response system. All three lead to the release of CRH. CRH in turn inhibits the GnRH pulse generator; in some cases its effects are mediated by the β-endorphin system, while in other cases, CRH may suppress GnRH release directly. Inhibition of GnRH release eventually results in suppression of gonadal function. Hence, reproductive stressors exert their effects by acting *centrally*, at the level of the hypothalamus.

PHYSICAL STRESS: EXERCISE AND REPRODUCTION

In the female

The relationship between strenuous physical activity and reproduction in the female has a long and complex history. The results of many studies, some dating back to the 1930s, have suggested a causal connection between strenuous physical activity and reproductive dysfunction. The image of the very lean female athlete (generally a runner) with reproductive problems, perhaps suffering from the so-called "runner's amenorrhea," has been a vivid one in the popular imagination. The causative relationships between physical activity and reproductive dysfunction have been very difficult to sort out. Training programs differ considerably in their intensity, and individual responses to the demands of training programs are quite different. It has also been difficult to pinpoint the critical physiological and endocrinological variables associated with reproductive dysfunction. Recent studies suggest that the nutritional and metabolic status of the individual is most likely the most important variable in predicting the effects of exercise.

The prepubertal female

Many studies have reported a delay in menarche when athletic training is begun before the onset of puberty. A summary of many of these studies is shown in Fig. 9.2. In general, high-intensity training programs result in a delay of about three years in menarche. Although the delay in menarche increases with the intensity of the athletic activity, ballet dancers and gymnasts consistently experience menarche later than

Figure 9.2
The effect of athletic activity on the age of menarche. Mean age at menarche of high school, college, olympic athletes, and ballet dancers compared to nonathletes (controls). Note the differences in the menarcheal ages in athletes trained before and after menarche. Ballet dancers consistently exhibit a longer delay in menarche than other athletes. (Reprinted from D. C. Cumming. 1990. *Seminars in Reproductive Endocrinology* 8[1], 15–24, with permission.)

other types of athletes, even those in high competition sports. These studies have generally been interpreted as indicating a causative relationship between athletic activity during late childhood and delay in menarche. Such a conclusion, although well-entrenched in popular stereotypes, is not strictly warranted, however. These studies have all relied on *retrospective* comparisons of the age of menarche in athletes and nonathletes. Such studies are statistically biased because most of the athletes began their training before menarche. Hence, the result itself cannot be used to demonstrate a causal connection between athletic activity and the age of menarche. What are needed are *prospective* studies in which individuals are followed in a randomized manner over a period of several years. At present no such studies have been carried out.

Some investigators have argued that the correlation between the delay of menarche and athletic activity may not be physiological at all. For example, since the long bones of the arms and legs continue to grow until the onset of menarche, girls who mature later may be encouraged to stay in athletics because they will perform better. Indeed, characteristics such as increased height, longer legs, narrower hips, and better neuromuscular coordination, all obvious assets in athletic activity, tend to occur more frequently in females with delayed puberty. Girls who mature earlier may not be as competitive in high-intensity athletic activities as those who mature later. The observed correlation between athletic activity and the delay in menarche may be a consequence of this difference.

The above theory not withstanding, some aspect of physical activity in the premenarcheally trained athletes does seem to be important. For example, the delay in menarche is proportional to the time in, and intensity of, training. Moreover, prompt onset of menarche occurs in ballet dancers who become inactive as the result of an injury or accident. These observations do indicate that strenuous physical activity in childhood does play a role in delaying menarche.

The postpubertal female

In the postpubertal female, many reports have noted a correlation between athletic activity and menstrual irregularities, with the incidence of irregularities increasing with the intensity or the type of training program. In the 1976 Montreal Olympics, for example, 59 percent of the female athletes reported some menstrual irregularity. Although inconsistencies and differences in reported outcomes in different studies have made it difficult to get a clear picture of the spectrum of dysfunction, the irregularities have been grouped into two reasonably well-characterized categories—*athletic amenorrhea* and *suppression of luteal function*. Let us consider athletic amenorrhea first.

Amenorrheic athletes typically display alterations in gonadotropin levels and/or pulsatility. A characteristic feature of amenorrheic athletes is very low LH and FSH levels and, as a consequence, low estradiol levels. In many cases, amenorrheic athletes may exhibit an irregular pattern of LH pulsatility, even while maintaining essentially normal mean LH levels. Two such examples in Fig. 9.3 show the LH patterns over a 24-hour period in a cyclic, sedentary female; a cyclic athlete; and two amenorrheic athletes. The LH profiles in the amenorrheic athletes are examples of two quite different, but typical, patterns. In the first case, the LH profile is quiescent during the daytime, and some LH pulsatility is observed during the sleep period. This pattern is, as we saw in Chapter 8, typical during the onset of puberty. In the other case, LH pulsatility is irregular during the day and absent during the night. In both cases, no follicular development takes place despite the fact that the mean LH levels may be similar to those in the cycling females.

220

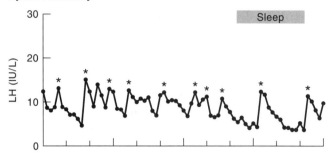

Cyclic sedentary

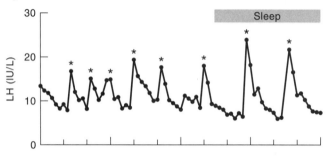

Cyclic athlete

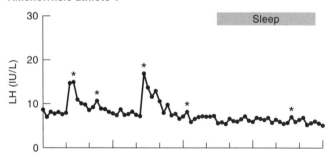

Amenorrheic athlete 1

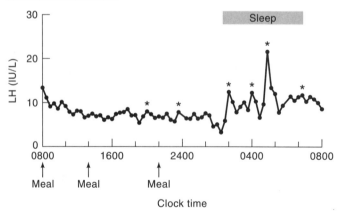

Amenorrheic athlete 2

Clock time

Figure 9.3 *(facing page)*
Effect of athletic activity on LH pulsatility. LH profiles during early follicular phase in a cyclic sedentary female, one cyclic athlete, and two noncyclic athletes. The asterisks indicate pulses identified by a pulse analysis program. Note the different sleep profiles in each. (Adapted from A. B. Loucks. 1990. Effects of exercise training on the menstrual cycle: Existence and mechanisms. *Medicine and Science in Sports and Medicine* 22[3], 275–280.)

What accounts for the low gonadotropin levels or altered LH pulsatility? The two possibilities are a defect either in the pituitary or in the hypothalamic function. The two can be distinguished by examining the effects of administering exogenous GnRH. If the pituitary is the problem, it will fail to respond to the exogenous GnRH. On the other hand, if exogenous GnRH stimulates the pituitary to secrete LH, then the defect must be in the hypothalamus. Experiments of this type have shown that pulsatile administration of GnRH to amenorrheic athletes results in pulsatile secretion of LH. This result shows that the abnormal LH pattern seen in amenorrheic athletes is not due to pituitary insensitivity to GnRH stimulation, but to insufficient endogenous GnRH.

The LH profiles shown in Fig. 9.3 also point out another important fact: *not all athletes are amenorrheic.* Nonetheless, subclinical manifestations of reproductive dysfunction may be present even if the individual is menstruating normally. Many regularly menstruating athletes are characterized by a condition generally referred to as *luteal suppression,* or corpus luteum dysfunction, resulting in reduced progesterone secretion and/or a shortened luteal phase. An illustration of this effect is shown in Fig. 9.4, which compares the monthly patterns of an estrogen and progesterone metabolite in the urine of cyclic sedentary (CS), cyclic athletic (CA), and amenorrheic athletic (AA) females. The extremely low levels of these metabolites in the AA group reflect the absence of any follicular development in the AA females. On the other hand, the reduced and abbreviated excretion of the progesterone metabolite indicates suppressed follicular and luteal development in AA compared to CA females. The CA females in this study were cycling normally, and they regarded their cycles as being perfectly regular. Such observations demonstrate that significant perturbations of the H-P-G axis can occur in a rigorous training program without necessarily leading to a complete suppression of ovarian function. Indeed, reduced progesterone secretion in the luteal phase has even been found in many recreational runners, showing that enhanced physical activity can produce minor, but detectable disturbances in ovarian function.

The incidence of luteal-phase disturbances and even anovulation can increase quite dramatically even in a short-term exercise program (Table 9.5). In this study, 28 young women with normal ovulatory cycles entered an exercise program (running from 4 to 10 miles/day) for a period of 4 weeks. The incidence of luteal-phase defects and anovulation (measured as loss of LH surge) in the 28 women was compared with those of the general population. Luteal-phase dysfunction increased from 15.8 percent to 33 percent in women who experienced no weight loss, and 63 percent in women who did experience weight loss. The incidence of anovulation increased from 10.6 percent to 42 percent and 81 percent, with no weight loss and weight loss, respectively.

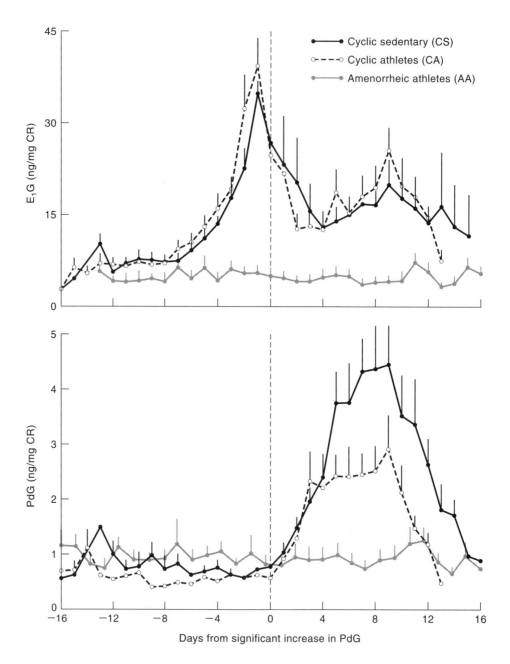

Figure 9.4
Steroid hormone patterns during the cycle in athletes. Patterns of estrogen (E₁G) and progesterone (PdG) metabolites in cyclic sedentary (CS), cyclic athletes (CA), and amenorrheic athletes (AA). Note the extremely low estrogen and progesterone levels in the AA group. There is no difference in the estrogen profile between the CS and CA groups. In contrast, the CA group has significantly lower progesterone levels during the luteal phase. (Adapted from A. B. Loucks. 1990. Effects of exercise training on the menstrual cycle: Existence and mechanisms. *Medicine and Science in Sports and Medicine* **22**[3], 275–280.)

Table 9.5
Effect of Short-Term Exercise on Ovarian Function

		Exercise	
Effect	General population	No weight loss	Weight loss
Luteal phase defects	15.8%	33%	63%
Anovulation	10.6%	42%	81%

Twenty-eight healthy young women (ages 20–30 years) participated in a study of progressively increasing exercise over a 4-week period. Some of the subjects experienced weight loss and others did not. The incidence of luteal-phase dysfunction and anovulation (measured by the absence of an LH surge) in both types of subjects was compared to that of the general population. (Adapted from Bullen *et al.*, 1985.)

Such observations demonstrate that significant perturbations of the H-P-G axis can take place even in short-term training programs without necessarily leading to a complete suppression of menstrual activity.

Luteal suppression appears to be an early response to the demands of increased physical activity. Why should the consequences of increased physical activity be manifested first in the luteal phase? A reasonable possibility is that luteal suppression represents a response of the body to an increased energy demand. The luteal phase of the ovulatory cycle requires about 8 to 10 percent more energy than the follicular phase. An increased demand for energy, as during an exercise program, will first affect the luteal phase. Hence, suppression of luteal progesterone levels might be the first of a series of energy-saving measures taken by the body in response to physical exertion. Luteal suppression may progress to full amenorrhea if the individual continues or intensifies the training program, and indeed, this is often the case. However, there are also athletes who have maintained the same menstrual condition for several years with no progression to amenorrhea. Hence, luteal suppression in athletes can be an *intermediate* or *end-point* in the body's adjustment to athletic training. Luteal suppression may be the end-point of a response to strenuous training in some, but not all individuals. In others, athletic training produces luteal suppression that progresses to amenorrhea.

Factors that affect the response to physical stress

What makes one individual more resistant to the demands of physical training than another? The critical-weight hypothesis discussed in the previous chapter proposes that crucial levels of body weight or fat content are required for normal reproductive function. Hence, one possibility is that differences in *somatotype*, or the difference in weight and proportion of body fat influences the response to physical activity. Runners and ballet dancers, for example, typically have a higher incidence of amenorrhea than do swimmers (Fig. 9.5). Runners are typically much leaner and lighter than swimmers. The proportion of body fat in high competition runners is about 15 percent, significantly lower than that of swimmers, which is about 21 percent. The critical-weight or critical-fat-ratio hypothesis can account for many cases of menstrual dysfunction, but it cannot account for all. For example, in athletes who cease their training programs after injury, menstrual function normalizes often before the critical

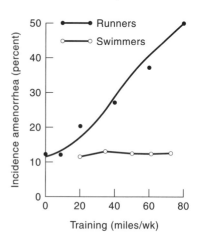

Figure 9.5
Amenorrhea in runners and swimmers. The incidence of amenorrhea increases significantly with the intensity of training in runners. In swimmers the incidence of amenorrhea appears to be independent of the intensity of training. (Adapted from C. Sanborn *et al.* 1982. *American Journal of Obstetrics and Gynecology* **143**[8], 859–861.)

weight or critical fat ratio is restored. This must mean that other parameters of metabolic status must play a role in exercise-related menstrual dysfunction.

Recent studies suggest that alterations in leptin release may be important. Recall from Chapter 8 that leptin appears to be necessary for triggering the onset of puberty. Leptin levels are significantly lower in athletes than in nonathletes, but the low mean leptin levels are not the critical factor. A recent comparison of leptin patterns in cycling and noncycling athletes indicates that a disturbance of the normal diurnal pattern of leptin release is associated with amenorrhea (Fig. 9.6). In this study, the mean leptin levels in both cycling (CA) and amenorrheic athletes (AA) were the same, but the normal diurnal leptin pattern has been lost in the amenorrheic athletes. The change in the diurnal leptin pattern must reflect a change in other metabolic parameters, not yet clearly defined, that distinguish the CA from the AA individuals.

Other studies suggest that severe disruptions in ovarian function occur in individuals who are predisposed to H-P-O axis dysfunction. Predisposition to menstrual dysfunction is not easy to measure precisely, and may involve a number of factors, including genetic ones, as well as the previous history of exposure to stress. For example, most athletes who experience menstrual dysfunction are younger than those who do not, are less likely to have had a previous pregnancy, and are more likely to have started serious athletic training at a younger age. The maturing H-P-O axis, encompassing roughly the 10 to 12 years after menarche, may be particularly sensitive to stress. Young females who are subject to high levels of physical, metabolic, and psychogenic stress during this critical period may develop a sensitivity to stress that later manifests itself as a predisposition to menstrual dysfunction. Those who experience stress at a more mature gynecological age suffer less dysfunction. Disturbances during the sensitive maturation process may well be partially irreversible, and H-P-G axis impairment and menstrual irregularity may be a long term result. The origin of the "predisposition" to menstrual dysfunction may lie in some cases in experiences of the young female during the maturational phase of the H-P-O axis. An important goal in any exercise or training program should take into account individual characteristics in order to minimize any long-term negative consequences on the H-P-O axis.

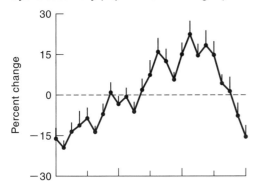

Cyclic sedentary (Leptin = 10.1 ± 1.3 ng/ml)

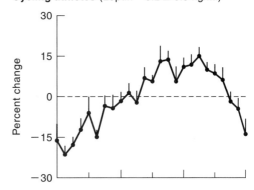

Cycling athletes (Leptin = 3.2 ± 0.5 ng/ml)

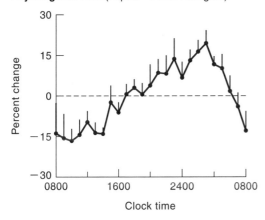

Psychogenic FHA (Leptin = 7.0 ± 1.5 ng/ml)

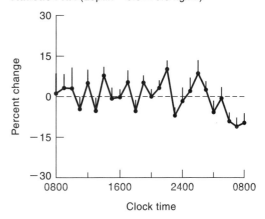

Athletic FHA (Leptin = 3.5 ± 0.5 ng/ml)

Figure 9.6

Leptin diurnal patterns in four groups of individuals. Both cycling and noncycling athletes have the same mean leptin levels. However, the amenorrheic athletes (athletic FHA) have lost the diurnal pattern of leptin secretion. Individuals suffering from psychogenic amenorrhea have intermediate mean leptin levels, but no disruption in the diurnal pattern. This finding suggests that there are neuroendocrine differences between exercise-induced and psychogenic amenorrheas. (Adapted from S. S. C. Yen. 1999. Chronic anovulation due to CNS-hypothalamic-pituitary dysfunction. In *Reproductive Endocrinology: Physiology, Pathophysiology, and Clinical Management*, 4th ed. S. S. C. Yen, R. B. Jaffe, and R. L. Barbieri, Eds. W. B. Saunders Co., Philadelphia, PA.)

In the male

Males in general appear to be more resistant to the demands of strenuous physical activity than females. There are no comparable data for boys that indicate any significant influence of exercise and strenuous training on the time of male puberty. It may be that because of their greater weight and lean muscle mass young boys are better prepared for the metabolic demands of puberty. Certainly, there is no evidence that conventional sports activities typically engaged in by adolescent males have any effect on the timing of puberty.

In the postpubertal male, chronic strenuous exercise, such as high mileage running or weight lifting, has been associated with some impairment of testicular function in some studies. Reduction in testosterone levels has been observed in runners, particularly those following a highly demanding training program (at least 50 miles per week), and in wrestlers and gymnasts as well. Changes in spermatogenesis, although less clearly defined, also appear to be found in some males with a high level of physical activity. In some studies, high physical activity donors to artificial insemination programs have been found to have lower pregnancy rates than donors with a normal activity profile. On the other hand, in other studies the sperm count in high mileage runners was found to be in the normal range, despite lowered testosterone levels. Lowered libido has been reported in some runners, but the importance of chronic fatigue has not been evaluated properly.

There is no convincing evidence for significant alterations in LH profiles in male athletes, suggesting that in males, inhibition of the GnRH pulse generator as a consequence of the multifaceted demands of strenuous training is much less likely to occur than in females. Despite a large number of studies involving many athletes, only two runners have been described with features that approach the male equivalent of "runner's amenorrhea." Elevation of cortisol levels, indicating an activation of the HPA axis, has been reported in instances of overtraining. These increases appear to have little effect on LH secretion, but there is some evidence that the higher cortisol levels may inhibit testosterone production by acting directly at the level of the testis.

SUMMARY

Physical activity in the prepubertal female can lead to a delay in the onset of puberty, while in the postpubertal female, it may lead to the suppression of ovarian and menstrual function. Reproductive dysfunction is due to inhibition of GnRH release. The adaptation of the individual female to the demands of physical exercise or a strenuous physical training program depends on a number of factors, including age, nutritional status, height and weight, body build, and emotional status. In general, the responses to physical stress occupy a continuum between luteal suppression and athletic amenorrhea. Luteal suppression refers to a disturbance in corpus luteum function manifested by lower levels of progesterone during the luteal phase. The female continues to menstruate more or less normally. Luteal suppression represents one of the first reasonably defined stages in the response of the body to physical activity, but it can also represent the end-point of a successful adaptation of the body to a strenuous physical training program. In other individuals, luteal suppression progresses to a complete inhibition of the H-P-G axis, culminating in amenorrhea. In most cases, the amenorrheic condition is reversible, and regular cycling returns once the intensity of the activity is decreased.

Males are much less sensitive to strenuous physical activity than females. While there have been reports of decreased sperm count or loss of libido, particularly in high-intensity runners and wrestlers, there does not appear to be compelling evidence that physical activity has a pronounced effect on reproductive activity in males.

METABOLIC STRESS: NUTRITION AND REPRODUCTION

A connection between nutritional status and reproductive function, particularly in females, has long been noted. We saw in Chapter 8 the important role that improved nutrition has played in the lowering of the age of menarche in many countries. The male reproductive axis is less sensitive to nutritional deficiency, and its effects are felt only when the malnutrition is quite severe. In the female, however, relatively minor nutritional deficiencies can have significant effects on ovulation. There are three levels at which we might consider that nutritional deficiency could affect reproductive function in females: gamete production, pregnancy, and lactation. Reproduction requires energy. It has been estimated that to produce a viable infant requires 50,000 to 80,000 calories, while lactation requires about 500 to 1000 calories per day, over and above what the caloric demand would be normally. Energy requirements for pregnancy and during lactation are quite similar. In contrast, much less energy is required for egg production. It may, therefore, be surprising at first sight that nutrient limitation in the female is manifested in the suppression of ovulation. A likely explanation is that if the outcome of a pregnancy is not predictable because of nutritional deficiencies, it does not make sense to produce an egg. We can imagine that females who continued to ovulate even if they were undernourished would probably not be able to carry a pregnancy to term, or would not even survive themselves. Therefore, the first response to nutritional limitation will be to suppress egg production.

The energy requirements for a successful pregnancy are so critical that we can surmise that our successful female ancestors must have been those who had the energy resources to sustain a pregnancy. One way to do this is to store fat, since fat is a very useful way to store energy. Evolution must have selected for women who could store fat. The pattern of fat accumulation in females and males is shown in Figure 9.7. During

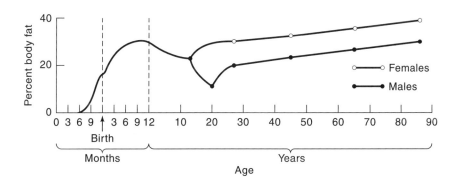

Figure 9.7
Changing fat content from late gestation through adult life in females and males. Note the significant difference in the amount of body fat in females and males at the onset of puberty. (Adapted from B. Friis-Hansen. 1965. Hydrometry of growth and aging. In *Human Body Composition: Approaches and Applications.* Symposia of the Society for the Study of Human Biology, Vol. VII. J. Brozek, Ed. Pergamon Press, New York, NY.)

gestation and early childhood, girls and boys accumulate fat at the same rate. At puberty, however, males experience a significant decrease in fat content, while the percentage of body fat in females increases significantly, and this difference is maintained throughout adulthood. In general, females attain a fat content equivalent to about 144,000 calories when they reach reproductive maturity. This is not necessarily reached at the time of menarche. A large fraction of this caloric reservoir is to be able to support a pregnancy and about four months of lactation. It is interesting to note also that too much fat is as much of a problem as too little fat. Obese women have the same problems with infertility as excessively lean women, although the reasons may be different.

Involuntary dietary restriction

Severe reduction in food intake can occur involuntarily in a number of circumstances, for example, in prisons, concentration camps, refugee camps, during famines, or under conditions where food availability varies with the season. Anthropological studies in tropical subsistence societies where seasonal variation in rainfall is the prime determinant of food abundance have provided valuable information about the relationship between nutrition and the control of ovarian function. Fig. 9.8 shows the results of a study carried out among the !Kung-San people of the Kalahari desert in South Africa. In this sparse environment the search for food is intense all year round, but even more so during the winter months. As a consequence, seasonal changes in nutrition and body weight are the norm. The women have a peak time for births occurring 9

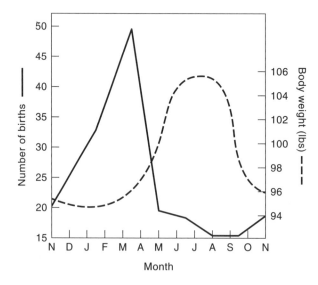

Figure 9.8
Involuntary food intake and fertility. Relationship between the time of births and mean body weight of females among the hunter/gatherer !Kung-San people of the Kalahari Desert. Note that peak weights (106 lb.) are reached during the rainy season in July and August. Births peak around March and April, about nine months after. (Adapted from L. A. Van der Walt, E. N. Wilmsen, and T. Jenkins. 1978. *Journal of Clinical Endocrinology and Metabolism* **46**, 658–664. © The Endocrine Society.)

months after the females reach their maximal weight. During the lean winter months, anovulation, luteal-phase defects, and amenorrhea are typical, while normal ovulatory cycles are restored during the time of maximum food availability. It is interesting to note that the maximal weight, 106 lb., is, as we saw in Chapter 8, essentially the same weight that has been considered to represent the critical weight for menarche to occur. Small increases in weight of 4 to 8 lb. around the critical level can restore ovulation and menstruation. These observations suggest that the concept of a critical threshold with respect to ovarian function may apply in both the prepubertal and postpubertal female. The factors that define the threshold have not been identified fully, but they presumably include weight, percentage of body fat, and other as-yet-unknown determinants of energy status.

Voluntary dietary restriction

Dieting

The availability of food is not usually considered to be an important factor in controlling reproductive function in the industrialized societies. On the other hand, the fashionable desire for leanness and increased emphasis on exercise and physical activity in Western societies have led to a plethora of dieting regimes. Even moderate food restriction and weight loss in normal cycling females has been found to result in reduced steroid hormone levels and suppression of ovulation. Figure 9.9 illustrates the effects of a short period of dieting on LH pulsatility in a healthy woman. In this example, the dieting woman lost 5 kg (11.25 lb.) during a 2.5-week dieting period, and then regained it afterward. The dieting period was characterized by complete inhibition of LH pulsatility due to suppression of the GnRH pulse generator. This example shows clearly that dynamic and reversible relationship between food deprivation and inhibition of the H-P-O axis.

More extreme dieting programs, especially if they are prolonged, may interfere with the normal homeostatic mechanisms that control appetite and weight. Disturbances of these control mechanisms can lead to pathological conditions (eating disorders) that in their consequences span a continuum from the very mild to the very severe. Eating disorders have been classified into three types: **anorexia nervosa** (AN), **bulimia nervosa** (BN), and a third type, a heterogeneous class that is defined by exclusion, i.e., neither AN nor BN. Eating disorders have long been found to be prevalent in athletes. They are more common among amenorrheic athletes than in **eumenorrheic** (menstruating normally) athletes.

Anorexia nervosa

Anorexia nervosa is an extreme form of voluntary nutritional restriction, in reality a complex neuroendocrine "self-starvation" syndrome (Fig. 9.10). Although AN was first defined as a clinical entity about 120 years ago, examination of the historical record indicates that it may date back to the pre-Christian era in the Mediterranean world. Adolescent females are very sensitive to food deprivation, the major effect of which is weight loss and reproductive dysfunction. AN is seen most frequently in adolescent females, occurring with a frequency of 0.5 to 1.0 percent in females from 15 to 30 years old.

Four criteria are used in diagnosis of AN: severe weight loss, amenorrhea, fear of gaining weight, and disturbance in body perception. There is also an unexplained and

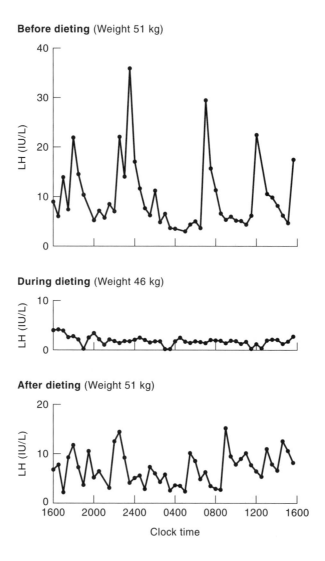

Figure 9.9
Effect of strict dieting on LH pulsatility. LH profile during the luteal phase before, during, and after two and a half weeks of dieting in a healthy young woman. The woman weighed 51 kg before and 46 kg during the dieting program. She quickly regained her original weight after stopping the diet. (Adapted from M. M. Fichter and K. M. Pirke. 1984. Hypothalamic-pituitary function in starving healthy subjects. In *The Psychobiology of Anorexia Nervosa*. K. M. Pirke and D. Ploog, Eds. Springer-Verlag GmbH Ltd.)

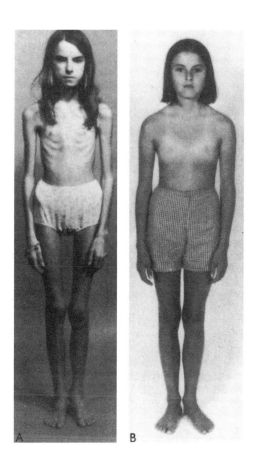

Figure 9.10
A 12-year-old girl with severe anorexia nervosa and marked generalized wasting before
(A) and after (B) recovery following 6 months of psychotherapy. (Courtesy of Dr. Dorothy
Hollingsworth, UCSD School of Medicine.)

pronounced sex difference in the incidence of anorexia nervosa: males account for at
most 8–10 percent of all anorectic patients. Despite much study, the etiology of AN
remains unknown, and it continues to have a poor prognosis, with a mortality rate of
up to 20 percent.

AN has commonly been viewed as being caused by a complex of psychological,
familial, and sociocultural factors, such as the "fear of fatness," or the struggle of a
young female to reconcile the complexities of the different roles demanded of females
in Western societies. However, a number of investigators have indicated that such sociocultural factors cannot account for all the epidemiological and clinical features of
the condition. For example, distorted body image and fear of fatness are not universal
traits, and are not generally seen in cases of AN in many non-Western societies. Moreover, the pronounced sex ratio (10 female to 1 male) in the incidence of AN is not easy
to fit into the sociocultural model. In both males and females, severe and progressive
weight loss is the defining feature of the condition. In the female, severe weight loss results in amenorrhea.

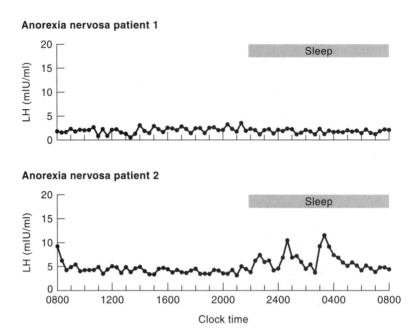

Figure 9.11

Two representative patterns of a 24-hour LH profile in two patients with anorexia nervosa. In one, the disorder leads to complete cessation of GnRH and LH release. In the other, there is some activation during the sleep period. Both patterns represent a regression to the prepubertal state. (Adapted from S. S. C. Yen. 1999. Chronic anovulation due to CNS-hypothalamic-pituitary dysfunction. In *Reproductive Endocrinology: Physiology, Pathophysiology, and Clinical Management*, 4th ed. S. S. C. Yen, R. B. Jaffe, and R. L. Barbieri, Eds., W. B. Saunders Co., Philadelphia, PA.)

The neuroendocrine picture of AN is complex, but it is clear that an essential aspect is marked suppression of the H-P-G axis (Fig. 9.11) and activation of the H-P-A axis. H-P-G axis suppression is due to H-P-A axis activation. Restoration of weight and fat content to the critical levels discussed above restores GnRH pulsatility. The H-P-A axis is activated even under mildly restricted feeding conditions, and a universal manifestation of self-starvation is hypercortisolism (high cortisol levels). CRH levels in anorectics are about 170 percent higher than in normal individuals. There appears to be a direct relationship between high CRH levels and reduced food intake. For example, injection of CRH into the brain of animals causes a marked reduction in feeding. The high levels of glucocorticoids due to activation of the H-P-A axis can cause euphoria and dependence in humans and animals. This has recently led to an interesting hypothesis for the etiology of AN. Initially, self-starvation is internally (unconsciously) rewarding because of the resulting high cortisol levels. CRH levels increase because of reduced food intake, and they, in turn, reduce food intake further. The internal reward increases, and so on. In most individuals, this progression can be interrupted without difficulty. In others, perhaps because of underlying neuroendocrine dysfunctions, this cycle of reduced food intake and reward may be strongly reinforced and self-sustaining, to the point that it becomes pathological, leading to AN.

Bulimia nervosa and other disorders

Bulimia nervosa appears superficially similar to AN. However, in contrast to AN, body weight is at or above normal. The characteristic feature of BN involves episodes of binge eating followed by some kind of purging, such as self-induced vomiting or the frequent use of diuretics, laxatives, or emetics. BN occurs with a frequency of 2–5 percent in females aged 15 to 30. Self-starvation, which is the primary and progressive feature of AN, occurs only over short periods in BN. As a consequence, reproductive dysfunction is less severe than in AN. Although less than 10 percent have normal menses, the incidence of amenorrhea (40 to 50 percent) is much less than in AN. BN, like AN, has tended to be seen as a sociological disorder, a reflection of our concern with being thin. However, recent studies suggest that BN may also be the consequence of an underlying neuroendocrine disorder.

Individuals that cannot be classified as anorectic or bulimic are placed in a third category, and these occur with about the same frequency as bulimics. Such individuals are of average weight, but typically have a pronounced preoccupation with body image and weight; binge eating is rare, but purging is frequent. Ovarian disturbances are less severe, and the incidence of amenorrhea is less than in bulimics.

SUMMARY

The nutritional/metabolic status of the female is of critical importance for normal ovulatory function. Minor restrictions in food intake can have important effects on ovarian function, and conditions or circumstances that lead to a significant nutritional deficit will result in the complete cessation of ovulatory activity. Restriction of food intake is common in athletes and in adolescent females, and amenorrhea a fairly common consequence. The reproductive effects are reversible in that normal ovulatory function is restored when normal food intake is resumed. Eating disorders, such as anorexia nervosa and bulimia nervosa, are pathological conditions that lead to reduced food intake and especially in the case of anorexia nervosa, result in suppression of ovulatory activity. The etiology of anorexia and bulimia nervosa is unknown, but the disorders probably arise from underlying neuroendocrine disorders that have not yet been fully defined. Suppression of ovulatory function is due to suppression of the GnRH pulse generator. What is not yet clear is how nutritional status is measured by the body and translated into a neuroendocrine signal controlling GnRH pulse generator activity.

PSYCHOGENIC STRESS

Cessation of menstruation in response to emotional stress (in the absence of organic disease, intense physical exercise, or obvious nutritional deprivation) is known as **psychogenic amenorrhea**. The stimulus that sets off the ovulatory dysfunction may be fear, failure, loss of a loved one, or some other type of emotional trauma. The condition has been recognized since the early 1800s, but it was in the 1940s, particularly from studies of women in the armed services during World War II, that the importance of psychogenic factors in menstrual function was articulated clearly.

Psychogenic amenorrhea is now considered to fall within a broad classification of amenorrheas, referred to as **functional hypothalamic amenorrheas** (FHAs), characterized by a deficiency in GnRH release, not associated with an organic lesion, and reversible. Exercise-induced and nutritionally induced amenorrheas discussed above represent two other types of FHAs. Functional hypothalamic amenorrhea is one of the most common types of amenorrhea, and is most prevalent between the ages of 20 and 30. Psychogenic amenorrhea develops in individuals who are basically psychologically healthy, and the interruption in menstrual cycling is temporary. The degree of suppression of GnRH pulsatility varies widely, from almost complete suppression of ovarian function leading to extremely low levels of steroid hormones, to somewhat reduced gonadotropin levels and near normal steroid hormone levels. Secretion of LH can revert to a prepubertal pattern, or in some cases, to the pubertal pattern with amplification of LH pulsatility during the sleep period (Fig. 9.12). Restoration of ovulation, and even achievement of pregnancy, can be accomplished by exogenous pulsatile administration of GnRH, demonstrating that the immediate underlying cause of amenorrhea is due to failure or impairment of the GnRH pulse generator. In most cases, once the source of stress is removed, normal reproductive activity returns.

The precise etiology of psychogenic amenorrhea remains unknown. It is not at all clear how emotional trauma is converted into a neuroendocrine signal that suppresses GnRH secretion. However, recent studies indicate that FHAs, such as psychogenic and exercise-induced amenorrhea, share a number of features that can be linked to nutritional/metabolic status. Low energy availability, for example, appears to be a common finding in both types of conditions. Compared to normal cycling women, patients with these two types of amenorrheas consume about 50 percent less fat and more carbohydrates and fiber, even though their caloric intakes are similar. In addition, other metabolic parameters, such as insulin, thyroid hormone, cortisol, and leptin levels, as well as other neuroendocrine factors, appear to be altered in characteristic ways. This finding suggests that both psychogenic and exercise-induced amenorrhea may be associated with metabolic deficits. Still, there are differences between the two types of amenorrheas. For example, individuals with psychogenic amenorrhea appear to have a normal leptin diurnal pattern, unlike individuals with athletic amenorrhea (Fig. 9.5). This finding suggests different factors may be involved in the origin of the amenorrhea. It is not yet clear whether psychogenic stress leads to subtle alterations in nutritional status or whether preexisting nutritional deficits make individuals less resistant to psychogenic stress. Despite these uncertainties, this evolving knowledge suggests that perhaps all reproductive stressors exert their effect by perturbing the metabolic and nutritional status.

SUMMARY

Emotional trauma, or psychogenic stress, sometimes leads to the suppression of ovulatory activity and amenorrhea in otherwise normally cycling females. Since menstrual cyclicity can be restored by exogenous administration of pulsatile GnRH, psychogenic stress can result in the inhibition of GnRH release. Psychogenic amenorrhea belongs to the class of functional hypothalamic amenorrheas (FHAs) are hypothalamic in origin, reversible, and not due to organic lesions. This class includes exercise-induced amenorrheas, as well as amenorrheas due to nutritional deprivation. Recent studies suggest that certain types of metabolic/nutritional deficits may cause FHA. Perhaps all reproductive stressors—strenuous physical activity, nutritional deficiency, and emotional trauma—induce metabolic changes, and the reproductive dysfunction that follows represents an adaptation to these nutritional deficits. It is not yet clear how metabolic status is translated into neuroendocrine signals that impinge on the hypothalamus.

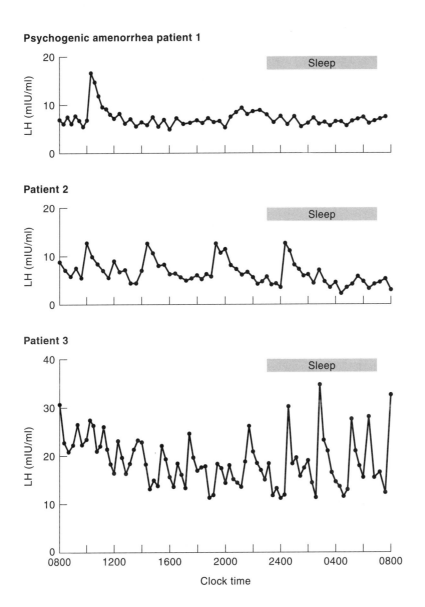

Figure 9.12

Representative 24-hour LH profiles in three patients with psychogenic amenorrhea. All three show marked reduction either in LH amplitude or LH frequency. Patient No. 3 shows marked LH enhancement during the sleep (S) period. (Adapted from S. S. C. Yen, 1986.)

QUESTIONS

1. Given that CRH is the stress hormone, can you think of a therapeutic approach for the relief of the symptoms and effects of stress?

2. Why is the critical-weight hypothesis incomplete as a model for the maintenance of reproductive function in females?

3. Amenorrhea and luteal suppression are common end-points in the response to the demands of physical training in females. What factors influence which end-point will be reached?

4. The responses of the body to physical, nutritional, and psychogenic stress have some common features. What are they, and how are they explained?

5. Why is anorexia nervosa more serious than other eating disorders?

6. Why are females more sensitive to stress than males?

7. Library project:
    a. Females appear to be more sensitive to different types of stress during the first 10 years after menarche. What accounts for this greater sensitivity during this period?
    b. What is the latest view regarding the metabolic parameters that may be important in generating the puberty trigger?

SUPPLEMENTARY READING

Bemporad, J. R. 1996. Self-starvation through the ages: Reflections on the pre-history of anorexia nervosa. *International Journal of Eating Disorders* **19**(1), 217–237.

Chroussos, G. P. 1998. Stressors, stress, and neuroendocrine integration of the adaptive response. *Annals of the New York Academy of Sciences* **851**, 355–363.

Dorn, L. D., and G. P. Chroussos. 1993. The endocrinology of stress and stress system disorders in adolescence. *Endocrine and Metabolic Clinics of North America* **22**(3), 685–697.

Frisch, R. E. 1988. Fatness and fertility. *Scientific American* **258**(3), 88–95.

Yen, S. S. C. 1999. Chronic anovulation due to CNS-hypothalamic-pituitary dysfunction. In *Reproductive Endocrinology: Physiology, Pathophysiology, and Clinical Management*, 4th ed. S. S. C. Yen, R. B. Jaffe, and R. L. Barbieri, Eds. W. B. Saunders Co., Philadelphia, PA.

ADVANCED TOPICS

Bergh, C., and P. Södersten. 1996. Anorexia nervosa, self-starvation and the reward of stress. *Nature Medicine* **2**(1), 21–22.

Bullen, B. A., *et al.* 1985. Induction of menstrual disorders by strenuous exercise in untrained women. *New England Journal of Medicine* **312**, 1349–1353.

Fichter, M. M., and K. M. Pirke. 1984. Hypothalamic-pituitary function in starving healthy subjects. In *The Psychobiology of Anorexia Nervosa*. K. M. Pirke and D. Ploog, Eds. Springer-Verlag, Berlin, Germany.

Hackney, A. C. 1996. The male reproductive system and endurance exercise. *Medicine and Science in Sports and Exercise* **28**(2), 180–189.

Laughlin, G. A., and S. S. C. Yen. 1997. Hypoleptinemia in women athletes: Absence of a diurnal rhythm with amenorrhea. *Journal of Clinical Endocrinology and Metabolism* **82**(1), 318–321.

Laughlin, G. A., C. E. Dominguez, and S. S. C. Yen. 1998. Nutritional and endocrine-metabolic aberrations in women with functional hypothalamic amenorrhea. *Journal of Clinical Endocrinology and Metabolism* **83**(1), 25–32.

Loucks, A. B. 1990. Effects of exercise training on the menstrual cycle: Existence and mechanisms. *Medicine and Science in Sports and Exercise* **22**(3), 275–280.

Sanborn, C. F., B. J. Martin, and W. W. Wagner. 1982. Is athletic amenorrhea specific to runners? *American Journal of Obstetrics and Gynecology* **143**, 859–862.

Van der Walt, L. A., E. N. Wilmsen, and T. Jenkins. 1978. Unusual sex hormone patterns among desert-dwelling hunter-gatherers. *Journal of Clinical Endocrinology and Metabolism* **46**, 658–668.

Yen, S. S. C. 1986. Chronic anovulation caused by peripheral endocrine disorders. In *Reproductive Endocrinology*, 2nd ed. S. S. C. Yen and R. B. Jaffe, Eds. W. B. Saunders Co., Philadelphia, PA.

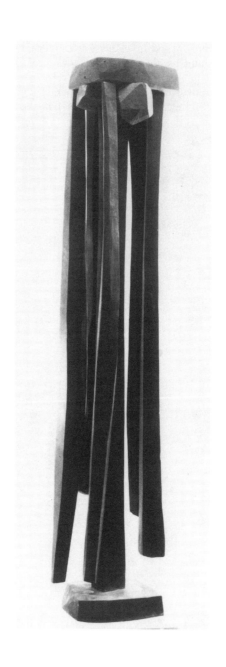

Isamu Noguchi, *Mortality*, 1962.
(Courtesy of the Isamu Noguchi Foundation, Inc.)

CHAPTER 10

Aging and Reproduction

When women go through menopause, where do men go?

—Title of video. 1996

I. V. Studios / Elizabeth Sher, Berkeley.

"The dangerous age" is marked by certain organic disturbances,
but what lends them importance is their symbolic significance.
The crisis of the "change of life" is felt much less keenly by
women who have not staked everything on their femininity.

—Simone de Beauvoir. 1953

The Second Sex. Translator and editor,
H. M. Parshley. Alfred A. Knopf, New York, p. 542.

The commonly held belief that aging routinely requires
pharmacological management has unfortunately led to neglect of
diet and lifestyle as the primary means to achieve healthy aging.
Now is an appropriate time to reassess this emphasis.

—W. C. Willett, G. Colditz, and M. Stampfer. 2000

Postmenopausal estrogens—Opposed, unopposed, or none of the above.
Journal of the American Medical Association 283(4), 534–535.

AGING IS NEITHER WELCOME nor pleasant, but there is a certain egalitarian quality to it: no one is immune. Rich or poor, famous or unknown, intelligent or not,
successful or unsuccessful, happy or unhappy—everyone ages. Some people age more
gracefully than others, and cosmetic surgery may mask some of its effects, but the inex-
orable progression of decline cannot be denied. There is, however, one aspect of aging
that does not appear equitable—this is reproductive aging. The human female experiences a relatively dramatic decline in her fertility beginning in her late 30s and terminating at about the age of 50, at which time she becomes permanently infertile. This
despite the fact that she will live on the average (in the industrialized societies) another
30 years. The human male, in contrast, does not experience an equivalent precipitous
decline in fertility. Although most men have completed their reproductive role by the
age of 50, they remain fertile until well into their 70s. What accounts for this differ-

ence? We cannot give a definitive answer to this question. We can, however, describe some of the features of reproductive aging in both sexes, and consider the hypotheses that attempt to understand the difference. In addition, we will examine the therapies that have been developed to deal with the consequences of reproductive aging, as well as the controversies that surround them.

REPRODUCTIVE AGING IN THE FEMALE

The perimenopause

The stage of life beginning in the late 30s and ending around age 50 has been commonly called the "change of life," but more recently, the more technical term *menopause*, introduced in the nineteenth century, has gained currency and is now used widely in the popular media. In a strict sense, the term "menopause" refers to the last menstrual period in a woman's life, and as such, can only be determined retrospectively. A better term is **perimenopause** (also known as the **climacteric**), defining the period before the menopause that is marked by changes in endocrinological parameters that signal the approach of menopause, as well as a period of time, generally a year, after menopause (Fig. 10.1). In common usage, however, the term "menopause" is often used to mean the perimenopause, as when a woman says, "I am going through menopause."

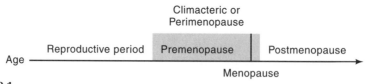

Figure 10.1
Reproductive aging in the female. The climacteric, or perimenopause, defines the period of physiological decline of the H-P-O axis. The age at the last menstrual cycle represents the menopause.

Characteristics

The perimenopause is marked by a number of changes in the female's reproductive physiology that may begin to be noted in her mid-30s. Notable changes include decreased fertility, higher risk of spontaneous abortion (miscarriage) and stillbirth, increased risk of conceiving a child with chromosome abnormalities, and a decrease in bone mass. Paradoxically, these changes may begin several years before significant changes in menstrual cyclicity can be noted in regularly cycling women.

Endocrinologically, several changes can be noted. An important one is the progressive decrease in the mean levels of the main ovarian estrogen, estradiol (Fig. 10.2). The decrease in mean estradiol levels does not tell the whole story, however. A more detailed analysis of ovarian endocrine function indicates that during the perimenopause, the cyclic patterns of estradiol and progesterone production change differently. Although daily progesterone levels begin to decrease after the age of 40, pulsatile progesterone production during the luteal stage remains stable. Daily estradiol levels also decrease during this period, but surprisingly the midcycle estradiol surge does not change (Fig. 10.3). Estradiol levels decline during the luteal phase, and this change can be noted even in 30- to 34-year-olds. The maintenance of estradiol surge levels implies

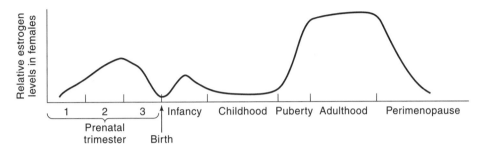

Figure 10.2
Mean estrogen levels in the female as a function of age. Estradiol levels peak during the postpubertal and adult period. They begin to decrease significantly over a 5-year period beginning at about age 40.

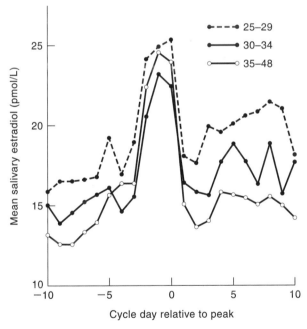

Figure 10.3
Mean levels of salivary estradiol for three age groups of females. For comparison, the patterns have been aligned on the day of the estradiol peak. Note that although the estradiol levels decrease with age, the estradiol peak remains the same. The lower levels during the luteal phase contribute to the lower fertility of older females. (Adapted from M. T. O'Rourke *et al.* 1996. Ovarian function in the latter half of the reproductive lifespan. *American Journal of Human Biology* 8, 751–759. © 1996, Wiley-Liss, Inc., a subsidiary of John Wiley and Sons, Inc.)

that the ability to generate an LH surge is also maintained until quite late in the perimenopause. We infer that the aging ovary continues to generate dominant follicles in a reasonably normal manner. These observations help us understand why many perimenopausal women continue to experience more or less normal menstrual cycles. The decline in estrogen and progesterone levels during the luteal phase, indicating altered corpus luteum function, may also contribute to the decline in the ability of perimenopausal women to carry a pregnancy to term.

Pituitary gonadotropins also begin to increase during the latter stages of the perimenopause. FSH levels begin to increase some 5 years before the menopause, eventually reaching levels 10- to 15-fold higher than during the follicular phase in a premenstrual woman. This increase is generally attributed to the decreasing estrogen and inhibin levels and to the relaxation of negative-feedback inhibition. LH levels increase about 3- to 4-fold, but they generally begin to increase after FSH.

These changes have suggested that the perimenopause, despite the maintenance of cyclic secretion of estradiol and progesterone, appears to be characterized by a gradual loss of coordination between the ovary and the pituitary and the ovary and the endometrium. We can think of the perimenopause as involving the dismantling of the fine-tuning of the H-P-O axis that was established some time after puberty. The length of the follicular phase begins to increase and the frequency of anovulatory cycles also increases. Menstrual cycles also become irregular, since the changing patterns of estrogen and progesterone no longer control endometrial development in a predictable way.

A number of somatic disturbances have also been associated with the perimenopause. Some of the most common are hot flushes (flashes), insomnia, loss of libido, nervousness, and depression. Hot flashes and insomnia are considered to reflect changes in the hypothalamic nuclei that regulate body temperature and sleep, in the latter case perhaps by disturbing the 24-hour biological clock that seems to govern many neuroendocrine secretions. The source of this 24-hour periodicity is the **suprachiasmatic nuclei** (SCN) in the hypothalamus. There have been suggestions that alteration in SCN signals may produce changes in the pulse frequency of the GnRH pulse generator during the perimenopause. However, in humans there is no definitive evidence for any change in the pulsatility characteristics of the GnRH pulse generator. Some aspects of reproductive function may depend on neural inputs that are synchronized by the SCN. However, the etiology (origin) of the majority of the symptoms associated with perimenopause remains controversial, particularly because these symptoms are experienced by some, but not all, women, and their severity is quite variable.

Causes

The perimenopause signals the physiological decline of the H-P-O axis. We do not yet understand the ultimate cause of this decline. There is general agreement that the immediate cause is the decline in ovarian follicular activity, but we do not yet have a clear understanding of the factors that initiate this decline. One hypothesis is that it is the depletion of primordial follicles beyond some critical value that precipitates the decline in follicular activity. Recall that loss of follicles that began during fetal life is continuous. The rate of loss from birth to about the age of 37–38 remains relatively constant, but after that the rate of loss doubles (Fig. 10.4). The number of follicles remaining in the ovary when the rate of loss increases is about 25,000. This number may represent a critical threshold, below which follicular recruitment and development is disturbed. The ovarian reserve, i. e., the number of primordial follicles at age 37–38, is

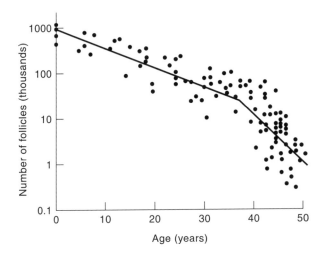

Figure 10.4
Decrease in the total number of follicles in human ovaries from birth to age 50. Note the
doubling of the rate of atretic loss beginning at about age 35–37 years. This change marks the
age at which perimenopausal symptoms begin to be manifested. (Adapted from M. J. Faddy,
R. G. Gosden, A. Gougeon, S. J. Richardson, J. F. Nelson. 1992. Accelerated disappearance of
ovarian follicles in mid-life: Implications for forecasting menopause. *Human Reproduction*
7(10), 1342–1346.)

quite variable, and this variability can have significant effects in fertility. For example,
women who continue to menstruate regularly after the age of 45 have been found to
have an ovarian reserve about 10-fold greater than a woman who experiences irregular menses. Corroborating evidence for the importance of the ovarian reserve has come
from animal experiments which show that removal of follicles from the ovary acceler-
ates the onset of full sterility. Hence, this view suggests that accelerated depletion of
the ovarian reserve, rather than age itself, may precipitate the changes associated with
the perimenopause.

An alternative, and perhaps complementary, hypothesis suggests that the disturbance in follicular dynamics is due to the decline in granulosa cell competence. Circumstantial evidence for this model comes from the comparison of granulosa cell function in granulosa cells isolated from follicles of old and young women. The follicles in
older women tend to have fewer granulosa cells at a comparable stage of development.
The granulosa cells themselves exhibit a progressive decline in steroid and protein production, decreased mitosis, and increased probability of dying. Thus, the decline in the
integrity of the granulosa-oocyte relationship may be due to a decline in granulosa cell
function. However, this pushes the problem another step back, since it is not at all clear
what precipitates the disturbance in granulosa cell function. Some investigators suggest that the origin lies not with the granulosa cell, but with the oocyte itself. The oocyte may turn out to be the ovarian pacemaker, perhaps because it has a limited lifetime. Decline in oocyte function could easily manifest itself in a loss of granulosa cell–
oocyte function. Many questions remain unanswered regarding this critical time in a
woman's reproductive period.

The menopause

The precise time of menopause, the last menses, is not easy to pinpoint. Menses can reoccur even after several months have passed without a period. For example, the likelihood of menses after six months without a period is 52 percent if the woman is between 45 and 49 years old, and 20 percent if she is over 53. Approximately 10 percent of women have a period more than one year after their last.

Estradiol production by the ovary ceases when the ovarian reserve reaches a low level, perhaps below 1000. Despite the loss of estradiol, estrogen levels in the perimenopausal or menopausal woman do not drop to zero. A weaker estrogen, *estrone*, which was always produced at low levels before, continues to be produced. Estrone, derived from a weak androgen, androstenedione, is produced by the adrenal glands, by adipose (fat) tissue, and by the stromal (nonfollicular) cells that remain active in the ovary. Estrone becomes the primary estrogen after menopause. Adipose tissue appears to be an important source of estrone because fat or obese postmenopausal women have higher estrone levels than leaner ones. Androgens, primarily in the form of androstenedione, increase, and the increase in the androgen/estrogen ratio leads to a variety of side effects, hirsutism (growth of facial or body hair), thinning hair, acne, deeper voice, and weight gain. These side effects vary greatly in severity and frequency. The menopause results in rather profound changes in the hormonal environment of the female. Much of the controversy surrounding menopause is to what extent we should interpret these changes as normal physiological changes, rather than as pathological ones.

SUMMARY

The menopause can be considered the end-point of a profound decline in the integrity of the H-P-O axis that begins between the ages of 35 and 40. We are not yet able to identify the initiating stimuli for this decline, but it is correlated with significant depletion of the ovarian reserve and progressive loss of granulosa cell function and is manifested in the progressive loss of follicular activity. The loss of follicular activity leads to the decrease in estrogen levels, increase in gonadotropins, loss of fertility, increased risk for osteoporosis, and a number of somatic disturbances whose etiology has been difficult to trace. It is quite clear that the loss of ovarian follicular activity is to a large extent separable from the normal aging process.

THE SIGNIFICANCE OF MENOPAUSE

Symbolic significance

Historically, the cessation of menstruation has been recognized and documented for at least 1500 years. It has probably been seen throughout most of that time as part of normal aging, of no special importance or consequence. Anthropological studies have noted that in most societies no rites of passage have celebrated or announced menopause as they have birth, puberty, marriage, or death. Things began to change in the eighteenth century, when the cessation of menstruation began to be linked to a number of organic and emotional problems in women. Menopause began to be seen as a "medical problem," a condition that required medical attention. Reproductive aging in the female has stimulated much more interest from both the medical profession and the public than reproductive aging in the male, and its study has benefited, and according to some, been burdened, by that attention.

One early voice who commented on the consequences of menopause was John Leake, who published his *Chronic or Slow Diseases Peculiar to Women* in 1777:

> *At this* critical time of life *the female sex are often visited with various diseases of the chronic kind . . . some are subject to pain and giddiness of the head, hysteric disorders, colic pains, and a female weakness . . . intolerable itching of the neck of the bladder, and contiguous parts are often very troublesome to others.*[1]

A not dissimilar point of view was expressed by the French physician de L'Isère in his *Treatise on the Diseases of Females* in 1845:

> *Compelled to yield to the power of time, women now cease to exist for the species, henceforward live only for themselves. Their features are stamped with the impress of age, and their genital organs are sealed with the signet of sterility . . . It is the dictate of prudence to avoid all such circumstances as might tend to awaken any erotic thought in the mind and reanimate a sentiment that ought rather to become extinct . . .*[2]

And in 1972, the American obstetrician R. A. Wilson carried this trend to a surprising extreme:

> *Estrogen deficiency is as much a disease as thyroid, pancreatic, or adrenal deficiency. No attempt will be made here to detail all of the unwholesome effects of this deficiency disease; a few will suffice, e.g. thinning of bones, dowager's hump, ugly body contours, flaccidity of the breast, atrophy of the genitalia . . . The estrogenic treatment of older women will inhibit osteoporosis and thus help to prevent fractures, as long as they continue healthful activities and appropriate diets. Breast and genital organs will not shrivel. Such women will be much more pleasant to live with and will not become dull and unattractive.*[3]

Common to these points of view, despite their 200-year span, is that the menopause is somehow more than the mere cessation of menstruation. In a very important way, the study of menopause has not been merely about endocrinological parameters, but also about its symbolic significance. F. Borner in his text *The Menopause*, published in 1887, had a sense about the larger canvas that contained this important period in a woman's life:

> *The perimenopause, or so-called change of life in women, presents, without question, one of the most interesting subjects offered to the physician, and*

---

[1]Leake, J. 1777. *Chronic or Slow Diseases Peculiar to Women.* Baldwin, London. Cited in Utian, W. H. 1980. *Menopause in Modern Perspective: A Guide to Clinical Practice.* Appleton-Century-Crofts, New York, NY.

[2]de L' Isère, Colombat. 1845. *Treatise on the Diseases of Females.* Translated by C. D. Meigs and L. Blanchard. Lea, Philadelphia, PA. Cited in Utian, 1980.

[3]Wilson, R. A., and T. A. Wilson. 1972. The basic philosophy of estrogen maintenance. *Journal of the American Geriatrics Society* **20**, 521–523.

especially to the gynecologist, in the practice of his profession. The phenomena of this period are so various and changeable . . . so ill-defined are the boundaries between the physiological and the pathological in this field of study, that it is highly desirable in the interest of our patients of the other sex, that the greatest possible light should be thrown upon this question.[4]

Perhaps more than in other areas of human reproduction, as the above statements suggest, menopause and its symptoms have been interpreted in ways that have reflected deeply held and generally unconscious views of the nature of the female. Hence, despite the universality of the physiological changes that characterize menopause, the significance of those changes has varied from one society to another.

Evolutionary significance

The postmenopausal infertile period of the human female constitutes about one-third of her total life span. The difference between the end of fertility and longevity in the human female stands in sharp contrast not only to that of the human male, but to that of females of other mammalian species. In other species, females appear to be fertile until they die. Although this conclusion has been questioned with the occasional discovery of infertile older females in the wild or in captivity, with perhaps only one exception (the pilot whale), there are no examples of species in the wild in which a substantial fraction of the females become sterile well before their death. Our closest primate relatives, chimpanzees and gorillas, apparently do not undergo menopause. Hence, the relatively long postmenopausal period typical of humans is extremely unusual. Moreover, reproductive aging in human females is also unusual because it occurs much earlier than that of other body functions (Fig. 10.5). Is this merely an accidental happenstance, or is there some deeper reason?

One possibility is that premature reproductive senescence has no particular significance and is simply a consequence of a fairly recent increase in human longevity. It may differ from the aging of other physiological functions for reasons that remain unspecified. Here we need to distinguish longevity (life span) from life expectancy. A significant increase in the latter has become most evident since the beginning of the twentieth century. Before then, the postmenopausal period was masked artefactually by a relatively short life expectancy due in large part to the inadequate control of infectious diseases. For example, in the early twentieth century, the life expectancy for women in the United States was about 48 years. Thus, in the early 1900s, relatively few women experienced menopause compared to those in the 1990s, when the life expectancy was close to 80 years. As far as can be determined from the surviving historical records, the age at menopause has remained remarkably constant at about 47–50 years for at least two thousand years. There does not appear to be any compelling evidence that early menopause is a consequence of a recent increase in the human life span.

An early menopause, or premature reproductive aging, presents an interesting puzzle for evolutionary biologists. Why should a woman cease reproducing halfway through her life span, especially when most other body functions have not yet begun to age? Would it not be better for her to continue reproducing at least until senescence really begins to take hold? We learned above that menopause is probably the result of

---

[4]Borner, F. 1887. *The Menopause: Cyclopedia of Obstetrics and Gynecology*, Vol. II. William and Wood, New York, NY. Cited in Utian, 1980.

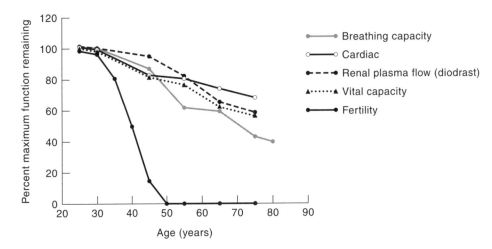

Figure 10.5
Decrease of physiologic function in humans as they age. Most body functions age gradually, and even at a late age, significant function remains. In contrast, fertility in females decreases rapidly and ceases by about age 50. (Adapted from K. Hill and A. M. Hurtado. 1991. The evolution of premature reproductive senescence and menopause in human females. *Human Nature* 2[4], 313–350, using data from Wood, 1990, and Mildvan and Strehler, 1960.)

the depletion of primordial follicles. But why was the human female not provided with a larger reservoir of follicles? This clearly happens in other mammalian species, such as the elephant and the baleen whale. A favored hypothesis suggests that menopause evolved to permit the female to switch her investment to close kin rather than to reproduction. Although there are different versions of this hypothesis, it often takes the form referred to as the "grandmother hypothesis." According to this hypothesis, the decision between an early or late menopause involves a cost-benefit comparison of the advantages of females continuing to give birth up until they die, or forgoing some births in order to survive to care for children and grandchildren. An important factor in this comparison is the prolonged care and rearing that human infants require in their development compared to infants of other species. This view can be summarized succinctly:

> As a woman ages, she can do more to increase the number of people bearing her genes by devoting herself to her existing children and grandchildren than by producing yet another child.[5]

The role of the postmenopausal woman has been explored in several anthropological studies carried out in existing hunter-gatherer societies. One well-known study examined the importance of foraging by women of different ages among the Hadza people, a hunter-gatherer society in Tanzania. The postmenopausal grandmothers, because they could work more hours unimpeded and were more efficient, were found to be better foragers than the younger women. The excess food they contributed was

[5]Diamond, J. 1996. Why women change (evolution of menopause). *Discover* 17(7), 130–138.

shared with their children and grandchildren. The food contributed by the postmeno-pausal grandmothers was critical precisely because it enhanced the probability of survival of their children and grandchildren. Hence, women gain more (in the sense of ensuring maximum survival of their genes) by withdrawing early from reproductive competition and devoting part of their postmenopausal period to the care of their children and grandchildren. The evolutionary significance of this observation was *"that menopause's benefits outweigh its costs, and that women can make more by making less."*[6]

More recent studies have not arrived at the same conclusion. Indeed, it has been difficult to replicate the results of the Tanzania study. For example, two recent studies of two hunter-gatherer societies in South America have not demonstrated the same evolutionary value of postmenopausal women as shown in the Hazda people. Available evidence from many studies suggests that fathers, not grandmothers, are more important for the survival of children.

In searching for alternative explanations anthropologists and evolutionary biologists have focused on characteristics that are uniquely human. An interesting possibility is that menopause is a consequence of the constraints of childbirth. The large head of human fetuses makes childbirth especially difficult for human females compared with other primates. The risks of child-bearing are greater in an older female than in a younger one. Moreover, because human infants require such prolonged care, it seems reasonable to suppose that evolution would favor limiting reproduction to younger women, for whom giving birth is relatively safer. The complementary advantage is that it would permit the mother to invest more time in the care of the young. Hence, different factors during human evolution may have contributed to the evolution of premature reproductive aging in the human female.

SUMMARY

Reproductive aging in the female appears to have acquired a significance and interest that go beyond its physiological characteristics. Particularly in the industrialized countries, menopause has tended to symbolize a fundamental transition in a woman's life, from a sexual to a nonsexual being. Menopause, or premature reproductive aging compared to the aging of most other body functions, also has evolutionary significance because it appears to be unique to humans. Although we are not yet certain that we understand the evolution of menopause, it may be a consequence of the increased risk of childbirth due to the large head of human infants, and the prolonged care that human infants require. Both constraints, unique to humans, may have provided the selective advantage to premature reproductive aging.

MENOPAUSE AS A MEDICAL CONDITION

A perspective

In the United States some 50 percent to 60 percent of women seek medical attention for symptoms thought to be associated with menopause. In a 1994 summary of the reasons for seeking attention, 70 percent of women reported physical symptoms, while 63 percent reported emotional symptoms. The physical symptoms generally in-

---

[6]Ibid.

volve physiological parameters, for example, loss of skin texture, hair loss, facial hair growth, reduction in breast size, weight gain, urogenital atrophy, osteoporosis, alterations in lipid metabolism and cardiovascular function. Many other symptoms, often characterized as psychosomatic or psychological, include hot flashes (flushes), cold sweats, numbness, backache, bloatedness, rheumatic pains, bowel disorders, fatigue, headache, palpitations, dizzy spells, irritability, nervousness, depression, excitability, loss of libido, insomnia, and panic attacks. The number and severity of the symptoms experienced vary enormously not only from one individual to another, but from one culture or society to another. For example, hot flashes (flushes), generally described as a collection of transient symptoms that include a sensation of being hot, sweating, increased heartbeat (tachycardia), anxiety, followed by rapid cooling or chilling, have generally been associated in the public mind as a universal menopausal symptom. However, only 50 percent to 75 percent of menopausal women in the United States and Europe report having experienced hot flashes. The percentage is lower in non-Western societies.

The *experience* of menopause involves not only the symptoms themselves, but also the way in which the symptoms are interpreted, and also on a complex mixture of social, cultural, and psychological factors. Some writers emphasize the psychological:

> *By far the greatest hazards of the menopause are psychogenical or culturally induced, and these are not so simply dispelled by a few pills. A psychiatrist working in China reported to me that she had never seen a menopausal psychosis in a Chinese woman. This she attributed to the fact that in China the older woman has a secure and coveted position.*[7]

Some are left perplexed:

> *The changes that affect both the body and the mind at and after menopause are immensely complicated. Not only are there hormonal changes, but also emotional, social, and family changes, and no one really knows why in fact some women, albeit the minority, pass through their fifties and sixties with little physical or emotional disturbance.*[8]

Some attempt to deny the significance of the symptoms:

> *There may be many symptoms and signs associated with menopause but, especially in the well-balanced, educated, contented woman who finds her family, sexual and professional life fulfilling, there may be no symptoms whatsoever.*[9]

Some try to be sensible and understanding:

> *All women have to incorporate and adjust to changes as best they can, and some have preferred to see them as an unremarkable matter of course rather than to appear awkward, self-important, neurotic, and so forth. We have seen, for example, that in general some individuals label their internal sensa-*

---

[7]Thompson, C. 1971. *On Women.* New American Library, New York, NY.
[8]Anderson, M. 1983. *The Menopause.* Faber and Faber, London, UK.
[9]Cooper, W. 1987. *No Change: A Biological Revolution for Women.* Arrow Books, London, UK.

*tions as important, and some do not. The point of interest is that profound
changes take place in all women, including those who do not report them.*[10]

Others (generally men) have taken a catastrophic view:

> *A large percentage of women . . . acquire a vapid cowlike feeling called a
> "negative state." It is a strange endogenous misery . . . the world appears as
> through a grey veil, and they live as docile, harmless creatures missing most
> of life's values.*[11]

> *The menopausal woman is not normal; she suffers from a deficiency disease
> with serious sequalae and needs treatment.*[12]

The point of view, expressed perhaps in an extreme way by the last two statements, that menopause is a pathological condition requiring medical intervention has probably been a prevalent view in most of the medical community in the United States and Europe during the last forty or fifty years. It has led, according to its critics, to the *medicalization* of menopause. Germaine Greer in her book *The Change: Woman, Aging, and the Menopause* captured the essence of this view:

> *It is not quite forty years since eliminating menopause was first mooted. The
> idea did not come from women, but from men who thought that the cessation
> of ovulation was a premature death, a tragedy.*[13]

Estrogen replacement therapy (ERT)

"Menopause as pathology" was probably first promoted in the popular media by the gynecologist R. A. Wilson with the publication of his book *Feminine Forever* in 1966. The message was simple, but powerful: the culprit, the cause of all the menopausal misery was estrogen deficiency. At the time, the idea seemed to make sense. The powerful and widespread effects of the steroid sex hormones, estrogens in the female, and androgens in the male, were beginning to be widely appreciated. It was a relatively easy leap of faith to consider that estrogen was the elixir of the "fountain of youth" for the female. The cure was simple as well: estrogen replacement therapy (ERT). Heavily promoted, ERT, otherwise referred to as the "youth pill," would cure or alleviate 26 symptoms, among them depression, frigidity, alcoholism, absent-mindedness, and suicide. The promotion worked: in the early 1970s, about 50 percent of the women between 55 and 64, and about 33 percent of women between 65 and 74 were on ERT. In retrospect, such extravagant hubris was bound to have its comeuppance.

The ERT bubble burst in 1975 with the publication of two studies that suggested that ERT was associated with an increased risk of endometrial cancer. These studies prompted a serious and more sober reexamination of the relationship between ERT

---

[10]Asso, D. 1983. *The Real Menstrual Cycle.* John Wiley and Sons, Chichester, NY.

[11]Wilson, R. A., and T. A. Wilson. 1963. The fate of the nontreated post-menopausal woman: A plea for the maintenance of adequate estrogen from puberty to the grave. *Journal of the American Geriatrics Society* **11**, 347–361.

[12]Wilson, R. A., R. E. Brevetti, and T. A. Wilson. 1963. Specific procedures for the elimination of the menopause. *Western Journal of Surgery, Obstetrics, and Gynecology* **71**, 110–116.

[13]Greer, G. 1992. *The Change: Woman, Aging, and the Menopause.* Alfred A. Knopf, New York, NY.

and menopausal symptoms. More than 50 studies analyzing this question have been published since 1982, but conclusions remain very controversial. The ERT issue spilled out of the medical community. It became and continues to be a public issue of great import, with supporters and skeptics reaching different conclusions from the same data. *What has been learned?* There is no easy answer to this question, nor is there consensus among the experts regarding the risks and benefits of ERT. What is clear is that despite the large number of studies, important questions remain unresolved. Our objective here is to try to summarize the more significant conclusions and observations, with respect to the potential benefits and risks of ERT and their significance for the health status of the postmenopausal female.

Benefits and risks of ERT

In general, we would expect that ERT would relieve menopausal symptoms that are due directly to estrogen deficiency. In fact, a direct and causal connection between estrogen deficiency and menopausal symptoms has been demonstrated for only a few of them. Hence, the major emphasis of all the recent studies has been on the two major benefits observed in the early ERT studies: prevention of coronary heart disease (CHD) and osteoporosis.

Coronary heart disease

Coronary heart disease (CHD) is the most common cause of death for women in the United States. It accounts for about 53 percent of all deaths in women over the age of 50 in the United States. Its toll is about 500,000 deaths per year, about eight-fold more than deaths due to breast cancer. Despite its high mortality, it may be ironic to note that the public does not view CHD with the same dread as it does breast cancer.

Estrogen has been considered to have a cardioprotective effect for several reasons. CHD is much more common in men than in women (2.5- to 4.5-fold increased risk of CHD death in men compared to women). CHD is uncommon in women before the age of 50 years. *In vivo* and *in vitro* studies in animals have reported at least 10 estrogen effects that might be expected to prevent or delay CHD. In the large Nurses' Health Study (1987), bilateral oophorectomy (removal of both ovaries) was associated with an increased risk of CHD, while no increased risk was observed in oophorectomized women treated with estrogen. The Postmenopausal Estrogen/Progestins Intervention (PEPI, 1995) trial examined the effects of estrogen on common heart disease risk factors. One consistent finding was that estrogen alone or in combination with a progestin (estrogen in combination with a progestin is known as hormone replacement therapy [HRT]) reduced LDL and increased HDL cholesterol compared to placebo. Low LDL and high HDL is considered to be cardioprotective. Further, more than 30 observational studies since the early 1970s have reported less CHD in postmenopausal women who used estrogen alone or with a progestin compared to those that did not. Indeed, from these studies the cardioprotective benefits of ERT suggested a 35 to 50 percent reduction in the risk for CHD.

Nevertheless, despite the consistency of the observational data and biological plausibility of the cardioprotective effects attributed to estrogen, none of these studies demonstrated definitively that hormone therapy in postmenopausal women prevented CHD. In particular, the observational studies were subject to significant bias since the participants who took estrogen tended to be younger, wealthier, more educated, and healthier than untreated participating women. Hence, the putative cardioprotective effects of estrogen may have been spurious because women who are less likely to develop CHD are more likely to be on estrogen. Explicit in this critique was that lifestyle factors, such as diet, exercise, smoking, and alcohol consumption, may play important roles in the development of CHD as well. The need for a properly controlled clinical trial was evident.

The first large, randomized, placebo-controlled trial of estrogen and CHD in women, known as the Heart and Estrogen/Progestin Replacement Study (HERS), was started in 1994. The HERS studied the effects of estrogen and progestin in women who already had CHD. Since patients with known CHD are at much higher risk for new events (such as a heart attack) than persons without CHD, the HERS design would make it possible to reveal any cardioprotective effects of estrogen in a short time. The HERS results, reported in 1998, showed that the severity of CHD rates did not differ between women receiving estrogen and progestin versus placebo. No benefit from HRT was seen despite an 11 percent decrease in LDL and a 10 percent increase in HDL cholesterol. A surprising finding was a 50 percent increase in CHD after one year of HRT, a difference that decreased to nonsignificance by 4 years.

The unexpected HERS finding, that is, the absence of an estrogen cardioprotective effect, has elicited much commentary and controversy. Two important explanations for the HERS results have been suggested. (1) The HERS result does not preclude the possibility that HRT may be useful for the primary prevention of CHD, but not for those who already have CHD. (2) The progestin component in the HERS study may have been responsible for the absence of any cardioprotective effect. To address these concerns, the Women's Health Initiative (WHI), a primary prevention trial involving more than 27,000 postmenopausal women, was initiated in 1998. Results are not expected until 2005.

Hence, at present, there is no reliable clinical trial data demonstrating that the risk of CHD in postmenopausal women is reduced by estrogen alone or in combination with a progestin. Nonetheless, because of the beneficial effects of estrogen on cardiovascular risk factors, it is possible that estrogen taken over the long term (10 years or more) will reduce the CHD risk. However, the observed increase in risk over the short term indicates that ERT or HRT should not be initiated solely to prevent new or recurrent heart disease. Alternatives to standard ERT or HRT using *selective estrogen receptor modulators* (SERMs) are being evaluated, and are discussed below.

Osteoporosis

Severe loss of bone mass, measured in grams of calcium in the bone, results in easily fractured bones and is known as osteoporosis (Fig. 10.6). Osteoporotic fractures contribute significantly to morbidity and mortality. In the United States, about 25 million people suffer from osteoporosis. Because most cases of osteoporosis are asymptomatic until a fracture occurs, this number is an underestimate of the true incidence of osteoporosis. The cost of treating osteoporosis and its consequences is enormous (estimated to be $10 billion), more than the cost of cardiovascular disease ($7.5 billion).

In both males and females bone mass increases during childhood and adolescence, reaching a peak at about 30 years of age (Fig. 10.7). The maximum bone mass reached depends on genetic and environmental factors. Genetic factors include race, stature, and age at menarche. Environmental factors include diet, exercise, alcohol consumption, and smoking. In females, bone mass begins to decrease rapidly at menopause and for about 10 years thereafter. Males normally reach a higher level and do not experience the abrupt decrease seen in females. The fracture threshold represents a value for

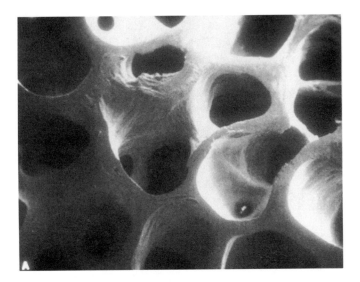

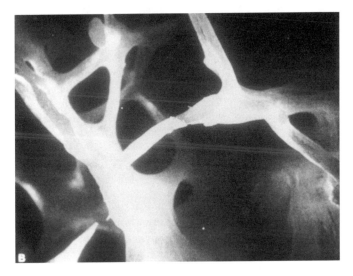

Figure 10.6
Comparison of bone ultrastructure (trabaculae) in normal (**A**) and osteoporotic bone (**B**). Trabecular bone is the innermost and most metabolically active part of bone. The trabaculae form a meshwork with intercommunicating spaces filled with bone marrow. (By permission from CompuMed, Los Angeles, CA.)

bone mass at which the bones fracture easily. Under normal conditions, females reach the fracture threshold at a much younger age than males (Fig. 10.7). In the United States, a woman at the age of 50 is estimated to have a 40 percent chance of experiencing an osteoporotic fracture at some time during her life. For this reason, osteoporosis has been considered a disease of women. Men also contract the disease, but the true incidence in men is unknown.

In a simplified sense, bone mass at any given age is determined by the balance between opposing actions of two types of bone cells—the *osteoblasts* and the *osteoclasts*.

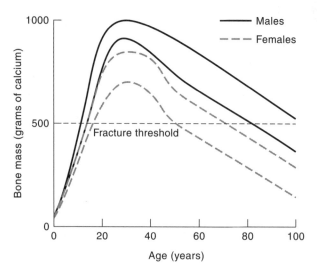

Figure 10.7

Variation of bone mass in females and males as a function of age. Bone mass, expressed as grams of calcium, increases through about the age of 30 in both females and males. In females, it decreases abruptly during the perimenopause and menopause, primarily because of the sudden decrease in estrogen levels. The age at which the fracture threshold is reached depends on the maximum level reached during the late 20s and early 30s. The higher level reached by males means that the fracture threshold is reached at a much greater age. (Adapted from G. R. Mundy. 1994. Calcium and common endocrine bone disorders. In *Clinical Endocrinology*, 2nd ed. Besser and Thorner, Eds. Mosby-Wolfe, London, UK.)

The activity of osteoblasts leads to bone formation (increase in bone mass), while osteoclastic activity leads to bone loss (decrease in bone mass). The balance between these two types of activity is influenced by many factors, but particularly by physical activity, nutrition (calcium and vitamin D intake, primarily), and estrogen. This means bone mass decreases when any of these three factors are limited. With proper levels of these factors, bone mass can be maintained at a level which minimizes fracture risk. Estrogen may have a beneficial effect on bone mass by inhibiting osteoclastic activity.

The abrupt decrease in bone mass in females is attributed to the decline in estrogen levels at menopause. The reduction in estrogen levels that accompanies the perimenopause and menopause is estimated to produce a 15 percent decline in bone mass compared to premenopausal levels (Fig. 10.8). It is important to emphasize that this loss cannot be prevented by diet, for example, by increasing calcium intake. On the other hand, additional decrease in bone mass can take place if physical activity declines or calcium intake is inadequate, both of which tend to occur as a person ages. The loss of bone mineralization due to loss of estrogen takes place over a 4- to 5-year period, after which the bone mass may stabilize at a new lower (15 percent less) level. This new level can be maintained only with proper calcium intake and exercise. If the premenopausal bone mass levels were below normal, a 15 percent loss can bring the person well into the fracture range. On the other hand, if the premenopausal levels were higher than normal, a 15 percent loss can be accommodated without incurring a risk of fracture.

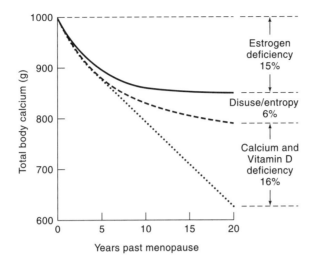

Figure 10.8

Factors that contribute to calcium depletion from bone in a postmenopausal woman. Loss of estrogen with menopause produces about a 15 percent depletion of bone calcium over a period of 5 years. Lack of physical or weight-bearing activity can account for an additional 6 percent depletion. Reduced calcium or vitamin D intake can lead to an additional 16 percent. Depending on the starting bone mass at the time of menopause, these combined losses will greatly increase the risk of bone fractures in postmenopausal women. (Adapted from R. P. Heaney, 1997.)

There is general agreement that ERT can not only stabilize bone mass irrespective of the age ERT is started (even if begun after the age of 70), but that under certain conditions it can result in a 5 to 10 percent increase in bone mass after 3–4 years of continuous use. Moreover, ERT may be useful for the management of fractures that have already taken place. ERT by itself, however, may not be sufficient to produce the full range of these beneficial results. Essential co-measures are consumption of at least 1200 mg/day of elemental calcium, avoidance of alcohol and tobacco, and maintenance of regular weight-bearing exercise (such as walking). Hence, as with the prevention of CHD, prevention of osteoporosis also depends on several lifestyle factors.

Cognitive functions

More recently, the effects of ERT and HRT on the decline of certain aspects of cognitive functioning in postmenopausal women who are otherwise healthy are beginning to be explored. These studies are inherently problematic because separating out the effects of normal aging and estrogen depletion is difficult. Nevertheless, despite the methodological shortcomings of the studies, two results appear to stand out. First, estrogen does not affect all aspects of cognitive functioning; in particular, no effect on perception, abstract reasoning, or higher-order intellectual activity has been noted. Estrogen seems to have no effect on visual or spatial memory, but it does appear to maintain or enhance verbal memory, as well as to maintain ability to learn new verbal material. This second result is interesting because it is consistent with the many reports that indicate that females have better verbal skills than males. Hence, ERT may be beneficial in slowing down verbal memory deficits that usually accompany the aging process.

Other menopausal symptoms

ERT is also prescribed for many other symptoms considered to be due to the loss of estrogen. Dry and itchy skin and urogenital atrophy (thinning of vaginal, urethral, and bladder walls, loss of vaginal secretions, shortening of the urethra) are examples of non-life-threatening menopausal symptoms, which in some studies have responded positively to ERT. The loss of vaginal texture and lubrication, for example, may make sexual intercourse more difficult or even painful. In some studies, ERT has been shown to be better than a placebo in relieving some of these symptoms. However, because there appears to be an age-dependent loss of estrogen receptors from these tissues, the benefits observed may be greatest in the early postmenopausal years, and less relief of these symptoms may be noted in the late postmenopausal period.

The causative factors underlying the many other menopausal symptoms associated with physiological, psychosomatic, psychological, and emotional disturbances remain essentially unknown. For example, the neuroendocrine factors or predisposing conditions that may be involved in generating the hot flash remain poorly understood, but many investigators have considered that estrogen deficiency may help trigger the hot flash. One model is that estrogen controls the temperature regulatory centers in the hypothalamus. Dysfunction of these centers resulting from estrogen depletion may trigger the transient temperature changes and sweating. A number of observations, however, have raised doubts about a direct connection between estrogen deficiency and the hot flash. For example, hot flashes, although a common perimenopausal symptom, are not a universal one. When they occur, they are transient, generally lasting 2 to 4 years during the early perimenopause. More importantly, however, ERT has not been uniformly successful in relieving the symptoms of hot flashes. In some studies, ERT is slightly better than a placebo, while in others, ERT *produces* hot flashes. Hence, the efficacy of ERT for relief of hot flashes has not been established unequivocally.

Endometrial and breast cancer

The doses of estrogen required to prevent CHD and osteoporosis tend to stimulate proliferation of endometrial tissue, and many studies have observed that continual stimulation (long-term, unopposed ERT) increases the risk of endometrial cancer. To reduce the risk, ERT has been supplemented with synthetic progestins because of the inhibitory effect that progestins have on endometrial proliferation, and indeed, addition of progestins to ERT has been shown to have a protective effect with respect to endometrial cancer. In most HRT protocols estrogen and progestins are given in a manner that approximates the pattern of estrogen and progesterone during the premenopausal period. Estrogen is given daily, while progestin is given only during 10 to 12 days of each month. Stoppage of the progestin results in what is termed "withdrawal bleeding" for 2 to 5 days. The withdrawal bleeding mimics normal menses, but is not a welcome occurrence. In some women, the bleeding may be prolonged and irregular. Uterine bleeding is probably the main reason HRT is discontinued.

The data on the risk of breast cancer with ERT or HRT is problematic. Short-term use of ERT is not considered to impose an increased risk of breast cancer. However, early studies suggested that long-term ERT use (more than 10 years) increased the risk by about 30 percent. Indeed, it is because of this increased risk that many women have chosen not to go on ERT or HRT. More recent ERT studies have reported similar values: 34 percent by the Collaborative Group on Hormonal Factors in Breast Cancer (CGHFBC) involving over 100,000 women, and 41 percent by the Nurses' Health Study involving 122,000 women. The significance of the increased risk can be kept in perspective by keeping in mind that it is relative to the absolute risk. The current lifetime risk of contracting breast cancer is about 12.5 percent in the industrialized countries. Hence, a 30 percent increase in risk would mean that the lifetime risk would increase to about 17 percent if a woman were to take HRT until her death.

The progestin component of HRT does not have a significant protective effect with respect to breast cancer as it does with endometrial cancer. In fact, most studies indicate that inclusion of a progestin component to ERT increases the risk of breast cancer substantially. For example, the CGHFBC study reported an increase of 53 percent over a ten-year period, and the Breast Cancer Detection Demonstration Project involving 46,000 women found an increase of 8 percent for each year of HRT over a ten-year period. With ERT alone, the increase was 1 percent per year over a ten-year period.

This increase in risk does not apply to all women, however, as has been shown by the two most comprehensive studies completed to date (the Nurses' Health Study involving 122,000 nurses over a period of 14 years, and the almost equally large Iowa Women's Health Study). These two studies report one very important conclusion: women who did not consume alcohol had *no increased risk* of developing breast cancer at all. The amount of alcohol that appears to be critical is 5 grams or more per day. Consumption of less than that had no effect on the breast cancer risk. This conclusion has extremely important implications. It indicates at the very least that the increase in breast cancer incidence associated with ERT or HRT is avoidable by abstaining from alcohol or keeping alcohol consumption low. Since none of the other studies stratified their data according to alcohol consumption, any effect of alcohol would have been missed.

The profound implications of this finding may have been ignored or not emphasized because it was difficult to explain the alcohol effect. However, a biologically plausible mechanism has been reported recently, and demonstrates that alcohol consumption by a postmenopausal woman on ERT increases the estrogen levels about 300 percent (three-fold) (Fig. 10.9). The alcohol effect is an interesting one. In postmenopausal women not on ERT, alcohol consumption has no effect on estrogen levels. On the other hand, in postmenopausal women *who are on ERT*, alcohol consumption results in a pronounced increase in estrogen levels. The levels of estrogen that are reached under these conditions are equivalent to the estrogen-surge levels during the follicular phase. The interpretation of these findings is that *ERT plus alcohol* results in an estrogen increase to levels that may be required for cancer-promoting effects. On the other hand, estrogen levels under *ERT without alcohol* are kept at levels below those that may be necessary for promoting breast cancer.

Other side effects

Several other adverse effects of ERT have been reported. Pulmonary embolism and venous thrombosis are seen in a small fraction of women. Uterine bleeding is probably the most common. Breast soreness is reported in the early stages of therapy, and

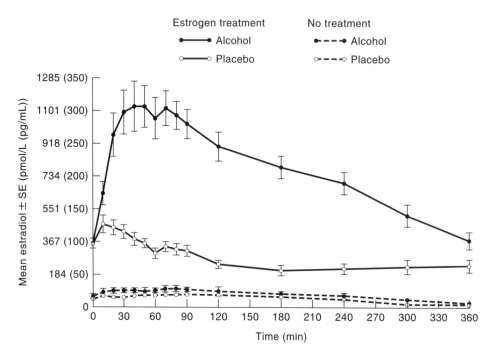

Figure 10.9

Effect of alcohol on estrogen levels in postmenopausal women. Women not on ERT have very low estrogen levels, and alcohol consumption (5 grams per day, equivalent to 3 ounces of wine) does not increase this level. This corresponds to the absence of any effect of alcohol in increasing breast cancer risk in women who are not on ERT. Women who are on ERT have estrogen levels similar to those found during the follicular phase. Consumption of alcohol produces an increase in estrogen levels that corresponds to the estrogen-surge levels during the ovulatory cycle. This effect lasts about 6 hours. This alcohol-associated rise in estrogen levels may account for the increased breast cancer risk of women who are on ERT and who consume alcohol. (Adapted from E. L. Ginsburg *et al.* 1996. Effects of alcohol ingestion on estrogens in postmenopausal women. *Journal of the American Medical Association* **276**, 1747–1751.)

appears to be most common in older women. Migraine headaches in women who have suffered from them premenopausally may be exacerbated with ERT, but the reason remains unknown. Gallstones appear to be associated with the use of oral estrogens commonly used in ERT. Fluid retention is a vexing problem in some women, particularly in women who are predisposed to *cyclic edema*, a condition characterized by facial bloating and puffiness and rapid gain and loss of weight of a few pounds. Some preliminary studies suggest that ERT may be contraindicated for a variety of conditions. The most important of these are indicated in Table 10.1. Although no conclusive data has yet been presented to justify these as absolute contraindications, the decision to start ERT has to be weighed carefully for those who present with any of these conditions.

Table 10.1
Suggested Guide for Beginning and Continuing ERT

Begin ERT?	
Yes	**No**
For menopausal symptoms	History of uterine or breast cancer
For prevention/treatment of osteoporosis	History of thromboembolism
For prevention of coronary disease	Abnormal liver function

Continue ERT?	
Yes	**No**
ERT well tolerated	ERT not tolerated—bleeding, sore
Bone density improved	breasts, fluid retention, weight
Menopause symptoms improved	gain, headaches
	Abnormal mammogram
	Other problems

Reference: Lufkin and Ory, 1995.

Alternatives to standard ERT or HRT

SERMs

Dissatisfaction with ERT/HRT, particularly with respect to the breast cancer risk, led to a search for compounds that might provide the beneficial effects of estrogen without its undesirable side effects. The result has been the synthesis of a class of compounds known as *selective estrogen receptor modulators* (SERMs). These compounds take advantage of the fact that there are two forms of the estrogen receptor, α and β. The α and β receptors are not distributed uniformly among estrogen-dependent tissues. Some tissues have only the α, and some have only the β. The SERMs bind with different affinity to the α and β receptors and they display estrogen agonist or antagonist effects in a tissue-dependent manner. A number of SERMs have been characterized, but at present three of them have been approved for human use: **tamoxifen** and **toremifene** for the treatment of breast cancer and **raloxifene** for the prevention of postmenopausal osteoporosis. In animal and human studies, both toremifene and tamoxifen have been shown to reduce cholesterol levels, maintain bone density, and protect against breast cancer, but long-term use of tamoxifen has been shown to lead to endometrial cancer, pulmonary embolism, and deep vein thrombosis. In contrast, the first randomized, placebo-controlled, clinical trial of raloxifene has had very promising results. The Multiple Outcome Raloxifene Evaluation (MORE), a short-term (about 3.5 years) trial involving about 5000 postmenopausal participants in 25 countries, demonstrated that raloxifene does not increase the risk of endometrial cancer, does not cause breast tenderness or pain, and decreased the risk of breast cancer by 76 percent. Hence, raloxifene appears to have the selectivity to separate the beneficial effects on bone and cardiovascular parameters from excess stimulation of endometrial and breast tissue. Since preclinical and clinical studies have shown that several SERMs have favorable effects on cardiovascular parameters, there is a real possibility that SERMs may reduce the risk of CHD. The Raloxifene Use for the Heart (RUTH) trial, involving about 10,000 women in 26 countries, begun in 1998 will determine the effect of raloxifene on cardiovascular morbidity and mortality. Raloxifene may turn out to be the SERM to replace standard ERT and HRT.

Bisphosphonates

Important progress has also been made in treating osteoporosis with a class of nonsteroidal compounds known as **bisphosphonates**. These compounds were first developed for the treatment of bone disease and bone cancer. Several of them (**pamidronate, alendronate, risendronate**) inhibit osteoclast activity, and in several clinical trials, alendronate therapy has been shown to result in an increase in bone mass and a reduction of fracture rates to approximately half of control rates at the spine, hip, and other sites in postmenopausal women. Alendronate, the only bisphosphonate approved for osteoporosis therapy, provides the best alternative to ERT for women in whom ERT is contraindicated or for women who choose not to take ERT.

The ERT skeptics

Despite what appear to be the important benefits of ERT/HRT, some investigators question the entire ERT approach to menopause. Hormonal dosing of large numbers of postmenopausal women somehow does not seem appropriate. A panacea in a pill seems insulting. Osteoporosis, for example, is not an inevitable consequence of menopause. It is seen much more frequently among Caucasians than among Blacks, where it is relatively rare. The traditional explanation is that Black women reach a higher bone mass while they are young. Hence, as we saw above, even with a 15 percent loss, their bone mass remains high enough to maintain suitably strong bones. Moreover, the increase in osteoporosis noted in many industrialized countries in recent years appears to reflect significant changes in lifestyle factors. British women today, for example, are twice as likely to suffer hip fractures as they were 30 years ago, a difference that is not due to differences in estrogen levels. Changes in diet, less exercise, decreased vitamin D and calcium intake, and increased incidence of alcohol consumption and cigarette smoking among younger females all clearly reduce the premenopausal bone-mass levels. Many investigators are convinced that attainment of high bone-mass values while young may be the best protection against osteoporosis.

The relationship of estrogen levels to CHD was questioned, even before the HERS results were reported. Japanese women tend to have lower levels of estrogen than women in the Western countries, yet their incidence of heart disease is much lower. The view that the rate of CHD is greatly accelerated at menopause has also been shown to be incorrect. CHD rates increase with age, but there is no change in the rate during the menopausal ages. Skeptics can point to many lifestyle factors that could equally well predispose women in the industrialized countries to CHD.

Those who are skeptical of the benefits of the use of ERT on a mass scale argue that we need to know more about those women who do not succumb to the more serious of menopausal symptoms. Germaine Greer argues that

> the obstacle to understanding here is the defect that disfigures all gynecological investigation; we do not know enough about the well woman to understand what has gone wrong with the sick one.[14]

---

[14]Ibid.

The skeptics contend that the benefits of ERT are relatively modest, and that most of the benefit may not due to estrogen itself, but to the other factors—diet, exercise, not smoking, etc.—that need to accompany ERT to see significant benefits. They argue further that changes in lifestyle factors are much less intrusive, and are likely to be more effective. They suggest that there are more than enough indications that lifestyle factors may account for most of the impact that menopause has on a woman. *But what happens if women do not change their lifestyle, or what about the women for whom it may be too late for lifestyle changes to have a significant effect?*

SUMMARY

The benefit/risk equation for standard ERT/HRT is not a simple one (Table 10.2). The most clearly demonstrated benefit is the positive effect on bone mass and reduction in the risk of osteoporosis and bone fractures. This positive effect is independent of the age at which the woman starts ERT/HRT. Less clear is whether ERT/HRT reduces the risk of CHD. There is no doubt that ERT/HRT is associated with an improvement in several risk factors used in assessing the risk of CHD, and there is much observational data to suggest that the risk of CHD is reduced. However, the only well-controlled clinical trial carried out to date failed to show that ERT/HRT had any beneficial effect on postmenopausal women who had CHD when the trial was started. A definitive trial to evaluate the effects of ERT/HRT in preventing CHD is under way, but its results will not be known for several years. Nevertheless, the beneficial effects of ERT/HRT on cardiovascular risk factors suggest that a reduction in the risk of CHD may be apparent only in the long-term.

Although not as well documented, ERT/HRT has been shown to have other beneficial effects, such as reduction in vaginal atrophy, preservation of skin elasticity, maintenance of verbal memory, and therapy for hot flashes.

The most serious risk of unopposed ERT is an important increase in the risk of breast and endometrial cancer. The risk for endometrial cancer can be eliminated if ERT is supplemented with a progestin, but the risk of breast cancer increases with HRT. However, two large-scale, long-term clinical trials have shown that the risk of breast cancer applies almost exclusively to women who consume more than a certain level of alcohol.

Table 10.2
Summary of Health Implications of ERT in Postmenopausal Women

Benefits	Risks
Decreased risk of cardiovascular disease	Possible increase in risk of breast cancer
Reduced risk of osteoporosis and osteoarthritis	Risk of thrombosis, pulmonary embolus
Preservation of skin elasticity	Risk of endometrial cancer if ERT not supplemented with progestin
Treatment of hot flashes	Risk of some types of liver tumors
Maintenance of verbal memory	
Possible reduction in risk of colon cancer	
Reduction in vaginal atrophy	

Not all of the listed benefits and risks have been documented equally well. See text for discussion.

ERT/HRT is currently contraindicated for women with a history of thromboembolism, certain types of liver abnormalities, a history of uterine or breast cancer, or abnormal mammograms, because of the risk of exacerbating the preexisting condition. ERT/HRT can also have undesirable, but not dangerous side effects, such as bloating, weight gain, sore breasts, and headaches, that cause many women to withdraw from the ERT/HRT regimen.

Alternatives to standard ERT/HRT that maintain the benefits of estrogen but reduce its risks have been developed. The bisphosphonate alendronate has been shown to be a reasonable alternative to estrogen in the maintenance and increase in bone mass in postmenopausal women. The SERM raloxifene has been found to reduce significantly the risk of breast and endometrial cancer, while maintaining the benefits to bone mass and cardiovascular parameters.

REPRODUCTIVE AGING IN THE MALE

Androgens and sexual function

The study of reproductive aging in the male has focused on its main symptom: the decline in sexual activity. Since antiquity, the decline of sexual functioning with age has been lamented by all men. We read in the King James translation of the Old Testament (Kings I:1–4) that King David, a prodigious womanizer most of his life, in his seventies "gat *no heat.*" The young woman who was sent to him to arouse him "*cherished the king and ministered to him: but the king knew her not.*" The desire to prevent or ameliorate this effect of aging was the impetus in the nineteenth century for a series of experiments and therapies that we would call now testosterone replacement therapy. We have discussed in Chapter 6 the experiments of Brown-Séquard in France and the vasoligation procedure of Eugen Steinach in Austria, heralded as a rejuvenation therapy in the 1920s. Sigmund Freud, W. B. Yeats, and many other eminent men of late-nineteenth- and early-twentieth-century European society underwent the Steinach operation. In the 1920s, Serge Voronoff took the concept of Brown-Séquard and Steinach to its logical conclusion when he transplanted chimpanzees' testes into men, and according to him "*put back human aging by 20 to 30 years.*" Although not recognized as such now, perhaps because of the decidedly commercial emphasis of their program, these men could be considered pioneers in the development of hormone replacement therapy. Their claims, almost poignant in their extravagance, for the prevention of aging in the male were no different from those who first promoted ERT for females in the 1960s. The concept on which the therapy was based was, given the knowledge at the time, a reasonable one.

There have been attempts in the recent past to define a "male perimenopause," a "male menopause," or an "andropause" by grouping symptoms that have been associated with aging in men: decline in sexual activity, insomnia, impaired memory, psychological depression, loss of concentration, and fatigue. The analogy with the female perimenopause or menopause is invalid because men do not lose fertility in the discontinuous way that women do. They remain fertile until way into their 70s. There is apparently even one documented case of paternity at 94 years. Age-related changes in reproductive parameters do take place in males, but none of the symptoms variously associated with the putative "andropause" have been found to be due to a physiological decline in the activity of the H-P-T axis *per se*, but appear to be a consequence of the overall aging process.

The age-related decrease in sexual functioning has been documented in a number of studies. In one well-known recent study, the 1994 Massachusetts Male Aging Study, involving 1709 men, 52 percent of the 1290 men who completed a sexual activity questionnaire experienced minimal, moderate, or complete erectile dysfunction. The prevalence of complete erectile dysfunction increased from 5 percent to 15 percent between 40 and 70 years. Another study of 183 healthy, married men aged between 60 and 80 years, shows the changes in sexual activity that accompany age (Table 10.3).

Table 10.3
Sexual Activity in One Year in 183 Older Married Men

Age (yr)	Low	Number of men	Moderate	Number of men	High	Number of men
60–64	0–21	20	25–30	16	51–136	19
65–69	0–2	13	3–30	16	31–200	16
70–74	0–2	21	3–19	20	20–125	20
75–79	0–3	7	4–27	8	28–76	7

Activity is measured as events leading to orgasm. The division into the low, moderate, and high categories is somewhat arbitrary. (Adapted from Tsitouras and Bulat, 1995, with permission.)

This and other studies show that aging men experience a decline in sexual interest and activity, marked by differing degrees of severity. Until recently, the decline in sexual functioning was thought to be due to a decline in androgen levels in aging men. This was certainly the rationale for the experiments of Brown-Séquard, Steinach, and Voronoff. The relationship between testosterone and erectile function was thought to be well understood from observing the consequences of removal of the testes in young men. Hence, it seemed natural to suppose that aging men were suffering from androgen deficiency and that the symptoms of aging were the result of that androgen deficiency.

Do androgen levels decline in aging men? The results of many recent well-designed studies have led to the conclusion that there is an age-related reduction in mean androgen levels. The decrease is not large, and many investigators do not consider the decrease to be physiologically meaningful for several reasons. First, the variability in androgen levels at all ages is large, and a large fraction of older men have androgen levels that are in the same range as young men. Second, several studies have shown that testosterone levels that are 60 percent of those found in younger men are sufficient to maintain normal sexual activity in men younger than 50 years. Third, there appears to be a very modest correlation between sexual activity and testosterone levels. The variation in sexual activity is quite large at each age (Table 10.3) but on average, older men with higher levels of sexual activity tended to have higher testosterone levels. However, as in the age-related changes in testosterone levels, there was a large overlap in testosterone levels with sexual activity. The slight correlation between testosterone

Table 10.4
Disorders Associated with Erectile Dysfunction

System affected	Disorder
Neurologic	Brain lesions, spinal cord lesions, pelvic nerve lesions, autonomic neuropathies
Endocrine	Diabetes, hypogonadism, hypothyroidism, hyperthyroidism, increased endogenous estrogen production (tumors)
Genital	Orchidectomy, radical prostectomy
Vascular	Sickle cell anemia
Other	Renal failure, hepatic cirrhosis, malignancies, chronic infections

Adapted from Tsitouras and Bulat, 1995, with permission.

and sexual activity remains unclear. It is quite possible that high sexual activity leads to higher testosterone, rather than the other way around. No relationship between smoking, obesity, muscle mass, or evidence of CHD and sexual activity was found. On the other hand, consumption of moderate amounts of alcohol (4 oz. per day) did not change testosterone levels but did lead to decreased levels of sexual activity. Fourth, it is not yet clear that testosterone supplementation in older males has been successful in increasing sexual activity.

As a general conclusion, for most men, the age-related decrease in sexual activity or age-related erectile dysfunction may not be due to androgen deficiency. Erectile dysfunction can result from a variety of disorders (Table 10.4). An important contributor to erectile dysfunction in older males is a side effect of medications (Table 10.5). Age itself appears to be the most dependable determinant of sexual activity, but the nature of the age-related factors responsible is not well understood.

Table 10.5
Some Common Therapeutic Drugs
Associated with Erectile Dysfunction

Drug group	Agent
Psychotropic	Barbiturates, alcohol (abuse), marijuana, lithium, phenothiazines, antidepressants, narcotic analgesics
Antihypertensives	Reserpine, methyldopa, clonidin
Other	Atropine, benzatropine, metaclopramide, estrogens, spironolactone, adrenal steroids (high doses), serotonin antagonists

Androgen replacement therapy

Despite the fact that the decline in androgen levels may not be physiologically important with respect to erectile function and libido, aging men do show clinical signs of androgen deficiency, in particular, loss of body mass and muscle strength (*sarcopenia*), and an increase in visceral (abdominal) fat. These features, when present in younger men, are known to be due to loss of testicular function, as a consequence of illness or physical trauma to the testes. In such cases testosterone replacement therapy is quite successful in restoring body mass and strength, vigor, and erectile function. However, the efficacy of testosterone replacement therapy in older men has not yet been demonstrated definitively. In some studies, it is relatively unsuccessful, while in others, some positive effects have been noted. The differences in response may depend on the general health of the individual, with healthy older men who maintain some level of physical activity responding better than others who may be in poor health and physical condition. The differences may also depend on the specific parameter used to measure response to the exogenous testosterone administration. In general, there appears to be some likelihood that with aging, androgen-dependent tissues respond less well to circulating androgens. This androgen resistance could be due to degeneration or loss of androgen receptors. Although neither of these possibilities has been studied extensively, an age-related decrease in androgen response, rather than a decline in androgen levels, may be more important in contributing to the symptoms of aging in males.

Androgen replacement therapy may also not be free of risks. The potential risks of androgen replacement therapy center around CHD and prostate cancer. Exogenous androgens may have deleterious effects on the cardiovascular system. The prostate is particularly sensitive to androgens. Here again, the data is not yet definitive. Some short-term studies of androgen replacement in older men who were not androgen deficient showed significant increases in prostate size. In other studies, testosterone replacement in older men who were androgen deficient did not result in increased prostate size. Many more studies are needed before we understand the complete risk/benefit equation for androgen replacement therapy in older men.

Aging and spermatogenesis

Although abundant anecdotal evidence has indicated that men remain fertile until quite old, relatively few studies have examined spermatogenetic parameters in aging men carefully. The data available indicate that the sperm count may not change significantly with age, but subtle changes in other measures of sperm quality may take place, such as smaller ejaculate volume, decreased sperm motility, and a higher fraction of abnormal spermatozoa. Again the variation in these parameters is large. Some postmortem examinations have also reported age-related reductions in the volume of the seminiferous tubules and the length of the tubules.

SUMMARY

Reproductive aging in the male is characterized by a decline in sexual activity and by the loss of body mass and muscle strength. Although androgen levels decline to some extent with age, there is no compelling evidence that the decline in sexual activity is associated with these modest changes in androgen levels. Rather, the reduction in sexual activity is probably due to as-yet-unidentified age-related factors that most likely work at the CNS level. The loss of body mass and strength, features that are androgen dependent, may be due to age-related androgen insensitivity in androgen-dependent tissues.

QUESTIONS

1. What observations in humans suggest that the ovarian reserve is an important parameter in determining the onset of the perimenopause?

2. How do you see the risk/benefit equation with respect to ERT? How would you explain it to your mother or other female relatives approaching the menopause?

3. How do you view the skeptics' response to the widespread promotion of ERT?

4. Which explanation/model for the early reproductive senescence in females do you find the most compelling?

5. Reproductive aging in the male can be due to androgen deficiency or androgen resistance. What is the difference, and how can they be distinguished operationally?

6. Do you agree with Simone de Beauvoir that the image of femininity determines the way in which menopause is experienced?

7. Library project:
   a. ERT does not always relieve many symptoms associated with menopause. What accounts for this lack of uniform response?
   b. Review the results of clinical studies that have examined the effects of androgen replacement therapy in older males.

SUPPLEMENTARY READING

Erickson, G. F. 1997. Perimenopause. In *Dissociation of Endocrine and Gametogenetic Ovarian Function*. R. A. Lobo, Ed. Springer-Verlag, Inc., New York, NY.

Diamond, J. 1996. Why women change (evolution of menopause). *Discover* 17(7), 130–138.

Ginsburg, E. L., *et al.* 1996. Effects of alcohol ingestion on estrogens in postmenopausal women. *Journal of the American Medical Association* 276, 1747–1751.

Gosden, R. G. 1985. *Biology of Menopause: The Causes and Consequences of Ovarian Aging*. Academic Press, London, UK.

Heaney, R. P. 1997. Osteoporosis. In *Preventive Nutrition: The Comprehensive Guide for Health Professionals*. A. Bendich and R. J. Deckelbaum, Eds. Humana Press Inc., Totowa, NJ, pp. 285–303.

Hill, K., and A. M. Hurtado. 1991. The evolution of premature reproductive senescence and menopause in human females: An evaluation of the "grandmother hypothesis." *Human Nature* 2(4), 313–350.

Tsitouras, P. D., and T. Bulat. 1995. The aging male reproductive system. *Endocrine and Metabolic Clinics of North America* 24(2), 297–315.

ADVANCED TOPICS

Agnusdei, D. 1999. Clinical efficacy of raloxifene in postmenopausal women. *European Journal of Obstetrics, Gynecology, and Reproductive Biology* 85(1), 43–46.

Barrett-Connor, E., C. A. Cox, and P. W. Anderson. 1999. The potential of SERMs for reducing the risk of coronary heart disease. *Trends in Endocrinology and Metabolism* 10(8), 320–325.

Barrett-Connor, E., and C. Stuenkel. 1999. Hormones and heart disease in women: Heart and estrogen/progestin replacement study in perspective. *Journal of Clinical Endocrinology and Metabolism* **84**(6), 1848–1853.

Colditz, G. A., *et al.* 1990. Prospective study of estrogen replacement therapy and risk of breast cancer in postmenopausal women. *Journal of the American Medical Association* **264**, 2648–2653.

Cummings, S. R., *et al.* 1999. The effect of raloxifene on risk of breast cancer in postmenopausal women: Results from the MORE randomized trial. Multiple Outcome of Raloxifene Evaluation. *Journal of the American Medical Association* **281**(23), 2189–2197.

Gapstur, S. M., J. D. Potter, T. A. Sellers, and A. R. Folsom. 1992. Increased risk of breast cancer with alcohol consumption in postmenopausal women. *American Journal of Epidemiology* **136**, 1221–1231.

Gooren, L. J. G. 1996. The age-related decline of androgen levels in men: Clinically significant? *British Journal of Urology* **78**, 763–768.

Hulley, S., *et al.* 1998. Randomized trial of estrogen plus progestin for secondary prevention of coronary heart disease in postmenopausal women. *Journal of the American Medical Association* **280**, 605–613.

Kirkwood, T. B. I. 1997. The origins of human aging. *Philosophical Transactions of the Royal Society, London* B **352**, 1765–1772.

Lufkin, E. G., and S. J. Ory. 1995. Postmenopausal estrogen therapy, 1995. *Trends in Endocrinology and Metabolism* **6**(2), 50–54.

Mildvan, A. S., and B. L. Strehler. 1960. A critique of theories of mortality. In *The Biology of Aging.* B. L. Strehler, Ed. American Institute of Biological Sciences, Washington, D.C.

Mundy, G. R. 1994. Calcium and common bone endocrine disorders. In *Clinical Endocrinology*, chapter 18, 2nd ed. G. M. Besser and M. O. Thorner, Eds. Mosby-Wolfe, London, UK.

O'Rourke, M. T., S. F. Lipson, and P. T. Ellison. 1996. Ovarian function in the latter half of the reproductive lifespan. *American Journal of Human Biology* **8**, 751–759.

Russell, R. G., and M. J. Rogers. 1999. Bisphosphonates: From the laboratory to the clinic and back again. *Bone* **25**(1), 97–106.

Wood, J. W. 1990. Fertility in anthropological populations. *Annual Review of Anthropology* **19**, 211–242.

Writing Group for the PEPI Trial. 1995. Effects of estrogen or estrogen/progestin regimens on heart disease risk factors in postmenopausal women: The Postmenopausal Estrogen/Progestin Interventions Trial. *Journal of the American Medical Association* **273**, 199–208.

Zumoff, B. 1997. Editorial: The critical role of alcohol consumption in determining the risk of breast cancer with postmenopausal estrogen administration. *Journal of Clinical Endocrinology and Metabolism* **82**(9), 1656–1658.

Constantin Brancusi, *The First Cry*, 1917.
(© Artists Rights Society [ARS], New York/ADAGP, Paris.)

PART

IV

A NEW LIFE

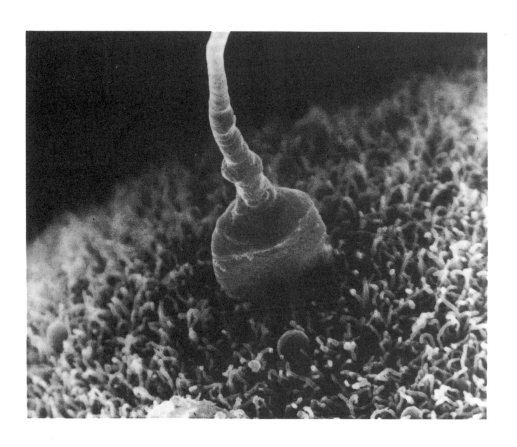

Human sperm penetrating an egg.
(© Photo Researchers, Inc. David Phillips, photographer.)

CHAPTER 11

Fertilization and Implantation

Selection pressure in the female reproductive tract permits the presentation of optimal sperm to the egg.

—P. M. Saling. 1996

Fertilization: Mammalian gamete interactions. In
Reproductive Endocrinology, Surgery, and Technology.
E. Y. Adashi, J. A. Rock, and Z. Rosenwaks, Eds. Lippincott-Raven Publishers.

Various processes have evolved that increase the ability of the embryo to access the blood supply of the mother. The main one is the invasion of the uterine wall by the embryo.

—C. A. Finn. 1998

Menstruation: A nonadaptive consequence of uterine evolution.
The Quarterly Review of Biology 73(2), 163–173.

FERTILIZATION, the fusion of the ovum with the spermatozoon, accomplishes three things: (1) the normal diploid complement of 46 chromosomes is restored; (2) the chromosomal sex of the embryo is determined; (3) embryogenesis is initiated. It is a remarkable process. Consider what is required. The ovulated oocyte, expelled from the ovary, must be "caught" by the fimbriae of the uterine tube and transported to the ampullary-isthmic region of the uterine tube (Fig. 11.1). The ovulated oocyte has a limited lifetime, perhaps no more than 24 hours, after which it degenerates and is resorbed by the tissue of the uterine tubes. The spermatozoa, deposited in the upper parts of the vagina, must traverse the cervical-vaginal junction, move through the length of the uterus and into the uterine tube, and finally to find and eventually penetrate the oocyte (Fig. 11.1). The functional life span of a spermatozoon after ejaculation is also short, perhaps no more than 24 hours. Each step appears precarious. If fertilization did not occur as frequently as it does, we would probably predict that the conditions under

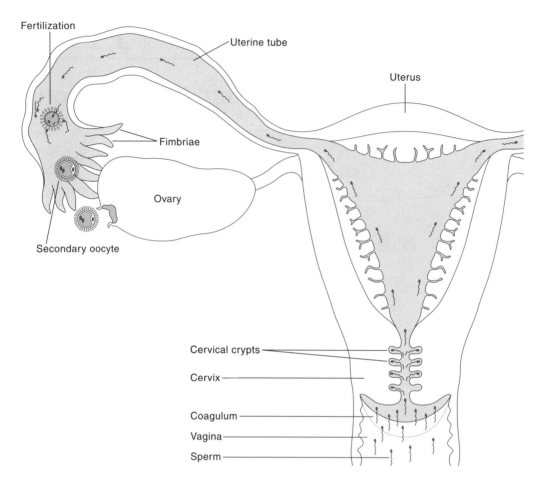

Figure 11.1
Transport of oocyte and sperm in the female reproductive tract. Sperm deposited during intercourse are trapped in a coagulum at the cervical-vaginal boundary and in small invaginations known as cervical crypts. Sperm released from the cervical crypts move through the uterus and into the uterine tube. The expelled secondary oocyte is caught by the fimbriae of the uterine tube and transported to the ampullary region of the uterine tube to await the arrival of sperm.

which the ovum and spermatozoon must meet are so special and unique that fertilization would be rare. In fact, the opposite is the case, thanks to millions of years of perfecting it. Each step is regulated, probably by the female genital tract and by the egg itself, thus ensuring that what appears to be a rather improbable event becomes highly probable.

Fertilization sets in motion a series of transformations that convert the fertilized egg into a viable and functional human being in the space of some 38 weeks. These transformations are of such complexity that even today with the considerably dispassionate approach of most modern biologists, we cannot but marvel at their intricacy.

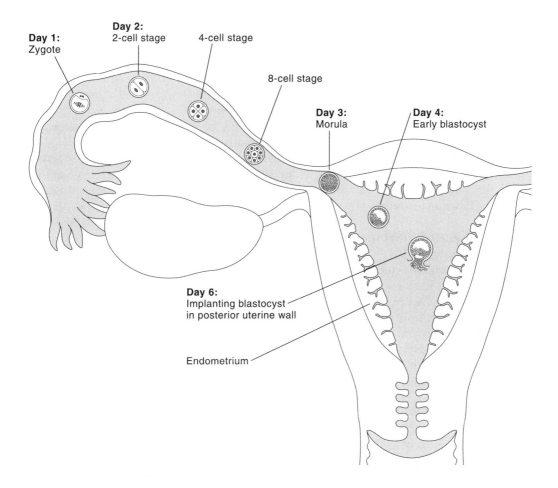

Figure 11.2
The preimplantation stage of human development. Day 1: Fertilization and formation of zygote. Day 2: Initiation of cleavage divisions and formation of the two-cell embryo. Day 3: The morula enters the uterus. Day 4: Formation of the blastocyst. Days 5–7: Implantation begins, commonly in the posterior uterine wall.

The first stage in the development of a new human being is known as the **preimplantation stage** of embryogenesis. Within a few hours after fertilization, the fertilized egg, the zygote, begins to undergo the **cleavage** divisions. Simultaneously, the developing embryo moves through the uterine tube, emerging into the uterus after 4 to 5 days. Around 5 to 7 days after fertilization, the developing embryo attaches to the uterine endometrium, and initiates **implantation** (Fig. 11.2). At this point, a pregnancy is considered to begin. In this chapter we will consider the requirements for fertilization, review the fertilization process itself, and finally examine the nature of implantation and the way in which the developing embryo signals its existence to the mother.

EGG AND SPERM

Oocyte transport

At ovulation, the secondary oocyte, which is arrested in the second meiotic division, is expelled from the ovary. The fimbriae of the uterine tube appear to move to cover the surface of the ovary, presumably to ensure that the extruded oocyte will land in the uterine tube. The oocyte, along with its adhering cumulus cells, is transported into the uterine tube by the beating of the cilia that line the interior of the uterine tube. Transport of the oocyte is blocked at the ampullary-isthmic junction (Fig. 11.2). The block is functional, rather than anatomical, since no clearly defined anatomical structure is present at this site. The block presumably permits the oocyte to remain at this junction to await the spermatozoa for fertilization. The oocyte remains functional at the junction for perhaps 24 hours.

Epididymal maturation

Spermatozoa taken directly from the testis, although fully formed in a morphological sense, have poor fertilizing ability compared to spermatozoa isolated from the distal end of the epididymis. This suggests that spermatozoa acquire fertilizing ability as they pass from the testis through the epididymis and into the vas deferens. Many changes in sperm characteristics have been identified during their 10- to 14-day sojourn through the epididymis. Three important changes whose significance for fertilization are rather clear are: (1) acquisition of motility, (2) acquisition of binding specificity to the zona pellucida (ZP), and (3) change from glucose to fructose as the main spermatozoon energy source. These and other changes experienced by the spermatozoa in the epididymis appear to be androgen dependent.

Motility refers to the ability of the spermatozoon to move in a forward direction by the motion of its flagellum. When examined in a culture dish, sperm taken directly from the testis remain relatively motionless. After passage through the epididymis sperm exhibit two alternating main types of motility—more or less forward motion interspersed with a gyrating motion in which the spermatozoon moves in circles (hyperactivated motility) (Fig. 11.3). Although the significance of these two types of motion remains unclear, both appear to be required for normal male fertility. At first glance it might appear that motility would be required at all stages after the spermatozoa have been ejaculated, but such is not the case. A number of studies have indicated that motility may be required only in passage from the vagina to the cervix and at the junction of the uterus and uterine tube. Movement through the uterus may not depend on motility.

The ability to bind to the ZP is, as we shall see below, absolutely essential for fertilization. It seems likely that this property is the result of an "unmasking" of the ZP binding sites on the sperm surface rather than their synthesis. The change from glucose to fructose metabolism is also essential because the sperm rely on components present in the seminal fluid for their metabolism and survival. In a sense, seminal fluid functions as a portable survival kit for the spermatozoon. Fructose, the main sugar found in honey, is the primary energy source present in seminal fluid.

Capacitation

For many mammalian species, spermatozoa taken from the epididymis still have poor fertilizing ability. Their fertilizing ability improves only after they have spent

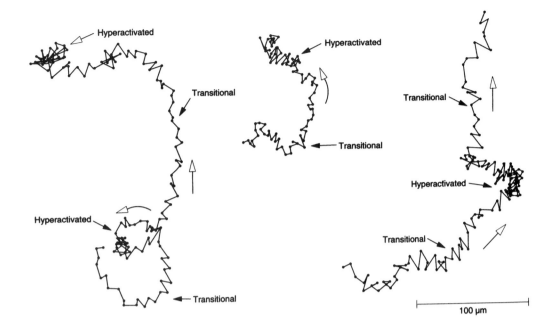

Figure 11.3
Multiphasic activity by human spermatozoa during which they spontaneously switch from
one form of movement (transitional) to another (hyperactivated). This is a common test for the
fertilizing potential of an ejaculate. (Reprinted from J. Aitken. 1995. Cell biology of human
spermatozoa. In *Gametes: The Spermatozoan*. J. G. Grudzinskas and J. L. Yovich, Eds.
Cambridge University Press, with permission).

some time in a suitably estrogen-primed uterus of a receptive female. The nature of
these uterine-dependent changes, generally referred to as **capacitation**, remains incompletely understood, but they include removal of components from the sperm surface,
which then permits the spermatozoon to penetrate the zona pellucida (ZP). Originally
discovered in 1951, capacitation appears to be necessary for fertilization in many, but
not all, mammalian species. In humans, although some early studies suggested that a
capacitation period of 5 to 7 hours was required to achieve optimum fertilization potential, more recent studies have questioned the need for capacitation. One important
argument against the need for capacitation in humans has come from the rapidly developing technology of fertilization *in vitro*, in which uncapacitated sperm suspended in
a salts solution have been shown to be quite capable of fertilizing an oocyte. The success of fertilization *in vitro* may indicate that exposure of the sperm to the salts solution
may produce the same results as exposure to uterine fluids. On the other hand, the conditions of fertilization *in vitro* are substantially different from those *in vivo*. In particular, under *in vitro* conditions the ratio of sperm to oocyte is at least 1,000 times greater
than under *in vivo* conditions. Because of this huge difference in sperm numbers, it cannot be said definitively that capacitation is not necessary in humans. For the present,
then, this remains an open question.

275

Semen analysis

The fertilizing ability of an ejaculate is assessed by several parameters. Table 11.1 provides a list of the most common ones. The sperm count, the concentration of sper-

Table 11.1
Normal Parameters for Evaluating Semen Quality

Semen parameters	Normal	Marginal	Abnormal
Volume (ml)	2–5	1–2	<1
Sperm count (10^6/ml)	20–250	10–20	<10
Sperm motility (%)	>50	40–50	<40
Normal sperm morphology	>50	40–50	<40

Reference: Overstreet *et al.*, 1992.

matozoa, expressed as the number of spermatozoa per milliliter (ml), is one of the most important measures of fertilization potential. Conventionally, normal fertility is said to require counts of at least 20 million per ml. Sperm counts can be quite variable, however, with counts of up to 150 million per ml not being unusual. A male is said to be subfertile if the sperm count is between 1 and 20 million per ml, and infertile if his sperm count is below 1 million per ml. Despite these definitions there are documented cases of fertility of males whose sperm count was 100,000 per ml. However, the couples in these examples tried to get pregnant for up to 12 years. These cases suggest that although the probability of conception per intercourse decreases as the sperm count decreases below 20 million/ml, over a long period of trials, a conception is possible.

Perhaps as important as the sperm count is the fraction of motile and morphologically normal sperm. Different types of abnormalities are seen even in normal ejaculates. Figure 11.4 illustrates some of the types of abnormalities commonly seen in normal ejaculates, for example, sperm with abnormally large or small heads, sperm with short tails, or sperm with two heads and/or two tails. Other morphologies are seen in ejaculates of men with *teratospermia* (characterized by a high frequency of abnormal sperm) (Fig. 11.5). The fraction of abnormal sperm that can be said to compromise the overall sperm quality enough to reduce fertility significantly has been difficult to establish in humans. In a species such as a bull, for example, 5 percent or less abnormal sperm is considered normal; higher values would disqualify a bull for stud purposes. In humans, a species not selected for high fertility, the fraction of abnormal sperm can be quite variable, and the quantitative relationship between the fraction of abnormal sperm and functional deficiency is not clear. Probably a value of greater than 50 percent abnormal sperm would be considered to compromise the sperm quality, but ejaculates with 30 percent and 40 percent abnormal sperm may not necessarily be associated with infertility.

The volume and composition of the ejaculate are also important parameters in determining overall fertility. Ejaculate volumes of 2 to 5 ml (average 3.5 ml) are considered normal for humans, and volumes less than 2 ml or greater than 5 ml are associated with subfertility. It is interesting to note that ejaculate volumes, as well as the number

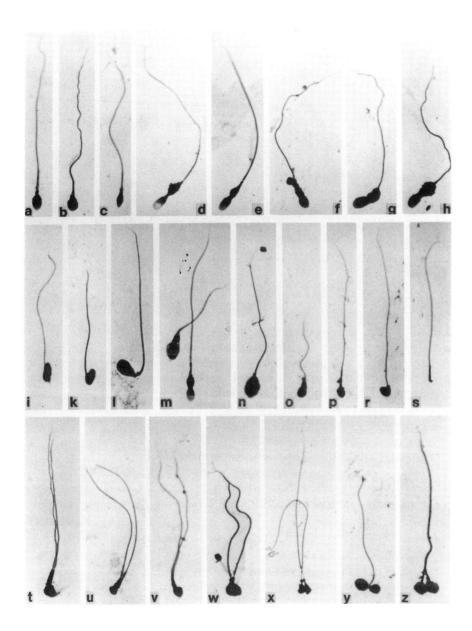

Figure 11.4
Examples of abnormal sperm from normal ejaculates: (a–c) normal sperm; (d–e) abnormal middle piece; (f–l) abnormal head, no middle piece; (m–r) abnormal; (s) no head; (t–w) two-tailed sperm; (x–z) two heads. (Reprinted from A. H. Holstein, E. C. Roosen-Runge, and C. Schirren, 1988. Courtesy of A. H. Holstein.)

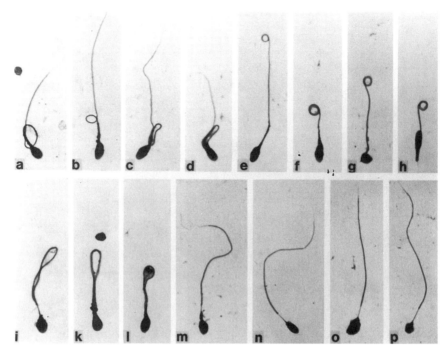

Figure 11.5
Abnormal sperm from teratospermic ejaculates (<40 percent normal sperm): (a–d) ring or
loop formation in the middle piece region; (e–h) loops at end of tail; (i–l) loops in principal
piece; (m–p) absence of middle piece. (Reprinted from A. H. Holstein, E. C. Roosen-Runge,
and C. Schirren, 1988. Courtesy of A. H. Holstein.)

of sperm per ejaculate, vary significantly among mammalian species (Table 11.2). Humans do not appear to enjoy any unusual status in this regard. The highest recorded
ejaculate volumes (500 ml) have been observed in the wild boar, a remarkable circumstance given its relatively small size. This may explain why the boar was such a power-

Table 11.2
Sperm Count and Ejaculate Volume
in Humans and Other Animals

Species	Sperm count (million/ml)	Volume (ml)
Ram	2000–5000	1–2
Human	50–150	2–5
Bull	300–2000	2–10
Dog	60–300	2–15
Stallion	30–800	30–100
Boar	20–300	150–500

Reference: Sharpe, 1994.

Table 11.3
Contributors to the Ejaculate

Source of secretion	% of ejaculate
Testes	5
Seminal vesicles	46–80
Prostate	13–33
Bulbourethral glands	2–5

Reference: Lundquist, 1949.

ful totem in many societies familiar with the species. The significance of this large variation in ejaculate volumes remains unknown.

Most of the ejaculate volume consists of seminal fluid contributed primarily by the seminal vesicles and the prostate, while the testes and the bulbourethral glands contribute less than 10 percent of the ejaculate volume (Table 11.3). Each of these four tissues contributes characteristic compounds. For example, the seminal vesicle marker is fructose. Hence, absence of a particular component in the ejaculate is indicative of abnormality in the tissue that provides that component. Because 40 to 80 percent of the ejaculate volume is contributed by the seminal vesicles, obstructions or dysfunction in the seminal vesicles are generally detected by a large reduction in ejaculate volume.

Concern about the sperm count

Several recent studies have suggested that the sperm count in some countries has been decreasing since World War II. In a comprehensive summary of 61 studies involving a total of 14,947 European men, the sperm count between 1938 and 1990 was found to have decreased from 113 million per ml to 66 million per ml. In a separate French study comparing the sperm count of healthy sperm donors in 1973 and 1992, a decline from 89 million per ml to 60 million per ml was reported. These studies have generated considerable controversy and have been criticized on many grounds. One of the perplexing observations from more recent studies is that the decrease in sperm count is not geographically uniform. Denmark, for example, appears to have suffered a greater decline than other European countries. Indeed, the first indications of significant changes in the sperm count came from studies carried out in Denmark. On the other hand, the sperm count in Finland, with a climate and level of socioeconomic development similar to that of Denmark, appears not to have undergone any change. The average sperm counts of Finnish men appears to be higher than Danish men.

The decrease in sperm count may be only one of several parameters whose changing values, in the eyes of some experts, reflect a deterioration of male reproductive health. The incidence of testicular cancer, cryptorchidism, and **hypospadias** (failure of the urethral folds to fuse properly during the development of the penis, resulting in the external opening of the urethra being on the underside of the penis, rather than at the tip) has also been reported to be increasing in England, Scotland, the Scandinavian and Baltic countries, Australia, New Zealand, and the United States. Denmark has the highest rate of testicular cancer in the world, where the lifetime risk of developing testicular cancer is 1 percent, over fourfold greater than in Finland or the United States.

The decline in male reproductive health suggested by some studies may not apply to all populations. Taken at face value, the studies indicate that a decline may be taking place in some but not all regions of the world. There is no general agreement about the nature of the factors that might account for the differences. The Danish investigators whose studies first brought this complex question to scientific attention have suggested that the decline could be due to exposure to a variety of toxic chemicals that have been shown to be weak estrogens or weak antiandrogens. The supposition is that such compounds may interfere with normal male sexual development during the fetal period. Denmark, which industrialized much before Finland, would have experienced a higher and longer exposure to such putative endocrine-interfering substances than Finland. However, other geographical regions that industrialized early do not appear to have experienced similar decreases in sperm count. Another recent suggestion is that maternal smoking may interfere with normal male development. Maternal smoking has been suggested as a possible way to account for the differences between Denmark and Finland. The smoking rates among Finnish women have been significantly lower than in other countries at least until the late 1980s. There is some evidence from studies with animals to suggest that maternal smoking can affect the fertility of both male and female offspring. No clear-cut data for humans is yet available.

SUMMARY

Ovulation and spermatogenesis are only the first steps in preparing for a successful fertilization. The ovulated oocyte must be captured by the fimbriae of the uterine tube at the moment of expulsion from the ovary. The captured oocyte is then propelled along the uterine tube to the ampullary-isthmic junction to await the arrival of the sperm. All this implies that the uterine tube is not simply an inert transport tube, but that it exerts essential physiological control over oocyte function and movement. The sperm must undergo a series of changes after they leave the testis to enable them to fertilize the egg. The first, known as epididymal maturation, are androgen dependent and confer motility on the sperm and the ability to penetrate the zona pellucida surrounding the oocyte. In some species, sperm must also be capacitated before they acquire optimal fertilizing potential. Capacitation requires that sperm be exposed to uterine fluids from an estrogen-primed uterus for a suitable period of time. In humans, it is not yet clear that capacitation is essential for fertilization.

The fertilizing potential of an ejaculate is determined by several parameters, the most important of which are a minimum sperm count (20 million per milliliter), a suitable ejaculate volume (2–5 milliliters), a low fraction of morphologically abnormal sperm, and sperm that exhibit normal and hyperactivated motility. There appears to be some preliminary evidence for a decline in male reproductive health, manifested in decreasing sperm count, increase in testicular cancer, cryptorchidism, and hypospadias, with some geographical regions being affected more severely than others, but the causes remain unknown. Moreover, not all measures of reproductive health are affected equally, so that for example, an increase in the incidence of testicular cancer may not be tightly correlated with a decline in sperm count. Much more research is needed before the extent and seriousness of this problem can be evaluated properly and its causes identified.

SPERM-EGG FUSION

Sperm transport in the female

After ejaculation and deposition of the ejaculate in the vagina, spermatozoa lead an extremely precarious existence. Because the vaginal secretions are much more acidic (pH 5.7) than the seminal fluid (pH 7.2–7.8), the ejaculate is converted into a gelatinous mass in the vagina which entraps over 99 percent of the spermatozoa. This is one point at which motility appears to be required for movement of the spermatozoa into the cervix. Motility presumably helps the spermatozoa escape the coagulation at the cervical-vaginal boundary. At the optimal part of the cycle (that is, at about the time of ovulation), cervical secretions are watery and alkaline, conditions that favor sperm penetration and viability. The cervix also contains small invaginations, known as **cervical crypts,** which appear to act as temporary reservoirs for sperm that escape the coagulum at the vaginal-cervical boundary (Fig. 11.6). Sperm may be released periodically from the crypts to continue their journey into the upper segments of the genital tract. The factors that regulate sperm release from the crypts in this more or less periodical fashion are unknown.

No more than 1 percent of the spermatozoa (and probably much less) in the ejaculate move into the uterus. Interestingly, although movement through the uterus can be extremely rapid (less than 5 minutes), motility may not be required. Nonmotile spermatozoa apparently are able to move as rapidly as motile ones. The mechanism by which sperm move through the uterus remains unclear. In mice, a sperm protein is re-

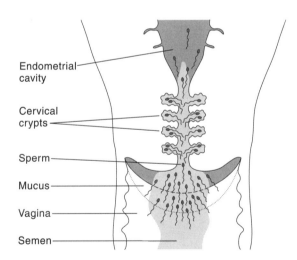

Endometrial cavity

Cervical crypts

Sperm

Mucus

Vagina

Semen

Figure 11.6
Sperm migration through the cervix. More than 99 percent of sperm are trapped in a coagulum at the cervical-vaginal boundary. Those that move into the cervix enter cervical crypts that appear to function as reservoirs from which sperm are released periodically to enter the uterine cavity. (Adapted from H. B. Croxatto. 1996. Gamete transport. In *Reproductive Endocrinology, Surgery, and Technology.* E. Y. Adashi, J. A. Rock, and Z. Rosenwaks, Eds. Lippincott-Raven Publishers, Philadelphia, PA.)

Table 11.4
Comparison of Sperm Dynamics in Different Species

Species	Site of deposition	Sperm/ejaculate (millions)	Sperm at fertilization site	Time to arrival at fertilization site
Human	Vagina	280	<200	5–68 min
Mouse	Uterus	50	<100	15 min
Rat	Uterus	58	500	15–30 min
Rabbit	Vagina	280	200–500	2–60 min
Cow	Vagina	3000	few	2–13 min
Sheep	Vagina	1000	600–700	6 min–5 hr
Pig	Uterus	8000	1000	15 min

Reference: Harper, 1982.

quired. Mice sperm that lack the protein *fertilin β* have a significantly reduced ability to move through the uterus. Fertilin β is synthesized during spermatogenesis and integrated into the sperm head membrane. The role of fertilin β in sperm transport through the uterus is unknown. An alternative and complementary possibility is that sperm transport is facilitated by small uterine contractions induced by prostaglandins in the seminal fluid. It is also not clear whether sperm move through the uterus as a cohort or individually. Motility, and in mice, the presence of fertilin β, appear to be required to traverse the junction of the uterus and uterine tube. The rate of attrition is high, however (Table 11.4). Of an estimated 100–300 million sperm deposited in the vagina, 100,000 to 1 million may reach the uterus, and probably no more than 200 reach the site of fertilization.

Sperm are subjected to a rigorous selection process once they are deposited in the female reproductive tract. The coagulum that forms at the cervical-vaginal boundary and the cervical crypts may have more than one function. First, the coagulum may provide a way to retain the sperm long enough so that they have a chance to move into the uterus. Without it, the sperm might simply fall back out the vagina. Second, the coagulum and the cervical crypts may also serve to select those sperm with the highest motility to ensure that they will, in principle, be able to reach the oocyte. Third, the physical barrier provided by the coagulum may also limit the number of sperm that make it into the uterus and eventually to the site of fertilization. This ensures that the likelihood of multiple fertilizations remains low (see below).

Sperm-oocyte interaction

Fertilization usually occurs in the ampullary region of the uterine tube (Fig. 11.1). The ovulated secondary oocyte has been primed to receive the sperm. Along the periphery of the egg plasma membrane, a large number of vesicles known as the **cortical granules** make their first appearance. The first polar body from the first meiotic division is contained in the **subzonal space**, the region between the egg membrane and the ZP (Fig. 11.7). There are four phases in the interaction of sperm and oocyte: (**1**) induction of the acrosome reaction (Fig. 11.8), (**2**) penetration of the ZP by the acrosome-

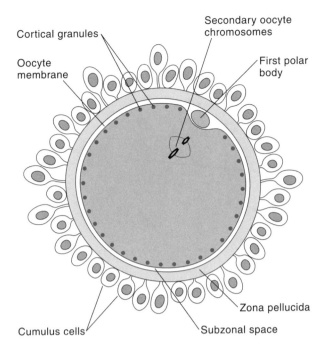

Figure 11.7
A secondary oocyte as it awaits fertilization. The second meiotic division has been arrested. The first polar body, the product of the first meiotic division, is found in the subzonal space between the oocyte membrane and the zona pellucida (ZP). The cortical granules line the interior periphery of the oocyte membrane.

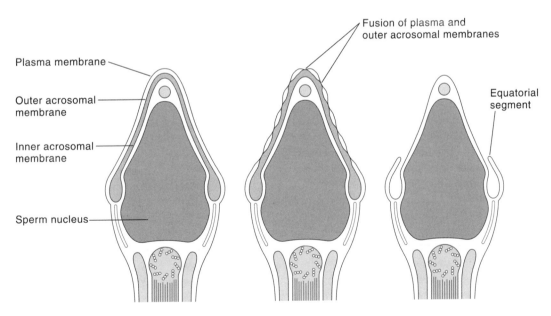

Figure 11.8
The acrosome reaction, initiated as the spermatozoon makes contact with the cumulus cells surrounding the oocyte, and completed before the sperm penetrates the ZP. Fusion of the outer acrosomal and sperm head plasma membrane at multiple, distinct points on a large portion of the sperm surface, followed by the disintegration of the acrosome, leaving a sperm retaining the equatorial segment of the sperm plasma membrane.

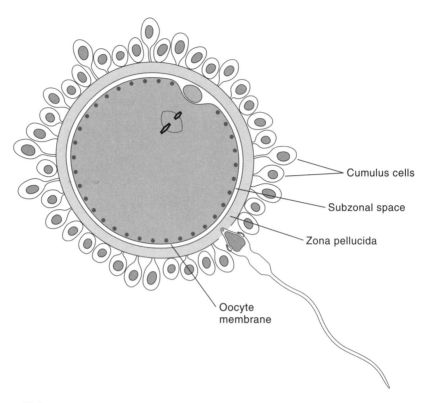

Figure 11.9
The initiation of fertilization. Movement of the sperm through the cumulus matrix, and after completion of the acrosome reaction, penetration through the zona pellucida.

reacted spermatozoon (Fig. 11.9), (**3**) fusion of the oocyte and sperm membranes (Fig. 11.10), and (**4**) induction of the cortical granule reaction and formation of the male and female pronuclei (Fig. 11.10).

The acrosome reaction begins probably when a spermatozoon comes in contact with the cumulus cells surrounding the oocyte. Recall that the cumulus cells are granulosa-derived cells that remained attached to the oocyte when the tertiary and Graafian follicles were formed and that remain with the oocyte at ovulation. The **acrosome reaction** involves the fusion of the sperm plasma membrane with the outer acrosomal membrane and the release of the acrosomal contents (Fig. 11.8). The acrosomal contents include hydrolytic enzymes that help the spermatozoon bore its way through the cumulus layer, and eventually through the ZP. There are probably several stages to the acrosome reaction, each contributing to the final goal, the penetration of the ZP. There are probably progressive triggers. For example, recent experiments indicate that progesterone secreted by the cumulus cells may be one of the first triggers of the acrosome reaction, which begins with the passage of the spermatozoon through the cumulus layer. The fraction of sperm capable of undergoing the acrosome reaction, even after epididymal maturation and capacitation, is variable, but rarely 100 percent. This means that not all sperm that reach the oocyte will in fact be able to fertilize it.

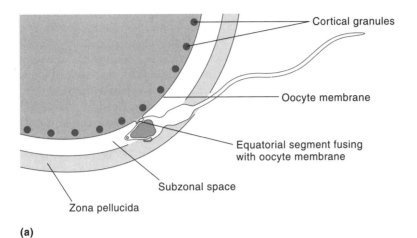

(a)

(b)

Figure 11.10
The completion of fertilization. (a) Fusion of the sperm plasma membrane at the equatorial region with the oocyte plasma membrane triggers the completion of the second meiotic division and the cortical granule reaction. (b) The second polar body is extruded into the subzonal space, and the maternal and paternal pronuclei begin to form. The cortical granule reaction, fusion of the cortical granules with the oocyte membrane, "hardens" the membrane, preventing polyspermy. The two pronuclei subsequently fuse to form the zygote, initiating embryogenesis.

Penetration of the ZP requires binding between one or two proteins of the ZP and specific receptors on the sperm surface. Binding of the sperm head to the ZP, then, is highly species specific. This specificity prevents fertilization of an oocyte of one species by the sperm of another. The mechanism of ZP penetration is not completely known. It probably involves enzymes released from the acrosome, which may help the spermatozoon digest its way through the ZP (Fig. 11.9). Hyperactivated motility may also be required. This property is manifested in the wild gyrations of the sperm flagellum after the sperm makes contact with the ZP. Under the conditions that are normally used for *in vitro* fertilization, hyperactivated motility is highly correlated with the fertilization rate of oocytes in couples undergoing assisted reproduction (see Chapter 16). Under normal *in vivo* conditions, several spermatozoa move through the cumulus cells and reach the ZP at about the same time, but only rarely does more than one penetrate the ZP completely. Under *in vitro* conditions penetration of the ZP takes 10 to 15 minutes.

After the spermatozoon penetrates the ZP, it enters the subzonal space, and the oocyte and sperm plasma membranes fuse (Fig. 11.10). Fusion of the two membranes triggers two important events: the cortical granule reaction and the resumption of the second meiotic division. The **cortical granule reaction** is characterized by the fusion of the cortical granules and the oocyte membrane. Presumably the granules empty their contents into the subzonal space. One important result of this reaction is that it "hardens" the oocyte membrane and prevents the fusion of any other sperm that may have also penetrated the ZP at about the same time. The cortical granule reaction limits the frequency of **polyspermy**, or the fertilization of an oocyte by more than one sperm. The reinitiation of the second meiotic division results in the formation and the extrusion of the second polar body (Fig. 11.10). The nucleus that remains in the secondary oocyte is referred to as the *female pronucleus*; it contains the haploid number of chromosomes (23), one member of each pair of homologous chromosomes. The sperm nucleus is converted to the *male pronucleus*; the two pronuclei remain separate and visible for several hours. Eventually, the pronuclei membranes disappear, the chromosomes replicate, and this marks the beginning of embryogenesis.

Summary

The different components of the female reproductive tract—the cervix, uterus, and uterine tubes—are not passive elements in the fertilization process. In fact, their active participation is absolutely required for fertilization. Cervical secretions around the periovulatory period may not only play an indispensable role in minimizing sperm loss after an ejaculate has been deposited at the cervical-vaginal boundary, but they may also provide a stringent selection filter permitting only those sperm with the requisite motility to move into the uterus. Uterine and uterine tube secretions around this time also provide an environment that facilitates sperm transport and survival. Sperm-oocyte fusion is orchestrated by signals provided by the oocyte. Perhaps the most obvious one is the induction of the acrosome reaction, which permits the sperm to penetrate the zona pellucida. The oocyte, through induction of the cortical granule reaction, also limits the frequency of polyspermy, the fertilization of an oocyte by more than one sperm. The barriers at the cervical-vaginal and uterus-uterine tube boundaries may not only serve as essential sperm selective agents, but may also work to limit the number of sperm that eventually reach the oocyte.

IMPLANTATION

The preimplantation embryo

Within a few hours after fertilization, the zygote initiates the **cleavage** divisions, going progressively in the next few days through the 2-cell, 4-cell, 8-cell, 16-cell stages, etc. A characteristic feature of the cleavage divisions is that the embryo does not increase in size, since the initial amount of cytoplasm is partitioned progressively into more and more (smaller and smaller) cells. The embryo at the latter cleavage stages is referred to as the **morula** (Latin for "berry") because of the solid berry-like appearance of the embryo at this stage (Fig. 11.11). The cells of the morula are loosely associated

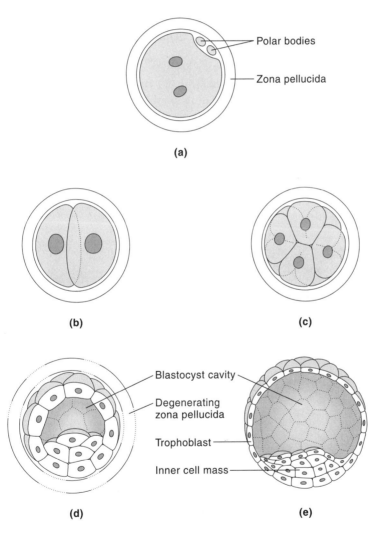

Figure 11.11
Preimplantation stages of human embryo. (a) Oocyte just after fertilization. (b) Two-cell stage embryo. (c) Eight-cell stage (morula). (d) Blastocyst formation begins about day 4, at which time the ZP begins to disappear. (e) At day 5–7, the mature blastocyst begins to implant.

Table 11.5
Derivatives of the Blastocyst

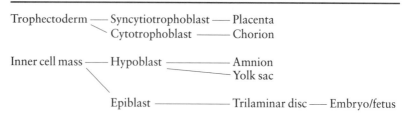

Trophectoderm —— Syncytiotrophoblast —— Placenta
          ＼ Cytotrophoblast ——— Chorion

Inner cell mass —— Hypoblast ————— Amnion
          ＼   ＼ Yolk sac

          Epiblast ————————— Trilaminar disc —— Embryo/fetus

and are held together by the zona pellucida (ZP). During the cleavage divisions, the embryo is transported from the site of fertilization through the uterine tube. The rate of transport is characteristic for each species. In humans, the developing embryo reaches the uterus in 5–6 days (Fig. 11.2).

At the 32- to 64-cell stage the morula begins to transform itself into the **blastocyst** (Fig. 11.11), which contains two functionally distinct tissues, the **trophectoderm (trophoblast)** and the **inner cell mass**. The trophoblast is the first of the **extraembryonic tissues** to appear, and it forms one of the progenitor tissues for the formation of the placenta, while the inner cell mass will give rise later to the embryo and fetus proper. Initially, the inner cell mass contains only about 8 cells, only about half of which are the progenitors of all of the tissues of the body. The other cells of the inner cell mass give rise to other extraembryonic tissues (Table 11.5). The blastocyst eventually "hatches" from the ZP (Figs. 11.11 and 11.12). The ZP has also had another very im-

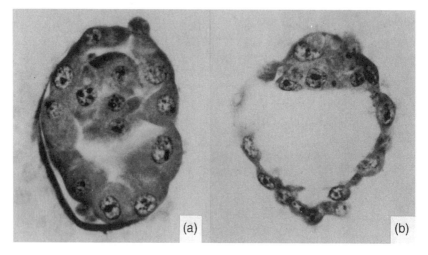

(a)   (b)

Figure 11.12
Photographs of two human blastocysts. (a) Section through a 58-cell blastocyst. The ZP is visible in the lower left hand, where a polar body can still be discerned. The ICM is just becoming distinct. (b) Section through a 107-cell blastocyst. The ZP is no longer present, and the ICM and the trophoblast are clearly defined. (Source: Specimens #8794 and #8663 from the Carnegie Collection, Carnegie Institution of Washington.)

portant function up to this stage: it prevents the implantation of the blastocyst until it has reached the uterus. Hence, the transport of the embryo along the uterine tubes follows a fairly tight schedule, the critical feature being that ZP degeneration should occur only after the embryo has reached the uterus.

Invasion of the maternal endometrium

During its preimplantation stages, the conceptus draws its nutritional support from the secretions of the uterine tubes and uterus. Eventually, however, these secretions are inadequate to support further growth, and it is at this stage that implantation begins. The blastocyst attaches to the uterine endometrium, a preferred site being the back wall of the uterine cavity. Attachment to and invasion of the endometrium by the blastocyst is guided by cells of the trophoblast. The trophoblast begins to differentiate into two distinct cell layers, the **cytotrophoblast** and the **syncytiotrophoblast** (Fig. 11.13), the latter being formed as the invading cells fuse together to form a multinucleate structure. The syncytiotrophoblast is responsible for the invasive properties of the developing blastocyst, secreting enzymes that digest the maternal tissues. The depth to which the blastocyst invades the maternal tissues varies with the species. In humans, implantation is highly invasive, and the blastocyst burrows its way completely not only into the uterine endometrium, but into the first third of the myometrium. By 12 days after fertilization the blastocyst is completely embedded in the maternal endometrium, and the syncytiotrophoblast has begun to make intimate contact with the maternal circulatory system through the formation of the lacunae, cavities filled with maternal blood (Fig. 11.12).

As implantation proceeds, the inner cell mass undergoes a series of transformations, forming first the **bilaminar disc**, consisting of two cell layers known as the **epiblast** and **hypoblast** (Fig. 11.13). The hypoblast forms the progenitor tissue for several extraembryonic tissues, some of which, together with the trophoblast, give rise to the fetal membranes, in particular to the *yolk sac, amnion, chorion*, and placenta (Table 11.5). The epiblast forms the *trilaminar disc*, which gives rise to all of the tissues and organs of the body (Table 11.6).

Table 11.6
Derivatives of the Trilaminar Disc

Layer	Derivatives (partial list)
Embryonic ectoderm	CNS, cranial and sensory nerves, retina, epidermis, hair, nails, inner ear, lens, mammary gland
Embryonic mesoderm	Connective tissue, muscles, blood, teeth, cardiovascular system, urogenital system, skeleton, teeth, connective tissue
Embryonic endoderm	Liver, pancreas, bladder, thyroid, tonsils, parathyroid, epithelial parts of lungs, bronchi, trachea

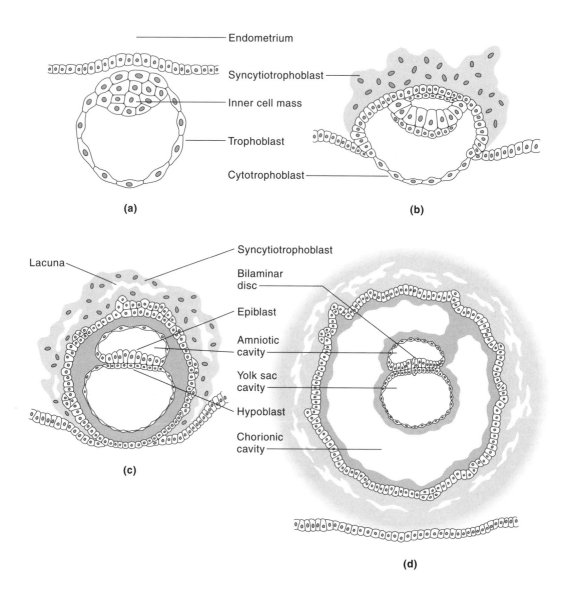

Figure 11.13

Progressive stages of implantation. (**a**) Initial contact between the blastocyst and uterine epithelium. (**b**) Early state of invasion, differentiation of the trophoblast into the cytotrophoblast and the syncytiotrophoblast, which provides the invasive potential of the embryo. (**c**) Further penetration of the blastocyst into endometrium; formation of the bilaminar disc, consisting of the epiblast and hypoblast, and amniotic cavity. (**d**) Completion of implantation characterized by the appearance of the lacunae, which represent establishment of intimate contact with the maternal circulatory system. Three cavities are present, amniotic, chorionic, and yolk sac. The yolk sac membrane developed from the hypoblast. The connecting stalk is the progenitor tissue of the umbilical cord. At about this stage, the epiblast begins its differentiation into the trilaminar disc.

Maternal recognition of the conceptus

A pregnancy is considered to be established at implantation. At this point, the mother begins to recognize that she is pregnant. In fact, it is also correct to say that the conceptus begins to take over the maternal H-P-O axis at this time. The need to do so is clear. Implantation occurs at about the time that the corpus luteum would start regressing if fertilization had not taken place. Since the corpus luteum maintains the endometrium, the tissue in which the conceptus is implanting, continued degeneration of the corpus luteum at this point would be disastrous for the pregnancy. Hence, what maintains corpus luteum function at this critical time? Or more generally, how is the cyclic nature of the maternal reproductive system converted to a noncyclic, pregnant pattern?

The conceptus itself suppresses corpus luteum regression by secreting the polypeptide hormone human chorionic gonadotropin (hCG). hCG production probably begins in the trophoblast even before implantation, but its levels increase dramatically once implantation begins. hCG, which is functionally equivalent to LH, stimulates the corpus luteum to continue producing progesterone and maintains the corpus luteum of pregnancy, which is much larger than the luteal phase corpus luteum. hCG is the earliest reliable biochemical indicator of pregnancy, and sensitive test kits that are easy to use at home are now widely available for the early detection of pregnancy. Hence, the initial stages of implantation depend on signals derived from among the first extraembryonic tissues of the developing conceptus. hCG levels increase rapidly until about the eighth week and then decline almost as precipitously, but remain detectable during the remaining months (Fig. 11.14). The corpus luteum remains throughout pregnancy, but its progesterone-secreting activity decreases significantly. Removal of the ovaries (and hence, the corpus luteum) during the first two months of pregnancy immediately

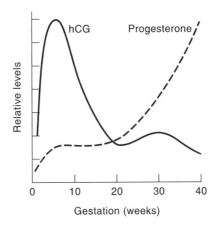

Figure 11.14
Maternal recognition of the implanting conceptus. hCG produced by the syncytiotrophoblast is the earliest easily assayed signal of implantation; it maintains corpus luteum function. hCG levels increase rapidly and reach a maximum about 8 weeks after fertilization. They begin to decrease when progesterone production by the developing placenta can maintain the stability of the endometrium. During the first 6 to 8 weeks of pregnancy, progesterone is contributed primarily by the corpus luteum.

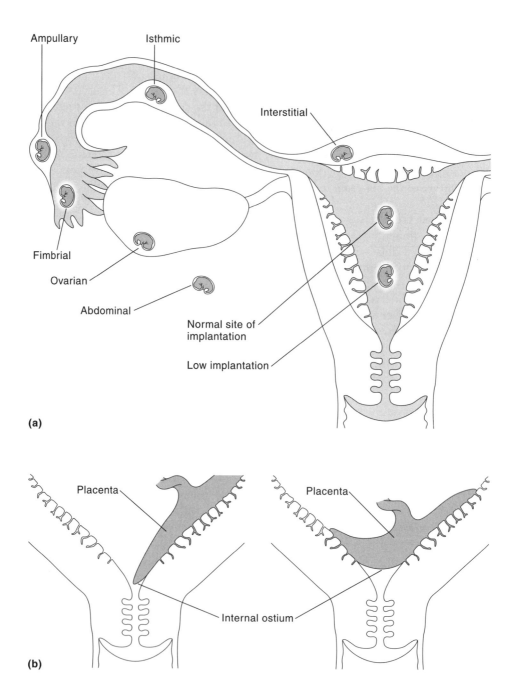

Figure 11.15
Ectopic implantation sites. (a) Implantations in the ampullary and the isthmic regions account
for about 80 percent of ectopic implantations. (b) Implantations in the internal ostium of the
cervix or close to cervical-uterine junction result in premature expulsion of the placenta, a
condition known as placenta previa.

terminates the pregnancy. However, removal during the last seven months has no effect on the pregnancy, indicating clearly that corpus luteum activity is no longer necessary.

The noncyclic, pregnant state is established and probably maintained mainly by progesterone, initially produced by the CL under hCG stimulation. When the activity of the corpus luteum of pregnancy begins to decrease at about the eighth week, progesterone production is taken over by the developing placenta. The placenta also begins to produce *estriol*, the main estrogen of pregnancy. The beginning of the fetal period coincides with the time at which the placenta begins to take over the function of the corpus luteum of pregnancy. Pregnancy is, in a sense, a prolongation of the luteal phase of the ovulatory cycle and is characterized by high levels of progesterone and estrogen, which together suppress cyclic ovarian function and menstruation.

Ectopic implantation

Although implantation usually takes place along the posterior or anterior uterine wall, infrequently the blastocyst can implant at other sites, and these are known as *ectopic* implantations. The most common of these ectopic sites are the uterine tubes (Fig. 11.15). Implantations in the uterine tubes give rise to **tubal pregnancies** (Fig. 11.16). Such pregnancies never reach term because growth of the fetus is severely limited. They can be quite dangerous because rupture of the uterine tube wall leads to severe bleeding which, if uncontrolled, can be lethal to the mother. A significant increase in tubal pregnancies in the last few years has been associated with the large increase in the incidence of sexually transmitted diseases (STDs) (see Chapter 17). The consequence of an STD

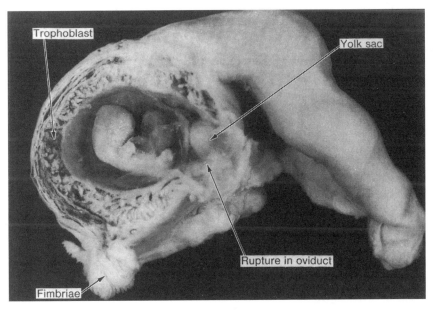

Figure 11.16
An ectopic implantation in the uterine tube, a tubal pregnancy. (Reprinted from T. H. Sadler. *Langmans Medical Embryology*, 7th ed. Williams and Wilkins, 1995, with permission from Lippincott Williams & Wilkins.)

is often the formation of scar tissue in the uterine tubes, which may impede the transport of the conceptus on its way to the uterus. Since the development of the conceptus follows its own internal clock, such a delay results in the formation of the blastocyst while the conceptus is still in the uterine tubes, and hence, implantation occurs in the uterine tubes. Implantation at other sites occurs with a much lower frequency. Implantations at the internal ostium or near the uterine-cervical boundary result in an abnormal development of the placenta, which results in severe bleeding during the latter stages of pregnancy (Fig. 11.15).

SUMMARY

The human conceptus has about 5–6 days of independent existence after being formed at fertilization. During this period, the developing embryo moves from the site of fertilization into the uterus, and as it does so, it transforms itself through a series of cleavage divisions and a fundamental internal cellular restructuring into the blastocyst. The blastocyst consists of two cell lineages with strikingly different developmental potential, the trophoblast and the inner cell mass. The inner cell mass (ICM) not only is the progenitor tissue for all of the somatic tissue of the body, but also contributes to the formation of extraembryonic tissues that later form part of the placenta. The cells of the ICM rearrange themselves to form the two layers of the bilaminar disc, the epiblast and hypoblast. In turn, the epiblast generates the trilaminar disc. The trophoblast is primarily responsible for implantation, the invasion of the uterus by which the embryo assures its further nutrition. Implantation is initiated by fusion of trophoblast cells at the site of attachment of the blastocyst to the endometrial epithelium to form the syncytiotrophoblast, which provides the invasive potential. The human conceptus implants completely by 10–12 days after fertilization. Trilaminar disc formation begins about 14 days after fertilization.

A clinical pregnancy, or maternal recognition of the conceptus, is marked by the appearance of hCG produced by the syncytiotrophoblast. hCG prevents the corpus luteum from regressing, thereby maintaining the progesterone production required to stabilize the endometrium so that the implanting embryo will not be expelled. Progesterone also suppresses the maternal H-P-O axis, thereby maintaining the female in a noncyclic, pregnant state. The developing placenta begins to take over the corpus luteum function at about 8 weeks gestation.

QUESTIONS

1. Why are such enormous rates of sperm production required for reproductive success?

2. Indicate the ways in which the female reproductive tract selects the sperm that will fertilize the egg.

3. What determines at what point the developing embryo is ready to implant?

4. What is the role of the corpus luteum in pregnancy?

5. Ectopic implantations are almost unknown in other mammalian species. Why are they so common in humans?

6. Library project:
    a. What is known about the factors that affect the volume and composition of the ejaculate?
    b. What is known about the functional lifetime of the ejaculated sperm and ovulated egg?

SUPPLEMENTARY READING

England, M. A. 1996. *Life Before Birth*, 2nd ed. Mosby-Wolfe, London, UK.

Holstein, A. H., E. C. Roosen-Runge, and C. Schirren. 1988. *Illustrated Pathology of Human Spermatogenesis*. Grosse Verlag, Berlin, Germany.

Moore, K. L., and T. V. N. Persaud. 1993. *The Developing Human*. W. B. Saunders Co., Philadelphia, PA.

O'Rahilly, R., and F. Müller. 1996. *Human Embryology and Teratology*, 2nd ed. Wiley-Liss, Inc., New York, NY.

Sadler, T. H. 1995. *Langmans Medical Embryology*, 7th ed. Williams and Wilkin, Baltimore, MD.

Saling, P. M. 1996. Fertilization: Mammalian gamete interactions. In *Reproductive Endocrinology, Surgery, and Technology*. E. Y. Adashi, J. A. Rock, and Z. Rosenwaks, Eds. Lippincott-Raven Publishers, Philadelphia, PA.

Wright, L. 1996. Silent sperm. *New Yorker* **71** (January 15), 42–50.

ADVANCED TOPICS

Aitken, J. 1995. Cell biology of human spermatozoa. In *Gametes: The Spermatozoon*, Chapter 11. J. G. Grudzinskas and J. L. Yovich, Eds. Cambridge University Press, Cambridge, UK.

Aitken, R. J. 1999. The human spermatozoon—a cell in crisis? *Journal of Reproduction and Fertility* **115**, 1–7.

Barratt, C. L. R., and J. D. Cooke. 1991. Sperm transport in the human female reproductive tract—a dynamic interaction. *International Journal of Andrology* **14**, 394–411.

Cho, C., D. O'Dell, J.-E. Faure, *et al.* 1998. Fertilization defects in sperm from mice lacking fertilin β. *Science* **281**, 1857–1859.

Cunningham, F. G., *et al.* 1993. The placental hormones. In *Williams Obstetrics*, 19th ed. Appleton and Lange, Norwalk, CT.

Glover, T. D., C. L. R. Barratt, and J. J. P. Taylor (Eds.) 1990. *Human Male Fertility and Semen Analysis*. Academic Press, New York, NY.

Harper, M. J. K. 1982. Sperm and egg transport. In *Reproduction in Mammals: Germ Cells and Fertilization*, 2nd ed. C. R. Austin and R. V. Short, Eds. Cambridge University Press, Cambridge, UK.

Keiding, N., and N. E. Skakkebaek. 1996. Sperm decline: real or artifact? *Fertility and Sterility* **65(2)**, 450–453.

Lipschultz, L. 1996. Debate over sperm quality. *Fertility and Sterility* **65(5)**, 909–911.

Lundquist, F. 1949. Aspects of the biochemistry of human semen. *Acta Physiologica Scandinavica* **19**, 7–105.

Overstreet, J. W., R. O. Davis, and D. F. Katz. 1992. Semen evaluation. *Infertility and Reproductive Medical Clinics of North America* **3**, 329–340.

Sharpe, R. M. 1994. Regulation of spermatogenesis. In *Physiology of Reproduction*, 2nd ed. E. Knobil and J. Neill, Eds. Raven Press, Philadelphia, PA.

Birth posture observed by United States Army surgeon in the mid-nineteenth century of a young Sioux woman who retired to the bank of a stream at the onset of labor. She sat cross-legged, thighs widely separated, arms folded, and head bowed, especially during labor pains, until birth occurred 40 minutes later. (G. J. Englemann. 1977. *Labor among Primitive Peoples*. AMS Press, New York, NY. Reprint of 1882 edition, published by J. H. Chambers, St. Louis.)

CHAPTER 12

Placentation, Parturition, and Lactation

The placenta has enormous genetic endowments, comparable to those of the brain and ovary. The molecular, cellular, vascular arrangements are designed for directing the traffic of biochemical and nutritional flow preferentially to the fetus . . . Thus, the placenta may be viewed as the third brain, which links the developed (maternal) and developing (fetal) brains.

—S. S. C. Yen. 1994

The placenta as a third brain. *Journal of Reproductive Medicine* 39(4), 277–280.

A hormone unexpectedly found in the human placenta turns out to influence the timing of delivery.

—R. Smith. 1999

The timing of birth. *Scientific American* 280(3), 68–75.

Although immaturity at birth can be assumed in most systems, no other organ presents such a dramatic change in size, shape, and function as does the breast during growth, puberty, and pregnancy and lactation.

—J. Russo and I. H. Russo. 1987

Development of the human mammary gland. In *The Mammary Gland. Development, Regulation, and Function*, Chapter 3. M. C. Neville and C. W. Daniel, Eds. Plenum Press, New York, NY.

WE REMAIN IGNORANT of most of the processes that guide human embryogenesis. However, through persistent, often ingenious, combinations of genetic, biochemical, and physiological approaches, the mysteries of prebirth human development are slowly being explained. To do justice to this critical stage in the life of every human being is beyond the scope of this book. For the details of human embryology, for example, the reader is referred to the standard embryology texts listed at the end of the chapter. Our objective here is to present an outline of the salient features of some key events during the prebirth stage of a human life. Our focus in this chapter is on (1) the formation and function of the placenta, and some of the better characterized aspects of the endocrinology of a human pregnancy, (2) the nature of the signals that precipitate parturition (birth), and (3) the development and functioning of the mammary glands.

PREGNANCY

Viviparity

Humans, like all placental mammals, are **viviparous** (Latin, *born alive*), which means that development of the fertilized egg takes place inside the mother. Birds and reptiles are *oviparous* organisms, in which fertilization occurs internally, but the young hatch from eggs deposited externally by the female. Both internal fertilization and viviparity are the consequence of changes that can be traced to about 400 million years ago when a conduit for passing gametes into the surrounding water appeared in some early fish. This conduit may have provided more control over gamete release, and hence, increased survivability. At some stage, many millions of years later, internal fertilization made its appearance in vertebrates. Internal fertilization was probably an early essential step in the transition from aquatic to terrestrial animals. Internal fertilization quite likely appeared in fishes, since a few extant fish and a few amphibian species exhibit it. However, in most of these species, the zygotes once formed internally are expelled into water where development is completed.

Viviparity is the consequence of many changes in the female reproductive tract, which eventually transformed it from a simple fertilization chamber into the structurally and functionally differentiated series of regions that constitute the cervix, the uterus, and uterine tubes found in modern mammals. Viviparity is not unique to mammals, since a few amphibian and reptile species are viviparous. Monotremes, the most primitive mammals, are oviparous and do not really possess a uterus. Marsupials show the first steps in the development of a uterus. Nutrition for the early stages of development is provided by secretions from the uterine glands, while the latter stages of development take place in an external pouch with the young attached to a nipple. A true uterus capable of supporting the development of the embryo to completion is a characteristic of placental mammals. But even in this group, there are significant differences in the way in which the blastocyst attaches to the uterine wall. In humans, the blastocyst is highly invasive in that it penetrates deeply into the underlying stroma of the uterus. It is probably this invasive property that accounts for the fact that ectopic implantations occur in human females but not in other animals.

Stages

It is convenient to divide the prebirth period of human development into three major stages—the preembryonic, embryonic, and fetal stages. The preembryonic stage covers the first two weeks after fertilization, and includes the preimplantation and the *periimplantation* periods, culminating in the formation of the bilaminar embryo. The embryonic stage, covering the third through the eighth week after fertilization, may be the most complex of these three stages, since all of the tissue and organ systems of the body are formed during this 5-week period. The fetal stage beginning at week 9 is characterized by continued growth and maturation of all the organs. Parturition (Latin, *travail*) represents the termination of the 9-month-long pregnancy (Fig. 12.1).

The normative time for a human pregnancy is considered to be about 266 days or 38 weeks after fertilization, and it is conventionally divided into three trimesters. The trimesters are measures of time, rather than being defined by important developmental stages. Clinically, pregnancy is measured from the date of the last menstrual period. Using this measure, the normal length of a pregnancy is about 280 days (40 weeks).

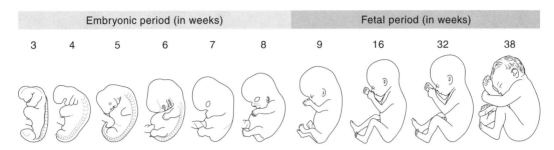

Embryonic period (in weeks)						Fetal period (in weeks)			
3	4	5	6	7	8	9	16	32	38

Figure 12.1
Embryonic and fetal development.

Hence, there is always a two-week difference between gestational age (measured from fertilization) and pregnancy age (measured from the last menstrual period). The standard way of estimating the expected date of delivery is to count back three calendar months from the first day of the last menstrual period, and then add a year and seven days. A moment's reflection should tell you why this method provides a reasonably accurate estimate of the expected date of delivery.

PLACENTATION

Placenta and fetal membranes

The placenta (Latin, *flat cake*) is commonly called the "after-birth" because it is expelled from the uterus after the fetus. It is disc-shaped with a diameter of about 7.5 in. and weighs about 1.5 lb. when fully formed. In many primitive societies the placenta was given special importance, and in some societies the ritual eating of the placenta was considered to bring good luck. Its function remained obscure, however, during all of classical antiquity. The fetal membranes associated with the placenta were recognized during Greek and Roman times, but no special term or appreciation of its function was evident. The term *placenta* did not appear until 1559, and it was perhaps a century later that its function began to be appreciated. Our view of the placenta has undergone significant modification in recent years. It was originally seen primarily as an organ for transporting nutrients from the mother and wastes from the fetus. More recently, its role as a complex neuroendocrine organ regulating all maternal-fetal interactions, regulating the growth of the embryo and fetus, and preparing the mother for the feeding of the infant after birth has been increasingly appreciated.

The placenta incorporates both fetal and maternal tissue. The fetal portion develops from the chorion, which develops from the cytotrophoblast and synctiotrophoblast. The maternal portion is a layer of the endometrium known as the **decidua** (Latin, *a falling off*) because this part separates from the uterus at the time of birth. The decidual cells are large cells connected to each other by tight junctions that form around the invading conceptus in a change known as the **decidual reaction**. As development proceeds, three decidual regions can be distinguished: the *decidual basalis*, marking the region in which the placenta will develop; the *decidua capsularis*, present in the early stages of implantation, but which eventually disappears; and the *decidua parietalis*,

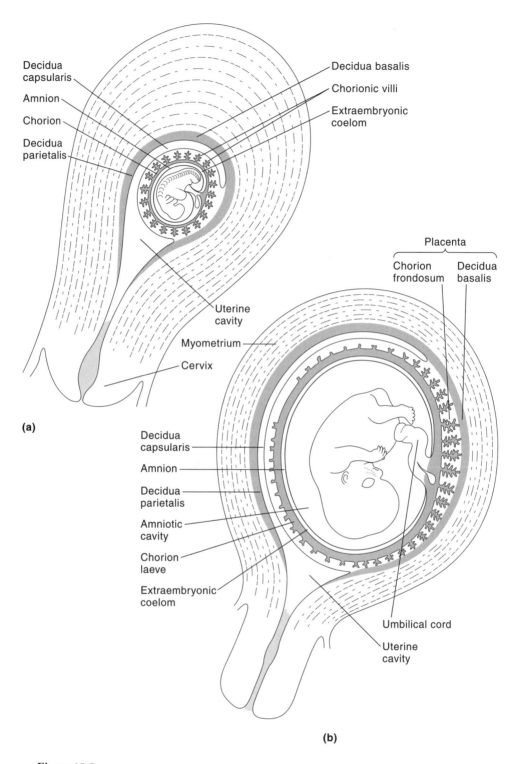

(a)

(b)

Figure 12.2
Development of the placenta and fetal membranes. (**a**) Chorionic sac at about 4 weeks. At this stage the chorionic villi are distributed uniformly around the sac. (**b**) Embryo and placenta at about 8 weeks. The decidua capsularis is becoming smooth, and the placenta proper is beginning to take form. By trimester 3, the amniotic cavity has expanded to take up the uterine cavity completely. (Adapted from R. O'Rahilly and F. Müller, 1996.)

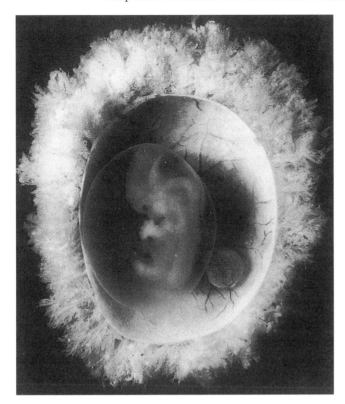

Figure 12.3
Photograph of a 7-week embryo in its amniotic and chorionic sac. Chorionic villi surround the chorionic sac. The chorion frondosum, marking the site where the placenta would form, is beginning to appear at the lower left. (Reprinted from R. O'Rahilly and F. Müller, 1996, with permission.)

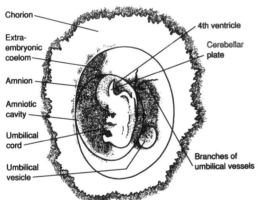

the endometrial regions far from the site of implantation (Fig. 12.2). Derivatives of the chorion make contact with the endometrium and form fingerlike projections known as the *chorionic villi* on its surface. The chorionic villi make the intimate connections with the endometrium and the maternal circulation (Fig. 12.3). However, the fetal and maternal circulatory systems remain separate. During the embryonic period, the chorionic villi appear uniformly around the chorionic sac so that both the decidual basalis and capsularis are penetrated by them. However, beginning in the fetal period, the decidua capsularis begins to degenerate and its surface becomes smooth (*chorion laeve*) (Figs. 12.2 and 12.3). The placenta is defined by the decidua basalis (maternal portion) and the chorion frondosum (fetal portion).

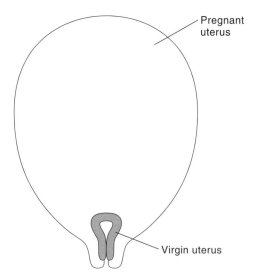

Figure 12.4
Size comparison of virginal and pregnant uterus at full term drawn to scale, about one-quarter of natural size.

The amniotic sac, the beginnings of which go back to the formation of the blastocyst, defines the region in which the embryo and later the fetus develops. The umbilical cord connects the embryo/fetus to the placenta, and its outer covering is the amnion (Fig. 12.3). The cord is 1 to 2 cm thick, and 30 to 90 cm long. Very long or very short cords are uncommon. Long cords have a tendency to coil around the fetus, and this may create serious problems during delivery if the cord is compressed as it passes through the cervix. Damage to the baby's brain may result due to cutting off of the blood supply.

The fetal period is characterized by a high rate of growth. At week 8, the human fetus is about 45 mm long; at week 12, 150 mm; at week 25, 375 mm; and at week 38, about 500 mm long. To accommodate this growth, the uterus also has to grow, eventually displacing the mother's internal organs and stretching the skin and the muscle of the interior abdominal wall (Figs. 12.4 and 12.5).

Maternal-fetal interactions

As the interface between the fetus and the mother, the placenta orchestrates all of the interactions between the fetal and maternal systems. Its secretions, more than those of the mother, dominate the hormonal milieu of the pregnant female. Pregnancy places complex demands on the maternal system, and the principal function of the placenta is to ensure the well-being of the developing fetus. It carries out its functions by regulating maternal protein and energy metabolism so that proper nutritional support for the rapidly developing fetus is provided continually. This requires that the pregnant female increase her dietary intake to ensure that the demands of fetal growth are met. In addition, the placenta provides for the transfer of nutrients and oxygen and removal of heat, carbon dioxide, and other wastes produced by the fetus. Moreover, it also begins to prepare the mammary glands so that the nutritional needs of the postnatal infant will be provided for. A variety of hormones are required for these complex functions, and the placenta is the primary source of many of them in the pregnant female. To discuss all of the identified placental hormones would go beyond the scope of this book. Nevertheless, a few merit some comment (Table 12.1).

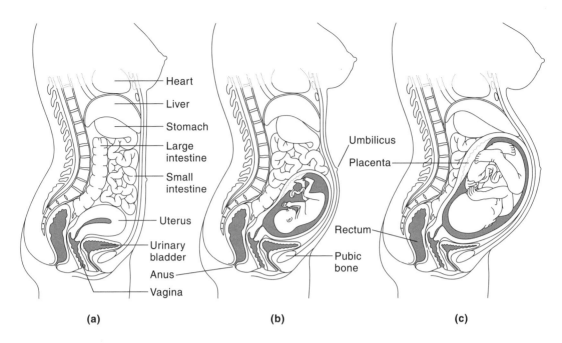

Figure 12.5
A woman's body before and during pregnancy: (**a**) not pregnant; (**b**) 20 weeks pregnant; and (**c**) 30 weeks pregnant. The uterus increases in size to accommodate the growing fetus. In (b), the uterus and fetus reach the level of the umbilicus; by 30 weeks, the uterus is displacing the maternal internal organs and stretching the skin and anterior abdominal muscular wall.

Table 12.1
Placental Hormones

Hormone	Function
Estrogens (estriol and estradiol)	Maintain estrogen-dependent tissues
Progesterone	Maintains endometrial stability; inhibits maternal H-P-O axis; mammary gland development
hCG	Maintains corpus luteum function during the first 8 weeks of pregnancy; may stimulate embryonic and fetal pituitary
hPL and hPGH	Regulate fetal growth, maternal and fetal energy utilization
CRH	Stimulates hCG production; regulates placental estrogen production; initiates parturition

This list includes the best known of the placental hormones. Many others, whose functions are less well known, are not listed here.

The feto-placental unit

The steroid hormone demands of the maternal and fetal systems are met not only by the placenta, but by the fetus as well through a series of coordinated interactions between the maternal, placental, and fetal tissues known as the **feto-placental unit**

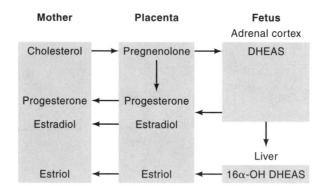

Figure 12.6
Simplified representation of the human feto-placental unit. Cholesterol from the mother is converted to progesterone in the placenta. Some of the progesterone is returned to the maternal system, and some is transferred to the fetal adrenal gland. The fetal adrenal gland converts progesterone to the weak androgen DHEAS, which is taken up by the fetal liver and converted to 16α-OH DEAS-DHEAS. This is transferred back to the placenta and converted to estriol, which serves as the main estrogen of pregnancy. Low levels of estradiol are also produced by the placenta by conversion of DHEAS into estradiol. The feto-placental unit is the primary source of steroid hormones in the pregnant female.

(Fig. 12.6). A highly simplified overview of this complex system can be summarized as follows:

The mother provides cholesterol from her diet to the placenta, which in turn converts it into progesterone. Some of this progesterone is diverted back to the maternal circulatory system, its main functions being the continued maintenance of a quiescent uterine endometrium so that the pregnancy can continue and suppression of the maternal H-P-O axis. Some of the progesterone passes through the blood vessels of the umbilical cord to the fetus. In the fetal adrenal glands, progesterone is converted to the weak androgen DHEAS, which is modified to 16α-OH-DHEAS by the fetal liver. 16α-OH-DHEAS is transferred back to the placenta, which converts it into estriol, quantitatively the major estrogen of pregnancy. Estradiol is also produced, but in lesser amounts.

Estrogens have multiple functions during pregnancy. They have been implicated in stimulating uterine growth and blood flow, relaxation of the pelvic ligaments, increasing cholesterol uptake, stimulating progesterone synthesis, and stimulating the growth of the breasts to ready them for milk production. Hence, in the pregnant woman, the functioning feto-placental unit, rather than the ovaries, determines and establishes the steroid hormone environment for the maternal system.

Other placental hormones

Nonsteroid hormones also play a part in the readjustment of the maternal physiology necessary to regulate the pregnancy, satisfy fetal needs, and prepare the mother for the postnatal needs of the infant (Table 12.1). These readjustments include the promotion and regulation of fetal growth; the shifting of maternal energy utilization in favor of fatty acid metabolism, while at the same time stabilizing glucose (carbohydrate) utilization (a much faster energy source) for fetal needs; and stimulation of mammary glands in preparation for milk production. Probably many different hormones take part in these functions, including the polypeptide hormone human placental lactogen (hPL), also known as chorionic somatomammotropin. hPL is a particularly interesting hormone because it binds to both the prolactin and growth hormone (GH) receptors, and as a result, it has both prolactin and GH activity. hPL is secreted by the syncytiotrophoblast as early as week 2 of gestation and increases continually during the rest of gestation (Fig. 12.7). Another recently identified hormone is the placental analog of growth hormone, and now referred to as human placental growth hormone (hPGH). Placental prolactin and hPL are important in preparing the mammary glands for milk production.

Placental hormones are important in helping the pregnant female to optimize her energy utilization. Many studies with primates in the wild indicate that pregnant females are quite judicious at food selection. For example, baboons in the wild can always balance their nutrient intake to what is seasonally available. In human societies, analysis of Australian and African tribes that to some extent still maintain a hunter-gatherer way of life, similar to that followed by our ancestors for thousands of years, has shown that food selection is quite purposeful, providing the energy-nutrient balance required for well-regulated fetal growth. Food aversions or preferences during pregnancy, commonly joked about and experienced by many pregnant women, may not be psychological whims, but may reflect placental hormone-mediated physiological effects.

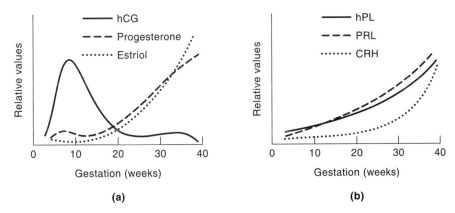

Figure 12.7
Placental hormones during gestation. (a) hCG and steroid hormones. The primary estrogen is estriol. (b) hPL, prolactin, and CRH appear early and increase continuously during gestation. CRH levels rise abruptly during the last few weeks of pregnancy.

The placenta also secretes GnRH, CRH, and β-endorphins. Placental GnRH is identical to hypothalamic GnRH and may be important for stimulating hCG secretion, as well as for regulating placental steroid hormone production. Placental CRH, also identical to its hypothalamic counterpart, appears at about 7 weeks and increases sharply during the last 5 weeks of gestation. We will see in the next section that recent studies suggest that placental CRH may be the "clock" that regulates the length of human pregnancy. Placental β-endorphin, as well as other endogenous opioid peptides (EOPs), may be important in controlling mood and appetite in the pregnant female. It is a common experience that many pregnant females have a sense of well-being, particularly before the physical discomfort of the latter stages of gestation become predominant. EOPs may contribute to this sense of well-being.

Multiple births

The most common types of multiple births are twins, of which there are two types, commonly referred to as fraternal and identical. Fraternal twins arise when two oocytes are ovulated simultaneously and each is fertilized, resulting in two zygotes, hence, the term **dizygotic twins**. Because they arise from two different zygotes, the twins may be of the same sex or of different sexes; genetically, they are related to each other as are normal sibs. Identical, or **monozygotic twins**, are thought to arise from a splitting of the inner cells mass into two groups of cells after the formation of the blastocyst. Each inner cell mass develops into a separate embryo, but because all the cells are derived from one zygote, both embryos carry the same genetic information, and hence, will be identical. Twinning rates vary significantly from one region of the world to another. Roughly three prevalence (low, intermediate, and high) twinning rates can be discerned, with Hawaii, Japan, and Taiwan having the lowest, and certain regions of Africa having the highest (Table 12.2). Since the monozygotic rates are fairly constant around the world, averaging 2–4 per 1000 births, the geographical variation in the twinning rates is due to variation in the dizygotic twinning rate. This is probably the result of a genetic disposition for multiple ovulations.

For dizygotic twins, placental and fetal membrane organization depends on how they implant in the endometrium. Most commonly, the two blastocysts implant separately, resulting in the formation of two placentas, two amnions, and two chorionic

Table 12.2
Twinning Rates

Prevalence	Rates (per 1000)	Regions
Low	2–7	Hawaii, Japan, Taiwan
Intermediate	7–20	Europe, most countries in Asia, the Americas, and Africa
High	>20	Parts of Africa, especially Nigeria, the Seychelles, Transvaal, Zimbabwe; in parts of the Americas with migrations from west Africa

Reference: MacGillivray *et al.*, 1988.

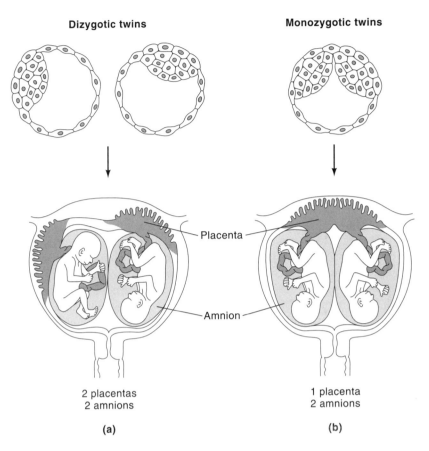

Dizygotic twins

Monozygotic twins

Placenta

Amnion

2 placentas
2 amnions

1 placenta
2 amnions

(a)

(b)

Figure 12.8
Twinning. (**a**) Development of dizygotic (DZ) twins. Two blastocysts implant separately, and two placentas develop independently. Occasionally, the two blastocysts may implant close together and the two placentas may fuse. (**b**) Development of monozygotic (MZ) twins. Two inner cell masses form, giving rise to two embryos, two amniotic cavities, but one placenta. MZ twins cannot be distinguished from DZ twins with one placenta by examination of the membranes alone.

sacs (Fig. 12.8). However, if the two blastocysts implant very close together, the placentas and chorionic sacs may fuse, but still retain two amnions. For monozygotic twins, there is generally one placenta and one chorionic sac derived from the original trophoblast, but two amniotic sacs, each originating from the two bilaminar discs (Fig. 12.8). Hence, monozygotic and dizygotic twins cannot be distinguished if they appear with one placenta and one chorionic sac. The cause of monozygotic twinning is unknown, and its incidence is relatively constant among different human populations (Table 12.2). In about 1 in 40 monozygotic twin pregnancies the separation of the two inner cell masses is not complete, and the two embryos do not develop independently because they share many cells and tissues. These unfortunate situations lead to the development of *conjoined twins*, commonly known as Siamese twins. The extent of fusion is variable, but in a large fraction of conjoined twins, the fusion of the two twins is extensive, leading to highly abnormal fetuses.

Triplets can arise from one, two, or three zygotes. One zygote would mean that the triplets were identical; two zygotes, two monozygotic twins, and a fraternal twin; three zygotes, all fraternal triplets. Higher-order multiples occur in various combinations. The incidence of multiple births has increased due to the use of hormone therapy for ovulatory failure.

The paradox of placentation

Placentation presents a particularly interesting and as-yet-unsolved puzzle in reproductive biology. In an immunological sense, both the sperm and the conceptus are foreign bodies from the maternal point of view. Vertebrates have developed different mechanisms for removing or neutralizing foreign cells or bodies. Generally, the first line of defense is an inflammation reaction, characterized by the mobilization of white blood cells that engulf and remove the foreign tissue. The inflammation reaction can be followed by activation of the immune system, which results in the formation of antibodies and activation of other cellular defense systems to remove the foreign cells if the inflammation response is not sufficient. For successful internal fertilization and viviparity, both the maternal inflammation and immune responses have to be suppressed. The importance of controlling the maternal inflammation and immune response can be appreciated by comparing fetal development in placental to that of nonplacental mammals. The nonplacental mammalian species are the monotremes (egg-laying mammals) and marsupials. Maternal-fetal contact is limited either by having the maternal-fetal interface separated by a maternally derived shell (monotremes) or by having very short gestation periods (marsupials). The extremely young marsupial fetus moves out of the uterine cavity and into an external maternal pouch to complete development externally rather than internally.

Placental mammals probably use several different mechanisms for regulating the maternal response to sperm and the conceptus. Progesterone is considered to have an important role in suppressing the maternal inflammation reaction. In animal studies, removal of progesterone during the very early stages of embryogenesis leads very quickly to a massive invasion of the uterus by white blood cells. Progesterone may also be important in permitting survival and passage of sperm, especially those deposited right after ovulation when progesterone levels begin to increase.

Other inhibitory mechanisms must come into play to prevent an immunological rejection of the implanting embryo. In fact, what happens is that the maternal immune system fails to respond to the implanting conceptus, and indeed, this lack of response is absolutely necessary for a normal pregnancy. The implanted conceptus is sometimes said to be in an immunologically privileged state. This special status applies only to the conceptus. It does not apply to most other types of foreign matter, including pathogens that may be trying to invade the maternal system. Failure of the regulatory mechanisms that maintain this privileged state is thought to account for a type of spontaneous abortion known as **preeclampsia**, or the premature expulsion of the placenta.

The mechanisms responsible for the localized suppression of the immune system that permits a pregnancy remain unclear. Some investigators have likened placentation to parasitic and viral infections that induce a localized and transient suppression of the host's immune system. The embryo can perhaps be likened to a parasite whose survival depends on its ability to suppress the host (maternal) immune surveillance system. Exploration of this analogy has led recently to a new hypothesis for the suppression of the maternal immune system. The *metavirus hypothesis* states that all placental mammals express a class of viruses known as *endogenous retroviruses* (ERVs) in trophoblast-derived tissue during implantation. ERVs are viruses encoded in the DNA of all mammals, and their expression results in the suppression of the immune system. According to this hypothesis, expression of the ERVs by the invasive syncytiotrophoblast results in a localized suppression of the maternal immune system so that implantation and placentation can take place. Another possibility is that the decidual interface that separates the fetus from the mother may, by limiting or controlling the extent of penetration by the embryo into the endometrium, help in insulating the embryo from the maternal surveillance system, and complement other mechanisms that inhibit the inflammation and immune response.

SUMMARY

The placenta provides the absolutely essential link between the fetus and the mother. It is probably fair to say that the placenta is responsible for orchestrating all the complex events of pregnancy. It informs the mother that she is pregnant and prepares the maternal system for the metabolic demands of the embryo and fetus during gestation and for its nutrition after birth. Placental hormones regulate fetal growth and are essential in preparing the fetus for existence outside the uterus. Placental progesterone has multiple functions. It suppresses the maternal H-P-O axis to prevent ovulation and maintain the mother in a noncycling state, and it maintains the uterus in a quiescent state so that labor will not be initiated prematurely. Placental progesterone also inhibits the maternal inflammation reaction that would normally be activated by the sperm and the implanting embryo. Tissues derived from the placenta probably also have a special role in conferring an immunologically privileged state on the embryo. This state prevents its rejection as a foreign body. The placenta also participates, as we shall see below, in transmitting the signal that initiates labor.

PARTURITION

Parturition defines the final gestation stage. Although the most dramatic aspect of this stage is the expulsion of the fetus from the uterus, preparation for birth begins a few weeks earlier.

Stages of labor

Labor (Latin, *toil* or *suffering*) refers to the sequence of events that begins with involuntary uterine contractions and ends with the expulsion of the fetus and placenta from the uterus. At least five physiological events can be distinguished: rupture of the fetal membranes (commonly referred to as "breaking the bag of waters"); cervical dilation, also known as cervical ripening; induction of contractions in the uterus and expulsion of the fetus; separation of the placenta from the uterine wall and its expulsion from the uterus; and involution of the uterus (contraction of the pregnant uterus to its former nonpregnant form). The stages of labor are diagrammed in Fig. 12.9.

The **dilation stage** begins with the onset of regular contractions of the uterus that are less than 10 minutes apart and culminates with the dilation of the cervix. This stage can be quite long, generally from 7 to 12 hours. The duration of this stage tends to be shorter for women who have given birth before, but it can be quite variable.

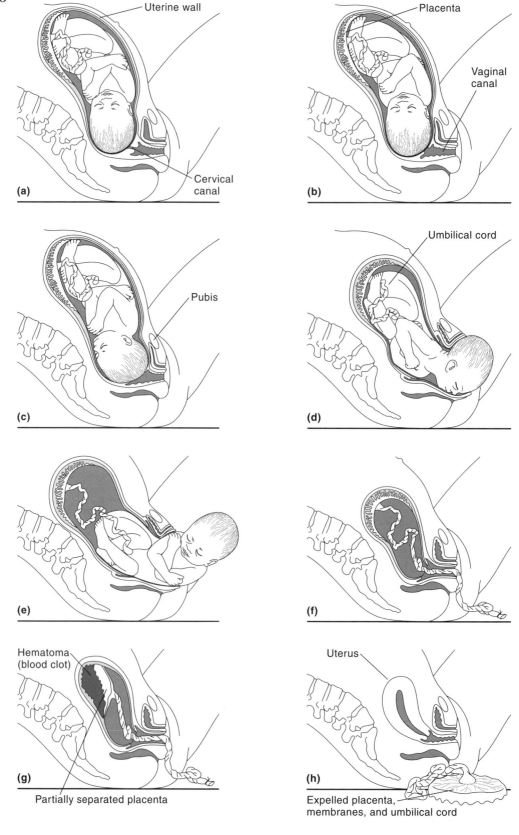

(a) Uterine wall, Cervical canal

(b) Placenta, Vaginal canal

(c) Pubis

(d) Umbilical cord

(e)

(f)

(g) Hematoma (blood clot), Partially separated placenta

(h) Uterus, Expelled placenta, membranes, and umbilical cord

Figure 12.9 *(facing page)*
Parturition. (a) and (b) The cervix begins to dilate during the first stage of labor. (c) to (e) The amniochorionic membrane ruptures; the fetus passes through the cervix and vagina and is delivered during the second stage of labor. (f) and (g) The uterus contracts during the third stage of labor, and the placenta separates from the uterine wall. This results in bleeding and formation of a large hematoma between the uterine wall and placenta. (h) As soon as the placenta and membranes are expelled, the recovery, or fourth stage, of labor begins. The uterus contracts, which constricts the endometrial arteries and prevents excessive bleeding.

The **expulsion stage** begins when the amniochorionic membrane ruptures and the fetus passes through the dilated cervix and vagina. This stage is much shorter than the dilation stage, generally about one hour.

The **placental stage** refers to the expulsion of the placenta. After the fetus is delivered, uterine contractions begin anew, resulting in the contraction of the uterus. As the uterus contracts, the placenta and fetal membranes separate from the uterine wall and are expelled from the uterus. Although expulsion of the placenta takes place in about 15 minutes in the majority of deliveries, in a few, expulsion may take much more time or may need to be manipulated. The separation of the placenta from the uterine wall results in bleeding and the formation of a large blood clot (hematoma) between the uterine wall and the placenta.

The **recovery stage** begins after the expulsion of the placenta. The uterus continues contracting and this restores it more or less to its prepregnancy form, and at the same time constricts the arteries to prevent excessive bleeding. This stage normally takes about 2 to 3 hours.

The timing of birth

What triggers birth? We might expect that under normal conditions birth is initiated when fetal organs have matured adequately to permit the baby to survive outside the uterus, and that labor is initiated by an internal signal that is generated when this maturational state has been reached. One early hypothesis was that the placenta was the source of the initiating signal. Given the undoubted importance of the placenta in regulating essentially all aspects of maternal-fetal interactions, a prominent placental role in the induction of labor was deemed eminently reasonable. It is interesting to note, however, that the ancient Greek physician Hippocrates suggested more than 2000 years ago that the fetus was responsible for initiating labor. He could provide no evidence to support his view, however.

The importance of the fetal brain in determining the length of gestation was first suggested in 1933 by the observation that human anencephalic fetuses (fetuses lacking major parts of the brain, in particular the hypothalamus) were born weeks late. In the 1950s, extremely long gestation times were noted in sheep grazing on skunk cabbage in mountain meadows in Idaho. Such sheep gave birth to one-eyed lambs that were twice the size of normal lambs. This deformity was attributed to abnormalities in the hypothalamus-pituitary axis. Many subsequent studies have shown that in sheep the fetal hypothalamus initiates parturition. In humans, the initiating signal appears to come from the placenta.

Parturition hormones

The proximate signal is considered to be the increase in the estrogen/progesterone ratio that begins to take place 2 to 4 weeks before birth. Recall that the predominant hormone in pregnancy in humans is estriol. Hence, the change in the estrogen/progesterone ratio really means a change in the estriol/progesterone ratio (Fig. 12.10). With this increase, the hormonal environment becomes estrogenic rather than progestonic. The increase in this ratio has the following consequences:

1. The uterus begins to become contractile. Paradoxically, the uterine contractions begin even in the face of very high progesterone levels. The importance of progesterone in maintaining a pregnancy has long been known. One of the most important functions of progesterone appears to be the suppression of uterine contractility. In animals, progesterone withdrawal precedes and heralds the initiation of labor. Interference with progesterone action, such as by administering progesterone receptor antagonists, triggers uterine contractions that result in the expulsion of the fetus. This is certainly the case in the earlier stages of human pregnancies. However, for reasons not well understood, this sensitivity begins to change in the latter part of pregnancy. As long as progesterone predominates, the uterus can be considered a bag of relaxed smooth muscle sealed at the bottom—the cervix—by inflexible collagen fibers. However, as estriol begins to

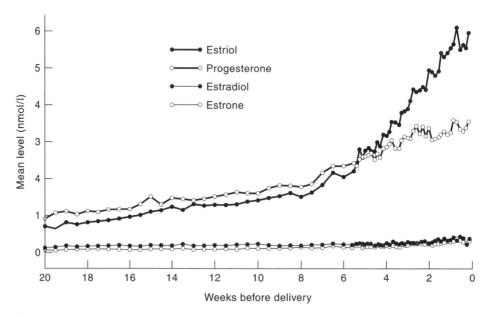

Figure 12.10
Mean levels of salivary progesterone, estriol, estrone, and estradiol during the last 20 weeks of pregnancy. The rapid estriol rise during the last 4 weeks produced by release of placental CRH may provide the triggering signal for the initiation of labor. (Adapted from J. Darne, A. H. G. McGarrigle, and G. C. L. Lachelin, 1987.)

predominate, cells of the myometrium begin to produce a protein called *connexin*. Connexin joins the uterine muscle cells together to form an electrically wired network of tissue which enables the uterus to undergo coordinated contractions despite the high progesterone levels.

2. The high estriol levels induce the myometrial cells to produce oxytocin receptors and the release of oxytocin from the maternal pituitary. Oxytocin, working on the oxytocin receptors, initiates uterine contractions and eventually induces labor. It is interesting to note that in baboons oxytocin release occurs at night. Oxytocin release may occur preferentially at night in humans as well, and this would explain why labor tends to be initiated at night.

3. Estriol also stimulates the synthesis of prostaglandins by the placental membranes covering the cervical area, and these in turn induce the production of enzymes that break down the collagen fibers that have maintained a sealed cervix. Removal of the collagen fibers permits the cervix to dilate progressively, and finally to open enough to permit the infant's head to pass through.

Placental CRH as the pregnancy clock

In humans, the relatively sharp increase in the estriol/progesterone ratio appears to be the proximate signal for initiating labor. Nevertheless, the crucial question is *What mechanism produces the change in the estriol/progesterone ratio?* The most recent studies indicate that the originating signal comes from the placenta. A number of studies have demonstrated that the timing of delivery is closely related to the rate at which placental CRH is released into the maternal and fetal circulations during the last few weeks of pregnancy (Fig. 12.7). Indeed, women with the highest CRH levels between 16 and 20 weeks of gestation were the most likely to deliver prematurely, while those with the lowest levels tended to deliver late.

How does placental CRH control the estriol/progesterone ratio? Current studies suggest that first, it stimulates the fetal pituitary to secrete ACTH, and second, both fetal ACTH and placental CRH stimulate the fetal adrenal to increase its conversion of DHEAS to estriol. Hence, when CRH levels reach a certain point, the steroid hormone environment shifts in favor of estrogens. Once estrogens begin to predominate, the changes that lead to birth begin to take place. The fetal adrenal also responds to the fetal ACTH by producing cortisol. Cortisol ensures that the infant's lungs undergo the final changes required for breathing air and helps to maintain CRH production by the placenta. Placental CRH may also work together with estrogen to stimulate the production and concentration of prostaglandins in the cervix, facilitating its softening and opening (Fig. 12.11).

Labor is triggered prematurely in about 8 percent of pregnancies. This means that under certain circumstances the triggering signal can be uncoupled from the maturation of critical organs, and the cascade of birth signals can be generated before the fetus is able to survive outside. A number of conditions are associated with preterm labor, but the specific causes remain unknown. There is a certain urgency in developing preventive or treatment strategies because preterm labor is a leading cause of fetal mortality. Our new understanding of the different factors that play a role in initiating labor should help us in devising intervention strategies for preventing premature labor. Exploratory studies with CRH receptor antagonists, oxytocin receptor antagonists, and prostaglandin blockers are under way. Moreover, work on developing simple and reliable ways of identifying women at risk for premature labor is also proceeding.

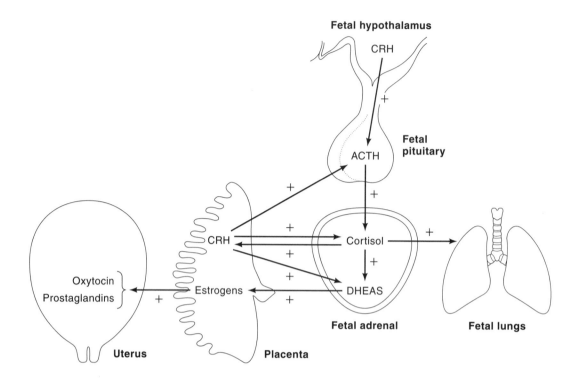

Figure 12.11
Hormonal interactions coordinating the timing of human parturition. A rapid rise in placental CRH near term results in stimulation of the fetal adrenal, an effect which includes activation of the fetal pituitary, leading to an increase in DHEAS and cortisol. Cortisol release by the fetal adrenal is necessary to prepare the fetal lungs for life after birth. Increased DHEAS leads to increased estrogen, increasing the estrogen/progesterone ratio. High estrogen levels initiate the cascade of signals that induce uterine contractions, cervical dilation, and oxytocin release that induces labor. (Adapted from R. Smith *et al.*, 1998.)

SUMMARY

The timing of birth in humans appears to controlled by a placental CRH clock. Placental CRH secretion beginning about week 12 initially stimulates the production of the weak androgen DHEAS by one part of the fetal adrenal. DHEAS is then converted by the placenta into estriol. CRH also stimulates another region of the fetal adrenal to produce cortisol. The fetal cortisol may be part of a positive-feedback system working at the placenta that keeps placental CRH production continually increasing. When placental CRH levels reach a critical threshold, the estriol/progesterone ratio increases dramatically, precipitating changes in uterine contractility, cervical dilation, oxytocin release, and other changes that lead to the initiation of labor. The rate of placental CRH production appears to function as the pregnancy clock.

LACTATION

In all mammalian species (except in humans since the adoption of breast-milk substitutes), lactation is as essential to reproductive success as gametogenesis and fertilization, since postnatal survival of the newborn depends upon the milk provided by the mother. Indeed, the term "mammal" refers to the mammary glands, the milk-producing organ that is characteristic of mammals. The basic features of the mammary glands in different species are quite similar, although the number, size, shape, and location can vary considerably among species. Although breast-milk substitutes have been widely available since the beginning of the twentieth century, their use began to be heavily promoted in the 1940s in the United States and several European countries. The consequence was that the percentage of breast-fed infants declined to 10–20 percent in the 1960s. Since then, however, at least among some segments of the population, there has been a revival of the practice of breast-feeding in many of the industrialized countries (Table 12.3). Unfortunately, in many parts of the urbanized underdeveloped world,

Table 12.3
Breast-feeding in the United States (Selected Years)

	1972–74	1978–80	1990–92	1993–94
All mothers (15–44 years)	30.1%	47.5%	54.2%	58.1%

Reference: National Center for Health Statistics. Health, United States, 1998.

breast-feeding has been undervalued not only as the most economical and beneficial source of infant nutrition, but as an important contributor to fertility control (recall the discussion of lactational infertility in Chapter 7). In this section, we will consider the anatomy and development of the mammary glands, and the factors that control milk production and release.

The mammary gland

Two general cellular compartments make up the normal, mature mammary gland: an adipose (fatty) stroma permeated by blood vessels and nerves and an epithelial compartment of **lactiferous ducts** and *lobules*, complex structures consisting of milk-secreting (alveoli) and nonsecretory epithelial cells (Fig. 12.12). The ducts extend radially out from the nipple toward the chest wall. Mammary gland development begins about the end of the fourth week of gestation, and fetal gland formation is not completed until about week 30 to 32 of gestation. At birth, the nipples are poorly formed and depressed (Fig. 12.13). At this stage, both the male and female glands consist of a few rudimentary ducts. Occasionally, within a few days after birth, the infant breast produces a milklike secretion known as "witch's milk," which subsides after about 3 weeks. Production of witch's milk is an indication of competence of the ductal tissue to produce milk. Essentially no further development takes place during infancy and childhood.

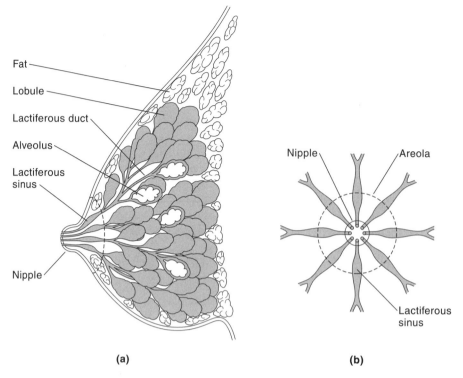

Figure 12.12
Developed mammary gland. (**a**) The ductular-alveolar network. The rounded, milk-producing alveoli empty through a system of smaller ducts into lactiferous ducts, which empty into milk reservoirs known as lactiferous sinuses. Each lactiferous duct runs through the nipple and to the outside. (**b**) The lactiferous ducts diverge radially from the nipple, and each duct empties 15–20 lobular networks.

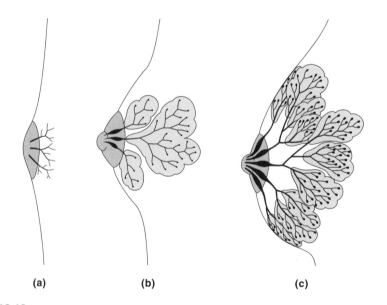

Figure 12.13
Illustrations of the progressive stages of postnatal development of the breast and the ductular network: (**a**) child; (**b**) young adult; (**c**) pregnancy. Growth during adolescence is due to deposition of fat and development of the ductular network.

At puberty in females, the ducts begin to grow and divide, forming club-shaped terminal end buds (Fig. 12.13). Breast size increases through fat deposition and the growth of fibrous connective tissue. The ducts, which drain different alveolar clusters, converge to form the *lactiferous sinuses*. The sinuses serve as small reservoirs for milk; they run through the nipple and directly onto its surface (Fig. 12.12). The areola and nipple become pigmented and wrinkled, particularly in the late postpubertal female. Full development of the alveolar structure occurs in pregnancy.

A number of hormones play a role in mammary gland development. During gestation and childhood, two anterior pituitary hormones, growth hormone (GH) and thyroid-stimulating hormone (TSH), appear to be required (Table 12.4). With the onset of puberty, under the action primarily of estrogen, the ducts begin to sprout and branch, and the precursors of the true alveoli begin to form at the end of the ductal

Table 12.4
Hormonal Control of Mammary Gland Development and Lactation

Stage	Hormones
Fetal and prepubertal	Possibly growth hormone and thyroid-stimulating hormone
Postpubertal	Estrogen, progesterone, adrenal corticosteroids, growth hormone
Early to midpregnancy	Placental estrogen, progesterone, corticosteroids; possibly hPL; mammary gland is fully competent to produce milk by the end of the fourth month of pregnancy
Late pregnancy	Placental estrogen, progesterone, hPL, prolactin; high levels of steroids and hPL suppress prolactin-dependent milk production
Postpartum First week: Colostrum	Abrupt decrease in placental hormones at birth removes inhibition to prolactin-stimulated milk production
Third week and after: Mature milk	Continued milk production depends on suckling-dependent prolactin release. Nipple stimulation activates neural pathways which may involve β-endorphin-mediated suppression of dopamine release. Decrease in dopamine permits prolactin release from the anterior pituitary. Milk release depends on oxytocin release from the posterior pituitary. Nipple stimulation and other nursing cues stimulate oxytocin release by neural pathways not well understood.
Weaning	Reduction in suckling intensity leads to synthesis of milk peptide that suppresses milk production

branches. With increasing age and successive ovulatory cycles, more complex lobular structures begin to appear. Progesterone and adrenal corticosteroids also contribute to the ductal growth and differentiation of the alveoli. Growth hormone and adrenal corticosteroids may also play a role in the deposition of fat and growth of connective tissue that determines the increase in breast size. During pregnancy, the previously developed ductal-lobular-alveolar system undergoes extensive growth under the action of placental estrogen, progesterone, prolactin, hPL, and possibly insulin. The mammary gland is fully developed for milk production by the end of the fourth month of gestation (Table 12.4).

Control of milk production and release (lactogenesis)

Milk production

What prevents milk production and release during the latter months of gestation? Milk production requires prolactin. During pregnancy, prolactin levels are high, but the milk-producing cells are not responsive to prolactin in the presence of high steroid hormone and hPL levels. At birth, the combined suppressive effects of placental hormones are removed. Prolactin levels also fall, but they are maintained at levels that support milk production if suckling is initiated. If suckling does not take place, prolactin levels decrease, and milk production slowly diminishes over a 3- or 4-week period.

The ultimate success of breast-feeding depends on the suckling intensity. Recall from Chapter 7 that the effectiveness of breast-feeding as a contraceptive depends on the frequency and duration of suckling. Copious milk production also depends on the intensity of suckling, since nipple stimulation is absolutely necessary for prolactin release from the anterior pituitary. Nipple stimulation during suckling activates neural pathways from the nipple up the spinal cord and via the brainstem to the hypothalamus (Fig. 12.14). Prolactin release in humans may take place by two mechanisms. One possibility is that the neural signals impinge on the β-endorphin-secreting neurons stimulating the release of β-endorphin. β-endorphin, in turn, suppresses dopamine release in the hypothalamus, and since dopamine normally inhibits prolactin secretion, the decrease in dopamine levels leads to increased prolactin secretion from the pituitary. Another possibility is that dopamine levels may be brought down directly by neural stimulation. Despite the uncertainty in the mechanism of prolactin release, there is no doubt that the amount of prolactin released depends on the intensity and duration of suckling. For example, suckling at both breasts simultaneously, as occurs when feeding twins, results in higher prolactin levels than suckling at one breast. Each time the infant suckles at the breast, prolactin levels increase abruptly (recall Fig. 7.16), inducing more milk production. Hence, each time the infant suckles, it not only releases milk stored in the ductal reservoirs, but the suckling stimulates more milk production. In effect, the infant is ordering its next meal.

Milk release

While prolactin is necessary for milk production, ejection of milk from alveoli into the lactiferous ducts requires another pituitary hormone—oxytocin (Table 12.4). The mechanism of oxytocin release from the posterior pituitary in humans is not completely clear. Stimulation of the nipple may lead directly to oxytocin release through neural signals transmitted through the spinal cord and stimulating the oxytocin-producing neurons. Another possibility is that β-endorphin may, as in the case of prolactin, also mediate the neural signals generated by nipple stimulation (Fig. 12.14). In

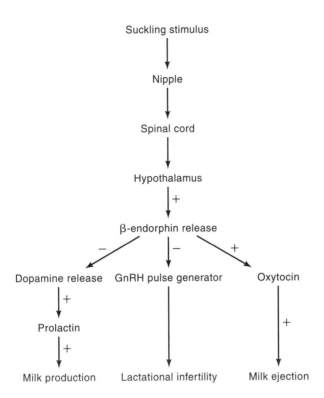

Figure 12.14
Control of lactogenesis. Stimulation of the nipple generates neural signals that go through the spinal cord and the hypothalamus. One model suggests that the neural signals activate the β-endorphin-secreting neurons. The increase in β-endorphin inhibits dopamine production, resulting in prolactin release, which is necessary for milk production. β-endorphin release stimulates oxytocin production, which in turn permits milk to be ejected from the alveoli. β-endorphin release also contributes to the inhibition of GnRH release which accounts for lactational infertility.

any case, oxytocin induces contraction of the epithelial cells that surround the alveoli, resulting in the ejection of milk into the ducts (known as the *milk let-down*). In contrast to prolactin release, however, oxytocin release does not always require nipple stimulation. Milk ejection often takes place without prior nipple stimulation, as a *conditioned reflex*. For example, milk let-down occurs frequently when the baby cries near feeding time, even before the infant begins suckling. If the breast is highly engorged, the milk will spurt out of the nipple. Oxytocin and milk release are particularly sensitive to physical and emotional stress. Discomfort or pain, especially if the breasts are engorged, may prevent milk release even when the breasts are full of milk. Worrying about breast-feeding itself may also inhibit milk release. Hence, the importance of psychological factors in the successful initiation and maintenance of breast-feeding should not be underestimated.

If the breast is not emptied of milk frequently, an inhibitory peptide produced in the milk will accumulate and suppress milk synthesis. This occurs normally as the infant is being weaned away from sole reliance on breast milk. With the decrease in suckling intensity, milk production gradually diminishes. Eventually, when suckling ceases altogether, the breast begins to involute, primarily because of structural changes in and disappearance of many of the alveoli. The basic ductal network remains, however. Breast volume is reduced, but the breasts invariably remain larger after lactation than before pregnancy. The breasts enter a period of inactivity similar to the one that prevailed before pregnancy. With another pregnancy, the alveoli will again be induced to develop. At menopause, the entire ductal apparatus begins to atrophy.

It may be necessary to suppress lactation for a variety of reasons. The woman may simply decide not to breast-feed. If the fetus aborts after 4 months or if it is stillborn, lactation will be unnecessary. If the mother is HIV-positive, breast-feeding is contraindicated. A number of methods have been used to suppress milk production. Before hormonal therapy became available, tight binding of the breasts was common. Sex steroids, estrogen or progesterone, which inhibit the release of prolactin have often been administered. Probably the most common therapy today is *bromocriptine*, a dopamine receptor agonist. Bromocriptine increases dopamine levels, which depresses prolactin release, and milk production decreases.

Characteristics of breast milk

Mature breast milk is a complex mixture of fat, protein, carbohydrates, vitamins, and minerals (Table 12.5). The primary energy source in breast milk is fat, but the fat is readily digestible because it is completely emulsified. The fat is also a good carrier of fat-soluble vitamins. Lactose, the main carbohydrate, is less sweet than table sugar and functions in promoting the proliferation of lactic acid-producing bacteria in the infant's intestines. One of the products of lactose breakdown, galactose, is important in forming the myelin sheath that envelopes nerve fibers. Mature breast milk begins to appear at around the end of the third week after birth. Before that, and particularly during the first week, a yellowish fluid called *colostrum* is produced. Colostrum has more protein, less fat and lactose, and a different mix of vitamins than mature milk.

Table 12.5
Composition of Breast Milk

Colostrum (appearing during first week postpartum)
  Differs from mature milk in that it has less fat and lactose, but more protein.

Transitional milk
  Transition from colostrum to mature milk

Mature milk (appearing about week 3 postpartum)
  Water
  Lactose
  Fat: Includes all essential, saturated and nonsaturated fatty acids
  Amino acids: All essential amino acids
  Proteins: Includes lactoglobulins and immunoglobulins
  Minerals: Includes calcium, iron, magnesium, potassium, sodium, phosphorus, sulfur
  Vitamins: Includes A, B complex, C, D, E, K

Both colostrum and mature milk are rich in immunoglobulins (maternal antibodies) and other defense factors that have a protective function. These include antimicrobial agents, anti-inflammatory factors, and antibodies that protect the breast-feeding infant against diarrhea, respiratory and urinary tract infections, and viral infections. The protection may improve with the duration of breast-feeding, especially because breast milk contains factors that may stimulate the development of the infant's own immune system. Hence, there is good evidence that breast milk not only confers passive protection against infections during lactation, but that it also stimulates the development of the infant's immune system so that protection against a variety of other diseases is enhanced.

SUMMARY

The development of the milk-producing elements of the mammary gland, the ductular-alveolar network, begins in the postpubertal female but is completed only during the first pregnancy. During the postpubertal period, ovarian and adrenal steroids are the primary hormones driving ductal differentiation, while placental hormones including estrogen, progesterone, prolactin, and hPL are responsible for the full maturation of the ductular-alveolar system. By the end of the fourth month of pregnancy, the mammary gland is fully capable of milk production. Production, however, is suppressed because of the high steroid hormone levels. Milk production begins at birth with the removal of placental hormones, and is maintained by prolactin release brought about by the continued nipple stimulation that takes place during suckling. Release of milk from the nipple requires the release of oxytocin from the posterior pituitary, which itself is dependent on nipple stimulation and other cues. The neural pathways that mediate oxytocin release in response to physiological and psychological factors remain incompletely understood. For the infant, the most significant benefit of breast-feeding is that it not only confers passive, short-term protection against infections during lactation, but that is also stimulates a long-lasting active immunity.

QUESTIONS

1. Why does it make sense to refer to the placenta as the "third brain"?

2. In what sense can pregnancy be considered to be an extended luteal phase?

3. Suppose you wanted to develop a simple test kit that would predict the day of birth reasonably accurately. What would be the basis of your test?

4. What is the ultimate trigger for the induction of labor?

5. For a large fraction of births, why does labor begin at night or during the very early morning hours?

6. Describe the role of hormones in lactogenesis.

7. Describe the important benefits of breast-feeding for the mother and infant.

8. Library project:
   a. There is much folklore surrounding conditions that might stimulate labor. Consider the ones you have heard or read about. Can any of them be reconciled with what is known about parturition?
   b. What conditions/factors affect the amount of milk produced by a lactating female, and her ability to breast-feed successfully?

SUPPLEMENTARY READING

Forsyth, I. A. 1991. The mammary gland. *Baillieres Clinical Endocrinology and Metabolism* 5(4), 809–832.

Hanson, L. A. 1998. Breastfeeding provides passive and likely long-lasting active immunity. *Annals of Allergy, Asthma, and Immunology* 81(6), 523–533.

Haynes, D. M. 1987. The human placenta: Historical considerations. In *The Human Placenta: Clinical Perspectives*. J. P. Lavery, Ed. Aspen Publishing Inc., Rockville, MD.

Johnson, M. H., and B. J. Everitt. 1995. *Essential Reproduction*, 4th ed. Blackwell Science Ltd., Oxford, UK.

Moore, K. L., and T. V. N. Persaud. 1993. *The Developing Human*. W. B. Saunders Co., Philadelphia, PA.

Schwartz, L. B. 1997. Commentary: Understanding human parturition. *Lancet* 350, 1792–1793.

Smith, R. 1999. The timing of birth. *Scientific American* 280(3), 68–75.

Wilson, L., Jr., and M. Parsons. 1996. Endocrinology of human gestation. In *Reproductive Endocrinology, Surgery, and Technology*, Vol. 1. E. Y. Adashi, J. A. Rock, and Z. Rosenwaks, Eds. Lippincott-Raven Publishers, Philadelphia, PA, pp. 452–475.

Yen, S. S. C. 1994. The placenta as a third brain. *Journal of Reproductive Medicine* 39(4), 277–280.

ADVANCED TOPICS

Darne, J., A. H. G. McGarrigle, and G. C. L. Lachelin. 1987. Saliva oestriol, oestradiol, oestrone and progesterone levels in pregnancy: Spontaneous labour at term is preceded by a rise in the saliva estriol/progesterone ratio. *British Journal of Obstetrics and Gynecology* 94, 227–235.

Finn, C. A. 1998. Menstruation: A nonadaptive consequence of uterine evolution. *The Quarterly Review of Biology* 73(2), 163–173.

Hall, J. G. 1996. Twinning: Mechanisms and genetic implications. *Current Opinion in Genetics and Development* 6, 343–347.

McLean, M., J. J. Davies, R. Woods, P. J. Lowry, and R. Smith. 1995. A placental clock controlling the length of human pregnancy. *Nature Medicine* 1(5), 460–463.

McMillen, I. C., I. D. Phillips, J. T. Ross, J. S. Robinson, and J. A. Owens. 1995. Chronic stress—the key to parturition. *Reproduction, Fertility, and Development* 7(3), 499–507.

Porter, D. G., R. B. Heap, and A. P. F. Flint. 1982. Endocrinology of the placenta and the evolution of viviparity. *Journal of Reproduction and Fertility* 31 (supplement), 113–138.

Smith, R., S. Mesiano, E.-C. Chan, S. Brown, and R. B. Jaffe. CRH directly and preferentially stimulates DHEAS secretion by human fetal adrenal cortical cells. *Journal of Clinical Endocrinology and Metabolism* 83(8), 2916–2920.

Villarreal, L. P. 1997. On viruses, sex, and motherhood. *Journal of Virology* 71(2), 859–865.

Wilde, C. J., *et al.* 1998. Autocrine regulation of milk secretion. *Biochemical Society Symposia* 63, 81–90.

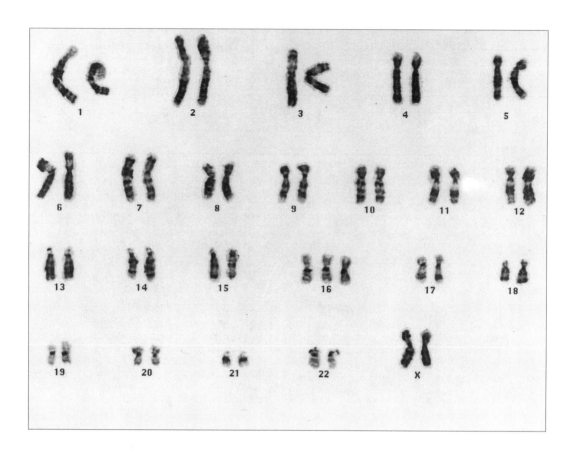

Trisomy-16 karyotype. Trisomy-16 is the most commonly observed trisomy in humans. Most trisomy-16 fetuses are lost spontaneously before birth, or are stillborn. (Courtesy of Dr. O. W. Jones, UCSD School of Medicine.)

CHAPTER 13

Fetal Loss and Birth Defects

We ought not to set them aside with idle thoughts or idle words about "curiosities" or "chances." Not one of them is without meaning; not one that might not become the beginning of excellent knowledge, if only we could answer the question— why is it rare, or being rare, why did it in this instance happen?

—J. Paget. 1882

Lancet 2, 1017, cited in K. L. Moore and T. V. N. Persaud. 1993. *The Developing Human*, Fifth edition, Chapter 8. W. B. Saunders Co., Philadelphia, PA.

If an abnormal child is born in a family or tribe, man insists on an explanation . . . ancient beliefs often plague parents even today, as they interpret them as portents or punishments.

—J. Warkany. 1971

Congenital Malformations: Notes and Comments. Year Book Medical Publishers, Inc., Chicago, IL, p. 19.

Tobacco is the most significant reproductive poison in current use.

—R. Forman, S. Gilmour-White, N. Forman. 1996

Drug-Induced Infertility and Sexual Dysfunction. Cambridge University Press, Cambridge, UK, p. 106.

NOT ALL CONCEPTIONS end with a live birth, and not all births result in a healthy, normal baby. In some cases, the newborn infant may be severely deformed, an occurrence that can have devastating consequences for the family. Human societies have long been aware of the births of deformed infants. Images of deformities in the form of drawings, carvings, or sculptures have been left in the ruins of many ancient societies. Perhaps the oldest known one, dated to 6500 BCE, and obtained from a Neolithic site in Çatal Hüyük in Turkey, depicts a two-headed female figure, whose significance re-

Figure 13.1
Ancient depiction of a deformity. White marble statuette of a two-headed figurine found in a neolithic shrine (6500 BCE) in Çatal Hüyük in southern Turkey. The significance of the figure is not clear. (Reference: J. Mellaart. 1963. Deities and shrines of neolithic Anatolia. *Archaeology* 16, 29–40.)

mains unknown (Fig. 13.1). In previous times, severely deformed infants were termed *monsters* (Latin, *monstrum*, meaning omen or warning), because these births were seen as a sign from the gods or God. In a book published in 1573 by A. Paré, titled *On Monsters and Marvels*, and which became widely known in Europe during the sixteenth and seventeenth centuries, the first two explanations for the birth of deformed infants were listed as the *grace of God* and the *wrath of God*, without apparent contradiction. Even today, as indicated in the epigraph by J. Warkany, many parents continue to view the birth of a deformed child as punishment.

Perhaps an unappreciated aspect of modern biology is that we have come to recognize that developmental defects, tragic and unfortunate though they might be, are nevertheless to be expected. A moment's reflection will convince us that the development of an individual from the zygote must be incredibly complex. There must be thousands of ways in which things can go wrong during the 9-month gestation period. Perhaps we should not be surprised when things go wrong, but instead should marvel that most people are born normal and healthy. In fact, as we will see below, fetal loss is very high in humans, and most conceptions are lost before birth. Our objective in this chapter is to review the incidence, nature, and our current understanding of the causes of fetal loss (loss before birth) and birth defects.

HUMAN TERATOLOGY

General causes

Fetal loss and birth defects are the result of disturbances in the developmental program that converts the zygote into a fully developed infant. These disturbances result in abnormalities of differing degrees of severity, in some cases leading to death before birth, or shortly thereafter. The study of developmental disorders is known as **teratology** (Greek, *teratos*, meaning "monster"). Disturbances in normal development can arise from many different factors. For our purposes, it is useful to group these into three categories: intrinsic (genetic), extrinsic (environmental), and maternal factors. **Intrinsic factors** are defects in the conceptus itself. Many such defects arise during oogenesis or fertilization and are the consequence of chromosomal aberrations. Others are due to mutations in single genes (**monogenic**), or the combined effects of several genes and their interactions with the environment (**polygenic/multifactorial**).

Extrinsic factors refer to a broad category of external conditions (with respect to the embryo or fetus) that can disturb normal development, for example, maternal nutritional deficiency, or exposure to environmental agents known as **teratogens**. Environmental agents may be *physical agents*, such as radiation; *chemical* or *biochemical agents* (drugs, chemicals, naturally occurring steroid hormones or their synthetic analogs); or *biological agents* (infectious organisms such as bacteria or viruses).

Maternal factors include a variety of conditions that result in the inability of the mother to support a pregnancy to term. This could be due to endometrial defects resulting in an unstable endometrium, endocrine disorders, malformations of the uterus, or immunological interactions between wife and husband. These conditions will be discussed in Chapter 16.

To appreciate the full complexity of embryonic and fetal development and the factors to which they are sensitive, it is important to recognize that different tissue and organ systems develop at different times during gestation. Hence, the sensitivity of a particular organ system to genetic or environmental factors will not be the same at all periods of gestation. During the first two weeks, encompassing the preimplantation and the periimplantation stages, the embryo is particularly sensitive to genetic factors, but relatively insensitive to environmental teratogens. Chromosome aberrations are considered to account for a very large percentage of spontaneous abortions that take place during this stage, most of them with minimum awareness by the female (Table 13.1).

Table 13.1
Causes of Prenatal Loss and Congenital Abnormalities

Cause	Prenatal loss	Congenital abnormalities
Unknown	40–50%	50–60%
Chromosome aberrations	50–60%	6–7%
Monogenic		7–8%
Polygenic/multifactorial		20–25%
Environmental toxins		7–10%

References: Wood, 1994; Moore and Persaud, 1993; Hook *et al.*, 1983.

Weeks 3–8 of gestation, during which all of the organ systems develop, is the period during which the embryo is most sensitive to teratogenic effects (Table 13.2). However, for each system, there are periods of major sensitivity followed by a period of lower sensitivity. Exposure to teratogens during the periods of major sensitivity may lead to severe abnormalities, while exposure at other times may have no effect or only minor consequences. The central nervous system, for example, appears to be continually sensitive to the effects of teratogens, with major sensitivity during the first 16 weeks of gestation, and less, but still significant sensitivity during the remaining time. This may account for the observation that defects of the brain and brain function account for the largest fraction of malformations of human organs (Table 13.3). Heart, kidney, and limb abnormalities account for a significant fraction of major abnormalities as well.

In most cases, we do not understand precisely how these different factors produce specific disorders. Nevertheless, even our limited understanding has had important consequences, since it has led us to recognize that certain types of disorders can be prevented. Not all disorders are preventable, however, nor will they be in the near future. Hence, even under the most ideal conditions, the incidence of fetal loss or birth defects cannot be reduced to zero. It is worth remembering that a measure of an advanced and humane society is how it treats its weakest and most defenseless members . . . *to contribute to the care of the physically and mentally handicapped is the price to be paid by those who have normal children—the price to be paid for overcoming the barbarism and cruelty inherent in societies that eliminate the weak and dispose of the deformed.*[1]

---

[1]Warkany, J. 1971. *Congenital Malformations: Notes and Comments.* Year Book Medical Publishers, Inc., Chicago, IL, p. 22.

Table 13.2
Malformations Arising During the Embryonic (Weeks 3–8)
and Fetal (After Week 8) Periods

Tissue	Period (weeks)	Major abnormalities
CNS	3–16	Neural tube defects; major brain malformations; mental retardation
	16–birth	Functional defects; cognitive deficits
Heart	3–8	Defects in formation of the atrial and ventricle chambers of the heart; defects in the aortic and pulmonary arteries; defects in the walls (septa) separating the different chambers of the heart
Upper limbs	4–6	Absence of one or more limbs; one or more deformed limbs
Lower limbs	4.5–6	Absence of one or more limbs; one or more deformed limbs
Ears	4.2–8.5	Low-set malformed ears; deafness
	8.5–20	Minor functional defects
Eyes	4.5–7.8	Abnormally small eyes; cataracts; glaucoma
	7.8–20	Functional defects
Teeth	6.9–8	Defective formation of enamel; stained teeth
Palate	6.9–8.2	Cleft palate
External genitalia	7.2–9.8	Masculinization of female genitalia
	9.8–birth	Minor anomalies

References: Moore and Persaud, 1993; O'Rahilly and Müller, 1996.

Table 13.3
Incidence of the Major Tissue Malformations

Tissue	Number/1000 births
Brain	10
Heart	8
Kidneys	4
Limbs	2
All others	6
Total	30

Reference: Connor and Ferguson-Smith, 1987.

Incidence

Fetal loss

Fetal loss refers to the termination of a pregnancy due to so-called "natural causes," resulting in a *spontaneous abortion*, or miscarriage, and clearly distinguishable from an *induced abortion*, which requires intervention on the part of the pregnant female herself or some other person. Reported spontaneous abortions give us a minimum estimate of postimplantation loss, or loss after a clinically recognized pregnancy has been established. Preimplantation loss can also occur, but this type of loss is very difficult to measure since no reliable measures of fertilization are yet available. Although postimplantation loss is in principle easier to measure (since hCG provides a reliable indicator of implantation), in practice, it is difficult to measure accurately since many losses go unreported or unnoticed, particularly if they occur very early in the implantation stages. Delayed menses in a woman who is sexually active may in fact represent a very early stage postimplantation loss, but is rarely reported as such.

Despite these limitations, current estimates suggest that preimplantation loss is quite high and that perhaps as many as 30–40 percent of human conceptuses are lost before or during implantation. Postimplantation loss varies with maternal age. The lowest rates (12 percent) are observed when the female is in her early 20s. The loss is greater before the 20s and increases significantly after the late 30s (Table 13.4). Com-

Table 13.4
Fetal Loss Increases with Maternal Age

Maternal age (years)	Probability of fetal loss (%)
<20	17
20–24	12
25–29	14
30–34	16
35–39	20
>40	28

Pooled data from nine different populations. Since loss was determined retrospectively, the level of loss is underestimated at all ages. (Reference: Wood, 1994.)

bining preimplantation and postimplantation loss suggests that 45–55 percent of human conceptuses are probably lost before birth.

Birth defects and perinatal loss

Many developmental abnormalities are not lethal for the embryo or fetus, but their effects are generally detected at the time of birth, and are referred to as **congenital abnormalities**, or birth defects. Such defects may involve structural, metabolic, functional, or behavioral anomalies. They may be of major or minor clinical significance. Some involve an easily noticeable anatomical, physiological, or endocrinological abnormality, while other defects involve subtle alterations that may not manifest themselves at birth, and many months can pass before the consequences of the defects can be established. The total incidence of birth defects depends on how they are defined. Major structural abnormalities (for example, *spina bifida* and *cleft lip*) are seen in about 2 to 3 percent of newborn infants. If we include minor, but yet medically significant anomalies, the figure increases to 6 to 8 percent. In addition, a certain fraction of infants die within a few months after birth. These **perinatal** deaths constitute about 1 to 2 percent of all births. Hence, about 8 to 10 percent of newborn infants are born with some type of significant defect.

INTRINSIC FACTORS

Intrinsic (genetic) defects are the most significant cause of fetal loss and birth defects. Most such defects are due to chromosome mutations (changes in chromosome number or large-scale changes in chromosome structure) or gene mutations (changes in the nucleotide sequence in one or several genes) which arise during gametogenesis. We will consider some of the consequences of these different mutations.

Chromosome mutations

Many different types of chromosome mutations have been observed. Chromosome mutations are usually lethal or result in severe congenital abnormalities. The most common are those involving changes in chromosome number.

Aneuploidy

Chromosome abnormalities, detected in 50 to 60 percent of spontaneous abortuses, are probably the single major source of fetal loss (Table 13.1). In contrast, they are responsible for a much smaller fraction of congenital abnormalities, about 6–7 percent. Most likely this is because chromosome mutations result in such severe abnormalities that very few of the embryos can survive to birth. Most chromosome abnormalities arise during gametogenesis, and most involve **nondisjunction**, which is the failure of homologous chromosomes to separate from each other during the first or second meiotic division. As a consequence, gametes either lacking one chromosome or having an extra chromosome are produced (Fig. 13.2). Fertilization involving these abnormal gametes results in an **aneuploid** embryo, or an embryo with a chromosome number different than 46.

An embryo whose cells have 45 chromosomes (rather than the normal 46) is said to be **monosomic** for the missing chromosome, while an embryo with 47 chromosomes (one chromosome appearing in three copies rather than the normal two) is

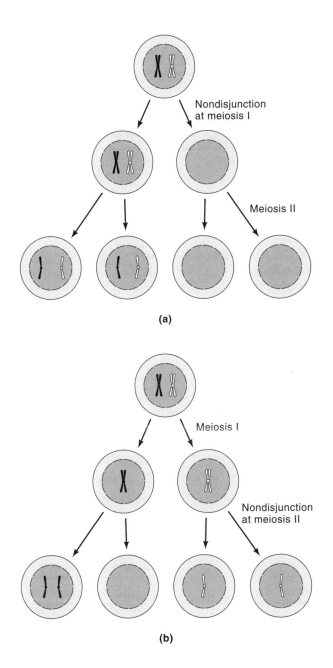

Figure 13.2

Nondisjunction of autosomes during meiosis. Nondisjunction is a failure of homologous chromosomes to separate from each other either in the first meiotic division (**a**) or in the second meiotic division (**b**). In either case, gametes containing both members of a homologous pair or lacking both members of a homologous pair will be formed. Fusion of the first type of gamete with a normal one results in a trisomic condition. Fusion of the second type of gamete with a normal one results in a monosomic condition. Nondisjunction can occur in either oogenesis or spermatogenesis.

Table 13.5
Trisomic Conditions Compatible with a Live Birth

Trisomy	Incidence/live births	Features
Trisomy-21 (Down syndrome)	1/800	Mental retardation, congenital heart disease
Trisomy-18	1/8000	Severe growth and mental retardation
Trisomy-13	1/25,000	Severe CNS malformations

Trisomy-21 accounts for 10–15% of mentally deficient institutionalized patients. About 75% of Trisomy-21 embryos abort spontaneously, and 20% are stillborn. Trisomy-18 and Trisomy-13 newborns rarely survive past six months after birth. (Reference: Moore and Persaud, 1993.)

known as **trisomy**. The incidence of aneuploidy is different for each chromosome pair. The most frequent type of trisomy observed is Trisomy-16, accounting for about 30 percent of all trisomic conditions. Although most Trisomy-16 embryos are aborted spontaneously, a small fraction survive to term, but rarely survive more than a few days after birth. Monosomic embryos have never been known to survive past the embryonic period, and are aborted very early in gestation. The trisomic condition is almost always lethal, again with most trisomic embryos being aborted during the embryonic period, or during the first trimester. However, a small fraction of fetuses trisomic for chromosomes 13, 18 or 21 are born alive. Newborns trisomic for chromosomes 13 or 18 are severely defective and die shortly after birth (Table 13.5).

Trisomy-21 (formerly known commonly as mongolism or Down syndrome) is the only trisomy compatible with life after infancy (Fig. 13.3). Trisomy-21 individuals are generally severely mentally retarded, but appear to have a normal life span. Trisomy-21 occurs with an overall incidence of 1 in 600–800 births, but it exhibits a very marked maternal age effect, with the incidence increasing almost 100-fold between the early 20s and the 40s (Table 13.6). Because of this age dependence, older mothers con-

Table 13.6
Incidence of Trisomy-21 as a Function of Maternal Age

Maternal age (years)	Incidence/live births
20–24	1/1400
25–29	1/1100
30–34	1/700
35	1/350
37	1/225
39	1/140
41	1/85
43	1/50
>45	1/25

Reference: Hook *et al.*, 1983.

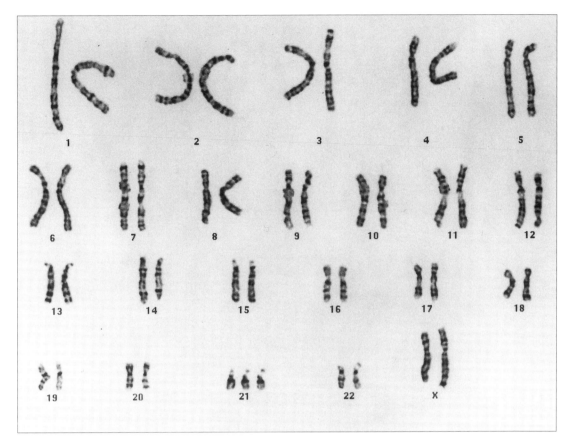

Figure 13.3
Example of an *XX*, Trisomy-21 karyotype. (Courtesy of Dr. O. W. Jones, UCSD School of Medicine.)

tribute a significant fraction of the total cases of Trisomy-21. The maternal age dependence of Trisomy-21 is not unique to chromosome 21, but is a special case of a general observation that the incidence of aneuploidy in embryos increases with maternal age. One common explanation for this observation suggests that the long arrest period that characterizes oogenesis in the female increases the likelihood of nondisjunction in female meiosis. Hence, for example, the oocytes that are being ovulated in a 40-year-old woman have been in an arrest stage for 40 years. Although plausible, it is not clear whether this explanation is valid.

More complex types of chromosomal abnormalities are also observed with a low frequency in spontaneous abortions. These abnormalities probably arise from anomalies at fertilization, and include, for example, **dispermy**, or the fertilization of one egg by two spermatozoa, or **digyny**, the fertilization of a diploid, rather than a haploid egg, leading to the formation of a **triploid** embryo (Fig. 13.4). **Tetraploid** embryos, characterized by three or four complete sets of chromosomes, are also seen. Both conditions are highly lethal, although such embryos can survive to birth.

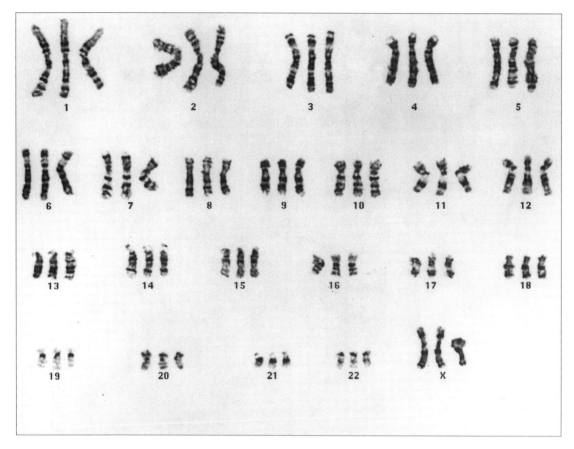

Figure 13.4
Example of an *XXX*, triploid karyotype. (Courtesy of Dr. O. W. Jones, UCSD School of Medicine.)

Ovarian teratomas and molar pregnancies

One type of unusual chromosome abnormality has come under significant scrutiny during the last 15 years. This is a condition in which the embryo has two sets of chromosomes from the same parent. It is of interest to review it here because of an important insight into human development that this condition has provided. Recall that normally the embryo starts with one set of maternal and one set of paternal chromosomes. Infrequently, by mechanisms that are poorly understood, embryos are generated that carry two sets of maternal or two sets of paternal chromosomes. These conditions lead to embryonic development known clinically as ovarian teratomas and hydatidiform moles. These two clinical conditions have been known for a long time, but it is only in the last 15 years or so that their chromosomal nature has been recognized.

Ovarian teratomas (Greek: *tera*—monster; *oma*—swelling), or **dermoid cysts,** have been recognized since the seventeenth century. They are growths that occur spontaneously in the ovary. Since they arise in maternal tissue, the cells in these growths carry only maternal chromosomes. Morphologically, teratomas are a disorganized mixture of mature tissues and generally contain bone, hair, neural tissue, skin, and fully developed teeth, and sometimes partial limb development. They are known now to be the result of spontaneous triggering of embryonic development in primary oocytes in the ovary. They can arise at any age in the female, even in infants, although the median age is about 30 years. They occur with an incidence of 1 to 2 percent. The ovarian teratoma is characterized by rather extensive development of tissues derived from the trilaminar disc, but extremely poor development of extraembryonic derivatives, so that no tissue resembling a placenta or chorion appears. Most teratomas are benign, but malignant degeneration (known as **teratocarcinomas**) occurs in about 2 percent of cases.

Hydatidiform moles (Greek: *hydatid*—watery; *mole*—shapeless mass) are severely abnormal conceptuses that contain two sets of paternal chromosomes; they were also recognized as clinical entities long before their chromosomal nature was understood. Molar pregnancies, at least in the early stages, are difficult to distinguish from normal ones. hCG levels, the first marker of implantation, are usually in the normal range. Moles, in contrast to ovarian teratomas, are characterized by extensive placental development, and little or no embryonic development. Molar development may continue for up to about two months before the tissue is aborted spontaneously.

Ovarian teratomas and moles are of special biological interest because they demonstrate clearly that normal development in humans (and in mammals in general) requires both the maternal and paternal set of chromosomes. The maternal chromosome set is not functionally equivalent to the paternal set, or vice versa, and they each appear to carry out an asymmetric and complementary division of labor during development. This means that in humans, virgin births (or **parthenogenesis**), in which no paternal contribution is present, are not possible. This is not the case in many nonmammalian species. For example, turkeys often reproduce parthenogenetically. It is known currently that the functional inequivalency of the maternal and paternal genomes is due to a set of genes that have been programmed (**imprinted**) during oogenesis and spermatogenesis to be expressed at different times and in different tissues during embryogenesis. As a consequence of imprinting, a maternally imprinted gene will be expressed differentially from its paternally imprinted homolog, and this differential expression seems to be necessary for normal development. Imprinted genes appear to be distributed throughout all the chromosomes. We do not yet know the number of imprinted genes, but estimates from studies in the mouse suggest that their number probably does not exceed 200. Much more work is required before we will have a good understanding of the nature of the imprinted genes, and the role that they play in development.

Gene mutations

Developmental defects can arise from mutations in one or several genes. At present, more than 4000 disorders linked to single- or multiple-gene mutations have been catalogued. Examples of monogenic disorders are sickle cell anemia, cystic fibrosis, Tay-Sachs disease, hemophilia, Duchenne muscular dystrophy, Huntington's disease, achondroplasia, neurofibromatosis, fragile X syndrome, familial breast and ovarian cancer, and phenylketonuria (PKU) (Table 13.7). The contribution of gene mutations, whether monogenic or polygenic, to prenatal loss cannot be estimated presently. On the other hand, monogenic disorders account for about 7–8 percent of congenital disorders (Table 13.1). For many single-gene disorders, we can make a direct connection between the disorder and a particular gene. There is no cure for most of the monogenic

Table 13.7
Examples of Monogenic Disorders

Disorder	Characteristics
Achondroplasia	Dwarfism; bone abnormalities
Camptobrachydactyly	Hand and foot malformations
Cystic fibrosis	Obstructive diseases of the lungs and pancreas
Duchenne muscular dystrophy	Neuromuscular degeneration
Fragile X syndrome	Severe mental retardation
Hemophilia	Absence of blood clotting
Huntington's disease	Progressive nervous system degeneration
Lesch-Nyhan syndrome	Compulsive self-mutilating behavior
Phenylketonuria (PKU)	Mental retardation

disorders, but palliative therapy is available for some: for example, clotting factor replacement for hemophilia, and dietary modification for PKU. Polygenic/multifactorial disorders account for 20–25 percent of congenital disorders (Table 13.1). Examples are neural tube defects, cleft lip, cleft palate, congenital heart disease, diabetes, and mental disorders such as schizophrenia and bipolar disease. In most cases the genes involved have not yet been identified.

Parental origin of mutations

Recall from Chapter 2 that the rate of germ line mutations is determined primarily by endogenous factors, most likely from oxidative damage to DNA from compounds generated during normal cell metabolism. The higher the metabolic rate, the more damage to DNA results. Although various antioxidants in the cells neutralize the effects of the oxidizing agents, some errors remain; if these errors are not removed by the error-correcting system of DNA replication, they will persist in the germ line. The germ line mutation rate depends not only on the metabolic rate of the gonad, but also on the number of cell divisions required to form the gametes. Consider the difference in gametogenesis in the two sexes. In the female, the oocytes are formed when she is still *in utero*. These oocytes do not divide anymore, and only undergo meiosis just prior to ovulation. Very few oocytes are ovulated. The number of cell divisions in the germ line lineage from the zygote to a PGC, to an oogonium, and finally to the oocyte is estimated to be about 24. In other words, an oocyte can be traced back 24 cell generations to the zygote. In the male, on the other hand, a stock of spermatogonia is produced during fetal life, and this stock serves as a self-generating source for spermatozoa that are produced continually once puberty begins. Recall from Chapter 6 that the type A (dark) spermatogonium provides the regenerating stem cell for spermatogenesis. As a consequence, more type A (dark) cell divisions have taken place in an older male than in a younger one. It is estimated that the number of cell divisions between the zygote and the spermatozoon is about 197 when the male is 20 years old, 427 when he is 30 years old, and 792 when he is 45 years old. With over 100 million spermatozoa being produced per day, the testis is a "hot spot" of metabolic activity. We might expect therefore that the male germ line should be more mutation prone than the female germ line, and therefore, that spermatozoa would be at a much higher risk of transmitting heritable mutations to offspring compared to oocytes.

There is substantial evidence indicating that there are significant differences in the mutation rate in males and females. The best-documented cases (because these are the easiest to measure accurately) have come from analysis of the spontaneous mutation rate of X-linked genes. All of these studies suggest that gene mutations (single-nucleotide substitutions) in the male germ line occur about 10 to 11 times more frequently than in the female germ line. One consequence of this difference is that monogenic disorders would be expected to have a higher frequency of paternal than maternal origin. Moreover, since the mutation rate increases with the age of the male, the incidence of monogenic disorders would be expected to be higher among children of older fathers compared with that of younger fathers. The evidence available at present is in agreement with these predictions.

The higher germ line mutation rate in the male provides an explanation for the observation that most of the DNA of the Y chromosome is noninformational. Mutations on the autosomes and the X chromosome have a chance of being repaired during meiosis. Recall from Chapter 2 that homologous chromosomes recombine with each other during the first division of meiosis. Associated with this exchange is a *recombination repair mechanism* that corrects errors that might have escaped the replication repair system. Mutations on the X chromosome may be repaired one cell generation later during female meiosis. The Y chromosome, in contrast, has no homolog, and there is no recombination repair mechanism to remove mutations on the Y. This means that over time, the Y accumulates mutations. Indeed, this is what is observed. The Y chromosome is loaded with nonfunctional *pseudogenes*, relics of what were once functional genes. Hence, the high mutation rate in the testis and the reduced ability for repair means that the Y chromosome is slowly being mutated into extinction.

The final positioning of the testes in scrotal species may be a way of reducing the high germ line mutation rate in males. In scrotal species, the testes are a few degrees cooler than body temperature. This lower temperature reduces the metabolic rate and therefore the mutation rate. Testes retained within the abdominal cavity are extremely susceptible to the development of testicular cancer, especially cancers of the germ line. Hence, the evolution of the scrotum may have provided a convenient way to keep the mutation rate at some acceptable level. In nonscrotal species, special venous cooling mechanisms have appeared to insulate the testis from the core body temperature. Hence, in scrotal and nonscrotal species, cooling mechanisms that reduce the metabolic rate appear to be an important aspect of testicular function.

The testis appears to have negotiated an uneasy Faustian bargain. In order to ensure the propagation of the species, it has to produce spermatozoa at a prodigious rate. A high price is paid for this, however: a much higher germ line mutation rate. This means that the male is responsible for a much larger fraction of genetic disease and birth defects than the female. The benefit of this complex bargain may be that the genetic variability generated by the testis provides new genotypes to be tested by natural selection.

SUMMARY

Chromosome mutations, of which the major type is aneuploidy, are the single most important contributor to fetal loss. They account for at least 50 percent of all spontaneous abortions. The contribution of gene mutations to fetal loss is more difficult to estimate, but all studies indicate that it is significantly less. In contrast, chromosome mutations account for only about 6–7 percent of congenital anomalies. Birth defects can be caused by a single gene (monogenic) or by multiple genes (polygenic), or in many cases, the defect can result from interactions between gene action and factors external to the developing embryo or fetus (multifactorial). Together these account for about 20 to 25 percent of congenital disorders. The causes of a significant fraction of fetal loss and of most congenital anomalies remain unknown (Table 13.1).

EXTRINSIC FACTORS

Normal embryonic or fetal development can be adversely affected by specific types of maternal deficits and by exposure of the pregnant female to chemical or physical agents (teratogens). In this section, we review the effects of the major classes of extrinsic factors.

Maternal nutritional deficiency

Pregnancy places special dietary demands on the mother. As suggested in the previous chapter, metabolic adjustments to pregnancy are met by adjustment in the food intake, or nowadays, by supplementation with vitamins and other essential dietary requirements. Deficiencies in certain compounds can lead to growth retardation and other, more serious disorders. One of the best documented examples of the association between a dietary compound and a serious disorder is that of **neural tube defects** (NTDs), disorders affecting the spinal cord and the brain. In the United States, the two most common NTDs are spina bifida (incomplete closing of the spine) and **anencephaly** (failure of the brain to form). Anencephalic infants die shortly before or after birth, while spina bifida infants will suffer from varying degrees of paralysis and disability. One important discovery of the last decade is that 50–70 percent of these two disorders can be prevented by folic acid supplementation during pregnancy. The U.S. Public Health Service recommendations are that all women of childbearing age consume 0.4 mg of folic acid daily because over half of the pregnancies in the United States are unplanned, and because these two defects occur about 3–4 weeks after fertilization, before most women are aware that they are pregnant.

Environmental agents

Over 3000 environmental agents have been tested for teratogenicity and developmental toxicity, categories that include significant tissue and organ malformations, embryo and fetal lethality, intrauterine growth retardation (IUGR), and different types of functional impairment. About 1200 of these agents are known to produce congenital anomalies in experimental animals, but only about 40 are known to cause defects in humans. The criteria for establishing an agent as a teratogen or developmental toxin are quite stringent, and the most important are: (1) proven exposure at a critical time during pregnancy (for example, prescriptions, physician's records, or other verifiable record); (2) consistent findings by two or more high-quality epidemiological studies; (3) detailed delineation of the clinical manifestation of the exposure; (4) an association between agent and effect that makes biological sense; (5) rare environmental exposure that should be associated with a rare effect. Verification of the effect in an experimental system is important but not necessary. It is estimated that environmental agents account for no more than 10 percent of human congenital abnormalities (Table 13.1). We consider here some of the environmental agents for which there is some evidence linking them to teratogenic effects in humans.

Biological agents—viruses and bacteria

The embryo and fetus are susceptible to infection by the microorganisms to which the mother is exposed. Fortunately, the maternal immune system is able to eliminate or inactivate the large majority of infectious agents. However, a few may escape destruction and enter the fetal circulatory system through the placenta (Table 13.8). The fe-

Table 13.8
Examples of Biological Teratogens

Teratogen	Major effects
Cytomegalovirus	Severe brain malformations; cerebral palsy
Rubella virus	Cardiac malformations; cataracts; glaucoma; IUGR; mental retardation
Human parvovirus B19	Eye defects; degeneration of fetal tissues
Human immunodeficiency virus (HIV)	Microcephaly; growth failure
Herpes simplex virus	Enlarged liver; hemolytic anemia
Varicella (chickenpox) virus	Cataracts; hydrocephaly; neurological defects; mental retardation
Venezuelan equine encephalitis virus	Microcephaly; CNS necrosis
Treponema pallidum	Congenital deafness; mental retardation; hydrocephaly
Toxoplasma gondii	Microcephaly; mental retardation

tal central nervous system appears to be particularly sensitive to the effects of these agents. The consequences of these infections are generally quite severe, and include spontaneous abortions, stillbirths, premature births, multiple tissue damage, congenital abnormalities, mental retardation, and sensitivity to disease after birth.

The effects of the rubella virus, the causative agent for German measles, was probably the first well-documented example of the teratogenic effects of a virus. The risk of fetal infection is about 20 percent if the mother is exposed to rubella. However, the sensitive period for teratogenic damage by rubella infections is between week 3 and week 12 of development, and the effects can vary greatly in severity. This example illustrates again the important principle that sensitivity to teratogenic agents is not uniform during gestation. Since a vaccine is available for the prevention of German measles, rubella-caused teratogenesis is very rare in the developed world.

Other infectious organisms have been consistently associated with developmental disorders. Several of these are organisms responsible for sexually transmitted diseases—herpes simplex virus, which causes genital herpes; human immunodeficiency virus (HIV), which causes AIDS; and *Treponema pallidum*, which causes syphilis. These will be discussed in more detail in Chapter 17.

Therapeutic and recreational drugs

A wide variety of chemicals can be transferred from the mother to the embryo or fetus. Important types of chemical teratogens are listed in Tables 13.9 and 13.10.

Table 13.9
Therapeutic Teratogens

Drug	Prescribed for	Major effects
Aminopterin	Antitumor	IUGR; skeletal defects
Isotretinoin (Accutane)	Acne	Neural tube defects; cardiovascular defects
Methotrexate	Antitumor	Multiple malformations
Dilantin	Epilepsy	Microcephaly; mental retardation
Thalidomide	Morning sickness	Abnormal limb development
Tetracycline	Antibiotic	Stained teeth
Valproic acid	Migraine; epilepsy	Craniofacial anomalies

Table 13.10
Major Recreational Teratogens

Agent	Major features
Alcohol	IUGR; microcephaly; fetal alcohol syndrome
Tobacco	Spontaneous abortions; IUGR
Cocaine	IUGR; neurobehavioral abnormalities
Caffeine	IUGR; prematurity

These fall into two main groups: those that are used *therapeutically* (prescribed for controlling or treating disease or pathologies), and licit or illicit drugs that are used *recreationally*. The placenta is not as effective a barrier to the passage of many chemicals as was once thought. The effect of the particular teratogen depends on the dose that the conceptus experiences, not the dose experienced by the maternal system, and also on period of gestation. Some teratogens produce quite severe effects, death, major malformations, mental retardation, and functional disorders. Others may not produce clear-cut morphological or physiological defects, but may result in **intrauterine growth retardation** (IUGR). IUGR is a consequence of disordered fetal growth that may have serious consequences postnatally. One important diagnostic criteria for IUGR is prematurity or lower-than-normal birth weight. IUGR, for example, is a common condition in infants whose mothers were on drugs during pregnancy. IUGR has been associated with varying degrees of neurodevelopmental impairment, from learning and cognitive defects and behavioral difficulties to epilepsy, cerebral palsy, and mental retardation, as well as being a risk factor for diabetes, hypertension, and cardiovascular disease. Assessing the consequences of IUGR remains a difficult enterprise.

Therapeutic drugs

The use of prescription and nonprescription drugs during pregnancy is surprisingly high. It is estimated that 40 to 90 percent of pregnant women take at least one drug, and some take as many as four, during the first trimester of pregnancy, the most sensitive period for teratogenesis. Table 13.9 lists a few therapeutic drugs that are known to have teratogenic effects. Methotrexate has been prescribed as an anticancer drug and for palliative therapy for rheumatic diseases. A few children with an uncommon and characteristic pattern of congenital anomalies have been born to women who received the drug during the first trimester of pregnancy. The effects of methotrexate have been attributed to the fact that it is a folic acid antagonist, meaning that its administration results in folic acid deficiency, which can have serious consequences. Phenytoin is one of several drugs prescribed for seizure disorders, and a characteristic pattern of congenital abnormalities (heart disorders, and facial clefts) has been seen among children of epileptic women treated with drugs of this type. Phenobarbitol, a barbiturate, is an anticonvulsant, and there is some limited evidence for teratogenic risk, but the risk is considered to be small to moderate.

One of the most recently recognized teratogens are the *retinoids*, vitamin A analogs, prescribed in the treatment of dermatologic diseases such as acne and psoriasis (Accutane, the trade name for *isotretinoin*, is one of the more popular retinoids). Children born to women treated with isotretinoin during the first trimester have been born with severe craniofacial abnormalities and cardiovascular defects. Based on the data collected, the teratogenic risk associated with retinoids is high.

Alcohol

The clinical characteristics that are now known as **fetal alcohol syndrome** (FAS) were first described by the French pediatrician Lemoine in 1968, and these included the following: growth deficiency, microcephaly (very small head due to lack of brain tissue development), anomalous facial characteristics, cardiac defects, limb deformities, hyperactivity, attention deficit disorder, delay in psychomotor and language development, poor visual memory, psychosocial maladjustment, low IQ, and mental retardation. FAS is a major societal health problem in the United States. FAS in its different forms accounts for a significant fraction of the cases of mental retardation in the United States. What is interesting about alcohol is that its effects have been suspected or known for a long time. Aristotle pointed out in his writings that women drunkards often gave birth to abnormal children. Drinking alcohol by the bride on her wedding night was prohibited in ancient Greek custom. The ancient Hebrews also understood the effects of alcohol. A proscription against it is evident in the following passage from the Old Testament (Judges 13:3–5):

> And the angel of the Lord appeared unto the woman, and said unto her,
> Behold now, thou are barren, and bearest not; but thou shalt conceive and a
> bear a son. Now therefore, beware, I pray thee, and drink not wine nor strong
> drink, and eat not any unclean thing; for lo, thou shalt conceive and bear a
> son.

More recently, during the English gin epidemic (1720–1750), a sharp decrease in the price of gin led to enormous problems in the health and well-being of infants. When the cause was identified, the English government raised the price of gin enough to control the amount of gin consumption. The National Institute of Medicine estimates that in the United States 20 percent of women who drink continue to do so while they are pregnant. As a result, about 1 infant in 1000 born has FAS.

Exposure to alcohol during the embryonic period is considered to produce the most severe defects. However, exposure to alcohol at all stages is dangerous, particularly since the central nervous system remains sensitive to teratogenic action throughout gestation. Although the mechanism of alcohol action in producing FAS remains unclear, important clues about the way in which alcohol damages brain development were reported recently. This study, carried out in rats, showed that alcohol induces apoptosis, or programmed cell death, when brain neurons are forming connections with each other. The equivalent vulnerable period in humans is not only during the early months of the pregnancy, but extends to the end of gestation and to several years after birth. This study indicates that late-pregnancy drinking carries serious risks for the fetus. Defects produced during the latter stages of pregnancy may not result in obvious anatomical abnormalities, but evidence suggests that cognitive ability may be impaired. Binge drinking is considered to be more harmful than one or two drinks daily. How much alcohol is safe? No precise answer can be given to this question. The safest course is to avoid it altogether.

Tobacco

Tobacco smoke contains over 3500 different compounds, many of which are mutagenic. The most abundant, nicotine, carbon monoxide, and hydrogen cyanide, have been considered to be the most toxic during pregnancy. Nicotine is water and lipid soluble, which means that it is readily taken in by all tissues. It is taken up preferentially by the brain and the adrenal gland. The precise mechanism of its effects is not well understood. In combination with carbon monoxide and hydrogen cyanide, one of its major effects is to reduce blood flow to the uterus, which restricts oxygen flow to the fetus. Because the fetus has a significantly reduced ability to metabolize nicotine, it will be exposed to the drug at least four times longer than an adult. It has been difficult to document the precise effects of maternal smoking on the developing embryo or fetus. Nevertheless, the risks are high. An estimated 100,000 spontaneous abortions, 5000 congenital abnormalities, and 200,000 cases of IUGR are attributed to maternal smoking.

Much less attention has been paid to paternal smoking, but it has been recognized that its effects may be no less significant. In a report published in 1986 by the National Research Council, paternal smoking was associated with increased risk of perinatal mortality, lower birth weight, increased risk of congenital malformations, and increased risk of childhood cancers. For example, the neonatal death rate for infants of smoking fathers was 17.2 per 1000 live births, while it was 11.9 per 1000 live births for infants of nonsmoking fathers. The rate of major malformations in newborns was 2.9 percent for smoking, and 0.8 percent, for nonsmoking fathers. Maternal smoking has little effect on the incidence of childhood cancers, while there appears to be a clear correlation with paternal smoking.

The paternal smoking effects are now considered to be due to mutagenic effects of tobacco smoke in the testis. Tobacco smoke contains many mutagenic compounds that are easily absorbed into the blood and, therefore, eventually reach the testes. The rapidly dividing germ cells are particularly susceptible to mutagenic damage. Moreover, tobacco smoke contains many compounds that potentiate the oxidizing effects of the normally produced free radicals. The continual production of spermatozoa during the male's reproductive lifetime means that paternal smoking may increase the germ line mutation rate. The risk that an embryo will be carrying genetic lesions with the potential to disturb normal development or that may result in childhood cancers is increased if the father is a smoker. The contribution cannot be estimated with great precision, but there is significant evidence that damage to the fetus may be as important as that of maternal smoking.

Cocaine

Cocaine is a topical anesthetic and a powerful CNS stimulant. No controlled study on the effects of cocaine on the fetus has been carried out. However, a variety of malformations (microcephaly, kidney defects, cardiac defects, and limb deformities) have been reported in children of mothers known to have used cocaine during pregnancy. A few case studies have also indicated that cocaine taken during pregnancy may lead to IUGR and behavioral abnormalities during infancy and childhood. The most reliable information suggests that premature expulsion of the placenta and cocaine-induced fetal CNS hemorrhaging may be the most important types of damage due to cocaine exposure.

Environmental chemicals

It is estimated that more than 60,000 chemicals are used in manufacturing processes, and 500 or more new ones are being introduced yearly. These include organic solvents, heavy metals, and pesticides (over 21,000 of these are registered) (Table 13.11). Exposure to most of these chemicals is generally confined to the workers in the

Table 13.11
Environmental Teratogens

Agent	Major effects
Heavy metals	Cerebral atrophy; mental retardation
PCBs	IUGR
DDT	Malformation of male internal and external genitalia

manufacturing plants that use them. However, a few may get dispersed widely, and in some circumstances can pose a possible threat to pregnant women. For example, the polychlorinated biphenyls (PCBs) used in the manufacture of plastics, paints, and other products have been found to be teratogenic in animal studies, and there is some evidence that PCB exposure during pregnancy is associated with IUGR. In Japan in 1968 and 1979, some batches of rice oil used in cooking were accidentally contaminated with PCBs. Women who ingested the contaminated rice oil gave birth to children with poor muscle tone and dark-brown staining of the skin; they also exhibited some impairment in learning. The staining disappeared a few months after birth. Similar effects were observed in the children of mothers from the Great Lakes region who ate PCB-contaminated fish. Although PCBs have been banned, they appear to have permeated the environment. They can be detected in the food chain and in almost all tissues of the body. The level of PCB exposure associated with the effects described above was much higher than the levels found in the environment. There is no clear evidence that the low levels of PCBs detected in the environment are teratogenic.

A number of environmental chemicals are weak estrogen agonists or weak androgen antagonists, which means that exposure to them at a critical time in gestation could disrupt normal development of the male internal and external genitalia. Effects of this type have been documented in animal studies, which usually employ high concentrations of these compounds. There is no agreement about the extent to which exposure to the much lower levels of these chemicals that most people might be exposed to is a significant problem in humans. One of the major difficulties in establishing well-defined cause-and-effect relationships is that humans consume many compounds in their food that are weak estrogens. These compounds, derived from plants that are part of the normal diet, are known as *phytoestrogens*. There is no compelling evidence that normal consumption of phytoestrogens should be a source of concern. In fact, the evidence indicates that phytoestrogens may exert a protective effect in the development of many cancers (see Chapter 18).

The number of environmental chemicals with teratogenic potential is large, but exposure of the general population to most of these is considered to be below the threshold needed to have significant developmental effects. There is widespread concern by the public that environmental chemicals are responsible for many developmental defects, functional impairment, childhood cancers, and even cognitive disorders, conditions for which no obvious cause can be found. Particularly troublesome have been the "cancer clusters" that have received much publicity over the last few years. These are cases in which an unusual number of cancers are seen in certain communities. Generally, some aspect of the environment, such as contaminated groundwater, the soil, the food, or the air, is blamed for the observed increase in cancers. However, despite intensive and exhaustive investigation by public health officials, not one of the cancer clusters found in the United States has been shown to be due to an environmental cause. In other parts of the world, in only a handful of clusters among the hundreds reported, has an environmental cause been convincingly identified.

Radiation

Ionizing radiation (X-rays, gamma radiation, radiation from radioactive elements) is of concern only after exposure to large doses, such as might be used in radiation therapy for cancer. Doses used for diagnostic purposes (medical X-rays) are not considered to be teratogenic.

Summary

Perhaps the most important lesson that can be gleamed from the study of teratogens is that the developing embryo and fetus are exceptionally sensitive to many different compounds and agents. Many of these produce very obvious and severe malformations or abnormalities, and we properly do everything we can to minimize exposures to such teratogens. Much concern has been expressed about the dangers of environmental chemicals. However, it has been very difficult to demonstrate convincingly that normal exposures to environmental chemicals contribute significantly to the overall rates of fetal loss or birth defects. On the other hand, the information we have at present indicates that alcohol and smoking may be responsible for the largest fraction of developmental abnormalities due to external agents. Many of the defects produced by alcohol and smoking may be subtle. The most insidious of these are those that perturb brain development and affect cognitive and psychological function. The cumulative cost of such defects to the individual and to the society is incalculable. It is ironic that up until recently we as a society have been much less concerned about alcohol and smoking, about whose effects in the general population there is no doubt, than we have been about environmental chemicals, whose contributions to developmental abnormalities are much less important.

PRENATAL DIAGNOSIS

A number of methods are now available for *prenatal diagnosis*, or for obtaining anatomical, chromosomal, physiological, and genetic information about the embryo and fetus. Prenatal diagnosis is valuable for several reasons. A long-term goal is to be able to identify an abnormal prenatal condition so that treatment for the condition *in utero* can be devised. It is also very useful in assessing the status of the fetus in cases where the method or timing of parturition needs to be planned. In other cases identifying embryos carrying chromosomal aberrations or other severe disorders early in gestation may give the parents the option of inducing an abortion.

Most prenatal diagnostic methods are invasive in that they require tissue from the fetus. Ultrasonography is the principal noninvasive procedure, and it is perhaps most useful during the second and third trimesters of pregnancy, when it is used to locate the placenta, determine fetal cardiac activity, detect multiple pregnancies, detect some types of structural abnormalities, and evaluate fetal well-being. Continuing technological improvements in ultrasonography are making it possible to obtain information about the developing conceptus at earlier and earlier ages. Another relatively noninvasive method is measuring *alpha-fetoprotein* levels in the maternal serum beginning early in the second trimester. Elevated or depressed levels of that protein are diagnostic for a number of developmental disorders, especially neural tube defects and some chromosomal aberrations.

There are several invasive diagnostic methods, including *amniocentesis*, *chorionic villus sampling* (CVS), *cordocentesis*, *fetal biopsy*, and *preimplantation embryo biopsy*. Amniocentesis involves puncturing the uterus and amniotic sac with a fine needle to remove a sample of amniotic fluid; it is used for analysis of chromosomes in fetal cells that are normally found in the amniotic fluid (Fig. 13.5). The earliest time in gestation that amniocentesis can be performed successfully is 16 to 18 weeks. CVS is a placental biopsy in which cells are taken for analysis from chorionic villi. Successful sampling can be done earlier than amniocentesis. Cordocentesis involves taking blood from the umbilical cord. This has been useful in evaluating the blood status of the fetus or detecting prenatal infections. Fetal biopsy involves removal of fetal tissue, usually skin, liver, or muscular tissue, that can be analyzed directly for several disorders. Pre-

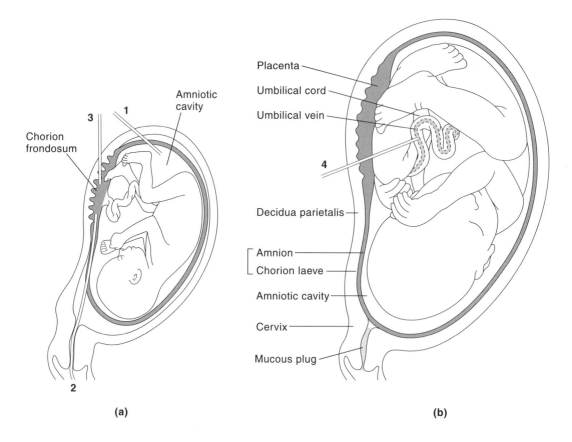

Figure 13.5
Four methods of prenatal sampling, (**a**) second trimester, (**b**) third trimester. The numerals indicate the routes of access: (1) amniocentesis, (2) transcervical chorionic sampling, (3) transabdominal chorionic sampling, and (4) cordocentesis. (Adapted from R. O'Rahilly and F. Müller, 1996.)

implantation embryo biopsy can be carried out as part of an *in vitro* fertilization program (see Chapter 16), and involves excising one or more cells from a morula stage embryo, or removing trophoblastic cells from a blastocyst. These cells can then be analyzed for chromosome aberrations.

QUESTIONS

1. Chromosome abnormalities are a major cause of fetal loss, but only a minor cause of congenital abnormalities. Why do you think this is the case?

2. The incidence of aneuploidy increases with maternal, but not paternal, age. How is this difference explained?

3. What do ovarian teratomas and hydatidiform moles have in common? How do they differ?

4. The mutation rate is higher in males than females. How is this difference explained, and what are the consequences?

5. Describe some of the effects of the most important human teratogens.

6. Library project:
    a. What is known about the mechanisms that give rise to ovarian teratomas and hydatidi-
       form moles?
    b. Describe advances in methods for prenatal diagnosis.

SUPPLEMENTARY READING

Forman, R., S. Gilmour-White, and N. Forman. 1996. *Drug-Induced Infertility and Sexual Dysfunction.* Cambridge University Press, Cambridge, UK.

Gawand, A. 1999. The cancer-cluster myth. *The New Yorker* (February 8), 34–38.

Moore, K. L., and T. V. N. Persaud. 1993. *The Developing Human.* W. B. Saunders Co., Philadelphia, PA.

O'Rahilly, R., and F. Müller. 1996. *Human Embryology and Teratology,* 2nd ed. Wiley-Liss, Inc., New York, NY.

Warkany, J. 1971. *Congenital Malformations: Notes and Comments.* Year Book Medical Publishers, Inc., Chicago, IL.

ADVANCED TOPICS

Bishop, J. B., K. L. Witt, and R. A. Sloane. 1997. Genetic toxicities of human teratogens. *Mutation Research* **396**, 9–43.

Conner, S. M., and M. A. Ferguson-Smith. 1987. *Essential Medical Genetics,* 2nd ed. Blackwell Scientific Publications, Oxford, UK.

Gadow, E. C., L. Otaño, and S. E. Lippold. 1996. Congenital malformations. *Current Opinion in Obstetrics and Gynecology* **8**(6), 412–416.

Golden, R. J., *et al.* 1998. Environmental endocrine modulators and human health: An assessment of the biological evidence. *Critical Reviews in Toxicology* **28**(2), 109–227.

Hook, E. B., P. K. Cross, and D. M. Schreiemachers. 1983. Chromosome abnormality rates at amniocentesis and in live-born infants. *Journal of the American Medical Association* **249**, 2034–2040.

Ikonomidou, C., *et al.* 2000. Ethanol-induced apoptotic neurodegeneration and fetal alcohol syndrome. *Science* **287**, 1056–1060.

Loebstein, R., and G. Koren. 1997. Pregnancy outcome and neurodevelopment of children exposed in utero to psychoactive drugs: The Motherisk experience. *Journal of Psychiatry and Neuroscience* **22**(3), 192–196.

MacKenzie, T. D., C. E. Bartecchi, and R. W. Schrier. 1994. The human costs of tobacco use. *New England Journal of Medicine* **330**, 975–980.

Nahmias, A. J., and A. P. Kourtis. 1997. The great balancing acts: The pregnant woman, placenta, fetus, and infectious agents. *Clinics in Perinatology* **24**(2), 497–521.

Shepard, T. H. 1995. *Catalog of Teratogenic Agents.* John Hopkins University Press, Baltimore, MD.

Smith, C. G., and R. H. Asch. 1987. Drug abuse and reproduction. *Fertility and Sterility* **48**(3), 355–373.

Wood, J. W. 1994. *Dynamics of Human Reproduction: Biology, Biometry, Demography.* Aldine de Gruyter, New York, NY.

Woodal, W. A., and B. N. Ames. 1997. Nutritional prevention of DNA damage to sperm and consequent risk reduction in birth defects and cancer in offspring. In *Preventive Nutrition: The Comprehensive Guide for Health Professionals.* A. Bendich and R. J. Deckelbaum, Eds. Humana Press, Totowa, NJ.

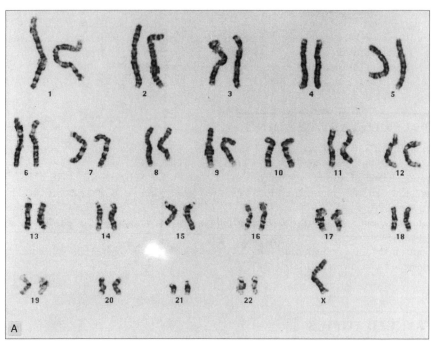

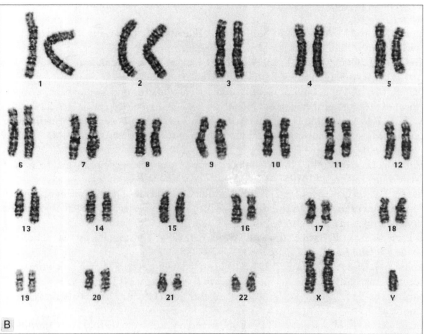

Sex chromosome aneuploidy. (A) *X0* (Turner syndrome) in a 17-year-old female with amenorrhea. (B) *XXY* (Klinefelter syndrome) in a 21-year-old male with hypogonadism. (Courtesy of Dr. O. W. Jones, UCSD School of Medicine.)

CHAPTER 14

Disorders of Sexual Development

. . . an individual who had been brought up as a girl, but who was attracted only by the female sex, had no menstruation, grew a mustache, had lately attempted coitus with a girl, and during the act claimed a discharge from the urethra of semen. The general lines of the body were feminine, the breasts were well developed . . . The hair on the pubis was of the female type. The external organs were those of a woman of normal development and presented no peculiarity.

—E. C. Dudley and H. M. Stowe. 1913

Gynecology. Practical Medicine Series, Vol. V.
The Year Book Publishers, Chicago, IL, p. 90.

The epigraph is taken from a text in common use in medical schools in the United States in the early part of the twentieth century. It is an excerpt from a description of an individual labeled as a "*hermaphrodite with functional testes and female generative organs.*" Descriptions of individuals with unusual or abnormal external genitalia are quite common in the clinical literature of that time and even much before that as well. However, a true understanding of the nature of those abnormalities was not really possible until the nature of sexual development itself was understood. It has been only in the last 20 years or so that we have developed a reasonably comprehensive view of the way in which the different elements of the reproductive system are formed. To a large extent, our knowledge has come from the analysis of cases in which sexual development is abnormal. Our objective in this chapter is to review the general principles by which we can understand the many different types of sexual abnormalities that are observed.

Recall from Chapter 4 that sexual development occurs in stages. Sex determination defines the gonadal sex of the individual, while formation of the internal and external genitalia defines the phenotypic sex of the individual. Finally, the full expression of phenotypic sex begins at puberty. Errors in sex determination lead to the clinical conditions of sex reversal and true hermaphroditism that we discussed in Chapter 4. In this chapter, we will consider the disorders of phenotypic sex and of puberty. Disorders of phenotypic sex result in a mismatch between gonadal and phenotypic sex, giving rise to a condition referred to in clinical literature as *pseudohermaphroditism*. The conditions that lead to this type of disorder are outlined in Table 14.1. Disorders of this type always present the individual, the family, and the clinician with the multifaceted problems of developing a suitable therapy and also trying to deal with the psychological and emotional trauma. Disorders of puberty affect the timing of puberty, and these can be classified as those that accelerate puberty and those that delay it. Many different conditions are known to disturb the timing of puberty (Table 14.2).

Table 14.1
Disorders that Affect the
Development of Phenotypic Sex

Sex chromosome abnormalities

Klinefelter syndrome, *XXY*
Turner syndrome, *X0*

Disorders of hormone synthesis

Testosterone deficiency
DHT deficiency
APH deficiency
Testosterone excess
Estrogen deficiency

Disorders of hormone action

Androgen insensitivity syndrome (AIS)
Estrogen insensitivity syndrome (EIS)
APH insensitivity

Table 14.2
Disorders that Affect the Timing of Puberty

Precocious puberty

GnRH-dependent: complete, or central, precocious puberty
    Idiopathic—sporadic or familial
    CNS lesions—tumors, congenital abnormalities

GnRH-independent: incomplete precocious puberty
    Premature thelarche
    Premature adrenarche
    Estrogen-secreting tumors
    Androgen-secreting tumors
    Congenital adrenal hyperplasia
    Gonadotropin-secreting tumors
    hCG-secreting tumors
    Constitutive LH receptor mutations

Delayed puberty

Hypothalamic or pituitary lesions: *hypogonadotropic hypogonadism*
    Idiopathic GnRH deficiency
    CNS tumors
    Hypothalamic delayed adolescence: chronic illness, malnutrition
    Hyperprolactinemia
    Impaired gonadotropin secretion

Gonadal lesions: *hypergonadotropic hypogonadism*
    Gonadal dysgenesis
    Ovarian failure: chemotherapy, radiation therapy, cancer
    Testicular failure: impaired androgen synthesis, defective LH receptor

DISORDERS OF PHENOTYPIC SEX

Disorders of phenotypic sex arise from changes in sex chromosome number, alterations in androgen, estrogen, or antiparamesonephric hormone (APH) levels, or mutations in the genes encoding the androgen or estrogen receptors. The defects that give rise to these disorders result in some cases in pronounced anatomical changes in the internal and external genitalia. Infertility due to the failure of gametogenesis is a consistent feature in all of these disorders.

Sex chromosome abnormalities

The most common sex chromosome aberrations that lead to a perturbation in the development of phenotypic sex are Klinefelter, first described in 1942, and Turner syndromes, first described in 1938 (Fig. 14.1). The classic phenotype of Klinefelter syn-

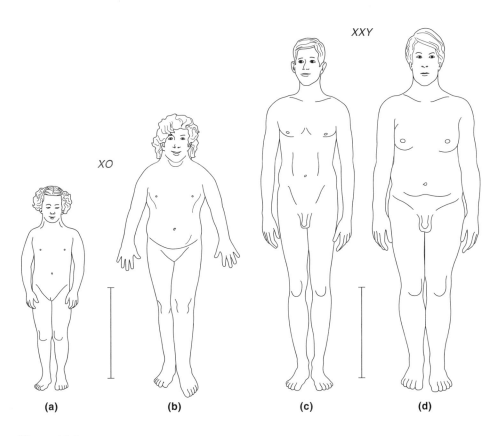

Figure 14.1
Turner and Klinefelter syndromes. (**a**) and (**b**) *X0* Turner syndrome in girls aged 4 and 14, respectively. The short stature and broad chest are characteristic. Webbing of the neck, also typical, is not always apparent. (**c**) and (**d**) *XXY* Klinefelter syndrome in a boy 16 and a man 21, respectively. The tall stature, long limbs, and body shape are characteristic. Breast development is often observed as well. (Adapted from R. O'Rahilly and F. Müller, 1996, with permission.)

drome, occurring with a frequency of about 1 in 800 males, generally develops after puberty. Klinefelter individuals have two X chromosomes in addition to a Y chromosome with testes (they are XXY males), although the testes are smaller compared to those of normal males. Androgen deficiency, seminiferous tubule disorganization, and disproportionately long limbs are the typical features observed. The consequences of androgen deficiency are lack of facial and body hair, a small penis, poor muscle development, and infertility and erectile dysfunction. Gynecomastia is also common. Other features seen in some but not all XXY individuals are mental retardation, learning difficulties, and a tendency to sociopathic behavior. The severity of the symptoms is quite variable. Androgen replacement is the main therapy and, if begun at puberty, has been successful in restoring muscle strength and libido in some individuals. It has not been possible to restore fertility, however, by androgen or other hormone therapy.

Turner syndrome, occurring with a frequency of about 1 in 2500 female births, is characterized by the $X0$ genotype, females with only one X chromosome, rather than the normal 2 Xs. The classic features of Turner syndrome include short, webbed neck, absence of pubic and axillary hair, and very small ovaries containing no follicles (*ovarian dysgenesis*). Follicles form at the normal time and in about the normal numbers in $X0$ embryos, but it appears that atretic loss is greatly accelerated, so that by birth the ovary is essentially empty of follicles. This finding may suggest that a gene or genes on the X chromosome may be important in regulating atretic loss. A number of other abnormalities may also be present, including cardiovascular and renal defects, osteoporosis, and inverted nipples. Hormonally, Turner syndrome individuals have low estrogen and androgen levels and elevated FSH and LH levels. Estrogen replacement therapy is required at puberty to develop the female body form, but fertility cannot be restored.

Both the $X0$ and XXY conditions are the result of errors in meiosis. For example, under normal conditions, an XY individual produces equal numbers of X- and Y-bearing spermatozoa. However, failure of the X and Y chromosomes to disjoin in the first meiotic division results in some spermatozoa that are XY-bearing and some that are *null*-bearing (not carrying a sex chromosome at all) (Fig. 14.2). When such sperm fertilize a normal egg, an XXY or $X0$ zygote is formed. Another possibility is that the failure occurs not in the first meiotic division, but in the second. In this case, XX and YY sperm are formed, which in turn can give rise to XXX and XYY zygotes (Fig. 14.2). Similar errors can occur in oogenesis, which results in either XX- or *null*-bearing eggs.

Hormone biosynthesis disorders

Recall from Chapter 4 the asymmetric requirement for development of the internal and external genitalia in males and females—androgens and APH in males, but no requirement for estrogen in females. Hence, deficiencies in androgens or APH would be expected to result in a lack of virilization of the genitalia in males, while excess androgens would be expected to result in the virilization of the genitalia in females. Let us consider how these different conditions come about.

Androgen deficiency

Androgen deficiency is most commonly caused by mutations in genes that encode the enzymes involved in testosterone and DHT synthesis. Defects in testosterone and DHT synthesis or in the response to them (see below) result in defects in the normal virilization of XY embryos, a condition referred to in the clinical literature as *male*

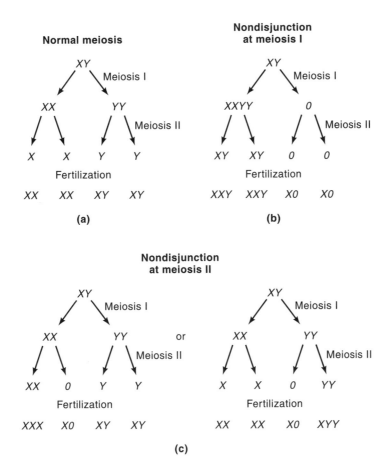

Figure 14.2
Nondisjunction in spermatogenesis. Consequences of nondisjunction of the X and the Y chromosomes in the first and second meiotic divisions in spermatogenesis. (**a**) Normal meiosis I and meiosis II. X- and Y-bearing sperm are produced in equal numbers, resulting in equal numbers of XX and XY fertilizations. (**b**) In a meiosis I error, the X and Y fail to disjoin, resulting in half the sperm carrying both an X and a Y, and the other half with no sex chromosomes. Equal numbers of XXY and X0 fertilizations would occur. (**c**) In a meiosis II error, the nondisjunction can affect the X or the Y chromosome at the second meiotic division, leading to indicated types of fertilizations. Similar errors can take place in oogenesis.

pseudohermaphroditism. This global term is not useful diagnostically, however, since the phenotype that it describes can arise in several different ways. Recall that the steroid hormone biosynthetic pathway is hierarchical, beginning with the synthesis of progestins from cholesterol, conversion of progestins to androgens, and finally, conversion of androgens to estrogens (Fig. 14.3). Steroid sex hormones are synthesized primarily by the gonads and secondarily by the adrenals. The adrenals specialize in the synthesis of the nonsex steroids, the glucocorticoids and the mineralocorticoids, both of which are dependent on the prior synthesis of progestins. The synthetic pathways of both sex and nonsex steroids are multistep conversions, each carried out by a specific enzyme. The enzyme is a protein, the product of a specific gene. Mutations in the gene that encodes an enzyme may lead to the absence of the enzyme, to a partially functional

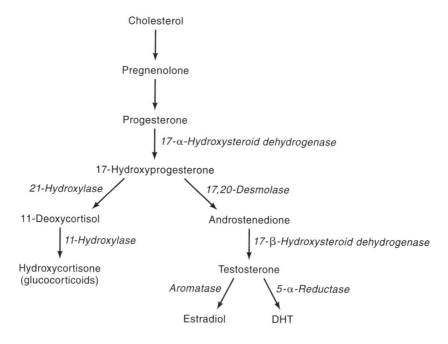

Figure 14.3
Highly simplified diagram of steroid hormone synthesis. Only a few steps in the biosynthetic pathways are shown. In the gonads, the enzyme *17-alpha-hydroxysteroid dehydrogenase* converts progesterone to 17-hydroxyprogesterone; *17,20-desmolase* converts 17-hydroxyprogesterone to a weak androgen, androstenedione, which in turn is converted to testosterone by *17-β-hydroxysteroid dehydrogenase*. Testosterone is converted to estradiol by *aromatase*, and in tissues that have the enzyme *5-alpha-reductase*, testosterone is converted to dihydrotestosterone (DHT). In the adrenal gland, 17-hydroxyprogesterone is converted to the glucocorticoids, successively by *21-hydroxylase* and *11-hydroxylase*. Deficiencies in these enzymes result in developmental disorders of phenotypic sex.

enzyme, or a nonfunctional enzyme. This in turn results in the deficiency of a given hormone.

The phenotype of the mutation, that is, the extent of the hormone deficiency, depends on the extent to which the mutation impairs the function of the particular enzyme. In general, deficiency of a particular hormone also leads to deficiencies of hormones that are downstream in the pathway. Deficiencies in testosterone, for example, can arise in several ways. For example, defects in the conversion of cholesterol to progestins affect both gonadal and adrenal hormone synthesis and result in deficiencies in sex and nonsex steroid hormones. These defects in general have severe consequences, since the glucocorticoid and mineralocorticoid hormones are essential for viability.

Defects in the synthesis of testoterone or DHT result in different phenotypes. This is because, as we learned in Chapter 4, testosterone and DHT have qualitatively different functions in the formation of the internal and external genitalia. A comparison of these expected differences is shown in Fig. 14.4. There is great variability in the pheno-

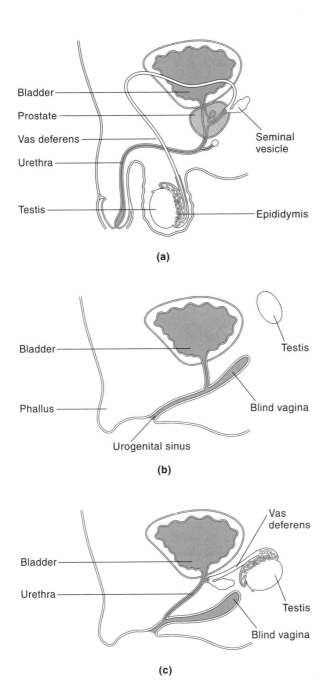

Figure 14.4

The consequences of testosterone or DHT deficiency. (a) Normal masculinization of the internal and external genitalia. (b) Testosterone deficiency prevents enlargement of phallus to form a normal penis. The testes remain in the abdominal position. The urogenital sinus and folds fail to fuse to form the scrotum, and instead they develop into a blind vagina and a urethra that does not empty through the phallus. No uterus forms because of APH suppression. This phenotype is seen in severe forms of testosterone deficiency. (c) DHT deficiency presents a similar, but not identical phenotype. The testis generally will move down from the original abdominal position into the inguinal canal. A vas deferens and seminal vesicle form since testosterone is present, but the prostate gland is missing. A small phallus is present, and urogenital sinus development is femalelike.

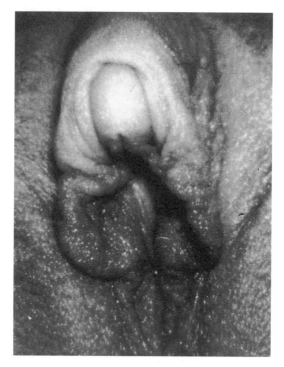

Figure 14.5
External genitalia of *X Y* individual with 17-alpha-hydroxysteroid dehydrogenase deficiency.
In this case, testosterone cannot be made, and as a consequence masculinization of the external
genitalia does not take place. An enlarged clitoris forms, the testes remain in the abdominal
position, and the scrotum is empty. (Reprinted from M. A. Damario and J. A. Rock, 1996,
with permission.)

types observed because the precise level of the androgens varies from one individual to another. For example, a mutation in the gene encoding the enzyme *17-alpha-*
hydroxysteroid dehydrogenase (see Fig. 14.3), resulting in the failure to synthesize
testosterone, can lead in an *X Y* individual to the phenotype shown in Fig. 14.5. External genitalia are female in appearance. The testes remain cryptorchid, and the scrotum
does not develop properly. The difficult questions in these cases is what "sex" the individual is considered to be.

DHT deficiency arises from a deficiency in *5-alpha-reductase*, the enzyme that
converts testosterone to dihydrotestosterone (DHT). The phenotype depends on the
extent of DHT deficiency. In males, DHT is required to form the prostate, penis, and
scrotum, and hence, a DHT deficiency would be expected to manifest itself in abnormalities in the internal and external genitalia. Internally, because testosterone levels
are normal, mesonephric duct development is almost normal except that no prostate
gland is formed (Fig. 14.4). The ejaculatory ducts empty into a blind vagina. No paramesonephric duct development takes place since APH levels are normal. The external
genitalia at birth have a pronounced femalelike character, with a clitorallike phallus,
no scrotum, a vaginal opening, and labia (Fig. 14.6). The testes generally remain in the
abdominal cavity, although they may be palpitated in the labia.

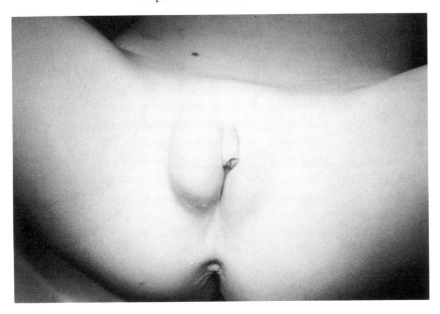

Figure 14.6
Appearance of external genitalia in a case of DHT deficiency. Note the small phallus, and vaginal opening. The testes in the inguinal position produce the bulging out, but a true scrotum does not form. (Courtesy of K. W. Jones, UCSD School of Medicine.)

DHT deficiency syndrome was originally characterized in 1961 in studies of large families in the Dominican Republic, Turkey, and Papua, New Guinea, in whom numerous members appeared to be female at birth, but who at puberty began to develop a penis and scrotum. In the Dominican Republic, this condition was often referred to as the "penis-at-12" syndrome. The external genitalia at birth can vary from being completely female with a blind vagina (complete DHT deficiency) to different levels of ambiguity (partial DHT deficiency). In the severe cases of the disorder (that is, those with female genitalia), the individuals are raised as females. At puberty, because of the very large increases in the levels of testosterone, masculinization of the external genitalia becomes possible, and limited enlargement of the phallus and descent of the testes into an abnormal scrotal pouch can take place. Hence, often a penis and a scrotum begin to develop in what was before considered to be a girl. DHT-deficient individuals are sterile, although a limited degree of spermatogenesis has been reported in some instances. It is also interesting that although many such individuals may have been raised as girls up through puberty, their psychosexual orientation after puberty is male. This syndrome illustrates clearly the different functions of testosterone and DHT and the very interesting and perplexing separation of function of these two androgenic hormones.

Testosterone elevation

Some mutations produce a deficiency in glucocorticoids and, at the same time, an increase in androgen production. One such condition is referred to as *congenital adrenal hyperplasia* (CAH), and it can arise from a deficiency in *21-hydroxylase* (see Fig.

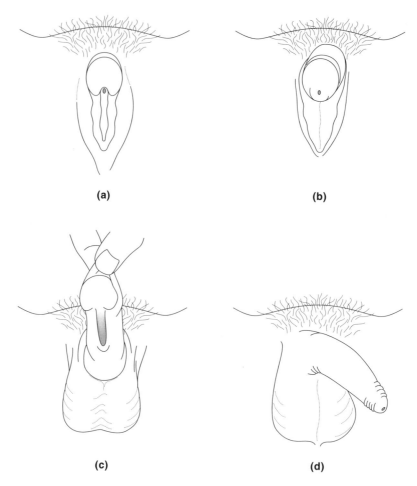

Figure 14.7
The appearance of external genitalia in *XX* individuals with congenital adrenal hyperplasia (CAH) with different degrees of severity. Malelike development of the external genitalia in an *XX* individual will depend on the extent of androgen production. In the examples shown, the least severe case (**a**) exhibits an enlarged clitoris, while the most severe (**d**) exhibits an almost normal penis and scrotum. The scrotum is empty, however, and ovaries are always present. (Adapted from M. H. Johnson and B. J. Everitt, 1995.)

14.3). Blockage of the glucocorticoid synthesis pathway shunts steroid hormone biosynthesis in the direction of androgen synthesis. In an *XX* embryo, the increased androgen levels result in the masculinization of internal and external genitalia. The degree of masculinization varies from simply an enlarged clitoris to the development of a more or less normal penis and (empty) scrotum (Figs. 14.7 and 14.8). Ovaries are present in all cases. A uterus and uterine tubes are also present, while mesonephric duct development is variable.

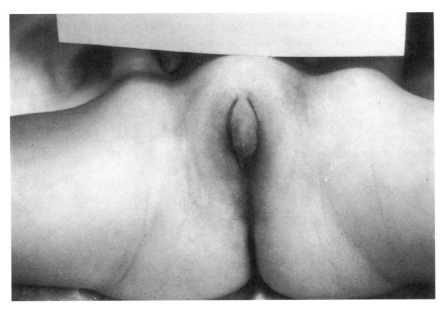

Figure 14.8
External genitalia in a case of CAH. Note extensive development of the clitoris and very small vaginal opening. (Courtesy of K. W. Jones, UCSD School of Medicine.)

APH deficiency

APH deficiency is caused by mutations in the gene encoding APH, which leads to the absence of APH activity. Such mutations are quite rare, and cases of APH deficiency tend to be clustered in families that are carriers of such mutations. Clinically, APH deficiency results in the **persistent Mullerian (paramesonephric) duct syndrome (PMDS)**, the main feature of which is the presence of a uterus and uterine tubes in males. The PMDS diagnosis is made generally when a uterus and uterine tubes are discovered in a boy during surgery for an inguinal hernia. External genitalia are normal, except that the testes are often cryptorchid. Affected individuals are often infertile and prone to the development of tumors of the germ cells. APH deficiency in females has not been described.

Estrogen deficiency

As indicated in Chapter 6, the first cases of estrogen deficiency were discovered in 1995. These involved two sibs, a sister and a brother, both of whom were shown to be deficient for aromatase, the enzyme that converts androgens to estrogens. In a female, inability to convert androgens to estrogens would mean that during fetal life the fetus would be exposed to higher androgen levels than normal. Accordingly, the girl sib exhibited significant virilization of the external genitalia. Surgery was required to reform the external genitalia. At puberty, due to the elevated testosterone levels, she developed obvious signs of virilization. Her FSH and LH levels were also quite high, indicating that the high testosterone levels were not by themselves capable of suppressing pituitary gonadotropin secretion. Estrogen replacement therapy was successful in inducing breast development, menstruation, and reducing the LH and FSH levels.

At 24 years of age, the male sib had a bone age of 14 years, was very tall (6 ft. 10 in.), and had osteoporosis. The low bone age and osteoporosis show that in males estrogens are required for skeletal maturation and bone maintenance. However, he had normal male external genitalia and secondary sex characteristics, indicating that estrogen plays no role in their development. Also very interesting was the fact that both sibs had appropriate gender identities and psychosexual orientation. This finding indicates that estrogen may not have an important role on the sexual differentiation of the brain as has been described in a number of nonprimate mammalian species.

Hormone receptor disorders

Since hormone action is mediated by receptors, we would expect that defects in androgen or estrogen receptor function would also result in phenotypic sex abnormalities. Such defects are characterized endocrinologically by insensitivity to either androgens or estrogens.

Androgen insensitivity syndrome (AIS)

The clinical features of AIS were described long before the molecular nature of the disorder was understood. AIS is a disorder arising from a mutation in the X-linked gene that encodes the androgen receptor. The defect in the androgen receptor means that although androgen levels are normal or above normal in affected individuals, androgen-dependent functions cannot be carried out. In particular, in XY embryos, mesonephric duct development does not take place so that affected individuals lack all mesonephric-derived structures. External genitalia are female, but with a blind vagina. Since APH is produced normally, no paramesonephric-derived structures are formed.

The AIS phenotype is variable. At one extreme is complete AIS (referred to in the older literature as *testicular feminization*), in which the phenotype is completely female at birth. The testes remain cryptorchid. At puberty, breast development takes place because of conversion of androgens to estrogens, and a female somatotype develops, except that there is no growth of pubic or axillary hair (Fig. 14.9). Very often it is only when the affected individual fails to menstruate that the disorder is discovered. Complete AIS reveals quite clearly the essentially female substratum of the human body plan. Partial AIS is characterized by a continuum of phenotypes, some with ambiguous genitalia, some with poorly developed male external genitalia, or some with normal male external genitalia but suffering from erectile dysfunction and infertility.

Estrogen insensitivity syndrome (EIS)

As of this writing only one case of EIS in humans has been described (see Chapter 6). EIS arises from a mutation in the autosomal gene encoding the estrogen receptor. As discussed in Chapter 6, the affected individual is male with normal male external genitalia and with normal androgen levels, and extremely tall (6 ft. 8 in.). Although 28 years old, his bone age is 15 years, and he has osteoporosis. Estrogen levels are very high, but there is no evidence of estrogen-dependent phenotypes, such as gynecomastia. The EIS phenotype is very similar to that of the male sib with aromatase deficiency described above.

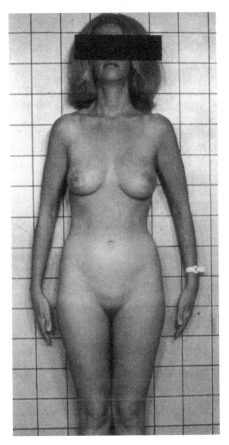

Figure 14.9
Adult *XY* individual with complete androgen insensitivity syndrome (AIS). Note the completely normal female body form. The testes remain in the abdominal cavity. No paramesonephric duct development takes place because APH activity is normal. External genitalia are indistinguishable from the normal female, but the vagina is blind-ended, since no uterus or cervix forms. Typically, pubic hair is sparse or absent, due also to the defect in the androgen receptor. (Reprinted from M. A. Damario and J. A. Rock, 1996, with permission.)

SUMMARY

In males, feminization of the internal and external genitalia can occur in several ways: sex chromosome abnormalities, deficiency in testosterone or DHT, androgen-receptor dysfunction, and APH or APH-receptor deficiency. In females, masculinization of the internal or external genitalia can occur in only one way, excess androgen production. This difference between males and females is simply a reflection of "induced" versus "default" nature of the fetal development of the internal and external genitalia (see Chapter 4). In some cases, surgical reconstruction and appropriate hormonal replacement have been successful in restoring external genitalia concordant with gonadal or chromosomal sex.

DISORDERS AFFECTING THE TIMING OF PUBERTY

Precocious puberty

Precocious puberty is defined as the appearance of secondary sex characteristics in girls younger than 8 years and in boys younger than 9 years. There are documented cases of puberty onset as early as 8 months of age, and there is even one case of a girl of 5 giving birth. For unknown reasons, precocious puberty occurs more frequently in girls than in boys. Two different classes of precocious puberty are recognized, *GnRH-dependent and GnRH-independent* (Table 14.2).

GnRH-dependent onset of puberty

GnRH-dependent precocious puberty is the result of premature activation of the GnRH pulse generator, a condition that clearly reflects a decoupling of the biological clock regulating the timing of puberty from its control of the GnRH pulse generator. This type of precocious puberty is referred to as **complete, or central, precocious puberty** (CPP), since as we saw in Chapter 8, reactivation of the GnRH pulse generator is sufficient to bring on all the changes that normally occur with the onset of puberty. Most cases of CPP are idiopathic (the cause is unknown), and no organic lesions can be identified. Other cases can be traced to tumors or other types of lesions in the hypothalamus or other sites in the central nervous system.

One of the remarkable success stories in modern reproductive biology has been the use of GnRH agonists in the treatment of CPP. Administration of long-acting GnRH agonists result in the down-regulation of the GnRH receptors in the pituitary gonadotrophs, and this in turn leads to suppression of gonadotropins and inhibition of gonadal function. The administration of GnRH agonists suppresses the onset of changes that typically occur during puberty. It is important to realize that endogenous GnRH secretion is not suppressed, but simply that its effects are counteracted by the GnRH agonist. Administration of the GnRH agonist can be continued until, for example, the mean age of puberty for boys is reached, at which time the endogenous GnRH reinitiates the onset puberty, which has been held in abeyance for several years. Hence, the GnRH agonist provides a method of postponing the onset of puberty.

GnRH-independent onset of puberty

GnRH-independent precocious puberty occurs when secondary sex characteristics appear without GnRH stimulation. This can occur in many different ways, some of which are indicated in Table 14.2. Such conditions can arise from isolated adrenal gland, pituitary, or gonadal activation, resulting in increased secretion of adrenal sex steroid hormones, gonadotropins, or gonadal steroid hormones. Because the constellation of changes that take place is not the same as those that occur when driven by GnRH activation, these types of early puberty are referred to as **incomplete precocious puberty** (IPP). A few examples will illustrate the general principles.

Premature development of breasts in young girls is caused by increased levels of estrogens, produced by either the adrenal glands or ovaries. Such increases occurring in very young girls (less than 8 years) are often transient. If the elevated levels are due to estrogen-secreting ovarian or adrenal tumors, the effects are not transient. In males, elevated testosterone levels can arise from androgen-producing adrenal or Leydig cell tumors. High testosterone levels lead to enlargement of the external genitalia and appearance of pubic and axillary hair, but without a concomitant increase in FSH, little or no spermatogenesis takes place. Recently, an interesting class of mutations leading to IPP has been discovered. These are mutations that result in constitutively active LH receptors in Leydig cells. Under normal conditions, LH receptors are activated by the binding of LH. However, some mutations in the gene encoding the LH receptor result in receptors that are always on; that is, it does not need to bind LH in order to be activated. Such constitutively active LH receptors lead to elevated levels of testosterone, which in turn produces the IPP phenotype. Prematurely elevated levels of estrogens and androgens can also result from isolated gonadotropin secretions from the pituitary, generally from pituitary tumors. The therapy for IPP, because of its multiple causes, cannot be as straightforward as it is for CPP.

Delayed puberty

We can easily guess by this stage that delay of puberty can be the result of hypothalamic, pituitary, or gonadal dysfunction. In all cases the end result is the same—gonadal deficiency, or **hypogonadism**. Diagnostically, two types of delayed puberty are recognized, hypogonadotropic hypogonadism and hypergonadotropic hypogonadism.

Hypogonadotropic hypogonadism

The diagnostic feature of **hypogonadotropic hypogonadism** is very low gonadotropin levels (hence, the term *hypogonadotropic*) due either to hypothalamic or pituitary defects. Hypothalamic dysfunction could be due either to failure of the GnRH pulse generator (idiopathic GnRH deficiency) or to suppression of the pulse generator by conditions such as CNS tumors, generalized stress, or hyperprolactinemia. Delay of puberty due to GnRH deficiency can be treated successfully by exogenous pulsatile GnRH administration. In males, fertility can be maintained, and in females, ovulatory cycles can be initiated and maintained. Pituitary dysfunction leading to gonadotropin deficiency can arise from many types of disorders. Such isolated LH or FSH deficiencies result in defective gonadal function, which in turn is manifested as a delay in puberty.

Hypergonadotropic hypogonadism

Hypergonadotropic hypogonadism is characterized by gonadal dysfunction in the presence of high gonadotropin levels (hence, the term *hypergonadotropic*), a condition similar to that of menopause. This condition is due to primary gonadal failure or dysfunction. Low gonadal steroids leads to a decrease in negative feedback, which in turn permits gonadotropin levels to rise. Many different types of gonadal lesions leading to this condition have been described, including primary lesions in Leydig or granulosa cell function and deficiencies in the enzymes of the steroid hormone biosynthesis pathway. Recently, a condition leading to LH resistance has been described. This is the result of mutations in the gene encoding the LH receptor that results in an aberrant LH receptor that does not respond to LH. In both males and females, the gonads fail to develop properly.

SUMMARY

The timing of puberty can be perturbed by hypothalamic, pituitary, or gonadal disorders, and these can either accelerate or delay the onset of puberty. Precocious puberty can result from premature activation of the GnRH pulse generator or from isolated pituitary or gonadal activation. The former leads to the clinically defined disorder known as complete or central precocious puberty because the pubertal changes that take place are the same as would normally occur in a much older child. The latter leads to a variety of conditions described as incomplete precocious puberty characterized by some, but not all, of the normal pubertal changes. Complete precocious puberty can be successfully treated with long-acting GnRH agonists that will desensitize the pituitary to GnRH stimulation. No general therapy is available for the treatment of incomplete precocious puberty.

Delayed puberty can be due to failure of the GnRH pulse generator to be activated at the normal time, or failure of the pituitary or the gonads to respond to hypothalamic or pituitary stimulation. Many different conditions lead to delayed puberty. Steroid sex hormone replacement has been a relatively successful therapy for many of these disorders.

QUESTIONS

1. Female pseudohermaphroditism arises in only one way, while male pseudohermaphroditism arises in several ways. How do you account for this circumstance?

2. A condition such as AIS raises serious ethical questions. For example, what should a young person diagnosed with AIS be told about her condition (remember that she is an adolescent)? Will the truth be so devastating that it may damage the person irreversibly? How should her condition be presented to her?

3. Indicate the phenotypes with respect to phenotypic sex you might expect to see as a consequence of a mutation in the gene that encodes the LH receptor that renders the LH receptor constitutive, i.e., the receptor does not require LH binding to be activated.

4. Why is it easier to treat GnRH-dependent than GnRH-independent precocious puberty?

5. Library project:
   a. Are there any problems associated with long-term usage of GnRH analogs in treating CPP?
   b. What factors are considered whenever surgical reconstruction of the external genitalia is contemplated for treating abnormalities of phenotypic sex?

SUPPLEMENTARY READING

Clark, P. A. 1998. Puberty: When it comes too soon—guidelines for the evaluation of sexual precocity. *Journal of the Kentucky Medical Association* 96(11), 440–447.

Damario, M. A., and J. A. Rock. 1996. Diagnostic approach to ambiguous genitalia. In *Reproductive Endocrinology, Surgery, and Technology*, Vol. 1, Chapter 42. E. Y. Adashi, J. A. Rock, and S. Rosenwaks, Eds. Lippincott-Raven Publishers, Philadelphia, PA.

Hughes, I. A. 1998. The masculinized female and investigation of abnormal sexual development. *Bailliere's Clinical Endocrinology and Metabolism* 12(1), 157–171.

Johnson, M. H., and B. J. Everitt. 1995. *Essential Reproduction*, 4th ed. Blackwell Science Ltd., Oxford, UK.

Malasanos, T. H. 1997. Sexual development of the fetus and pubertal child. *Clinical Obstetrics and Gynecology* 40(1), 153–167.

O'Rahilly, R., and F. Müller. 1996. *Human Embryology and Teratology*, 2nd ed. Wiley-Liss, Inc., New York, NY.

Zajac, J. D., and G. L. Warne. 1995. Disorders of sexual development. *Bailliere's Clinical Endocrine Metabolism* 9(3), 555–579.

ADVANCED TOPICS

Morishima, A., *et al.* 1995. Aromatase deficiency in male and female siblings caused by a novel mutation and the physiological role of estrogens. *Journal of Clinical and Endocrinological Metabolism* 80(12), 3689–3698.

Smith, E. P., *et al.* 1994. Estrogen resistance caused by a mutation in the estrogen-receptor gene in a man. *New England Journal Medicine* 331, 1056–1061.

Swerdloff, R. S., and C. Wang. 1998. Influence of pituitary disease on sexual development and functioning. *Psychotherapy and Psychosomatics* 67(3), 173–180.

Wiener, J. S., J. L. Teague, D. R. Roth, E. T. Gonzales, Jr., and D. J. Lamb. 1997. Molecular biology and function of the androgen receptor in genital development. *Journal of Urology* 157(4), 1377–1386.

Lee Krasner, *Memory of Love*, 1966.
(© 1999 Pollock-Krasner Foundation/Artists Rights Society [ARS], New York.)

SOCIETAL ISSUES

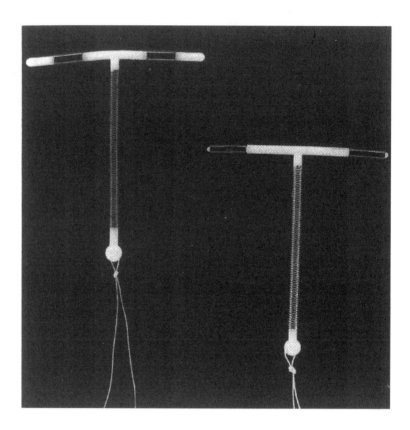

Intrauterine device (IUD). On a worldwide basis, the IUD is the most commonly used birth control device. (Left) The TCu-380 is a flexible polyethylene T-shaped tube with a copper collar on each of its arms. (Right) The TCu-380Ag, identical to the TCu-380 except that the copper wire has a silver core, was developed in Finland. (*Intrauterine Devices*. 1997. World Health Organization, Geneva, Switzerland, with permission.)

CHAPTER 15

Artificial Control of Fertility

*Today, more than one million women of many countries
are taking the pills, not simply because of their great effectiveness
but also because they provide a natural means of fertility control
such as nature uses after ovulation and during pregnancy.*

—J. Rock. 1963

The Time Has Come: A Catholic Doctor's Proposals to End the Battle
Over Birth Control. Alfred A. Knopf, New York, NY, p. 167.

*In terms of its impact on fertility, it may be claimed
that education is the best contraceptive.*

—S. J. Segal. 1996

Contraceptive development and better family planning.
Bulletin of the New York Academy of Sciences 73(1), 92–104.

THE CONTROL OF FERTILITY has probably been a constant preocupation in human affairs. Since ancient times human societies appear to have been interested in controlling or limiting fertility, and all societies have fashioned their own potions and devices for limiting the number of births. Documents recovered in archeological excavations reveal a variety of means to prevent undesired births. Prescriptions for contraceptives and **abortifacients** (agents that induce an abortion) have been found in surviving Egyptian, Chinese, and Indian texts. Most such prescriptions were probably not very effective, but some make sense. For example, part of an Egyptian text found in the Ebers Papyrus (1550 BCE) describes a medicated tampon made with ground acacia seed. When placed in the female, the acacia seed would ferment, releasing lactic acid, which is toxic to spermatozoa. Lactic acid has been used in many commercially available **spermicides** (or "sperm killers"). Hence, the Egyptian tampon functioned as a spermicide, although the designers of the tampon had no idea how it worked. The Romans made a significant contribution to the field of birth control by developing the condom, which was fashioned out of goat or fish bladders. In addition, hundreds of herbal recipes to prevent conception or to induce abortion were available in the Roman pharmacopoeia. **Coitus interruptus**, or withdrawal before ejaculation, has probably been the most universal method of birth control. Infanticide, the killing of the unwanted newborn, has been used by a number of societies as well.

Effective methods of fertility control are quite new. The major breakthrough came in the late 1950s when the first birth control pill was introduced. Since then, a number of innovations have been introduced. Our objective in this chapter is to review the different modern methods for controlling fertility. We will also consider some of the newer, but still experimental, methods that are in different stages of efficacy trials.

BIRTH CONTROL STRATEGIES

There are four different strategies for the artificial control of fertility: (1) prevention of gamete formation or release; (2) prevention of fertilization; (3) prevention of implantation; and (4) prevention of a live birth (Table 15.1). The commonly used term *contra-*

Table 15.1
Modern Birth Control

Strategy	Method
Prevention of gamete formation	
Females	Combined oral contraceptives (COCs)
	Progesterone-only contraceptives (POCs)
	Progesterone-only or combined injectables
	Progesterone analog implants
	GnRH analog/estrogen pills (experimental)
Males	Androgen injectables (in trials)
	Androgen/progesterone injectables (experimental)
	GnRH analog/androgen implants (experimental)
	Anti-FSH vaccine (in trials)
Prevention of fertilization	
	Sterilization
	Tubal ligation in the female
	Vasectomy in the male
	Barriers
	Male condom
	Female condom
	Diaphragm
	Spermicides
	Specific sperm inactivators
	Calcium blockers (experimental)
	Mifepristone (experimental)
	Intrauterine device (IUD)
	Antisperm vaccines (experimental)
	Periodic abstinence (rhythm)
Prevention of implantation	
	Intrauterine device (IUD)
	Postcoital estrogen/progestin
	Mifepristone
	Anti-hCG vaccines (in trials)
Prevention of live birth	
	Medical abortion
	Surgical abortion

ception encompasses strategies (1) and (2). Most of the methods available today—for example, sterilization, the intrauterine device (IUD), the diaphragm, the condom, spermicides, and rhythm—have strategy (2) as their rationale, while hormonal contraceptives (the "pill") employ strategy (1). A newly coined term, **interception**, encompasses the methods of strategy (3), and most of these are still in different stages of development. The IUD is unusual in that it utilizes strategies (2) and (3). Strategy (4) involves inducing an abortion and has been legally available in the United States since 1973, but is still proscribed in many countries in the world. Except in a few countries, induced abortion has not been used as a widespread birth control method. The degree of general acceptability of these strategies varies considerably. Strategies (1) and (2) are widely accepted in the United States, although they may be proscribed by certain religious groups; while strategies (3) and (4), which rely on interfering with development after fertilization, are much more controversial. For example, the licensing of RU-486, a progesterone antagonist widely used in Europe for early term abortions, was denied in the United States until very recently, while abortions were legal.

Modern birth control methods have found wide acceptability on a worldwide basis. According to a 1994 United Nations report, about 60 percent of couples use some form of birth control. For a variety of reasons, however, in some regions of the world, especially in Africa, the use of modern birth control methods is much lower (Table 15.2). The IUD is perhaps the most widely used reversible birth control method, prin-

Table 15.2
Prevalence of Contraceptive Use by World Region

Major area/region	% Couples using	
	Any method	Modern methods[a]
World	57	49
Less developed regions	53	48
Africa	18	14
Northern Africa	38	35
Sub-Saharan Africa	12	8
Asia/Oceania[b]	58	54
Eastern Asia[b]	79	79
Latin America	58	49
More developed regions	72	50
Asia: Japan	64	57
Northern America	74	69
Europe		
Eastern Europe	70	29
Northern Europe	80	76
Southern Europe	68	31
Western Europe	76	69
Australia-New Zealand	74	70

Survey data with 1990 as the average date. [a]Includes sterilization, COCs, IUDs, injectables, condoms, and vaginal barrier methods. [b]Excludes Japan, Australia, and New Zealand. (Reference: *Levels and Trends of Contraceptive Use as Assessed in 1994*. United Nations, New York, 1996.)

Table 15.3
Estimated Number of Users of Modern Reversible
Contraceptive Methods, 1990 (in Millions)

Method	World	Developed countries	Underdeveloped countries
IUD	109	13	96
Pill	93	49	44
Condom	50	31	19
Injectables and implants	11	1	10
Vaginal barriers	8	6	2

Reference: *Levels and Trends of Contraceptive Use as Assessed in 1994.* United Nations, New York, 1996.

cipally because it is the primary method used in China (Table 15.3). If China is excluded, oral contraceptives are the most commonly used contraceptive method, followed by the condom, injectables, levonorgestrel implants, and vaginal barrier methods. Demographic surveys in the United States indicate, perhaps surprisingly, that female and male sterilization is the most common method used by Americans. In a 1995 survey of 15- to 44-year-olds, sterilization accounted for about 40 percent of those who used birth control (Table 15.4). The sterilization option, however, depends strongly on the age of the user. As might be expected, it is more frequent among older users who have already completed their families. Oral contraceptives are most prevalent among the young. Use of the IUD is relatively uncommon in the United States, primarily because of litigation problems that have complicated its marketing. Both the IUD and the diaphragm are used much more frequently in other countries, particularly

Table 15.4
Birth Control Methods in the United States, 1995

Method	% Usage by age				
	15–44	15–19	20–24	25–34	35–44
All methods	64.2	29.8	63.5	71.1	72.3
Female sterilization	27.7	0.3	4.0	23.8	45.0
Male sterilization	10.9	0.0	1.1	7.8	19.4
Pill	26.9	43.8	52.1	33.3	8.7
IUD	0.8	0.0	0.3	1.7	1.1
Diaphragm	1.9	0.1	0.6	1.7	2.8
Condom	20.4	36.7	26.4	21.1	14.7
Other	11.4	19.1	15.5	10.6	8.3

The category "Other" in Table 15.4 includes the use of spermicidal jellies, foams, and sponges, all of which act by killing or inactivating sperm. Also included in this category is the rhythm method, in which intercourse is avoided during the days before and after ovulation. (Reference: National Center for Health Statistics. Health, United States, 1998, with Socioeconomic Status and Health Chartbook.)

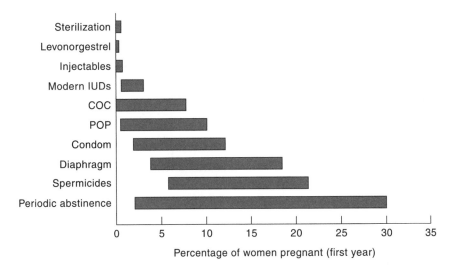

Figure 15.1
Estimates of birth control failure rates. (Reference: International Planned Parenthood
Foundation. IMAP, 1994. Statement on contraceptive efficacy.)

in the underdeveloped countries. Condom use is still quite high, perhaps bolstered by
campaigns that emphasize the importance of the condom in helping prevent the transmission of HIV and other sexually transmitted diseases (STDs).

Based on actual use, sterilization, levonorgestrel implants, and injectables are the
most effective methods of birth control (Fig. 15.1). The actual failure rates for these
three methods are 1 to 2 percent. Least effective birth control methods are the condom,
diaphragm, spermicides, and periodic abstinence. The failure rates for the IUD, COC,
and POP range from 1 to 10 percent despite the fact that the theoretical or optimal failure rates for these three methods are less than 1 percent. The difference is due to the
failure to use the methods consistently or properly, and this depends on the age of the
user. The overall failure rate in the United States for oral contraceptives is 2 to 3 percent, but among those younger than 20, the failure rate is 8 to 18 percent. These relatively high failure rates contribute significantly to the number of unintended pregnancies. This high rate of failure contributes to the much higher pregnancy rate in the
United States for those younger than 20 compared to their counterparts in other countries where the pattern of sexual behavior is the same as in the United States (Table
15.5). The pregnancy rates in this age group are twice that in Canada, and almost 10
times that of the Netherlands. It is estimated that half of these pregnancies are unintended and due to the failure of the birth control method. These high failure rates are
considered to be due to lack of information about the proper way to use the method,
lack of counseling, and the greater likelihood that a younger person will not follow the
birth control regimen optimally.

There is no doubt that the availability of modern birth control methods has had
enormous beneficial effects in all societies in which they have been introduced. Libera-
tion of women from the burden of unwanted pregnancies, the separation of sexuality
from reproduction, and the reduction in morbidity and in the number of unsafe abortions have led to unparalleled improvement in the quality of life. A significant number

Table 15.5
Pregnancy Rates for Women under the Age of 20 in the United States and Some European Countries

Country	Pregnancy rates per 1000 women
United States	97
England & Wales	46
Canada	40
Sweden	35
Denmark	25
Holland	10

Levels of adolescent sexual activity are approximately the same in these countries. (Reference: E. F. Jones *et al.* 1986. *Teenage Pregnancies in Industrialized Countries.* Yale University Press, New Haven, CT.)

of the problems that continually beset people in some underdeveloped countries can be traced to the unavailability, cost, or lack of information about modern birth control options. In countries like the United States, modern birth control methods have substantial economic benefits to the society at large. As shown in Fig. 15.2, an analysis of the 5-year costs of 15 different birth control methods leaves no doubt that birth control is cost-effective primarily because it reduces the rate of unintended pregnancies. The

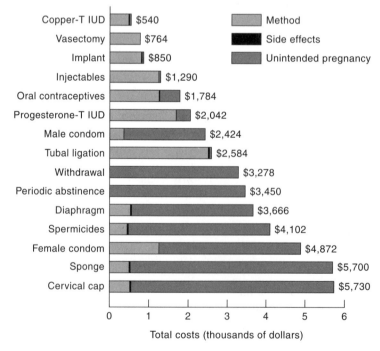

Figure 15.2
Comparison of 5-year costs associated with different birth control methods. Comparison of three types of costs that affect medical costs in a managed payment model typical in the United States: method used, adverse and beneficial side effects, and unintended pregnancies. (Reference: J. Trussell *et al.*, 1995, with permission.)

most cost-effective methods are the IUD, implants, and injectables, with most of the cost related to the method itself. The cost increases as the effectiveness of the method decreases. The barrier methods are the least cost-effective because their failure rate is so high. Hence, although the least effective methods may initially appear to offer cost savings, the cost of unintended pregnancy drives up their true cost. On the other hand, barrier methods, the male condom in particular, provide important protection against sexually transmitted diseases.

PREVENTION OF GAMETE FORMATION

In terms of its almost revolutionary impact on society, the development in the 1950s of an effective hormonal method to suppress ovulation initiated the modern era of birth control research. Three scientists in the United States, Gregory Pincus, John A. Rock, and C.-R. Garcia, are credited with the development of the first effective "oral contraceptive." In the United States and most countries of Europe, contraception is associated with the birth control pill, which made its appearance in 1960. What may not be appreciated is that the development of hormonal contraception can be traced back to experimental observations made in the 1920s.

In the female

The first suggestion for a hormonal contraceptive appeared in 1921 in a report describing the effects of transplanting the ovaries of pregnant rabbits into nonpregnant animals. The recipient rabbits became temporarily sterile. The author suggested that this effect could be developed into a method of sterilizing women for short periods. Hormonal contraception was discussed publicly for the first time at the Seventh International Birth Control Conference in Zurich, Switzerland, in 1930. These initiatives were premature, primarily because sex hormones were at that time still poorly understood. The 1920s through the 1950s were devoted to the systematic study of the steroid sex hormones; the focus of most of this work was on the use of the sex hormones, particularly estrogen and progesterone, as therapeutic drugs for menstrual and menopausal dysfunction. A new impetus for reconsidering hormonal contraception came after the end of World War II, when three social movements coalesced—birth control, population control, and eugenics. The common element in all three movements was the desire to control human reproduction by separating sexuality from reproduction.

Combined oral contraceptive (COC)

In work that began in 1951, knowledge of the ovulation-inhibiting effects of estrogen and progesterone (first discovered in the 1930s) and the ability to produce synthetic estrogen and progesterone analogs dating to the 1940s led Pincus, Rock, and Garcia to market the first effective oral contraceptive in 1956. The Federal Drug Administration (FDA) approved the first birth control pill for mass use in the United States in 1960. Today, millions of women use the pill. There is no doubt that the introduction of hormonal contraceptives has contributed significantly to the health and welfare of many people.

By today's standards, the estrogen doses delivered by the early generations of pills were high. Later, synthetic progesterone analogs were added to increase the efficacy and decrease the side effects. Pills containing both estrogen and progesterone became known as the **combined oral contraceptives** (COCs). Early COC formulations, because of the relatively high estrogen content, had significant side effects. Estrogen-related side effects included nausea, bloating, fluid retention, weight gain, irritability, nervousness, headaches, breast tenderness, increased blood pressure, and increased risk of blood clot formation. The major progesterone-related side effects included increased appetite, erratic menstrual bleeding, and breast shrinkage. These side effects generated so much controversy that even today many women still refrain from using COCs for fear of the side effects. However, the new generation of COCs containing the lowest levels of estrogens and progestins compatible with high efficacy has been shown in extensive studies involving many thousands of women to be safe for the average woman. Having been taken by millions of women, COCs have been more thoroughly tested than most other drugs. But not every woman is able to use them. They are contraindicated in a small percentage of women, for example, those with thromboembolic or cardiovascular disorders, impaired liver function due to hepatitis, deep varicose veins, known or suspected breast cancer, hypertension, epilepsy, and for smokers over the age of 35. Regular use of COCs has been found to be useful in the prevention of iron deficiency anemia, benign breast disease, pelvic inflammatory disease, and cancers of the endometrium and ovary.

The efficacy of COCs lies in their suppression of ovulation, which in turn depends on the combined action of estrogen and progestin. The primary effect of progestin is to suppress LH secretion, while the estrogen suppresses FSH secretion. Estrogen stabilizes the endometrium to minimize irregular shedding and breakthrough bleeding of the endometrial tissue. Because estrogen stimulates the synthesis of progesterone receptors, estrogen potentiates progesterone action. The new formulations are said to be multiphasic because the estrogen and progestin dosage is varied periodically during the cycle. For example, the biphasic COC provides altering doses of estrogen and progestins during the first half of the cycle, while only progestins at higher levels are given during the second half of the cycle. The dosage regimes of the triphasic formulations alter the estrogen and progestin concentrations throughout the cycle. Both are generally prescribed in 28-day packs, which deliver the hormones for 21 days, and these are followed by 7 days of inert tablets, containing iron and/or vitamins.

Injectables

Hormonal contraceptives administered by intramuscular injection were developed soon after COCs. Initially, the injections contained a progesterone analog and provided protection for about three months. One important advantage was that the injectable dispensed with the problem of daily pill taking. The newer formulations, which contain progesterone and estrogen analogs and provide protection for one month, have been shown to provide a safe option for most women. Their failure rate is less than that of COCs. During the last few years, their use has been increasing in many developing countries.

Progesterone-only contraceptives

Progesterone-only contraceptives were developed initially for women for whom estrogen was contraindicated. Three types of formulations are available. Progesterone injectables, usually containing the progesterone analog **depo-provera**, were developed first. A single high-dose injection of the synthetic progesterone provided sufficient protection for about three months. The need for repeated injections could be a disadvantage for those who dislike injections. On the other hand, these three-month injectables were an advantage in that the problem of compliance with daily pill taking was dispensed with. In addition, the necessity to visit a doctor's office every three months provides a built-in monitoring system. However, one of the major side effects of the initial formulations was their disruption of menstrual bleeding patterns, and this was the main reason for their discontinuation by users. The newer injectables with a combination of estrogen and progesterone cause much less disruption in bleeding and have minimal side effects.

Alternatives to progesterone injectables are progesterone-only pills (POPs), also referred to as "mini-pills." The mini-pills provide a low-dose-alternative to the progesterone-only regimes, and reduce progesterone-associated side effects. They are also quite effective. Studies have shown that ovulation is suppressed regularly only about 50 percent of the time, and their high efficacy is attributed in addition to the resulting changes in the cervical mucus, making it thicker and more viscous and hence a very effective barrier to sperm penetration. There are important differences between COCs and POPs, and these should be clearly understood when a particular method of contraception is being comtemplated (Table 15.6).

Table 15.6
Comparisons of COCs and POPs

	COCs	POPs
Efficacy	0.1–8.0% failure rate during the first 12 months of use	0.5–10.0% failure rate during the first 12 months of use
Mode of action	Ovulation inhibition	Primarily by thickening cervical mucus
Use	Taken each day in 3-week cycles with a 1-week break	Taken every day without a break
Eligibility	Contraindicated for women with estrogen sensitivity	Indicated for women sensitive to estrogen
Bleeding	Usually good cycle control	Irregular bleeding quite common
Compliance	Pill should not be missed for more than 24 hours	Pill should not be missed for more than 3 hours
Breast-feeding	Not suitable	Suitable

Reference: International Planned Parenthood Federation. IMAP, 1998. Statement on steroidal oral contraception.

Levonorgestrel implants

The newest version of the progesterone-only formulations is the subdermal implants, which were first approved in Finland in 1983. Since then their use has spread to most countries where contraceptives are available. They are particularly popular in Europe. They require surgical implantation, usually in the underarm (Fig. 15.3). The

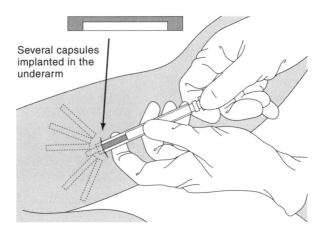

Several capsules implanted in the underarm

Figure 15.3
Schematic illustration of subdermal capsules and their implantation in the underarm.

first approved implant, Norplant I, consists of six capsules containing the synthetic progesterone analog *levonorgestrel*; the capsules release the hormone slowly over a period that can be as long as 5 years. The practical advantage of an implant is that it avoids the necessity of having to take a pill every day. The slow, constant release of the hormone avoids large daily fluctuations. The implants have been approved in many countries, including the United States, on the basis of comprehensive, clinical trials. The World Health Organization (WHO) has also been monitoring over 8000 women for 5 years to see if adverse reactions might show up in longer-term trials. At the present time, some 70,000 woman-years of experience have been accumulated. The most recent report from WHO confirms the effectiveness of Norplant I. The pregnancy rate is similar to that with surgical sterilization, and no unexpected adverse side effects have been noted beyond those noted in the initial clinical trials. Adverse reactions typically noted are those associated with progesterone-only formulations. Breakthrough bleeding generally disappears after a few months. A new generation of implant, Norplant II, is undergoing clinical trials. Preliminary results indicate that users of these second-generation implants experience fewer disturbances in menstrual bleeding.

GnRH analogs

Both GnRH agonists and antagonists have potential for use as contraceptives. Long-acting GnRH agonists desensitize the pituitary gonadotrophs (recall from Chapter 7 that continuous GnRH leads to a large decrease in LH and FSH production), eventually suppressing ovulation. To maintain estrogen- and progesterone-dependent tissues, low doses of estrogen and progesterone analogs are administered periodically. This method has been found to be highly effective in preliminary trials. An added benefit may be that reducing endogenous estrogen levels reduces the risks of breast, ovarian, and endometrial cancers (see Chapter 18). A GnRH-agonist-based contraceptive may be available in the near future.

GnRH antagonists compete directly with endogenous GnRH and prevent GnRH-dependent release of LH and FSH from the pituitary. A GnRH antagonist given in a large dose at midcycle or in a smaller dose on a weekly basis has been shown to suppress ovulation. However, extensive clinical trials using GnRH antagonists have not yet been conducted, and it will be several years before they could be used.

In the male

Androgen contraceptives

Until recently, most contraceptive research has been focused on the female; much less has been done in trying to develop appropriate contraceptives for males. This will probably change in the future as the suitability of newer methods is explored systematically (Table 15.7). Most research has focused on the development of steroid hormone contraception. Recall from Chapter 6 that testosterone is necessary not only for spermatogenesis, but also for maintenance of erectile function and libido in the male. Interference with testosterone synthesis or testosterone function would inevitably affect erectile function, which makes it an unacceptable basis for a contraceptive. For this reason, significant effort has been devoted to searching for and testing compounds that inhibit spermatogenesis specifically. Many compounds have been screened to determine whether their effects on spermatogenesis can be separated from their effects on testosterone function. Some have been found, but those which affect spermatogenesis specifically also tend to be toxic. A good example is *gossypol*, derived from cottonseed, and used as a contraceptive in China. The effects of gossypol in the short term and at the proper dose appear to be real. Interruption of spermatogenesis and impaired motility of surviving sperm have been observed, but in the long term, gossypol is cytotoxic (kills cells). This clearly prevents its widespread use as a male contraceptive.

Table 15.7
Status of Male Contraceptives

Method	Delivery	Status
Hormonal		
Testosterone analogs	Weekly injections	Effective in short-term trials. Widespread use unlikely because of weekly injections
Testosterone/progestin combination	Monthly injections	Human trials planned
GnRH analog/testosterone combination	Implant or patch	Human trials initiated
Nonhormonal		
Nifedipine (Calcium channel blocker)	Pill	Human trials planned
Mifepristone (RU-486)	Pill	Animal studies planned
Anti-FSH vaccine	Injection	Human trials under way
Antisperm vaccine	Injection	Initial success in animal studies

While decreasing testosterone levels will not work, the converse—increasing testosterone—is more successful. The principal idea is that exogenous administration of testosterone leads to the suppression of LH and FSH secretion because of the negative feedback effects of testosterone. (Recall, however, that we now know that most of this suppressive effect is probably due not to testosterone, but to estrogen derived from testosterone.) Suppression of LH secretion leads to a significant decrease in the intratesticular testosterone levels, high levels of which are required for spermatogenesis. Libido and erectile function are not generally affected because they can be maintained by the lower systemic testosterone provided by the exogenous testosterone. Suppression of FSH levels probably also plays a role in the inhibition of spermatogenesis. Several short-term pilot studies (1 year) have shown that spermatogenesis can be suppressed sufficiently by this procedure to be effective as a contraceptive. However, the most effective protocols required weekly testosterone injections, which significantly reduced their likelihood of acceptance. Other testosterone injectables in combination with progesterone analogs that can be administered once a month or once every three months are being tested, but their possible use is still some years away (Table 15.7).

GnRH analogs

The nonsteroid alternatives currently being considered for male contraception are of three types (Table 15.7). One uses a long-acting GnRH agonist in combination with a testosterone analog. The long-acting GnRH agonist down-regulates the pituitary gonadotrophs, leading to a decrease in LH and FSH production and suppression of spermatogenesis. The testosterone analog provides sufficient systemic androgen levels to maintain erectile and other androgen-dependent functions. Early trials of an implant delivery system have begun, but it will be several years before the effectiveness of this method is known.

Nonhormonal methods

Another class of possible male contraceptives is known as *calcium channel blockers*, because they inhibit the entry of calcium into cells. Sperm cells exposed to these compounds quickly lose their mobility, and hence, their ability to fertilize. Two types of compounds have been proposed for studies in animals and humans—mifepristone (RU-486), an antagonist at the progesterone receptor, and nifidepine, a drug normally used in the treatment of hypertension and migraines (Table 15.7). As a progesterone antagonist, RU-486 may also work by inhibiting the acrosome reaction required for sperm penetration of the zona pellucida. The most effective method of delivery for either compound has not yet been worked out.

Two other promising possibilities include an anti-FSH and antisperm vaccines (Table 15.7). Both of these are examples of several **immunocontraception** methods under development (see below). The rationale is to use specific types of antibodies to inactivate different proteins required for spermatogenesis (anti-FSH) or fertilization (antisperm vaccine). An anti-FSH vaccine is already undergoing human trials, and preliminary success in developing an antisperm vaccine in guinea pigs suggests that extension of the initial studies to other animals is warranted. Possible use in humans is still some years away.

SUMMARY

The development of hormonal formulations to suppress ovulation in the 1950s was the beginning of a revolution in the application of endocrinology to the control of human fertility. The "pill" in its various forms represents the most frequently used form of birth control in most countries of the world. Since their original formulation, hormonal contraceptives, whether as combinations of estrogen and progesterone analogs or progesterone-only preparations, have undergone progressive improvements that increase efficacy and reduce side effects. Currently, the most effective steroidal contraceptives are the levonorgestrel implants and the injectables. These contraceptives, together with the COCs, represent perhaps the most thoroughly tested drugs in use in the world. The development of male contraceptives has lagged significantly compared with female contraceptives. Androgen-based contraceptives for males have been shown to be quite effective in suppressing spermatogenesis, but their method of delivery (weekly injections) prevents their widespread use. A number of newer male contraceptive methods are being tested, but it is still too early to tell whether they will find wide acceptability.

Nonsteroidal alternatives using GnRH analogs are the most promising of the newer methods that rely on suppressing gametogenesis. Although not yet available on a mass basis, GnRH-agonist-based formulations for suppressing ovulation appear to be highly effective and easy to use. An added advantage is that endogenous estrogen levels can be reduced to minimize the risk of breast, ovarian, and endometrial cancers. Similar formulations for use in males are being developed and tested. It seems reasonable to expect that new and effective nonsteroidal formulations will be available for both females and males in the near future.

PREVENTION OF FERTILIZATION

The prevention of fertilization is probably the oldest form of birth control. The methods that employ this strategy use surgical, physical, or chemical means to prevent the sperm from reaching the egg.

Sterilization

The term *sterilization* refers to a **tubal ligation** in the female and a **vasectomy** in the male (Figs. 15.4 and 15.5). In both cases, the passage of spermatozoa is prevented in females by ligating, or cutting, the uterine tube and in males, by cutting the vas deferens. Both are highly effective, and together they constitute the major contraceptive method in the United States. Because both can be considered effectively irreversible, these procedures are carried out on individuals who are no longer interested in having children. Some more recently developed sterilization procedures are touted as being more reversible, but in the absence of large-scale studies, optimism about the reversibility is probably unwarranted.

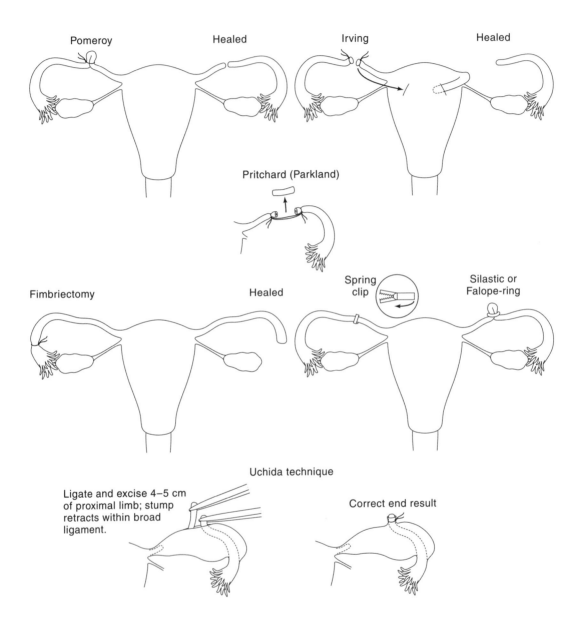

Figure 15.4
Tubal sterilization procedures. The uterine tubes are approached through a small incision in the abdomen. Sterilization is accomplished by ligation, clamping with clips or rings, or electrocoagulation. (Adapted from R. A. Hatcher *et al.*, 1994.)

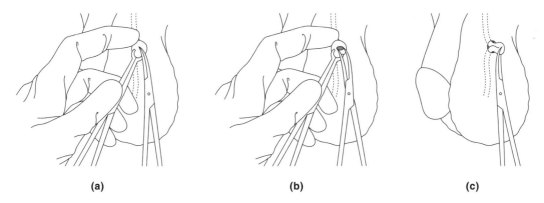

Figure 15.5
"No-scalpel" vasectomy. The vas deferens (dotted line) is grasped by special ring forceps and the skin is pierced by special dissecting forceps (**a**). The forceps stretch an opening (**b**), and vas is lifted out and cut or clipped (**c**). (Adapted from Population Information Program.)

Barrier methods

As the name indicates, these methods involve a barrier that prevents the passage of spermatozoa. The barrier can be *physical* or *chemical*. The barrier methods are in general problematic, with significant disadvantages, not the least of which are high failure rates. In males, the condom is the principal barrier method. In terms of limiting fertility, the effectiveness of condoms in actual use is considerably less than in theory. However, one important advantage of the condom is that it reduces the risk of contracting a sexually transmitted disease.

In females, a number of devices are used singly or in combination. These include spermicides, diaphragms, cervical caps, and the recently introduced female condom. These act as both physical and chemical barriers. Spermicides, as the name implies, contain substances that kill or inactivate sperm. Many such compounds have been used. Lactic acid has been a common component of many spermicides. More recently, compounds known as surfactants have been introduced. Surfactants disrupt or dissolve the sperm surface membrane. Spermicides come in many forms, as tablets, creams, jellies, foams, sponges, and suppositories. They are deposited high in the vagina at the external os of the cervix one hour or less before coitus. Contraceptive sponges are designed to be a depository for semen. The sponge, impregnated with a spermicide, entraps and inactivates sperm. Its failure rate is relatively high also. The failure rate of spermicides is highly variable, as might be expected, given the fact that the positioning of the spermicide is critical for its effectiveness. Although relatively inexpensive, spermicides cannot be considered to be very effective.

The diaphragm is a thin rubber cup stretched over a thin wire ring. It is placed in the upper vagina so that it covers the external os of the cervix. It is fitted by a physician and is generally used with a spermicide. It is relatively safe, but not terribly effective, in large part because of the difficulty of insertion.

The cervical cap is a variation of the diaphragm, but it is held in place by suction rather than by the wire ring. Its failure rate is similar to that of the diaphragm, and it is considered to be more difficult to insert.

383

SUMMARY

Physical and chemical barrier methods are probably the oldest birth control methods. There have been significant improvements in their effectiveness, but they are poor substitutes for the hormone-based contraceptives. Nevertheless, they are still widely used around the world. Their primary advantage is that they reduce the risk of sexually transmitted diseases. Sterilization, vasectomy in the male and tubal ligation in the female, is the most effective nonreversible birth control method available today. Its use is limited primarily to older individuals who no longer desire to have children.

PREVENTION OF IMPLANTATION

This birth control strategy, referred to as interception, relies on generating a nonreceptive or "hostile" endometrium, thereby interfering with a normal implantation. Although modern interceptive methods are quite effective, their use has been quite controversial, particularly in the United States.

Intrauterine device (IUD)

The first IUD for human use was developed about a century ago, and since then a number of types have been marketed. The idea, however, is a very old one, originating according to tradition in the ancient Middle East. In order to keep their camels from becoming pregnant during the long caravan trips between the Middle East and the Far East, the camel drivers put small, smooth pebbles in the camel's uterus. The development of the modern generation of IUDs began in 1963 with a comprehensive study of IUDs initiated by the Population Council of New York. One important outcome of this study was the introduction of plastic IUDs in the shape of a T in 1967, and in 1968, plastic IUDs containing copper. The first copper IUD was introduced in 1976, and an improved version, the Copper T 380A, was approved for marketing in the United States in 1984. Since then more than 25 million have been distributed in over 70 countries. It is considered to be one of the most effective, long-acting reversible contraceptives available. The latest versions provide protection for up to 10 years. Variations of this basic design are being marketed in many countries (Fig. 15.6). Generally a 4–5-month period of adjustment is required. Bleeding during the first few months is common, but will disappear afterwards. The major risks with the older model IUDs were pelvic inflammatory disease, ectopic pregnancy, and even uterine perforation, if not fitted properly. In the United States these risks, either real or bolstered in the heat of litigation, were such that IUDs are no longer routinely prescribed and fitted. A female who wishes to be fitted with one must sign a form in which she effectively waives her rights to sue the manufacturer.

The newer IUD models have significantly reduced the risk associated with their use, and on a worldwide basis, IUDs are the most commonly used birth control method (Table 15.3), even in many developed countries. In Sweden about 30 percent of women use the IUD, since it is cheaper than, and just as effective as, the COCs. Some of the newer models also have a progesterone-releasing capability, in which the progesterone analog levonorgestrel is delivered continuously in small doses. The data available indicates that these new IUDs significantly reduce the initial blood loss, making them equivalent to the low-dose, progesterone-containing, subdermal implants. The

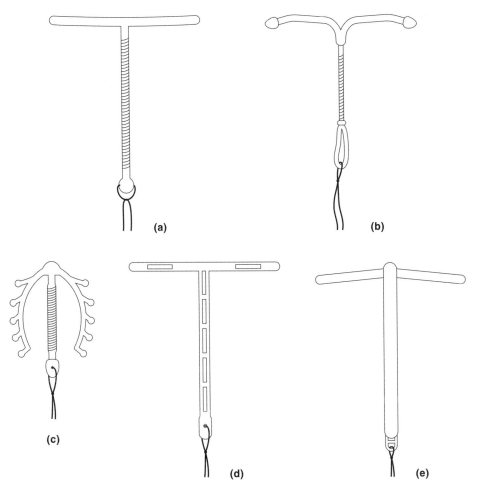

Figure 15.6
Available types of IUDs, variations on a theme. (a) TCu-200, length 36 mm, width 32 mm;
(b) Nova T, length 32 mm, width 32 mm; (c) MLCu-250, length 29 mm, width 18 mm;
(d) TCu-220, length 36 mm, width 32 mm; (e) Progesterone-releasing model, length 36 mm,
width 32 mm. (Adapted from *Intrauterine Devices: Technical and Managerial Guidelines for
Services*. 1977. World Health Organization, Geneva.)

IUD has many advantages, but its use is contraindicated under certain circumstances
(Table 15.8).

The mechanism of action of IUDs is still being debated. The early hypothesis proposed that an IUD interfered with implantation by producing changes in the endometrium. More recent studies suggest that the primary effect of the IUD is on sperm and
egg transport. As a foreign body, the IUD induces a uterine inflammatory response that
changes the characteristics of the uterine and uterine tube fluids. These changes are
thought to impair sperm and egg transport and decrease sperm viability, thus reducing

Table 15.8
Indications and Contraindications for IUD Use

Indications
  Inability to, or desire not to, use a hormonal method
  No history of pelvic inflammatory disease
  No history of uterine or uterine tube infections
  Minimal risk of contracting a sexually transmitted disease
  Desire to use a long-term reversible contraceptive that does not require daily pills

Contraindications
  Pregnancy or suspicion of pregnancy
  Suspected malignancy of the uterus or uterine tubes
  Uterine abnormality
  Undiagnosed bleeding
  Sexually transmitted disease
  Painful or long menstrual periods
  Severe anemia
  Unresolved abnormal pap smear
  History of ectopic pregnancy

Reference: Information about the IUD, Population Council, www.popcouncil.org.

the probability of fertilization. The copper in the copper-containing IUDs may enhance the contraceptive effect, but the precise mechanism of action is not yet clear. In contrast, the efficacy of the progesterone-releasing IUDs is considered to depend on the effect of progesterone in thickening the cervical mucus so as to act as an effective barrier to the passage of sperm.

Emergency contraception

Postcoital pill

These are commonly referred to as "morning-after" pills. The most popular regime is known as the Yuzpe regime developed in the early 1970s. These pills provide a very high dose of estrogen or a mixture of estrogen and progesterone and are taken within 72 hours after intercourse. The estrogen disrupts the pattern of endometrial development, generating a "hostile" environment for implantation. The postcoital pill can be quite effective, but it is not recommended for long-term use, and it is certainly contraindicated for women who are sensitive to estrogen. Postcoital pills have been available in most European countries for more than a decade, where they are sold in special packages for emergency contraceptive use. In the United States, postcoital pills were endorsed by the FDA only in 1997, and approved for distribution in 1998, but can be dispensed only with a prescription.

Mifepristone (RU-486)

Mifepristone, commonly referred to in the media in the United States as the "abortion pill," is an antagonist at the progesterone receptor. Hence, it interferes with the action of progesterone. In an interceptive sense, therefore, RU-486 can be used either before or after implantation. In the preimplantation mode, RU-486 can be used as a postcoital pill or as a once-a-month pill. In both cases, it impairs endometrial support for implantation. In Europe, where use of RU-486 is most widespread, this type of use is quite common, particularly among older women. The advantages are obvious. Daily pill taking can be dispensed with. We will see below that if taken after implantation has occurred, and up to 6 to 7 weeks after fertilization, RU-486 is very effective in inducing an abortion. Its use as an abortifacient under these conditions (hence, the name "abortion pill") has been tested widely and intensively in Europe. In combination with prostaglandin, it has been found to be very effective and safe as an abortifacient for early terminations.

SUMMARY

The IUD is the most widely used birth control method around the world. Its effectiveness is based on its interference with sperm transport and the prevention of implantation. Newer progesterone-releasing devices, because of their effectiveness, safety, and cost, have found widespread use in many developing countries. In the United States, because of a history of liability problems, IUDs are not routinely prescribed. A very recent development in birth control methods that rely on the prevention of implantation is RU-486, a progesterone-receptor antagonist. RU-486 has been found to be effective and safe, but its use is limited because of its cost. RU-486 is not routinely available in the United States because of the controversy regarding its use as an abortifacient.

The methods described above, using three different strategies for controlling fertility, are by far the most widely used. They differ in cost, effectiveness, ease of use, and compliance. Important features with respect to these properties are summarized in Table 15.9. It seems clear that currently the most effective and convenient methods are the injectables and implants.

Table 15.9
Convenience of Modern Reversible Contraceptive Methods

	Resupply requirements	Used at intercourse	Degree of client compliance required	Delivery approach
Orals	One cycle per month	No	High	Clinical and nonclinical
Injectables	One injection every 1–3 months	No	Low	Clinical, potential for nonclinical
Implants	Every 5 years	No	None	Clinical
IUDs	3–10 years	No	None	Clinical
Condoms/ spermicides	Every act of intercourse	Yes	Very high	Nonclinical

Reference: C. M. Huezo. 1998. Current reversible contraceptive methods: A global perspective. *International Journal of Gynecology and Obstetrics* **62 Suppl. 1**, S3–S15. Copyright 1998, with permission from Elsevier Science.

PREVENTION OF A LIVE BIRTH

Induced abortion is the termination of a pregnancy by artificial means. All human societies, as far as we know, have practiced abortion, whether sanctioned by law or not. The large majority of abortions have been and continue to be induced because the pregnancy is unintended and unwanted, even if in many cases the real reasons are masked by invoking medical complications. It is probably the case that throughout human history, and especially before modern contraceptive methods became available, induced abortion was widely used to control fertility.

A perspective

Despite its ubiquity, abortion remains controversial. Since the 1970s, some 50 million abortions have been induced annually, and it is estimated that about 40 percent of these have been illegal. Many have been carried out by untrained individuals, resulting in about 100,000 preventable deaths each year, with the overwhelming number of deaths occurring in poor, undeveloped countries.

The annual worldwide abortion rate is between 32 and 46 abortions per 1000 women aged 15–44, but the rate varies considerably between countries (Fig. 15.7). The Netherlands has the lowest rate (5 per 1000), while the former Soviet Union reported the highest (112 per 1000 in the 1980s), although the actual (unofficial) rates are considered to have been close to 200 per 1000. The mean number of abortions undergone by the average woman during her reproductive lifetime in the countries of the former Soviet Union was, officially, about *4*, but it was not uncommon for a woman to have *10 or more* abortions. The reason for the high abortion rate in countries like the former Soviet Union was that modern contraceptive methods were not available. From multiple studies around the world, it has become clear that the abortion rate is a function not only of the effectiveness of contraceptives, but also their cost and availability. It is estimated that the average woman would undergo *10* abortions during her lifetime in a society that does not provide accessible alternatives to abortion to control fertility. On the other hand, about *0.7* abortions per lifetime would be ex-

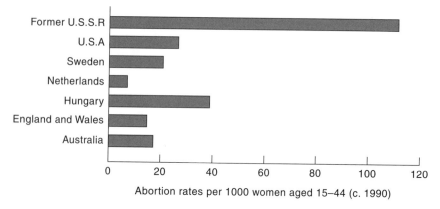

Abortion rates per 1000 women aged 15–44 (c. 1990)

Figure 15.7
Comparison of abortion rates in selected countries. (Adapted from A. Kulczycki *et al.*, 1996.)

Table 15.10
Unplanned Pregnancies Resulting from
Contraceptive Failure, United States, 1988

Method	Number of unplanned pregnancies (thousands)
Sterilization	28
IUD	18
Pill	324
Condom	612
Diaphragm	220
Rhythm	160
Others	378
Total	1740

Reference: Segal, 1996.

pected in a society that *does* provide highly effective contraceptive alternatives. The Netherlands, where free contraceptive counseling is available to all ages and, importantly, where the availability is noncontroversial, comes closest to the lower figure. The abortion rates in the United States reflect in part the fact that contraceptive counseling is controversial and limited in its accessibility.

Abortion rates will never be zero for a variety of reasons. First, no contraceptive regime is 100 percent effective, and second, some couples, even if contraceptives are easily available, do not use any form of contraception. In the United States, it is estimated that about 10 percent of the couples who use no contraceptive method contribute somewhat over 50 percent of the unintended pregnancies, while the other 50 percent is contributed by contraceptive failures in the 90 percent of couples who do use some type of birth control regime. As we saw in Fig. 15.1, the failure rate for different birth control methods varies considerably. In 1988, for example, contraceptive failure contributed about 1.7 million unintended pregnancies (Table 15.10). About half of these pregnancies were terminated by abortion. It is estimated that unintended pregnancies account for more than half of the total number of pregnancies in this country. The failure rate varies with age: most of the unintended pregnancies occur in women younger than 25. Hence, the availability and adoption of more effective contraceptive methods by young couples would substantially reduce both unplanned pregnancies and the abortion rate.

Although women and couples seek abortions for many reasons, these reasons can be grouped into two broad categories: *therapeutic* and *elective*. A therapeutic abortion may be required because the pregnancy poses significant risk to the physical well-being of the mother. In other cases, the mental well-being of the mother could justify a therapeutic abortion. Abortions to prevent the birth of a severely deformed or handicapped child could fall under the category of therapeutic abortions. Until recently, the evidence for a malformed fetus was based on the results of amniocentesis. However, a drawback of this procedure is that it could not be carried out successfully until week 16 of gestation, that is, not until the second trimester. Abortions carried out during the second or third trimester pose significant risks for the mother. Recently, chorionic villus (extraembryonic tissue that forms in the early stages of placenta formation) biopsy permits sampling fetal cells several weeks earlier, which means that the decision to induce an abortion can be made much earlier, and therefore, the risk to the mother would be much less. However, only significant chromosome abnormalities are generally detectable by these methods. Most single-gene mutations that give rise to an abnormal fetus or lead to developmental defects after birth would not be detected. On a worldwide basis, probably no more than 40 percent of abortions are induced for therapeutic (very broadly defined) reasons.

Elective abortions, termination of an unwanted pregnancy for nontherapeutic reasons, account for 60 percent or more of abortions. Elective abortions tend to arouse more controversy than therapeutic abortions. For example, is an abortion justified if the fetus is the undesired sex? There are no easy or universal solutions to these dilemmas. But perhaps the example of the Netherlands is instructive for those who would like to keep the abortion rate as low as possible. Their example shows that where the latest and most effective contraceptive options are available free or at low cost, where counseling and sex education is widespread, easily available, and informative, where the society promotes reproductive responsibility in a positive and open manner, and where modern abortion services are readily available at low cost, abortion rates and abortion-related maternal mortality can be kept at very low levels.

Abortion methods

Before the introduction of modern abortion techniques, many procedures were used to induce abortions, but they generally involved physical trauma to the uterus (for example, by sharp blows to the abdomen) or the introduction of sharp objects or toxic substances into the uterus. All of these procedures posed grave risks to the mother, and hence, maternal mortality rates were quite high. With the introduction of modern surgical techniques mortality rates have dropped significantly. The introduction of medical abortion techniques, interventions that do not require mechanical removal of the fetus, has provided, at least in the view of some, more acceptable alternatives to surgical abortions. Despite the relative efficacy of the newer medical alternatives, surgical abortions are still considered a last resort, something to try when all else has failed.

The risk to the mother associated with an abortion varies significantly with gestation time. Abortions during the first trimester (up to 12 weeks), performed according to current medical standards, carry a small risk. The abortion maternal mortality rate is lower than that of normal childbirth. For example, during the period 1979 to 1986, the mortality rate from pregnancy or childbirth (9.1 per 100,000 live births) was 15 times higher than the mortality rate from legal first trimester abortions (0.4–0.6 deaths per 100,000 procedures). Abortions during the second (12 to 24 weeks) and third trimester (after 24 weeks) carry a higher risk, and in many cases, the mortality rate is higher than that for normal childbirth (see below).

First trimester abortions

In the United States the large majority of abortions are carried out during the first trimester (Table 15.11). A number of different surgical or medical methods are available for inducing abortions during the first trimester (Table 15.12). The frequency with which they are used differs from country to country. The major surgical method, **vacuum aspiration** (VA), also called *suction curettage*, was introduced into the United States in 1967, and is now the most widely used method (over 90 percent of abortions are performed by VA). It can be performed on an outpatient basis and is considered to

Table 15.11
Characteristics of Legal Abortions, United States, Selected Years, 1972–1989

| | % Distribution | | | |
	1972	1978	1984	1989
Age				
<19	32.6	30.0	26.4	24.2
20–24	32.5	35.0	35.3	32.6
>25	34.9	34.9	38.3	43.2
Marital status				
Married	29.7	26.4	20.5	20.1
Unmarried	70.3	73.6	79.5	79.9
Weeks of gestation				
<8	34.0	52.2	50.5	49.8
9–10	30.7	26.9	26.4	25.8
11–12	17.5	12.3	12.6	12.6
13–25	8.4	4.0	5.8	4.2
16–20	8.2	3.7	3.9	4.2
>20	1.3	0.9	0.8	1.01
	Total Numbers			
Reported legal abortions	586,760	1,157,776	1,333,521	1,396,658

Reference: Hatcher *et al.*, 1994.

Table 15.12
Modern Abortion Methods

Method	Time in Gestation
Medical abortion	
Prostaglandins alone	Through week 3
Mifepristone/misoprostol	Through weeks 7–8
Methotrexate/misoprostol	Through week 9
Mifepristone/gemeprost	Through week 19
Surgical abortion	
Vacuum aspiration	Through weeks 7–8
Dilation and evacuation	Second trimester
Instillation	Second trimester
Hysterotomy	Third trimester

Misoprostol and gemeprost are prostaglandin analogs. Instillation refers to the injection of toxic compounds into the amniotic sac.

be a very effective and safe way of inducing early, first trimester abortions. (The mortality rate is 0.4 per 100,000.) A suction tube, whose diameter increases with the weeks of gestation, is connected to a vacuum pump, and inserted into the uterus. The fetus and endometrial tissue are suctioned out. Different methods are used to dilate the cervix to introduce the suction tube into the uterus. Vacuum aspiration can be performed in a physician's office with only a local anesthetic up through about the ninth week. After that, and up through the fourteenth week, it is advisable to carry it out in a hospital. Uterine bleeding and severe cramping can often accompany the procedure.

Medical alternatives to surgical methods came into their own with the introduction of mifepristone (RU-486), first marketed in 1988 in France, and now available in Britain, Sweden, other European countries, and China. Its use in the United States has been severely restricted. Although RU-486 by itself can work, recent extensive studies indicate that a combination of RU-486 and the prostaglandin analog *misoprostol* is more effective and safer. Prostaglandins, which produce uterine contractions, have been used alone effectively up to about week 3. The studies involving RU-486 and misoprostol indicate that the combination is 95 percent effective in inducing abortion up through weeks 8 or 9. The normal procedure involves taking two doses of RU-486 36 to 48 hours apart, with the second dose supplemented with a low dose of prostaglandin. Expulsion generally begins 4 hours later. Side effects include nausea, abdominal cramps, and bleeding that can last 10 to 14 days. A smaller study in the United States involving 17 sites in different states reports results very similar to the European ones.

Because RU-486 is expensive and not available in many countries, alternatives have been sought. A number of recent studies have shown that a combination of *methotrexate* (a cytotoxic compound sometimes used for tumor therapy) and misoprostol is 90 to 95 percent effective in inducing an abortion up through weeks 7 to 8. An advantage is that both of these compounds are inexpensive and easily available in most countries.

Second trimester abortions

Second trimester abortions are considered to be more risky and complicated than first trimester abortions. The older procedures, referred to as *instillation* techniques, involved injecting substances toxic to the fetus into the amniotic sac. One standard method used a concentrated saline solution. The high salt concentration kills the fetus and induces expulsion after 24 to 48 hours. Alternatives to salt, such as urea and glucose, have also been used. Injection of prostaglandins, especially up to week 19, has been used in the United States. A safer and more effective alternative to the prostaglandin option is the use of RU-486. This involves a high dose taken orally and followed 36 hours later with multiple doses of a prostaglandin analog, *gemeprost*, given vaginally, over a period of 24 hours. This is now the standard method used in Britain for second trimester abortions up to about week 19. It induces expulsion faster and has fewer complications. The maternal death rate is relatively low (7–8 per 100,000), but higher than that of vacuum aspirations. Instillation methods can have significant complications, the most serious being severe uterine hemorrhage, failure of the placenta to detach, and the fetus being born alive. (After expulsion, the fetus dies quickly due to respiratory failure.)

Dilation and evacuation (D & E) is the most common procedure used in the United States; it is generally performed between 12 and 16 weeks gestation, although some proponents use this method up through 20 weeks. Since the fetus is much larger in the second trimester, surgical crushing instruments are needed and the cervix must be dilated to a greater extent. Among the several methods available for dilating the cervix is the use of laminaria tents, small cylinders of dried and sterilized seaweed. The tent is placed in the cervix. It absorbs water readily and swells to a diameter 3 to 5 times greater than its original size. The second part of the procedure, the evacuation, involves using a metal scraper to remove the fetus and endometrium and more powerful suction pumps to remove the crushed tissue. A D & E carries an eightfold greater risk than a vacuum aspiration, can produce severe discomfort and even pain, and should always be performed in a clinic or hospital.

Third trimester abortions

Abortions during the late second trimester and third trimester require a **hysterotomy**, which can be likened to a cesarean section. The fetus is delivered alive and must be left to die, which often takes several hours depending on the age of the fetus. An alternative, and controversial, method in use in the United States is the so-called *partial-birth abortion*, in which labor is induced, and the fetus is destroyed as it emerges from the birth canal. The mortality rate from a hysterotomy is about 50 per 100,000 procedures.

Summary

Induced abortion is the birth control measure of last resort. Abortion rates will never be zero primarily because many couples do not use effective birth control methods, and secondarily because most methods are not 100 percent effective. Abortion rates, however, can be kept to a minimum if modern birth control methods are easily available, free or inexpensive, and if birth control information and advice is easily obtainable and noncontroversial. Modern abortion methods, whether surgical or medical, are extremely safe, especially if performed during the first trimester of pregnancy.

NEWER DEVELOPMENTS

Despite the impressive successes of the steroid contraceptives, contraceptive research is not a high priority item, and funding levels are quite low. The pharmaceutical industry, which supported much of the previous research, is not inclined to continue, given the high costs of bringing a product to market and the previous litigious history of contraceptive methods. The United States, which used to be a leader in contraceptive development, is no longer a key player. Most of the research programs are being supported by the WHO and the governments of India and China.

Although there is a pervasive lack of funding, small studies seeking to improve the existing birth control methods or to design newer ones are in progress. One approach is based on immunocontraceptives, or the use of antibodies to inactivate highly specific protein targets. Such an approach would be safer and have fewer side effects, particularly if the target inactivated did not have general systemic repercussions. COCs, for example, could be considered a shotgun approach since a large and complex system, regulation of ovarian function, is the target. Hence, the objective in these new approaches is to inactivate a very specific target necessary only for fertilization or implantation. The long-term objective in this research is not to develop one "universal" birth control regime (which probably does not exist), but to develop effective and safe

Table 15.13
Contraceptive Usage Increases as Availability
of Modern Contraceptive Methods Increases:
Data from 36 Developing Countries

Number of methods available	Contraceptive usage
0	8%
1–2	27%
3–4	43%
5–6	62%

Reference: Adapted from PATH, 1992.

alternatives to the ones currently available. The worldwide availability of several effective alternatives cannot be anything but beneficial to all societies. Studies from 36 different countries have shown that the prevalence of contraceptive use increases with the number of contraceptive methods available (Table 15.13). All of these new approaches are generally in the early stages of development. Hence, none of them is available on a mass basis, and it will be probably some years before any of them will be. Nevertheless, it is of interest to describe the essential features of each type to indicate their rationale.

Anti-FSH vaccine

FSH in the male does not appear to have any function other than to support spermatogenesis. A vaccine directed against FSH would not affect testosterone-dependent functions because testosterone synthesis depends on LH. The validity of this conclusion has been confirmed by a recently completed 5-year study of the effects of an anti-FSH vaccine in monkeys. The vaccine, developed in India, was shown to be efficacious and reversible. The trials to test its efficacy in humans were begun in India in 1995.

Anti-ZP vaccine

An antibody directed against the zona pellucida proteins would bind to the ZP proteins, thereby preventing the ZP from participating in the acrosome reaction. This, in turn, would inhibit penetration of the ZP by the spermatozoon. An anti-ZP vaccine has been developed and is being tested in experimental animals. Results are encouraging in the sense that the vaccine does prevent fertilization. However, the vaccine apparently also disturbs follicle development. Hence, much more work is necessary before such a vaccine could be considered for humans.

Anti-hCG vaccine

A vaccine directed against hCG could, in principle, be very effective. As we saw in Chapter 11, hCG is produced by the syncytiotrophoblast and is necessary to maintain secretion of progesterone by the corpus luteum. Inactivation of hCG would be expected to inhibit implantation. An anti-hCG vaccine is not easy to make. Recall that hCG and LH are extremely similar molecules, and hence, most antibodies directed against hCG would also recognize LH. Nevertheless, several vaccines highly specific for hCG have been made, and some have been tested successfully in animals. One vaccine is in the early phase of human trials.

Other possible vaccines

Antiacrosomal and antiepididymal protein vaccines are also being developed. In both cases the vaccines are directed against the spermatozoa, inactivating the sperm in different ways to prevent fertilization. The antiacrosomal vaccine is further along in its development, and trials in monkeys and baboons have been initiated. For those who believe that the human being begins at fertilization, such vaccines would be more acceptable since they would prevent fertilization. An antitrophoblast vaccine, directed against syncytiotrophoblast cells, is also in the early stages of development. Like the anti-hCG vaccine, the antitrophoblast vaccine would inhibit implantation.

SUMMARY

Future birth control methods will probably rely on the inactivation of specific target proteins necessary for gametogenesis, fertilization, or implantation. This is the rationale for the current focus on the development of immunocontraception methods. Preliminary tests for some of these new contraceptive vaccines are quite promising, but their availability is still some years away. Sadly, the future of much of this research and development may be in jeopardy because of liability concerns by pharmaceutical companies that took the lead in the development of hormonal contraceptives, and because of substantial decreases in financial support by the United States.

QUESTIONS

1. Menstrual cycling is not abolished with COCs despite the fact that ovulation does not take place. How do you explain this fact?

2. Compare the indications and contraindications for COCs and POCs.

3. The use of COCs is associated with a number of nonreproductive benefits. List some of these, and explain why COCs are responsible for the benefit. (Include information from Chapter 18.)

4. Mifepristone (RU-486) is effective as an abortifacient and birth control method. Explain why this is the case.

5. What types of emergency birth control methods are available, and how do they work?

6. What conditions/circumstances are important in reducing the induced abortion rate? Explain why that is the case.

7. Library project:
   a. The FDA has not yet approved the widespread use of mifepristone (RU-486) in the United States. What scientific reasons have they presented to support their position?
   b. The first postcoital pill (under the trade name Preven) has recently been approved in the United States. What is its mechanism of action and how does it differ from RU-486?

SUPPLEMENTARY READING

Burnhill, M. S. 1998. Contraceptive use: The U.S. perspective. *International Journal of Gynecology & Obstetrics* **62 Suppl. 1**, S17–S23.

Ewart, W. R., and B. Winikoff. 1998. Toward safe and effective medical abortion. *Science* **281**, 520–521.

Hatcher, R. A., *et al.* 1998. *Contraceptive Technology*, 17th revised ed. Ardent Media, Inc., New York, NY.

Intrauterine Devices: Technical and Managerial Guidelines for Services. 1997. World Health Organization, Geneva, Switzerland.

Ory, H. W. 1982. The non-contraceptive health benefits from oral contraception. *Family Planning Perspectives* **14**(4), 182–184.

Oudshoorn, N. 1994. *Beyond the Natural Body: An Archeology of Sex Hormones.* Routledge, London, UK.

Reifsnider, E. 1997. On the horizon: New options for contraception. *Journal of Obstetric, Gynecologic, and Neonatal Nursing* **26**(1), 91–100.

Segal, S. J. 1996. Contraceptive development and better family planning. *Bulletin of the New York Academy of Sciences* **93**(1), 92–104.

INTERNET SITES

Internet sites with information about fertility control

Contraceptive Research & Development Program (CONRAD): http://www.conrad.org

Population Council: http://www.popcouncil.org

Office of Population Research at Princeton University: http://opr.princeton.edu/ec/

ADVANCED TOPICS

Creinin, M. D. 1996. Oral methotrexate and vaginal misoprostol for early abortion. *Contraception* 54, 15–18.

Edwards, R. G. 1994. Implantation, interception and contraception. *Human Reproduction* 9(6), 985–995.

Fraser, H. M. 1993. GnRH analogues for contraception. *British Medical Bulletin* 49, 62–72.

Huezo, C. M. 1998. Current reversible contraceptive methods: A global perspective. *International Journal of Gynecology & Obstetrics* **62 Suppl. 1,** S3–S15.

International Planned Parenthood Federation (IPPF). 1994. International Medical Advisory Panel (IMAP) statement on contraceptive efficacy. *IPPF Medical Bulletin* 28, 1–2.

IPPF. 1995. IMAP statement on intrauterine devices. *IPPF Medical Bulletin* 29, 6–20.

IPPF. 1998. IMAP statement on steroidal oral contraception: Comparison of COCs and POPs. *IPPF Medical Bulletin* **32(6),** 20–30.

Kulczycki, A., M. Potts, and A. Rosenfield. 1996. Abortion and fertility regulation. *Lancet* 347, 1663–1668.

Levels and Trends of Contraceptive Use as Assessed in 1994. 1996. United Nations, New York.

Program for Appropriate Technology in Health (PATH). 1992. Contraceptive method mix: The importance of ensuring client choice. *Outlook* 10(1), 1–8.

Thong, K. J., P. Lynch, and D. T. Baird. 1996. A randomised study of two doses of gemeprost in combination with mifepristone for induction of abortion in the second trimester of pregnancy. *Contraception* 54, 97–100.

Trussell, J., J. A. Leveque, J. D. Koenig, R. London, S. Borden, and J. Henneberry. 1995. The economic value of contraception: A comparison of 15 methods. *American Journal of Public Health* 85, 494–503.

Two very different approaches to the problem of infertility. (Top) *Female Figurine.* Tlatilco, Mexico, 1300–1800 BCE. Probably used as a fertility amulet. (Cited in H. Speert. 1973. *Iconographica Gyniatrica.* F. A. Davis Company, Philadelphia, PA. Reprinted with permission of The Brooklyn Museum of Art, Brooklyn, New York.) (Bottom) A collection of advertisements that appeared in *Parents Magazine*, San Diego, March 1999.

CHAPTER 16

Infertility and Assisted Reproduction

*It is most important to recognize that with the exception
of cases in which there is an absolute barrier to fertility,
combined factors in both partners are frequently operating.*

—H. W. G. Baker. 1994

Male infertility. *Endocrine and Metabolic
Clinics of North America* 23(4), 783–793.

*The thoughtful clinician must be humble given the knowledge
that a large proportion of couples will eventually conceive on their
own without medical intervention and should explain this to
couples who do not have an absolute barrier to pregnancy.*

—D. Guzick. 1996

Human infertility: An introduction. In *Reproductive Endocrinology,
Surgery, and Technology*, Vol. 2, Chapter 101. E. Y. Adashi, J. A. Rock,
and S. Rosenwaks, Eds. Lippincott-Raven Publishers, Philadelphia, PA.

It [alcohol] provokes the desire but takes away the performance.

—W. Shakespeare.

Macbeth. Act II, scene iii, 34.

IN EVERY SOCIETY there has been a tension between the desire and need on the part of some members of the society to limit fertility, and the desire on the part of another segment of the population to be fertile. Until quite recently most of our attention has been devoted to developing effective and safe ways to limit fertility. This is largely because the control of fertility has been seen as an urgent personal and societal problem. Accordingly, as we saw in the last chapter, the effort invested in addressing this problem has paid off remarkably well. Infertility, on the other hand, has been seen as a regrettable, but personal, tragedy, without important societal implications. Infertility was also a problem about which little could be done. The historical turning point in the way we think about infertility came in 1978, when the British team of Steptoe and Edwards reported the first birth following human *in vitro* fertilization. Infertility could now be seen as a medical problem, something fixable, rather than as something to be regretted. Since then thousands of pregnancies have been achieved worldwide by a variety of procedures generally referred to as **assisted reproductive techniques** (ARTs). The development of these techniques has been so rapid that discussion of the social, ethical, and moral issues that they raise take place increasingly after the fact, which, to the consternation of many, makes the discussions almost irrelevant.

Our understanding of the factors that influence the fertility of couples has increased tremendously during the last twenty years. However, it is important to emphasize that relatively few definitive causes of infertility have been identified. This means that in many cases, to the disappointment of the couple, an infertility evaluation does not provide a specific etiologic diagnosis. The infertility may remain unexplained, or it may be the result of a combination of factors that vary from couple to couple. Our objective in this chapter is to review some of the factors that affect fertility and some of the known causes of infertility. The causes of infertility can be grouped into three major categories. First, infertility, especially in the female, can arise from normal, nonpathological processes that are associated with the perimenopause. We have considered this aspect already to some extent in Chapter 11. Second, infertility may be the result of intrinsic endocrinological, physiological, or neurological disturbances that interfere with normal gamete formation, sexual functioning, or the ability of the female to sustain a pregnancy to term. Third, the infertility may be affected by lifestyle factors and behaviors, such as smoking or drinking, or the use of therapeutic or recreational drugs, extremes of weight or exercise, or exposure to environmental toxicants. Finally, we will describe briefly the rapidly developing field of ARTs, which provide hope for many couples for whom no hope was available before.

FERTILITY IN HUMAN POPULATIONS

The fertility of any society depends primarily on the fertility of its females. The birth rate, however, depends primarily not only on the ability of its females to conceive and give birth to live children, but also on their desire to do so. This is a theoretical distinction that, in the history of most human societies, was not very important, but that is having significant practical consequences in most developed countries. There is also, as we learned in Chapters 8 and 10, a natural rhythm to fertility. Both affect the number and pattern of births.

Birth rate versus fertility

The birth rate in the United States has decreased dramatically in the last two hundred years. For example, in 1790, the birth rate in the newly formed United States was 55 per 1000 population, averaging about 8 births per female. In 1990, the birth rate was 15.5 per 1000, or 1.8 births per female. Except for the post–World War II "baby boom" resulting in 3.8 births per female, the birth rate has gone down consistently in the United States since its founding. Similar changes have been seen over the last 100 years in the nations that constitute the developed world.

There are several reasons for these dramatic changes in birth rates, but in essence they all reflect profound changes in the roles and aspirations for women in these societies, the most important of which have been the postponement of marriage, the delay of childbearing, and the increased use of birth control methods. Concern about the environment and overpopulation, unfavorable economic conditions, and perhaps liberalized abortion laws have also contributed to the decline in the birth rate.

In the United States, for example, 28 percent of females between 20 and 24 years old in 1960 were single. In 1985, that percentage had gone up to 58.5 percent. Similarly, 10 percent of those between 25 and 29 years were single in 1960, and the fraction had increased to 26 percent by 1985. Overall, it is estimated that about 16 percent of the decline in the birth rate is accounted for by the increase in the average age at first marriage, and the remainder by decisions of couples to limit the number of births by different means.

The decline in the birth rate has often been referred to as a decline in fertility. It should be clear, however, that the two terms refer to two different aspects of population growth. Even the term "infertility" is used imprecisely in most contexts to refer to the inability to have children. Two terms are used in the scientific literature when discussing inability to have children—**infertility** and **infecundity**, the first referring to the inability to conceive, and the second, to the inability to carry the conceptus to term successfully. However, quite often "infertility" is used to cover both types of conditions, because in many cases it may not be possible to make that distinction. We will use "infertility" in that double sense here, unless it is important to make the distinction in the context under discussion.

With these distinctions in mind, it is reasonably clear that the decrease in birth rate does not reflect a generalized increase in infertility. For example, the proportion of married couples that are infertile has remained the same since 1965, with only one qualification: in 1965, 3.6 percent of females between 20 to 24 years were infertile, while 10.6 percent in the same age range were infertile in 1982. The large increase in infertility in that age range has been attributed to the very high rate of sexually transmitted diseases (STDs) (discussed in Chapter 17) in young females in the period between 1965 and 1982, an increase considered to be due in large part to the increase in the number of sexual partners in this age range.

Age-dependent changes in fertility

One of the important consequences of postponing birth to a later age is that we have become much more aware of the decrease in fertility with maternal age. This relationship cannot be demonstrated easily in contemporary populations because of the widespread use of birth control. However, the decline in fertility can be seen from the way in which birth rates decline in population groups that have not practiced any form of birth control. One such example is taken from Mormon genealogical data from the middle of the nineteenth century (Table 16.1). In this group, fertility remained more or

Table 16.1
Age and Fertility in a Population with No Birth Control

Age (years)	Relative rate	
	Wife	Husband
15–19	0.96	0.90
20–24	1.00	1.00
25–29	1.03	0.99
30–34	0.99	1.04
35–39	0.90	0.97
40–44	0.62	0.83
45–49	0.14	0.82
50–54		0.73
55–59		0.48

Relative fertility rates according to maternal and paternal age calculated from Mormon genealogical data, with wives born between 1840 and 1859. Rates are measured relative to the age range 20 to 24. (Reference: Menken *et al.*, 1986.)

less constant up through the late 30s in females, declining significantly after that. In contrast, fertility in males declined much more slowly. The difference between the two indicates that the decline in fertility in couples is due primarily to the aging of females. Other measures of the decline in female fertility are consistent with this result. For example, in a large-scale French study of women with *azoospermic* (a sperm count of zero) husbands and who sought sperm donors, the pregnancy rate for women under 31 years was 74 percent, between 31 to 35, 62 percent, and for those over 35 years, it had declined to 54 percent. Measures of infecundity are also available. Consider, for example, the increase of the spontaneous abortion rate with maternal age. In the United States, this is about 10 percent for women 30 years and under, 18 percent for those between 30 and 40 years, and 34 percent for those between 40 and 44 years. Moreover, an age-dependent decline in the pregnancy success rate using ARTs is also seen: 17 percent for women younger than 30, 13 percent for those between 31 to 35, and 10 percent for those 36 to 40 years (based on a recent study in England).

Summary

Fertility declines with maternal age. The decline is clearly manifested beginning in the late 30s and is more marked after 40. Studies in different populations in the industrialized world suggest that 30 to 40 percent of women who defer pregnancy until the mid to late 30s will have significant infertility problems. Indeed, women in this age group represent a major fraction of the total infertility population. Not only do women in this age group have a lower pregnancy rate, but they also have a higher incidence of pregnancy loss. This finding should be taken into account by couples who for a variety of reasons wish to postpone having children.

INTRINSIC INFERTILITY IN HUMAN POPULATIONS

In this section we will consider infertility arising from endocrinological, physiological, or anatomical defects that are not age-dependent or due to the effects of external agents. It is important to keep in mind that questions of infertility use the couple as their reference point. In general, a couple is said to be subfertile or infertile if they have not achieved a pregnancy or live birth after 12 months of unprotected intercourse. This time period is arbitrary since the time required for conception depends on the months of exposure (Table 16.2). The probability of becoming pregnant after 1 year of unprotected intercourse is 85 percent, and increases marginally to 93 percent after 2 years. By convention, a 1-year period is considered to be stringent enough for a diagnosis of infertility, although not all experts agree on the reliability of this definition.

Using this convention, in the industrialized countries, about 15 to 20 percent of all couples will experience some infertility during their reproductive lives. About half of these either will be unable to have any children or will not have the number that they desire. Attempts to ascertain the causes of infertility in couples who seek treatment have yielded somewhat disparate results. For example, a large, comprehensive study in England found that in 28 percent of infertile couples, the cause of infertility was unexplained (Table 16.3). In about 26 percent of infertile couples, the cause was attributed to the male, while in about 41 percent, the cause was attributed to the female. The remaining 5 percent was due to coital problems attributed in large part to the male.

Table 16.2
Time Required for Conception

Months of exposure	% Pregnant
1 month	23
3 months	57
1 year	85
2 years	93

Reference: A. F. Guttmacher. 1956. *Journal of the American Medical Association* **161**, 855–866. Copyright 1956, American Medical Association, with permission.

Table 16.3
Causes and Frequency of Infertility in Couples

Cause	Frequency (among infertile couples)
Unexplained	28%
Male	
Sperm defects	21%
Other problems	3%
Coital problems	5%
Female	
Ovulatory failure	18%
Uterine tube defects	14%
Endometriosis	6%
Cervix/uterus	3%

Data from a comprehensive study in England. (Reference: Hull *et al.*, 1985.)

The results from another group of six large studies are shown in Table 16.4. In this case, the cause was unknown in 18 percent of couples, while infertility could be attributed to the female in 57 percent and to the male in 21 percent of couples. These differences probably arise in part because of the way in which different parameters are measured. For example, the fraction attributed to male infertility, due overwhelmingly

Table 16.4
Causes of Infertility in Couples

Cause	Frequency
Unexplained	18%
Female factors	
Ovulation disorders	29%
Uterine tube defects	16%
Cervix/uterus	5%
Endometriosis	7%
Other	4%
Male factors	
Sperm count and sperm defects	21%
Both male and female	4%

Pooled data from seven epidemiological studies in industrialized countries between 1962 and 1983. (Reference: Spira, 1986.)

to problems of sperm quantity and quality, varies from one country to another, depending on how the sperm parameters are assessed. These numbers suggest that in about 50 percent of infertile couples, the infertility is primarily due to problems in the female. In about 25 percent of couples the cause of infertility is unknown, and the remainder to problems primarily in the male.

Female infertility

Table 16.5 lists the more common categories of dysfunction that lead to infertility in women. It is very difficult to give precise values to the relative incidence of each of

Table 16.5
Major Causes of Infertility in Women

Cause	Examples of disorders
Ovulatory failure	Polycystic ovarian syndrome Hypothalamic anovulation Gonadal dysgenesis Resistant ovarian syndrome Hypogonadotropic hypogonadism
Impaired gamete or zygote transport	Impaired oocyte capture Uterine tube cilia defects Uterine tube defects Endometriosis Antisperm antibodies Pelvic inflammatory disease Cervical mucus
Implantation defects and recurrent spontaneous abortion	Polycystic ovarian syndrome Chromosome abnormalities Uterine and endometrial anomalies Corpus luteum defects

Adapted from Healy *et al.*, 1994, with permission.

these causes. For example, a World Health Organization (WHO) study published in 1985 showed the following causes of infertility in women: unexplained, 40 percent; tubal damage, 36 percent; ovulatory disorders, 33 percent; endometriosis, 6 percent. A similar distribution was seen in Asia, Latin America, and the Middle East, while in Africa, the overwhelming contribution to infertility among women was tubal damage.

Ovulatory failure

Many conditions can contribute to ovulatory failure. A few general classes of disorders are listed in Table 16.5. We have had occasion to discuss some of these in previous chapters. For example, we mentioned gonadal dysgenesis in Turner syndrome in

Chapter 14, hypothalamic failure due to stress in Chapter 9, and hypogonadotropic hypogonadism as a condition that delays puberty. Pituitary failure can also occur in the postpubertal female. **Polycystic ovarian syndrome** (PCOS) is considered to be one of the leading causes of anovulatory infertility. PCOS is a multifaceted disorder with a complex array of symptoms. The precise defect in PCOS remains unknown, although some disturbance in the estrogen-to-androgen ratio may be important in the origin of the disorder. PCOS females appear to have higher androgen levels or higher androgen-to-estrogen ratios than non-PCOS females. One consequence is that follicular dynamics is disturbed, and ovulation does not take place. The disturbance in estrogen synthesis may account for the observation that large numbers of Graafian-stage follicles accumulate in the ovary (the large follicles were originally referred to as cysts, hence, the origin of the name PCOS). Associated with PCOS is hirsutism (growth of body hair), obesity, and increased risk of endometrial cancer. Other less-well-understood disorders include resistant ovarian syndrome and luteinized unruptured follicle. In **resistant ovarian syndrome**, the ovary is insensitive to gonadotropin stimulation. **Luteinized unruptured follicle** refers to the formation of corpus luteum–like follicle before ovulation. Both of these disorders are thought to reflect serious disturbance in intraovarian factors that are necessary for ovulation.

Impaired sperm or zygote transport

Conditions that affect the ability of the sperm to reach the site of fertilization or that affect the ability of the conceptus to move into the uterus contribute significantly to female infertility. For example, unusually viscous cervical mucus acts as a very effective barrier to sperm movement through the cervix and into the uterus. A small fraction of infertile women produce such mucus. Pelvic inflammatory disease (PID), an infection of the uterine tubes caused primarily by sexually transmitted microorganisms, is the main cause of tubal disorders and tubal infertility. PID is also associated with a three- to eightfold greater risk of ectopic pregnancies. Among such organisms, chlamydial infections (see Chapter 17) account for about 75 percent of the cases of tubal infertility. Not all chlamydial infections result in infertility, however. In some studies, 25 percent of fertile women appear seropositive for chlamydia, that is, show evidence of having been infected with chlamydia.

Endometriosis is a condition in which endometrial tissue grows at ectopic sites. The condition was first described in the 1800s, but the term itself was not coined until 1927 in a paper that presented the standard hypothesis for the cause of endometriosis. According to this view, abnormal flow of endometrial tissue into the uterine tubes and into the abdominal cavity during menstruation leaves small deposits of endometrial tissue in these ectopic sites. The tissue begins to grow at these sites, and once it has done so, undergoes the same changes that would occur in the uterus. Although the favored ectopic sites are the uterine tubes and abdominal cavity, consistent with the standard hypothesis, endometriosis can occur in almost every organ of the body. In these unusual cases, it is presumed that endometrial cells are transported by the circulatory system and deposited at distant sites. There are even a few reports of endometriosis in men who had received estrogen therapy, indicating that endometriosis might have multiple causes. The menstrual-flow hypothesis has been debated vigorously, and the simplest way to summarize our knowledge of endometriosis is to say that the cause remains unknown.

The relationship between infertility and endometriosis is complex. A number of studies suggest that 20 to 40 percent of infertile women suffer from endometriosis, and, indeed, endometriosis is suspected in any woman complaining of infertility. On the other hand, 10 percent of fertile women have endometriosis. Endometriosis is an important disorder in women: about 4 per 1000 women in the 15 to 64 age range in the United States are hospitalized with endometriosis each year, slightly more than are hospitalized with breast cancer. Severe endometriosis in the uterine tubes, the ovaries, or in the pelvic regions can certainly compromise fertility, but in a number of controlled studies, medical suppression of endometriosis or surgical excision of the tissue has not improved fertility.

Implantation defects and recurrent spontaneous abortion

Impaired uterine receptivity can result either in the failure of implantation or in pregnancy loss at a later stage in gestation. Traditionally, lack of appropriate progesterone stimulation was considered to be the main cause of implantation failures, but implantation also depends on many factors other than progesterone. Recurrent spontaneous abortions are associated with uterine cavity abnormalities that prevent normal completion of gestation. About one-third of women with recurrent abortions also experience difficulties in conceiving, and 40 to 50 percent of these women also suffer from PCOS.

Male infertility

The true extent of male infertility remains controversial. Over the past 50 years the perception of the degree to which males were involved in the infertility of the couple has oscillated. Initially, infertility was seen as primarily a female problem; later the pendulum swung to the other side, and some studies in the United States indicated that 40 percent of all infertility was due to the male. More recently, a downward trend in the male contribution has been noted, but precise and reliable numbers are very difficult to obtain. Due to the redefinition of the lower limit of the "normal" sperm count, many men who would previously have been characterized as infertile or subfertile are now considered normal.

Male infertility arises from two different types of problems: *semen quality* and *coital impairment*. Semen quality includes a broad category of characteristics, such as sperm count, sperm morphology, sperm motility, ejaculate volume, and composition of the seminal fluid; coital problems result from erectile dysfunction, retrograde ejaculation, and/or failure to ejaculate. Male infertility falls into three treatment categories: untreatable sterility (12.5 percent); potentially treatable conditions (12.5 percent); and untreatable subfertility (75 percent).

These statistics are discouraging in that they indicate that most cases of male infertility are untreatable by conventional methods, which do not involve ARTs. Some of the conditions that fall into each of these categories are indicated in Tables 16.6. A few are considered briefly below.

Untreatable sterility

This treatment category encompasses severe and persistent impairment in the ability to produce sperm, which is referred to as **male factor infertility**. This condition can come about in a number of different ways. Primary seminiferous tubule failure refers to a complex of disorders that profoundly disturb seminiferous tubule function, for ex-

Table 16.6
Infertility in Males

Conditions resulting in untreatable sterility	Seminiferous tube failure
	Chromosome disorders—Klinefelter syndrome
	Cryptorchidism
	Testicular atrophy
	Idiopathic
Potentially treatable conditions	Genital tract obstructions
	Congenital
	STDs or other types of infections
	Trauma
	Sperm autoimmunity
	Idiopathic
	Trauma
	Gonadotropin deficiency
	Hypothalamic dysfunction
	Pituitary disorders
	Hyperprolactinemia
	Coital disorders
	Impotence
	Retrograde ejaculation
	Failure of ejaculation
Untreatable subfertility	Oligospermia—sperm count less than 20 million/ml
	Asthenospermia—less than 25% motile sperm
	Teratospermia—greater than 85% abnormal sperm
	Nomospermia—normal-looking sperm, but with impaired ability to fertilize

Adapted from Baker, 1994, with permission.

ample, to Sertoli cell failure. Klinefelter syndrome, as we saw before, is characterized by essentially no sperm production. Recent studies of azoospermic males have shown that some of them show small deletions from the Y-chromosome that result in a severe impairment in the first stage of spermatogenesis. The genes that are deleted must be those that direct the spermatogonial stages of spermatogenesis. Reduction in spermatogenesis due to testicular atrophy resulting from trauma, disease, or radiation therapy is, in most cases, irreversible. A large fraction of conditions that affect sperm number or quality are idiopathic.

Potentially treatable disorders

The most common potentially treatable conditions are listed in Table 16.6. Genital tract obstructions in the vas deferens, the epididymis, or the ejaculatory duct can be congenital or they can develop as a consequence of STD, tuberculosis, trauma, or from a number of other conditions. Many are potentially treatable by surgical intervention to remove the blockage, or in congenital cases, reconstructive surgery to repair the malformation. For example, intervention to remove epididymal and vas deferens blocks are moderately successful, with about 70 percent of men achieving substantial sperm output within a year after the surgery. The success rate in other types of treatments varies considerably.

An autoimmune attack on sperm is a common type of condition affecting sperm function, and it is also perhaps the most common medically treatable condition seen in men. The reason for the production of sperm antibodies is not usually known. Most cases arise spontaneously; others are due to trauma to the testis or to obstructions that result in breakage of the Sertoli cell barrier. The condition is characterized by the presence of antibodies to sperm in seminal fluid or even in the circulatory system, or by proteins of the immune system (immunoglobulins) coating the sperm. Sperm antibodies are often found in fertile men, but presumably the concentration is too low to impair fertility. Therapy involves the use of immunosuppressive drugs, such as the steroid prednisolone, taken over a period of 4 to 6 months. About 25 percent of couples are able to establish a pregnancy during such a regimen.

Gonadotropin deficiency can arise from a number of different causes, such as disorders of the GnRH pulse generator, pituitary dysfunction, suppression of gonadotropin secretion by drugs, hyperprolactinemia, illness, and malnutrition. In general, gonadotropin deficiency is a rare cause of male infertility, and in most cases, it can be treated by administration of LH and/or FSH or by pulsatile GnRH therapy.

Organic defects that result in erectile dysfunction, failure of ejaculation, or retrograde ejaculation (ejaculation of the sperm into the bladder rather than into the urethra) are generally infrequent in the industrialized countries. Some types of organic defects can be treated surgically. Short periods of psychogenic erectile dysfunction are seen, but these cases can usually be treated with psychotherapy.

Untreatable subfertility

Most cases of male infertility or subfertility are due to sperm counts that are lower than normal or that contain a substantial fraction of abnormal or dysfunctional sperm (Table 16.6). Many treatments have been tried over many years—treatment with gonadotropins, androgens, antiestrogens, nutritional supplements, anti-inflammatory drugs—but none has been consistently or uniformly effective. If the sperm count is not too low, continual attempts to conceive over several years have been successful in some cases. No treatment has been shown to increase semen quality and fertility in this group.

Summary

Much progress in identifying intrinsic conditions that affect fertility adversely in females and males has been made in the last 10 to 15 years. Technological improvements have helped significantly in the evaluation of infertile couples. However, in most cases the precise mechanisms at work remain unknown. Polycystic ovarian disease is perhaps the best example of a condition commonly associated with infertility in women that has been studied extensively, but whose etiology remains unknown and whose treatment remains unsatisfactory. Cause-and-effect relationships for most conditions associated with infertility remain ambiguous, and this has prevented the development of effective therapy for most infertility conditions.

EXTRINSIC FACTORS—LIFESTYLE

A number of individual lifestyle factors can compromise fertility. The most important of these factors are intensive exercise, extremes of weight, promiscuous and unprotected sexual activity, and immoderate drug use. For example, in susceptible individuals, intensive exercise can have dramatic consequences for reproductive function as we saw in Chapter 9. Extremes of weight can also have pronounced effects on fertility (Table 16.7). We had considered the effects of very low weight before in Chapter 9, but a

Table 16.7
Relative Risk of Infertility in Females as a Function of Departure from the Ideal Body Weight*

% of ideal body weight	Relative risk
86–119%	1.00
<86%	4.70
>119%	2.20

*A relative risk of 4.70 means that females who weigh less than 86% of their ideal body weight are 4.70 times as likely to suffer from infertility than if their weight was in the normal range (86–119%). (Reference: Green *et al.*, 1988.)

weight greater than 120 percent of the ideal body weight also increases the risk of infertility substantially. For females in particular, risky sexual activity that increases the likelihood of acquiring a sexually transmitted disease increases the chances of pelvic inflammatory disease (PID) discussed in the previous section. Less well documented, but probably no less real, are the consequences of drug use. We will now consider the effects of drugs on fertility.

Recreational drugs

Immoderate use of tobacco, alcohol, and other recreational drugs has been considered to have important effects on fertility or fecundity for some time, but their effects have not been easily quantifiable, since often more than one drug is used at the same time. Research on the reproductive effects of recreational drugs in humans, such as nicotine, marijuana, alcohol, cocaine, and other mood-altering compounds, is relatively recent. It is also difficult to carry out. Ethical constraints do not permit controlled human experimentation, and hence, the information we have available comes from scattered case reports, a few studies of volunteers, and epidemiological and retrospective studies. Moreover, because many drug users suffer from a variety of health problems, it is generally difficult to separate the drug's effects from the general health condition of the subject. Nonetheless, despite interpretive problems and some inconsistencies in findings that leave a number of important questions unanswered, there is substantial agreement that drug abuse poses serious risks to reproduction and fertility. For any given drug, the degree of risk depends on the amount taken and the frequency

Table 16.8
Classes of Abused Drugs

CNS stimulants	Amphetamines
	Caffeine
	Cocaine
	Nicotine
Hallucinogens	LSD
	Mescaline
	Penylclidine (PCP)
CNS depressants	Barbiturates
	Alcohol
	Marijuana
	Tranquilizers
Opiates	Heroin
	Morphine
	Codeine
	Dihydromorphine

Adapted from Smith and Asch, 1987, with permission.

of use. Sporadic use carries much less risk than chronic use. Equally importantly, the age of the user is being seen as an extremely important risk factor, particularly with respect to long-term consequences.

Most of these drugs belong to a large class of neuroactive substances that exert their effects in the CNS (Table 16.8) by altering the function of a number of neurotransmitter systems. The effects of most of these drugs are multifaceted. In some cases they disrupt the H-P-G axis by acting *centrally*, that is, by altering the GnRH pulse generator or its stimulation of the pituitary. In other cases, they may disrupt gonadal function directly. Unless drug use is chronic and long-term, the disruptive effects in otherwise healthy and reproductively mature individuals is usually transient; the effects may even be reduced by the development of tolerance to a given drug. On the other hand, chronic use of drugs during the critical years of H-P-O axis maturation in females may lead to lesions that may compromise reproductive function in an irreversible and permanent manner. Since drug use and abuse is most prevalent among the young (Table 16.9), the consequences of such abuse are potentially more serious.

Alcohol

Alcohol is one of the most commonly used drugs in the United States, and its effects on reproductive function depend on whether the alcohol intake is acute or chronic. The acute responses are directly related to the alcohol itself, while chronic effects are complicated by secondary systemic disturbances produced by long-term use of alcohol. In females, chronic use is associated with a large array of disorders, including amenorrhea, anovulation, luteal-phase dysfunction, ovarian pathology, and hyperprolactinemia. The most severe disorder, amenorrhea, is a common characteris-

Table 16.9
Drug Use by Age Group

Drug	Age group	(%)
Alcohol	12–17	31.5
	18–25	71.5
	26+	60.7
Marijuana	12–17	12.3
	18–25	21.9
	26+	6.2
Stimulants	12–17	1.8
	18–25	4.0
	26+	0.7
Cocaine	12–17	1.8
	18–25	7.7
	26+	2.1
Hallucinogens	12–17	1.1
	18–25	1.6
	26+	<0.5

Reference: National Household Survey of Drug Abuse, 1985. National Institution on Drug Abuse, Washington, D.C.

tic in alcohol-dependent women in the United States, Europe, and Japan. Most, but not all, such individuals are characterized by the low estrogen and high gonadotropin levels typically found in menopause. Other individuals exhibit essentially normal estrogen levels. This finding suggests that the cause of the amenorrhea may be different in different alcoholics. Although severe amenorrhea could be due to a suppression of the GnRH pulse generator, this has not been found to be the case. However, not all alcohol-dependent women are amenorrheic or, for that matter, infertile. Many cases of alcoholic women conceiving and giving birth have been documented, demonstrating that in some women the H-P-G axis can develop tolerance to the chronic effects of alcohol. Unfortunately, as we saw in Chapter 13, the children born to such women have a high probability of disabling physical and developmental problems.

Less severe disorders in the reproductive system have been noted in nonalcoholic women, moderate or social drinkers who are otherwise healthy and well-nourished. These disorders, including pain just prior to or during menstruation, luteal-phase dysfunction, disturbances in ovulation, lower-than-normal estrogen levels, and hyperprolactinemia, appear to be alcohol dose–dependent and of short-term duration if alcohol intake is reduced. In one particular study comparing the effects of different amounts of alcohol in well-nourished subjects, occasional or very moderate social drinkers (less than average of three drinks per day over a three-week period) did not experience any significant ovarian or menstrual disturbance. On the other hand, those who averaged four drinks per day developed significant ovarian or menstrual dysfunction.

Hyperprolactinemia is perhaps one of the most consistent features of alcohol intake, both in the alcohol-dependent female and in the moderate and social drinker. Elevated levels of prolactin often persist for long periods during sobriety, especially in alcoholics. Although high levels of prolactin are normally associated with ovarian and menstrual dysfunction, hyperprolactinemia with normal menstrual cycles has been observed in alcohol-dependent women. The mechanism of alcohol-induced hyperprolactinemia remains unclear, and the contribution of hyperprolactinemia to reproductive dysfunction is controversial.

In contrast to the multifaceted effects of alcohol in females, the primary effect of alcohol in the male is on testicular function, resulting in a fairly consistent picture of decreased libido, erectile dysfunction, and infertility depending again on whether the dose is acute or chronic. Acute doses of alcohol increase the subjective impression of sexual arousal, perhaps because alcohol reduces inhibitions, but objective measures of arousal do not always follow. Alcohol suppresses testosterone synthesis in the testis indirectly by its conversion to acetaldehyde. In addition, it increases the metabolism of testosterone by the liver. When liver function is impaired significantly, the pattern of circulating sex steroids changes, resulting in increased conversion of androgens to estrogens. As a result, hypogonadism and gynecomastia are common in alcoholic men. Erectile dysfunction is due not only to the lowering of testosterone levels, but also to the direct effects of alcohol on the penis in suppressing erections. Chronic alcohol abuse often results in more or less irreversible impotency, which can persist even after years of sobriety.

Sperm morphology and sperm count are affected by both acute and chronic alcohol use. Sperm without heads or with abnormal flagella were found in a study of nonalcoholic men who had consumed 0.4 to 0.8 gm/kg of alcohol. A higher fraction of morphologically abnormal sperm is consistently found in alcoholic men.

Although alcohol has been shown to impair both pituitary and hypothalamic function in animals, such effects have been difficult to document in humans. LH levels appear to be normal in alcoholic men, suggesting that pituitary and hypothalamic function in human males are relatively resistant to the effects of alcohol.

Tobacco

As we saw in Chapter 13, tobacco smoke has multiple and serious effects on the developing fetus. It also has a number of effects on H-P-G axis activity, generally attributed to nicotine. However, the neuroendocrine changes that accompany nicotine administration are complex and varied, and the effects of acute and chronic nicotine ingestion differ quite significantly. Chronic nicotine use, in contrast to acute, results in an increase in β-endorphin, which in turn inhibits GnRH secretion. The most consistent effects seen in habitual smokers are suppression of the H-P-G axis activity, inhibition of prolactin secretion, and moderate stimulation of the H-P-A axis.

In males, increased morphological abnormalities in sperm, decreased sperm count, decreased testosterone levels, and smaller ejaculate volumes have been reported in smokers compared with nonsmokers. The best-documented studies indicate that sperm count in smokers is 13 to 17 percent lower than in nonsmokers. In females, fertility decreases with the numbers of cigarettes smoked per day; menopause occurs at an earlier age in smoking than in nonsmoking women; menstrual irregularities appear to be more common among smokers than nonsmokers; and oocytes obtained from smoking women have significantly decreased *in vitro* fertilization capacity (see below). Although the specific mechanisms by which tobacco smoke interferes with reproductive function remain to be worked out, epidemiological studies since the 1980s have consistently shown that smoking adversely affects fertility. The most recent, methodologically sound, population-based study from eight geographic areas of the United States showed that smoking a pack of cigarettes per day leads to a significant increase in infertility in females.

Marijuana

The neurological effects of marijuana have been known for at least 2000 years. Indian ascetics used marijuana to "destroy the sexual appetite." Modern studies on the reproductive effects of marijuana are quite recent. A few controlled studies of males are available, but none of females. The principal neuroactive component of marijuana, Δ^9-tetrahydrocannabinol (THC), has been shown to inhibit the secretion of LH, FSH, and PRL in laboratory animals, including the Rhesus monkey. In all cases, the inhibition of gonadotropin levels occurs at the hypothalamus. In a Rhesus monkey treated with THC, for example, LH and FSH levels can be restored by exogenous administration of GnRH. The effects of THC, particularly in the female Rhesus monkey, tend to be long-lived, with resumption of a normal ovulatory cycle taking several months.

In human females, chronic marijuana use is associated with infertility, but the mechanism remains unknown. In both monkeys and humans, continued use leads to the development of tolerance, characterized by the return to apparently normal hormone levels in individuals who smoke marijuana regularly. The development of tolerance varies considerably from one individual to another, however. The effects of marijuana on males are more controversial because of conflicting reports. Lowering of testosterone levels has been reported in some studies, but not in others, although it is not clear what role the development of tolerance played in the levels of testosterone measured. On the other hand, decreases in sperm count appear to be a consistent finding. The effects on sperm count may not be permanent, however, since they may return to more or less normal levels 1 to 2 months after marijuana intake ceases. Marijuana use is generally considered a factor that may contribute to unexplained infertility in men. In animal studies, significant changes in sperm morphology have also been documented, although the strength of the effect depends on the extent of use.

Caffeine

No large-scale controlled studies on the effects of caffeine have been carried out. However, some observational studies suggest surprisingly that more than 2 cups of regular brewed coffee (not instant or decaffeinated) is associated with a 50 percent reduction in the chance of conception, but the infertility is not related to anovulation. However, it is not really clear that the effect is due to caffeine itself. Regular coffee contains a number of other compounds, in particular tannins, which are also found in caffeinated soft drinks. Some investigators suggest that the fertility effects may be due to the tannins, and not the caffeine.

Cocaine

Recent studies of cocaine usage indicate that 30 percent of males and 20 percent of females between the ages of 24 and 34 have used it more than once. Cocaine is a powerful CNS stimulant, exerting its effects through modulation of norepinephrine and dopamine concentrations. Part of the mythology of cocaine rests on anecdotal reports claiming that cocaine heightens sexual powers and sexual arousal temporarily, although males report this effect much more than females. If it has an effect as an aphrodisiac it is probably because, like other CNS drugs, it suppresses inhibitions. Some reports have indicated that cocaine applied topically to the male genitalia increases sexual performance, but if true, that effect is probably due to the anesthetic action of cocaine, prolonging erection and delaying ejaculation. In males, chronic use is associated with erectile dysfunction, decreased sperm count, and increased numbers of abnormal sperm. Chronic use in females is associated with amenorrhea, menstrual cycle disturbances, and 11 times greater risk of tubal infertility.

Amphetamines

The effects of amphetamines on reproductive function are not clearly defined, but the effects appear to depend on whether the usage is acute or chronic. Most studies have examined effects on males. The reported effects (these have not been well-controlled studies) for low doses are stimulation of sexual desire. On the other hand, chronic use of amphetamines is associated with loss of libido and erectile dysfunction.

Therapeutic drugs

Therapeutic drugs are prescribed for the prevention of disease or as therapy for existing disease. Many affect reproductive function. The drug classes for which we have the most information are listed in Table 16.10. Antihypertensives, prescribed to reduce blood pressure, are used by millions of people each year. It is estimated that over 35 million people in the United States are hypertensive, and about 2 million die each year from the consequences of hypertension. Since a very large fraction of hypertensives are middle-aged and old men, they are the main users of hypertensive drugs. With respect to reproductive function, men appear to be the most sensitive as well. The major effects reported are loss of libido, inability to maintain an erection, delayed or absent orgasm, and inability to ejaculate. There are several classes of antihypertensive drugs, and their effects vary. In addition, drug sensitivity varies considerably from person to person. The use of hypertensives, as we saw in Chapter 10, is an important factor in the loss of erectile function experienced by aging men.

Table 16.10
Therapeutic Drugs and Fertility

Class of drug	Effects
Antihypertensives	Decreased libido, erectile dysfunction, delayed or absent orgasm
Psychotropic and CNS drugs	
Antipsychotics	Decreased libido, erectile and ejaculatory dysfunction, hyperprolactinemia
Antidepressants	Exacerbate sexual dysfunction
Cancer chemotherapy	Amenorrhea, inhibits spermatogenesis

Reference: Forman *et al.*, 1996.

Psychotropic and CNS drugs are another large category of frequently prescribed drugs. The two major classes are the antipsychotics and the antidepressants. The effects of these drugs are not easy to measure because the illness itself may affect reproductive function. Nevertheless, decreased libido, erectile and ejaculatory dysfunction in males, and hyperprolactinemia in females have been reported with the use of antipsychotics. The effects of antidepressants are even less well documented. Anecdotal data have reported both stimulation and inhibition of sexual desire. Hence, at this stage, it seems impossible to provide definitive information about the effects of this category of drugs.

Chemotherapeutic drugs for the treatment of cancer are also widely prescribed. Their main effect is on rapidly dividing tissue. In males, the primary effect is to suppress spermatogenesis. Depending on the drug and the length of treatment, the cessation of spermatogenesis may be reversible, and sperm production may resume some months after the chemotherapy is stopped. In females, ovarian function appears to be particularly sensitive to the effects of these drugs, and very often the consequence of cancer chemotherapy is to render the female permanently infertile.

SUMMARY

Lifestyle factors have an important, but not easily quantifiable, effect on human fertility. The effects of factors such as risky sexual activity on the part of the female, intensive exercise programs, or extremes of weight have been well documented and are understood reasonably well. The effects of recreational and therapeutic drugs are less well understood. Nevertheless, there is little doubt that immoderate use of alcohol and tobacco, the two most commonly used recreational drugs, has important adverse effects on both female and male fertility. Some of these effects may be transient, but other effects may be only partially irreversible, especially for females who are heavy drug users during the maturational period of the H-P-O axis.

EXTRINSIC FACTORS— ENVIRONMENTAL TOXICANTS

As we saw in Chapter 13, reproductive toxins are responsible for spontaneous abortions, congenital abnormalities, prematurity, low birth weight, and stillbirth. Many of these toxins also interfere with ovarian and testicular function, libido, erectile function, or the ability to establish a pregnancy. These toxins may act by damaging cells or tissues; by mimicking normal hormones, either as agonists or antagonists; by interfering with cellular reactions; or they may be metabolized to toxic or even more toxic compounds.

Although there are many thousands of reproductive toxins, only a relative few have been tested in animals (Table 16.11). For obvious reasons, data on their effects on humans has come from accidental exposures, not as the result of well-controlled studies. This means that cause-and-effect relationships have been particularly difficult to establish in humans. Extrapolation from animal studies to humans, where animals have often been exposed to large doses over extended periods of time, may not be valid.

In animal studies, heavy metals (mercury and lead) have been shown to be toxic for testicular and ovarian function, resulting in suppressed spermatogenesis, altered sperm morphology, and decreased implantation rates. A few organic solvents have

Table 16.11
Environmental Toxins and Fertility

Class of toxin	Effects
Heavy metals	Inhibit spermatogenesis; affect sperm morphology; toxic in ovaries; interfere with implantation
Pesticides	Cause sterility; are anti-androgenic; cause defective internal and external male genitalia
PCBs	Cause testicular disorders; act as weak estrogens; cause defective masculinization

Reference: Forman *et al.*, 1996.

been shown to lead to testicular damage. A number of pesticides function as weak estrogens or as weak anti-androgens. Perhaps the best-documented example is DDT, which, until its use was banned in the developed countries in the late 1970s, was used in the control of malaria-causing mosquitos and other pests. It was shown in 1995 that the major breakdown metabolite of DTT, known as DDE, is an androgen-receptor antagonist. DDE interferes with the development of the internal and external male genitalia, which are dependent on androgen function.

The use of DDT has been associated with numerous examples of abnormal male genital development. One well-known study examined the significant reduction in the birth rate and penis size of alligators in several lakes in Florida. This massive effect on the alligator population was associated with a spill of DDT in the affected areas. Although DDT is no longer used in the developed world, it is still used in large amounts in many developing countries. The half-life of DDT is about 100 years, which means that because of heavy prior use, DDT remains widely dispersed in the environment. The levels considered to be anti-androgenic are 64 parts per billion (ppb). Higher levels are still found in many regions, particularly where DDT was used intensively. A number of investigators have suggested that residual DDT levels, and perhaps other pesticides with similar characteristics, may have contributed to the decrease in sperm count and the increased incidence of male reproductive disorders that have been reported in some countries. Pesticides with weak estrogenic activity have also been blamed for the decline in sperm parameters, and they could also interfere with normal male genital development. It has been suggested that another source of estrogenic compounds is the recycled synthetic estrogens from widespread use of oral contraceptives. These compounds are excreted into the urine, and because they tend to be highly resistant to breakdown, they may eventually be distributed into the water supply. Although this scenario seems far-fetched, there is preliminary data showing that these compounds have been detected in drinking water. It is clear that reproductive toxicology, the study of environmental toxicants, is still in its beginning stages.

SUMMARY

Exposure to heavy metals, industrial solvents, pesticides, and other environmental toxicants has been linked to reproductive dysfunction, including pregnancy loss, fetal loss, birth defects, and infertility. However, the causal relationships between toxicant exposure and adverse reproductive outcomes in humans are equivocal and controversial. Some studies show an association, while others do not. The reasons for the discrepancies are not known, but quite likely are due to differences in methodology, imprecision in measuring exposures accurately, differences in sample sizes, and other unknown confounding factors. More studies are needed to clarify the mechanisms by which toxicants work and to identify the important toxicants that affect human reproduction.

ASSISTED REPRODUCTIVE TECHNIQUES (ARTs)

The first ART, *in vitro* fertilization (IVF), was developed by Steptoe and Edwards in England in 1978. Since then, thousands of ARTs (both the original IVF and its modifications) have been performed. In 1994, for example, in the United States and Canada, 39,390 cycles of ARTs were carried out. In 1995, the number of ARTS reported were 59,142 cycles. The development and application of ARTs is one of the main growth industries in human reproductive biology, and today the different ARTs are the only alternatives for many cases of infertility in males and females. ARTs are technology driven, and since the original introduction of IVF, several modifications have increased the applicability of ARTs to different types of infertility.

In vitro fertilization (IVF) and embryo transfer (ET)

IVF, as the name implies, is the fertilization of human oocytes outside the body. IVF was originally developed for patients with tubal infertility that could not be corrected surgically. Conceptually, the procedure is simple:

1. Collect oocytes from the woman after artificial stimulation of follicular development.

2. Mix the oocyte with sperm collected from the husband or a donor, allowing the oocyte to be fertilized.

3. Culture the zygote in a culture dish through cleavage and the morula stage.

4. Transplant the very early embryo into the uterus of the recipient female (usually the donor of the oocytes), whose endometrium had been primed with estrogen and progesterone to prepare it for implantation (Fig. 16.1).

The probability of establishing a pregnancy increases with the number of embryos transferred: about 9 percent with one embryo, 18 percent for two, 29 percent for three, and 32 percent for four. In most programs three or four embryos are transferred. Transferring more does not significantly improve the chances of establishing a pregnancy. Moreover, multiple pregnancies carry significant risks for the mother and fetuses. All IVF procedures utilize some type of ovarian stimulation in order to obtain several oocytes for fertilization. The ovarian stimulation regimes have become quite complex, but fundamentally involve the use of FSH and LH (or hCG). The newer methods of ovarian stimulation generally include suppression of endogenous hormone production by GnRH agonists, followed by carefully controlled ovarian hyperstimulation by human FSH and LH. The use of IVF and ET has been expanded to pa-

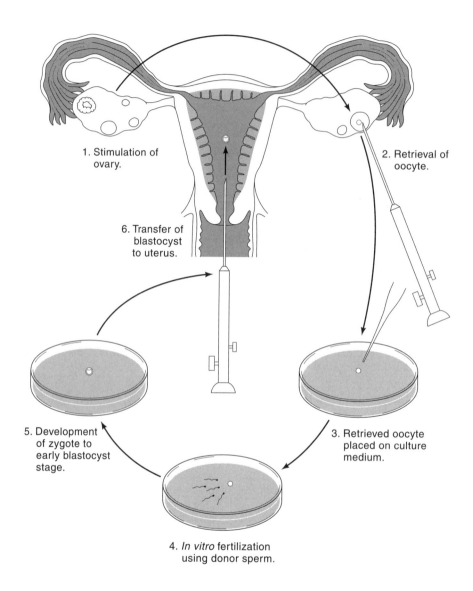

1. Stimulation of ovary.

2. Retrieval of oocyte.

6. Transfer of blastocyst to uterus.

5. Development of zygote to early blastocyst stage.

3. Retrieved oocyte placed on culture medium.

4. *In vitro* fertilization using donor sperm.

Figure 16.1
Standard IVF and ET protocol.

tients with infertility problems other than tubal infertility, such as endometriosis, some types of male infertility, and many cases of idiopathic infertility.

The success rate varies according to the skill and experience of the clinicians, and has improved since the early years. Perhaps the most important measure of success for the couple is the "baby take home rate," which in the early years was about 12–14 percent. A 1994 summary of IVF and ET procedures carried out in 249 ART clinics in the United States and Canada indicated a mean success rate of about 21 percent (deliveries

Table 16.12
Comparison of Reported Outcomes for ART Procedures

Variable	IVF	GIFT	ZIFT	ICSI
Number of procedures	36,035	3741	1078	5052
Transfers per retrieval (%)	89.7	98.5	91.7	92.5
Pregnancies	8299	1142	322	1461
Pregnancy loss (%)	18.6	21.5	18.3	18.9
Deliveries	6754	896	263	1185
Deliveries per retrieval (%)	22.3	27.0	27.9	23.5
Single births (%)	64.1	65.3	68.8	64.1
Ectopic pregnancies per transfer (%)	0.9	1.2	1.4	0.3
Birth defects per birth (%)	0.7	0.3	0.6	1.1

Data from 1995 and compiled from 281 clinics in the United States and Canada. (Adapted from ARTs in 1995, 1998, with permission.)

per oocyte retrieval), while in 1995, the overall success rate reported by 281 clinics increased to about 27 percent (Table 16.12). Of these, 63.7 percent were single births, 28.3 percent were twins, 5.9 percent were triplets, and 0.6 percent were higher-order multiple births. 1.2 percent of the pregnancies were ectopic, and 2.7 percent of the births had birth defects.

It has generally been the case that more embryos are generated than are implanted. The extra embryos are saved by freezing them at very low temperatures. Until recently, clinicians had hesitated to use the frozen embryos for fear that freezing or thawing might produce serious defects in the embryos. However, the first successful pregnancy generated from a thawed-out embryo has now been reported. It is not yet clear whether this procedure will have general applicability, but it is a measure of how far we have come in improving ARTs.

Other ARTs

Modifications of the original IVF and ET protocols have been introduced in attempts to simplify the procedure, to improve the success rate, and to extend it to patients for whom standard IVF procedures were inadequate. One of the earliest successful modifications of IVF was **gamete intrafallopian transfer** (GIFT). GIFT uses IVF protocols for ovarian stimulation and oocyte retrieval, but it differs in that fertilization is accomplished by transferring both the oocytes and spermatozoa directly into the uterine tubes. A potential advantage is that fertilization takes place in a more "normal" environment, that is, in the uterine tube, rather than in a laboratory dish. However, uterine tube function is required. Hence, GIFT is not indicated in cases of tubal absence or tubal defects. In general, greater skill on the part of the clinician is required also. The success rate for GIFT procedures is a little higher than in the standard IVF and ET (28.5 percent versus 21.1 percent) (Table 16.12). **Zygote intrafallopian transfer** (ZIFT) differs from standard IVF and ET protocols in that embryo transfer is carried out earlier, at the pronuclear stage, shortly after fertilization when the male and female pronuclei are still separate. Instead of waiting for the cleavage divisions to take place in a culture dish, the newly formed zygote is transferred so that these divisions

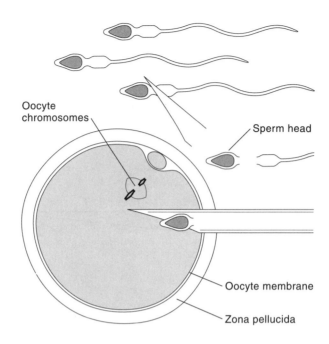

Figure 16.2
ICSI procedure. Removal of the sperm tail by an injection pipette, and injection of an isolated sperm head into the oocyte.

take place in the uterine tube, as happens normally. Success rates with ZIFT are similar to those with GIFT (Table 16.12).

The newest ART procedure is known as **intracytoplasmic sperm injection** (ICSI), first applied to humans in 1992, after first being tested in animals since 1966. The three ART procedures described above require that sperm be able to penetrate the zona pellucida in order to fertilize the oocyte. Perhaps for this reason they have not been very successful in cases where the sperm count was extremely low or when sperm quality was poor. ICSI involves the introduction of the sperm nuclei or intact spermatozoon directly into the oocyte (Figs. 16.2 and 16.3). ICSI, therefore, dispenses with the normal route of fertilization. The success rate of ICSI in treating couples with male infertility for whom the other ARTs were not possible is comparable to that of IVF techniques in couples without male infertility. The major advantage of ICSI is that it provides a treatment alternative for the types of male infertility that constitute the majority and for which no other therapy was available. The success in these cases suggests that ICSI can be applied with equal success to couples without male infertility.

New frontiers

The ARTs described above have, despite relatively low success rates, enabled many couples to have children. However, they are not the ideal solution to the problem of female and male infertility. With females, the biggest problem and limiting factor is the scarcity of fertilizable eggs. ARTs have generated a huge demand for eggs. Unlike

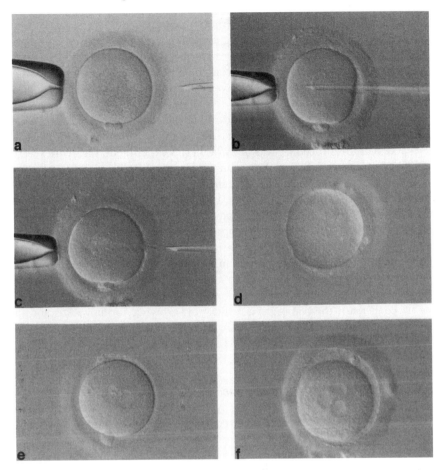

Figure 16.3
Photomicrographs of the ICSI protocol. (a) An oocyte after removal of its corona cells is held on a holding pipette with its first polar body at the 6 o'clock position. The injection pipette containing the spermatozoon is coming from the right. (b) The injection pipette has penetrated deep into the oocyte, after which the spermatozoon is expelled from it. (c) Removal of the injection pipette leaving the spermatozoon behind. (d) Oocyte with one pronucleus and two polar bodies. (e) The two pronuclei have moved together in preparation for fusion. The second polar body is at the 1 o'clock position on the edge of the zygote, and the first polar body is at 6 o'clock. (f) An unusual case, a zygote with three pronuclei. The third pronucleus probably arose after the second polar body was retained in the oocyte, so that two female and one male pronuclei were formed. The first polar body is at 1 o'clock. (Reprinted from A. Van Steirteghem *et al.* 1996. The development of ICSI. *Human Reproduction* **11** [**Suppl. 1**], 59–72, by permission of Oxford University Press.)

sperm, which are available in dozens of sperm banks around the country and for which there is an essentially limitless supply, human eggs are an extremely rare commodity. Indeed, the human egg is the rarest cell in the body. A recent advertisement in a student newspaper offered $50,000 for the egg of a "tall, at least 5 feet 10 inches, athletic coed with no family history of serious disease and with SAT scores of at least 1400." In the United States, a number of egg donor programs have been established. The donors selected for these programs can earn several thousand dollars for their participation, but they must also submit themselves to demanding ovarian hormone stimulation protocols lasting 1 to 4 months. The protocols are not entirely without risk. Problems reported are nausea, headaches, bloating, production of ovarian cysts, and possibly even an increase in the risk of ovarian cancer. Even if the procedure is successful, only a few eggs can be harvested.

Two alternative solutions to the egg scarcity problem are being explored. The furthest along in demonstrating its potential is ovarian tissue transplantation. The first reported success of this procedure was in 1994. In this experiment, the ovaries were removed from young ewes. Strips of ovarian tissue containing the cortex of the ovary, which is rich in primordial follicles, were removed from the extirpated ovaries, frozen, and then later transplanted into the same animal. The ewes afterwards were able to resume ovulatory cycles; some of them mated, and gave birth. The success of these initial experiments has suggested that this approach may be successful in humans as well. Experiments in transplanting human ovarian tissue into mice have given promising results. A human experiment was begun in England in 1995. Ovarian tissue was removed from a 3-year-old girl who was to undergo radiation therapy for cancer. The radiation would render her permanently sterile. The therapy was successful in eliminating the cancer, and the young girl is healthy and growing normally. At some future date, if the young woman so desires, her frozen ovarian tissue, stored for many years, could be reimplanted into her ovaries. The chances for success seem reasonable given the success with animals.

Transplanting ovarian tissue from one person to another would be much more difficult because the normal immune reaction against foreign tissue would have to suppressed. Immunosuppressive methods are available, but they are not completely effective, and they carry their own risk. However, if effective immunosuppressive therapies can be developed, transplantation between different individuals would be possible. Moreover, ovarian tissue could be derived from persons who have just died or even from fetal ovaries. This latter possibility has been carried out in mice, with the fertility of adult mice being restored by transplanted mouse fetal ovarian tissue.

The second alternative to egg scarcity is to generate them *in vitro*, that is, to develop an artificial ovary, so that primordial follicles can be developed into mature ones in a test tube. This is clearly a formidable challenge, but some initial steps are being taken in exploratory studies in the mouse. The most difficult problem is finding incubation conditions that will keep the ovary functional long enough for the follicles to develop. The signals that guide follicular development are poorly understood; hence, it has not been possible to generate fertilizable eggs from primordial follicles. Some success has been reported with tertiary follicles, but even there the success rate is very low. Despite the failures, optimists suggest that within 20 years it may be possible to generate eggs from human primordial follicles. The possibilities that can be imagined are so far-reaching that we may find them repellent. For example, an aborted female fetus with its millions of primordial follicles could become the biological mother of literally thousands of children. This capability would be useful in preserving a species that was on the brink of extinction, but it is much more difficult to imagine acceptable applications for humans. A much more complex world may be just around the corner. Our problem is that we have not yet assimilated the current ARTs, and soon we may be asked to think about developments for which there is no precedent in human society.

SUMMARY

The success of the different ARTs has generated a number of important questions. Debate on these issues continues, but no general consensus exists yet about most of the issues raised. The following is a list of some of the concerns and questions most commonly mentioned.

1. ARTs are expensive, are still considered experimental or elective procedures, and therefore are not covered by many health insurance plans. Their availability is limited primarily to those able to afford them.

2. The success rate is still somewhat low. How many times should the process be repeated before giving up, especially if it is covered by insurance?

3. It is possible, although again with low success rates, to have women in their 40s or even in their 50s conceive and bear children. At what maternal age should ARTs not be performed at all?

4. In cases where the female that carries the baby has not contributed the oocyte, what rights does the donor female have?

5. What rights does a surrogate mother (the female that contributes the uterus, but not the oocytes) have?

6. Often more fertilizations are carried out than embryo transfers are made. The unused embryos are frozen and stored. How long should they be kept? Who owns them? Do the embryos have any rights? Who retains the rights over the embryos if the couple dies?

7. By making ARTs more widely available, are we propagating mutations in the population that should be maintained at the lowest possible levels? This question has been raised most recently about the use of ICSI in certain types of male infertility that are known to be due to mutations in genes on the Y chromosome. Since the Y chromosome is passed from father to son, ICSI makes possible the transmission of the defective Y chromosomes to the sons. Under normal circumstances, men who carried such mutations would not have children, and their mutations would die with them. However, if ICSI were carried out in a completely unrestricted way, such mutations, relatively rare in the population, would be propagated more widely. Hence, more males would carry the mutations, and the frequency of male infertility would increase.

QUESTIONS

1. Compare the contributions of infertility and infecundity to the age-related decrease in reproductive potential in human females.

2. Female infertility appears in general to be more susceptible to treatment than male infertility. Why do you think this is the case?

3. What are some of the diverse symptoms of PCOS and endometriosis?

4. What therapies are available for treating female infertility?

5. Draw up a list of fertility disorders and the types of ARTs that could be used in their treatment.

6. What stand do you take with regard to ARTs that permit the transmission of infertility from parents to progeny?

7. What arguments can you offer *pro* and *con* the use of ARTs for women and men over 50 years old?

8. Library project:
   a. What therapies are available for treating endometriosis, and how successful are they?
   b. Describe the latest advances in generating Graafian-stage follicles *in vitro*.
   c. What are potential risks in undergoing ovarian stimulation for egg donation?

SUPPLEMENTARY READING

Assisted reproductive technology in the United States and Canada: 1995 results generated from the American Society for Reproductive Medicine/Society for Assisted Reproductive Technology Registry. 1998. *Fertility and Sterility* 69(3), 389–396.

Bradley, K. A., *et al.* 1998. Medical risks for women who drink alcohol. *Journal of General Internal Medicine* 13(9), 627–639.

Carson, D. S., and K. K. Bucci. 1998. Infertility in women: An update. *Journal of the American Pharmaceutical Association* 38(4), 480–488.

Forman, R., S. Gilmour-White, and N. Forman. 1996. *Drug-Induced Infertility and Sexual Dysfunction*. Cambridge University Press, Cambridge, UK.

Forti, G., and C. Krausz. 1998. Clinical review 100: Evaluation and treatment of the infertile couple. *Journal of Clinical Endocrinology and Metabolism* 83(12), 4177–4188.

Green, B. B., N. S. Weiss, and J. R. Daling. 1988. Risk of ovulatory infertility in relation to body weight. *Fertility and Sterility* 50, 721–726.

Guzick, D. S. 1996. Human infertility: An introduction. In *Reproductive Endocrinology, Surgery, and Technology*, Vol. 2, Chapter 101. E. Y. Adashi, J. A. Rock, and S. Rosenwaks, Eds. Lippincott-Raven Publishers, Philadelphia, PA.

Ingamells, S., and E. J. Thomas. 1998. Endometriosis: A continuing enigma? *Hospital Medicine* 59(6), 437–441.

Male infertility update. 1998. The ESHRE Capri Workshop Group. *Human Reproduction* 13(7), 2025–2032.

Travis, J. 1997. Brave new egg. *Discover* (April 1997), 76–80.

ADVANCED TOPICS

Baker, H. W. G. 1994. Male infertility. *Endocrine and Metabolic Clinics of North America* 23(4), 783–793.

Chuang, A. T., and S. S. Howards. 1998. Male infertility: Evaluation and nonsurgical therapy. *Urologic Clinics of North America* 25(4), 703–713.

Chuttinguis, S., M. R. Forman, H. W. Berendes, and L. Isotalo. 1992. Delayed childbearing and risk of adverse perinatal outcome. *Journal of the American Medical Association* 268, 7–15.

Falcone, T. 1999. New technology and new challenges for assisted reproduction (clinical conference). *Cleveland Clinic Journal of Medicine* 66(2), 78–82.

Healy, D. L., A. O. Trounson, and A. N. Andersen. 1994. Female infertility: Causes and treatment. *New England Journal of Medicine* 343, 1539–1544.

Kiedrich, K., and R. Felberbaum. 1998. New approaches to ovarian stimulation. *Human Reproduction* 13 Suppl. 3, 1–13.

Laurent, S. L., *et al.* 1992. An epidemiologic study of smoking and primary infertility in women. *Fertility and Sterility* 57, 565–572.

Marantides, D. 1997. Management of polycystic ovary syndrome. *Nurse Practitioner* 22(12), 34–38.

Menken, J., J. Trussell, and U. Larsen. 1986. Age and infertility. *Science* **233**, 1389–1394.

Palermo, G. D., J. Cohen, and Z. Rosenwaks. 1996. Intracytoplasmic sperm injection: A powerful tool to overcome fertilization failure. *Fertility and Sterility* **65**(5), 899–908.

Sharara, F. I., D. B. Seifer, and J. A. Flaws. 1998. Environmental toxicants and female reproduction. *Fertility and Sterility* **70**(4), 613–622.

Spira, A. 1986. Epidemiology of human reproduction. *Human Reproduction* **1**(2), 111–115.

French army poster from World War I warning soldiers of the dangers of STDs. The text was written on a tombstone and appeals to soldiers to "resist the seductions of the street" that will lead to a disease "as dangerous as war" and leading to a "useless death without honor." (S. D. Gilman. 1989. *Sexuality: An Illustrated History Representing the Sexual in Medicine and Culture*. Plate 315. John Wiley and Sons, Inc., Chichester, NY, with permission.)

CHAPTER 17

Sexually Transmitted Diseases

Venereal disease is bad like other disease,
but especially because it corrupts sex.

—T. Rosebury. 1971

Microbes and Morals.
Viking Press, New York, NY, p. xvii.

Who made the world I cannot tell;
'Tis made, and here I am in hell.
My hand, though now my knuckles bleed,
I never soiled with such a deed.

—A. E. Housman. 1936

No. XIX in *More Poems.* A. A. Knopf, New York, NY.

SEXUALLY TRANSMITTED DISEASES (STDs), unlike other infectious diseases, have had a twilight existence in the public consciousness. Whereas information about other infectious diseases is disseminated easily and without controversy, information about STDs has tended to be significantly less public, at least in the United States. The connection of STDs with what is seen as illicit, illegal, or promiscuous sexual behavior has certainly been the reason for this state of affairs. From a scientific and medical point of view, however, STDs have not been neglected. The study of STDs has been an active part of medical microbiology, and it has provided us with important insights into the nature of the microorganisms that cause disease, the way they evolve, their virulence, and pathogenicity. STDs have also provided the medical community and public health officials with important challenges in designing therapies and ways of controlling the spread of STDs. Our objective in this chapter will be to review the nature and the major characteristics of STDs and their often terrible consequences, and to describe the generally silent STD epidemic experienced by the United States and most parts of the world. If the old adage that "knowledge is power" has any validity, it certainly applies to knowledge of STDs, for only with widely shared knowledge will STDs be contained.

THE NATURE OF STDs

STDs were formerly known as *venereal* (derived from Venus, the goddess of love) diseases. The classical venereal diseases, **syphilis** and **gonorrhea**, were first described clearly in the sixteenth century. However, descriptions of diseases or conditions with symptoms similar to those of syphilis and gonorrhea appear in ancient Egyptian and Chinese writings, as well as in the Old Testament and Greek and Roman documents. In the Old Testament, for example, discharges from the genitals in both men and women, which could represent gonorrhea, are described in Leviticus 15, and rules for isolation of the individual and prevention of contamination of other individuals are set down. Although none of these descriptions can be considered definitive with respect to a known STD, they do suggest that some forms of STDs were probably present in those societies.

During the Middle Ages, numerous descriptions of discharges from the penis and vagina appear in writings of the time, in many cases clearly connected to prostitution. A public edict in Avignon in southern France that appeared in 1347 stated that

> *a surgeon appointed by the authorities examine every Saturday all the whores in the house of prostitution and if one is found, who has contracted a disease from coitus, she shall be separated from the rest and live apart in order that she may not distribute her favours and may thus be prevented from conveying the disease to the youth.*[1]

From the surviving descriptions of the disease it seems probable that the disease being described was gonorrhea. Indeed, the term "clap," coined in 1378 to define the "*certain inward heat and excoriation of the urethra,*"[2] is still in popular use today for gonorrhea. In contrast, according to many authorities, no definitive descriptions of syphilis have been found in ancient or medieval literature. Moreover, examination of skeletal remains from burial sites in Europe and the Middle East has not uncovered evidence of the lesions produced by the late stage of syphilis. Hence, many investigators believe that syphilis did not exist in Europe or the Middle East prior to 1493.

The first clear descriptions of syphilis appear in a dramatic fashion in 1495 during the siege of Naples by the mercenary army of Charles VIII of France. The army eventually had to withdraw because so many of the soldiers had become afflicted with what was described as the "pox." The new pox spread quickly through Europe, to the Americas, and other parts of the world. The new pox was given the name *syphilis* in 1530, but the term did not become common in medical usage until the middle of the nineteenth century. The origin of syphilis remains unclear. One hypothesis, proposed soon after the epidemic started, suggested that syphilis originated in the Americas and was brought back to Europe in 1493 by Spanish sailors who had voyaged with Columbus. However, no compelling evidence of syphilis in pre-Columbian America has been produced. The more likely possibility is that the organism responsible for syphilis, *Treponema pallidum*, had existed in Europe in a less virulent form, probably for hundreds of years. It may have been brought to Europe from the Middle East and ultimately from sub-Saharan Africa, where other treponemal diseases, such as *yaws*, had been common since antiquity. It may be that the organism acquired a new virulence as it became

---

[1]Cited in M. A. Waugh. 1990. History of clinical developments in sexually transmitted diseases. In *Sexually Transmitted Diseases*, 2nd ed. K. K. Holmes *et al.*, Eds. McGraw-Hill, Inc., New York, NY.
   [2]Ibid.

adapted to transmission by sexual contact. Whatever its origins, its study and the search for treatment changed the face of medicine and public health in fundamental ways.

The sudden emergence of syphilis in the sixteenth century is an instructive lesson for us because of the unexpected emergence of new types of STDs during the middle and latter part of the twentieth century. Genital herpes simplex virus and human papillomavirus infections began to be recognized in the 1960s, and eight new STDs have been identified since 1980, the best known being HIV, the virus that causes AIDS. The term *STD* denotes a disease caused by any of more than 25 infectious organisms that are transmitted primarily through sexual activity. Sexually transmitted pathogens are continuously changing and evolving, and we can therefore expect that new STDs will emerge and possibly become established. This is facilitated by increasing global travel, making possible the transfer of potentially pathogenic organisms from one ecosystem to another. New detection techniques have permitted the identification of previously unknown pathogenic organisms, which has facilitated their connection to diseases of previously unknown cause.

Modes of transmission

The main route of infection of STDs is person to person by sexual contact. Transmission occurs most efficiently by anal and vaginal intercourse, and less efficiently by oral intercourse. The likelihood of transmission increases with the number of sexual encounters with different partners. Other routes of infection are also possible under certain circumstances, however. Nonsexual, person-to-person transmission can also occur with some organisms. **Parenteral transmission** (blood to blood) is seen among drug users through sharing of contaminated needles. The relative contribution of sexual versus parenteral transmission varies depending on the pathogen, the risk behavior of the population, and other factors. Finally, transmission can occur from mother to fetus or mother to infant. A pregnant woman with an STD can transmit the infection to the fetus before or during birth.

Classes of STDs

STDs can be classified by the type of causative microorganism and also by the type of disease that the microorganism produces. *Bacteria, viruses,* and *protozoa* are the agents responsible for STDs (Table 17.1). Bacterial and protozoal STDs produce inflammatory responses, which are generally characterized either by discharges from the vagina, cervix, or urethra, or they produce ulcerative responses, in which ulcerations of the external and internal genitalia are the chief diagnostic feature. Syphilis, for example, produces genital ulcers (this is why it was known as a pox for a long time), but, in contrast to gonorrhea and chlamydia, it does not produce genital discharges. Viral STDs, except for rectal bleeding and discharge produced by herpes simplex virus (HSV), rarely produce genital discharges. Some STDs may remain localized to the site of infection, but others can progress to systemic and highly lethal infections (particularly syphilis, HIV, and hepatitis B). Bacterial and protozoal STDs can generally be cured with an appropriate antibiotic regimen. On the other hand, no curative therapy is available for viral STDs. Treatment for viral STDs is palliative, meaning that the therapy relieves the symptoms or may retard the progression of the disease, but does not eliminate the virus. Viral infections can become latent with no outward manifestations of disease, but may reappear at later times.

Table 17.1
Common STDs, Causative Agents, and Primary Features

STD	Causative agent	Primary features
Bacterial		
Syphilis	*Treponema pallidum*	Painless ulcers; rash; fever; pain; swollen lymph nodes
Gonorrhea	*Nesseria gonorrohoeae*	Discharge (vaginal, urethral); rectal bleeding; fever; pain; lower abdominal pain (females)
Chlamydia Trachoma	*Chlamydia trachomatis*	Discharge (vaginal, urethral); rectal bleeding/discharge; lower abdominal pain (females)
Lymphogranuloma venereum (LGV)	*Chlamydia trachomatis*	Blisters; fever; pain; swollen lymph nodes
Chancroid	*Haemophilus ducreyi*	Painful ulcers; swollen lymph nodes
Protozoan		
Trichomoniasis	*Trichomonas vaginalis*	Discharge (vaginal, urethral)
Viral		
Genital herpes	HSV-1 and HSV-2	Painful ulcers; rectal bleeding and discharge; swollen lymph nodes
Human papillomavirus	Several types	Growths (warts, tumors)
Hepatitis B	Hepatitis B virus	Rash; fever; pain
AIDS	HIV-1 and HIV-2	Rash; fever; pain; swollen lymph nodes; destruction of immune system

Adapted with permission from *Research Issues in Human Behavior and STDs in the AIDS Era.* Copyright 1991 by the American Society for Microbiology.

The difference in curative potential for the different STDs means that strategies for long-term control also differ for the different types of STDs. In the long run, prevention of initial infection is the most effective way to control viral STDs, and this requires the development of vaccines. Presently, the only effective vaccine available is for the hepatitis B virus. Surveillance of viral STDs must be of longer term because of the persistent and latent nature of viral infections. Programs to modify behavior to reduce risk of transmission may be very valuable if designed properly, as has been shown by the HIV/AIDS public education programs. However, such programs need to be ongoing. The control of bacterial and protozoal STDs would also be improved by public education programs. Until recently, the need was not felt to be as acute precisely because curative therapies were available. The emergence of antibiotic-resistant varieties of different STDs signals the importance of public education programs that stress prevention as the most important strategy for controlling the spread of STDs.

SUMMARY

Sexually transmitted diseases (STDs) are highly infectious diseases transmitted primarily by person-to-person sexual contact. STDs can also be transmitted by person-to-person nonsexual contact, parenterally (blood to blood), and by mother to fetus or mother to infant. Bacteria, viruses, and protozoa are the agents responsible for STDs. These agents infect the internal and external genitalia. Bacterial and protozoal STDs are generally characterized by inflammatory responses of the infected tissues, while ulcerative responses are typical of viral STDs. Viral STDs cannot be cured, and the virus can become latent, but may reappear at later times. Bacterial and protozoan STDs can generally be cured by antibiotic therapy.

BACTERIAL STDs

Syphilis

Syphilis is a bacterial STD that produces genital ulcers. Compared to other STDs, its incidence is relatively low (Table 17.2). The incidence of syphilis declined all during the 1970s and 1980s, then increased significantly during the early 1990s; however, it may be decreasing once again. The increase during the early 1990s is attributed to the reemergence of syphilis among users of illegal drugs (especially crack cocaine) and their sex partners. Because of the long latent period characteristic of syphilis infections, its prevalence is not known with any certainty.

Table 17.2
Estimated Annual Incidence and Prevalence of Common STDs in the United States, 1994

STD	Incidence	Prevalence
Chlamydia	4,000,000	Not available
Gonorrhea	800,000	Not available
Syphilis	101,000	Not available
Papillomavirus	500,000–1,000,000	24,000,000
Genital herpes	200,000–500,000	31,000,000
Hepatitis B (sexually transmitted)	50,000–100,000	Not available
HIV	Not available	630,000–897,000
AIDS	79,897	Not available
Trichomoniasis	3,000,000	Not available
Pelvic inflammatory disease	>1,000,000	Not available

Adapted with permission from *The Hidden Epidemic: Confronting STDs*. Copyright 1997 by the National Academy of Sciences. Courtesy of the National Academy Press, Washington, D.C.

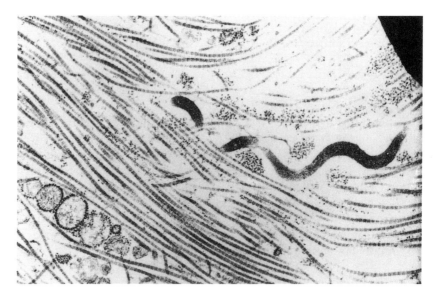

Figure 17.1
Electron micrograph of *Treponema pallidum*, the causative agent of syphilis, in human tissue. (Reprinted from S. A. Morse *et al.* [Eds.]. *Atlas of Sexually Transmitted Diseases and AIDS*. 1996. Mosby-Wolfe, London, UK, by permission of the publisher.)

Syphilis is caused by the bacterial spirochete *Treponema pallidum*, which enters the body through a break in the skin or mucous membrane (Fig. 17.1). Humans are apparently the only natural host for this organism. The incubation period varies from 10 days to 10 weeks, with an average of 3 weeks. A red spot on the skin at the site of infection appears some time after infection, evolves into a bump, and eventually ulcerates, forming what is called a *chancre*. In heterosexual men, the chancre generally forms on the penis, while chancres in the anus are common in homosexual men. In women, chancres form in the external genitalia or in the cervix. Chancres can form in the mouth and on the breast as well. The chancre is painless and disappears after 3 to 6 weeks. The chancre stage is referred to as the primary stage of syphilis.

If untreated, the *T. pallidum* organisms disperse from the original site of infection to almost every organ in the body, becoming a systemic infection. Its progression, however, is relatively slow and may produce few obvious symptoms. During this period the infected individual is infective. In some cases, this second stage of syphilis can manifest itself with fatigue, weight loss, headache, muscle ache, and fever. In addition, wartlike growths may appear in the external genitalia, the groin, and around the anus. These symptoms can persist for several weeks or months, but eventually they disappear as well. At this time, the disease enters a prolonged latent phase with no symptoms, but it can be detected by appropriate laboratory tests. The secondary stage of syphilis generally appears when the infected individual has not received antibiotic therapy during the primary stage. The infected individual is highly infectious during the first 2 to 4 years after infection. Antibiotic therapy is also generally effective during the secondary stage.

The lethal consequences of syphilis follow from the complications of the tertiary stage of syphilis, which may begin 15 to 30 years after the initial infection. Cardiovascular and neurological damage leading to heart attack, stroke, dementia, and spinal cord damage are the usual features of this stage. Syphilis in women can have drastic consequences for the fetus, depending on the stage of syphilis. Pregnancy occurring during the primary or secondary stage of syphilis results in a child that has syphilis, is premature, or is stillborn. Inexplicably, pregnancy during the tertiary stage may result in a healthy newborn.

Gonorrhea

Gonorrhea is caused by the bacterial species *Nesseria gonorrhoeae*, which appears to be highly adapted to humans since it does not infect other primates or other animal species (Fig. 17.2). *N. gonorrhoeae* is highly sensitive to dryness and survives only in the host. The primary route of transmission is through sexual intercourse. It is estimated that the chance of a woman becoming infected by a man who has gonorrhea is between 50 and 90 percent after one act of coitus. The chance of a man being infected by a woman after a single exposure is about 20 percent. In both cases, the likelihood of infection rapidly increases with multiple exposures.

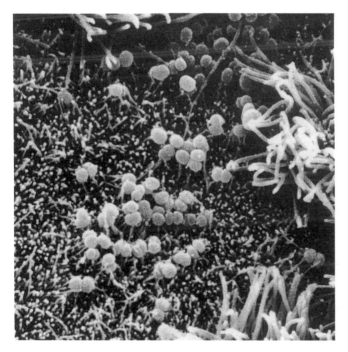

Figure 17.2
Scanning electron micrograph of *Nesseria gonorrhoeae* (seen as small spheres) attaching to the microvilli of cultured human uterine tube tissue. (Reprinted from S. A. Morse *et al.* [Eds.]. *Atlas of Sexually Transmitted Diseases and AIDS*. 1996. Mosby-Wolfe, London, UK, by permission of the publisher.)

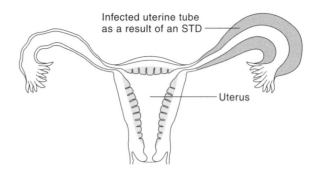

Figure 17.3
Pelvic inflammatory disease (PID). Infected uterine tube as a consequence of a bacterial STD.

About 800,000 new cases of gonorrheal infections were reported in 1994 (Table 17.2). Although gonorrhea has been considered easy to treat with penicillin or tetracycline, several new strains of *N. gonorrhoeae* have been appearing in the last few years that are resistant to the older antibiotic therapy. In 1976, all gonorrheal infections could be cured easily with penicillin, but in 1987, 2 percent were resistant, and in 1994, 30 percent were resistant not only to penicillin but to tetracycline, the two traditional antibiotics used to treat gonorrhea. Resistance of gonorrheal infections to the newer generation of antibiotics has been documented in Southeast Asia, the Western Pacific, and several states in the United States. For these new strains, newer and much more expensive antibiotics are required for effective therapy. Successful antibiotic treatment does not prevent infections at a later date, and in fact reinfections are quite common.

In men, the usual symptom of gonorrhea is *urethritis*, or inflammation of the urethra, which results in painful urination and discharges from the penis. The rectum can also be infected, resulting in painful defecation and spontaneous rectal discharges. In women, usual sites of infection are the cervix, uterus, and uterine tubes, conditions referred to as pelvic inflammatory disease (PID) (Fig. 17.3). The symptoms are painful urination, bleeding, fever, and lower abdominal pain. Gonorrhea is an important cause of PID. Interestingly, some 50 percent of infected women are asymptomatic, while only 1 to 3 percent of infected men are.

Gonorrhea can seriously affect the fertility of a woman, even if successfully treated with antibiotics. Irreversible scarring of the uterine tubes may have occurred before antibiotic therapy, resulting in infertility or increasing significantly the likelihood of having an ectopic pregnancy. Another serious complication is mother-to-infant transmission during birth, which often results in gonococcal eye infections that can cause blindness. Spontaneous abortions, prematurity, and other complications of pregnancy have been associated with gonorrhea. In men, the major complication of untreated gonorrhea is epididymitis, resulting in severe pain in the scrotum.

Chlamydia

Four million cases of chlamydial infections are reported annually (of which 2.6 million occur in women), making it the most common bacterial STD in the United States (Table 17.2). Transmission almost invariably occurs person to person through

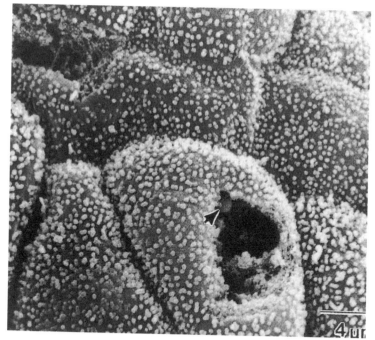

Figure 17.4
Scanning electron micrograph of human uterine tube tissue in culture infected with *Chlamydia trachomatis*. Note the large hole where a cell has been ruptured. The arrow points to a *C. trachomatis* cell that remains within the interior of the human cell. (Reprinted from S. A. Morse *et al.* [Eds.]. *Atlas of Sexually Transmitted Diseases and AIDS.* 1996. Mosby-Wolfe, London, UK, by permission of the publisher.)

sexual intercourse, but can also take place at birth, or by hand-to-eye transfer of genital secretions. It is estimated that two-thirds of women and one-third of men are infected by a single encounter with an infected partner. A large fraction of these infections are asymptomatic, perhaps as high as 85 percent in women and about 40 percent in men. The major sites of infection are the external genitalia, particularly the urethra in males and vagina in females. Much more problematic are infections in the cervix, uterus, and uterine tubes, many of which remain unrecognized and untreated. Such infections contribute significantly to the more than 1 million cases of PID reported each year. Detection of chlamydial infections is not as straightforward as with other STDs, but early and uncomplicated infections are very responsive to antibiotic therapy.

The causative agent is *Chlamydia trachomatis*, an intracellular parasitic bacterium that infects and multiplies within the cells that it infects (Fig. 17.4). Two forms of *C. trachomatis*, known as **biovars**, infect humans. These are **trachoma** and **lymphogranuloma venereum** (LGV). LGV infections are uncommon in the United States, but are widespread in Asia and Africa. The trachoma biovar favors the same sites as *N. gonorrhoeae*, but in general, the symptoms are milder. Indeed, a large fraction of infected women are asymptomatic. The most common symptom is a burning sensation while urinating, symptoms that appear 7 to 10 days after infection. Vaginal or urethral discharge may also be present. Rectal infections are commonly found in homosexual men. An important complication for women is PID, which may go unnoticed in many cases. Indeed, the first indication of a chlamydial infection is difficulty in getting pregnant or an ectopic pregnancy. Premature birth is a problem common in women who get infected with trachoma after they get pregnant. Some 70 percent of infants born to infected women are infected at the time of birth. Pneumonia and eye problems are very common in such infants.

Infection with the LGV biovar follows a different course. A small blister forms a few days after infection, usually on the penis or in the vagina. Because of its small size and lack of pain, the initial lesion may go unnoticed. The second stage of the infection begins with the infection of the lymph nodes in the groin or deep inside in the pelvis. In men, the swollen nodes eventually rupture, releasing pus. Flulike symptoms accompany this stage—fever, chills, nausea, headache, and muscular and joint pain. In women, the breakdown of the nodes may result in abscesses in the pelvis, the rectum, or even in massive enlargement of the external genitalia due to blockage of the lymph vessels.

Simple and reliable diagnostic tests for chlamydial infections have only recently become available. The less aggressive nature of chlamydial infections, as well as the lack of appropriately sensitive and relatively inexpensive detection methods, has meant that many infections go undetected, making control efforts more difficult and less effective, and the consequences of infection difficult to prevent. Antibiotic therapy is available for both types of infections, but the treatment program is more complicated than with gonorrhea. Multiple-dose antibiotic regimens are normally required, and they must be followed exactly for the infection to be cured. Reinfections are also very frequent. *C. trachomatis*, like *N. gonorrhoeae*, is very sensitive to dryness and does not survive very long outside a host.

Chancroid

Chancroid, characterized by painful genital ulcers, is caused by the bacterium *Haemophilus ducreyi*, first identified in 1900 (Table 17.1). Chancroid is common in parts of Africa, Asia, and Latin America, but very uncommon in the United States. Five states, Florida, Georgia, Louisiana, New York, and Texas, account for over 80 percent of the reported chancroid cases. Less than 1000 cases of chancroid were reported in the United States in 1994. The majority of cases occur in males.

Chancroid ulcers are often difficult to distinguish from syphilis and genital herpes ulcers, and definitive identification requires culturing the organism extracted from the ulcer. The incubation period is usually 4–10 days, although longer incubation periods are not uncommon. The lesion begins as a soft pustule, which progresses into a deep, painful ulcer. The penis and vulva are the most common sites of infection. Penicillin and sulfonamide therapy have been the traditional treatment method. However, during the past decade, antibiotic resistant *H. ducreyi* strains have appeared. Proper treatment requires the use of newer and more expensive antibiotics.

SUMMARY

Chlamydia is the most common bacterial STD in the United States. Most chlamydial infections are asymptomatic and difficult to diagnose definitively. Chlamydial infections account for a large fraction of cases of pelvic inflammatory disease (PID), ectopic pregnancies, and infertility. Gonorrhea is much less common than chlamydia, but gonorrheal infections also contribute to PID and infertility in women. The appearance of new gonorrheas resistant to traditional antibiotics in Southeast Asia and in several states in the United States has stimulated the development of new antibiotics to treat such infections. Syphilitic infections proceed through three stages, the first two of which are susceptible to antibiotic therapy. Tertiary-stage syphilis begins 15 to 30 years after the initial infection and is characterized by severe cardiovascular and neurological damage. Chancroid, characterized by painful genital ulcers, is relatively uncommon in the United States.

PROTOZOAN STDs

Protozoa are free-living unicellular eukaryotes. Many different species can infect humans, and many of these are intestinal parasites. Very few species are considered to be transmitted sexually. Perhaps for that reason protozoan STDs are generally less well known than bacterial and viral STDs.

Trichomoniasis

The most common protozoal STD is **trichomoniasis**, caused by the unicellular parasite *Trichomonas vaginalis*, which most frequently infects the vagina, urethra, and associated genital tissue (Fig. 17.5). It is an exclusive parasite of humans, and sexual

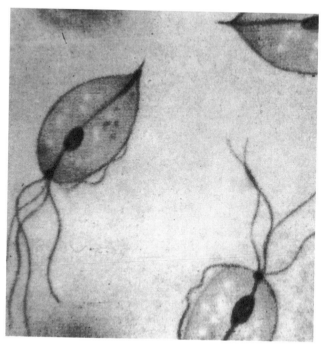

Figure 17.5
Trichomonas vaginalis, the causative agent for trichomoniasis. (Reprinted from S. A. Morse *et al.* [Eds.]. *Atlas of Sexually Transmitted Diseases and AIDS.* 1996. Mosby-Wolfe, London, UK, by permission of the publisher.)

Table 17.3
Worldwide Reported Incidence of Curable
Sexually Transmitted Infections in 1996

STD	Number of cases (in millions)
Trichomoniasis	170
Chlamydia	89
Gonorrhea	62
Syphilis	12
Total	333

Reference: WHO Fact Sheet N.110. Internet:
http://www.who.org.

intercourse appears to be the sole method of transmission. With about 3 million new cases each year (Table 17.2), trichomoniasis is a surprisingly common STD.

On a worldwide basis, trichomoniasis is the most common curable STD (Table 17.3). In men, infections tends to be asymptomatic, although mild urethral discharge and a burning sensation during urination may be present. Moreover, infections in men are self-limiting, so that men do not remain infective for long periods of time. In women, the primary symptoms are vaginal discharge, vulval itching and irritation, burning urination, and painful intercourse. Chronic infections are very common in women, and consequently, women are the primary infection reservoir. Trichomoniasis can be treated very effectively with the antibiotic *metronidazole*, if both sexual partners are treated simultaneously. However, metronidazole-resistant strains of *T. vaginalis* have been identified recently in cases of persistent trichomoniasis. The important complications occur in pregnant women, where trichomoniasis is associated with premature delivery and low birth weight. For unknown reasons, the use of oral contraceptives reduces the risk of infections.

SUMMARY

Trichomoniasis is the most common curable STD in the world. In the United States, the incidence of trichomoniasis approaches that of chlamydia. The urethra and vagina are the primary infection sites. Chronic infections are very common in women, and consequently, women are the primary infection reservoir. Antibiotic therapy is very effective if both sexual partners are treated simultaneously.

VIRAL STDs

Viral infections are responsible for many different types of disease in humans. Only a few viruses are transmitted sexually, but these are responsible for the majority of STDs.

Genital herpes

Genital herpes is caused by herpes simplex virus type 1 (HSV-1) or type 2 (HSV-2). HSV-1 and HSV-2 are members of the human herpes virus family that includes at least eight different viruses, each with different tissue specificity (Table 17.4). The herpes virus particle is a fairly large (diameter 200 to 250 nanometers) and complex proteinaceous structure, consisting of a DNA core embedded in an icosahedral protein capsid, and an envelope (Fig. 17.6). Although HSV-1 is more commonly associated

Table 17.4
Human Herpes Viruses

Virus	Principal diseases
Herpes simplex virus (HSV-1)	Skin and mucosal ulcers, especially oral and genital
Varicella zoster virus (VZV)	Chickenpox, shingles
Epstein-Barr virus (EBV)	Infectious mononucleosis
Cytomegalovirus (CMV)	Disease in immunosuppressed patients
Human herpes virus 6 (HHV-6)	Roseola infantum, nonrash illness in young children and possibly pneumonia in immunosuppressed patients
Human herpes virus 7 (HHV-7)	Some cases of roseola infantum
Human herpes virus 8 (HHV-8)	Associated with Karposi's sarcoma and some lymphomas

Reference: *Atlas of Sexually Transmitted Diseases and AIDS*, 1996.

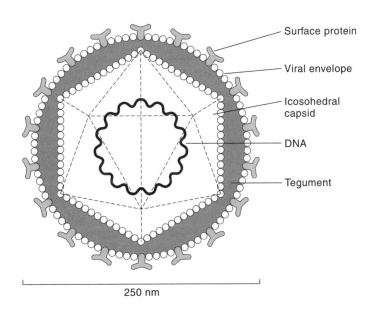

Figure 17.6
Structure of herpes simplex virus (cross-sectional view). The viral envelope is studded with proteins required for infection. At least five different protein species are recognized. An icosohedral protein capsid houses the viral DNA core. The region between the envelope and the capsid is called the tegument and consists of fibrous proteins.

with cold sores that appear in the mouth or throat, and HSV-2 with genital blisters, both types can infect either site. Infection is manifested by the appearance of small blisters that, over the course of 3 to 4 weeks, ulcerate. The ulcers may fuse to form larger ulcers that can be quite painful. The skin ulcers eventually form a scab and fall off. Ulcers can appear in sites other than the site of infection, for example, the thighs and buttocks. Cervical infections are quite common in women who exhibit ulcers in the external genitalia. Inflammation of the urethra, leading to painful urination, is common in both men and women. Fever, headache, muscle weakness, and fatigue frequently accompany the appearance of the blisters. In a few cases, more serious reactions to the infection can take place, for example, meningitis or neurological abnormalities that require hospitalization.

Three to four weeks after the initial infection the disease passes into a latent stage of variable duration. During this stage, infection develops in the sensory nerves in the vicinity of the original site of the infection. Periodically, symptoms return with the appearance of new lesions, often accompanied by tingling sensations or severe pains in the buttocks, hips, or legs. HSV-2 infections are more likely to recur than HSV-1 infections (90 percent versus 50 percent). The recurrence rate is about 4 to 5 episodes per year, although it may diminish with time. The infected individual is infective during the primary infection and during each recurrence. Recent studies indicate that transmission can occur even during the latent periods when no outward manifestation is apparent. Consequently, individuals can infect, and be infected, without realizing it.

HSV infections may be the most prevalent STD in the United States, with an estimated 31 million people already infected, and 200,000 to 500,000 new cases of genital herpes reported annually (Table 17.2). These estimates suggest that at least one in four women and one in five men will become infected with HSV during their lifetime. The complications of genital herpes extend beyond the infected individual. Infection of an infant just prior to or at birth can have terrible consequences—death or severe mental retardation for the infant. A primary infection during the early stages of gestation can cause spontaneous abortion, severe malformations, and intrauterine growth retardation. Infection of the fetus during a recurrent episode does not appear to have as severe consequences as infection during the primary episode. Some 73 percent of women who deliver babies infected with herpes have no apparent herpes lesions, so that is clear that fetuses can be infected during the latent periods of the disease.

There is no cure for genital herpes. The antiviral drug *acyclovir* relieves the symptoms of the primary infection, and suppresses recurrent episodes to some extent. Recurrence rates return to the same values when the drug is discontinued.

Human papillomavirus

Papillomaviruses constitute a large family of viruses, numbering more than 60 types, of which about 20 have the anal-genital region as their preferred site of infection. Papillomaviruses have a three-dimensional structure similar to that of the herpes viruses (Fig. 17.7). They also have a DNA core, an icosahedral protein capsid, and an envelope, but they are much smaller (diameter 50 nanometers), and consist of fewer proteins. The primary symptom of a papillomavirus infection is the appearance of warts (papillomas). One relatively common form of papilloma infection leads to *plantar warts*, which form in the soles of the feet. Papillomavirus infections of the anal-genital region are most commonly on the penis and scrotum in males and the labia, vagina, or cervix in females.

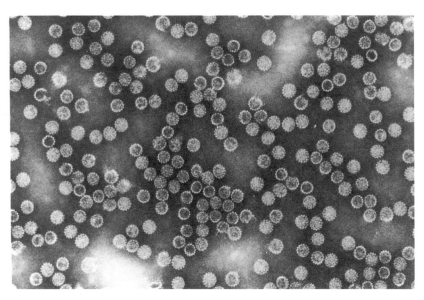

Figure 17.7
Electron micrograph of human papillomaviruses extracted from a plantar wart. (K. K. Holmes *et al. Sexually Transmitted Diseases*, 2nd ed. 1990. McGraw-Hill, New York, NY, reproduced with permission of The McGraw-Hill Companies.)

Papillomaviruses tend to infect the innermost layers of skin or mucous membranes, where they multiply. The virus migrates to the surface, and virus particles are shed, available to infect another person. Some viruses also stimulate cell division in infected cells, producing a wart, a precancerous lesion, or a cancer. Hence, depending on the virus, infection may be asymptomatic, may result in warts appearing in the anal-genital region, or much less frequently in a cancer. In men, the penis, scrotum, or urethra appear to be preferred sites for warts, while in women, warts tend to form in the labia, vagina, and cervix. Cancerous lesions tend to be found in the cervix. In both sexes, warts may form in the anus. If large enough, they can interfere with intercourse or with defecation. Diagnosis of papillomavirus infections is difficult and expensive, and treatment is less than satisfactory. There is no cure. The warts or lesions can be removed surgically, but this treatment is not curative. At best, removal of the lesions may help reduce transmission to another sexual partner.

Papillomaviruses are the second most prevalent STD organisms in the United States. Recent estimates of the prevalence of papillomavirus infections suggest that as many as 24 million Americans are already infected with papillomavirus, and close to 1 million new infections are occurring every year (Table 17.2). The prevalence among sexually active college-age men and women is particularly high, approaching 50 percent. Over 80 percent of prostitutes are infected. The efficiency of transmission per exposure is not known with any precision, but in some studies, over two-thirds of men who have had sexual intercourse with women with cervical warts have become infected. The incubation period can vary from 3 to 8 months.

The major complication of papillomavirus infection is the development of a genital or rectal cancer. However, induction of cancer depends on the type of virus and other unknown factors. Hence, it is difficult to estimate the likelihood that an infection will lead to a cancer. Transmission of the virus from mother to infant at birth is another important complication and is estimated at about 1 chance in 400 if the mother has genital warts. An infected infant develops warts in the genital area or infrequently in the respiratory tract. In the latter case, the infection can spread to the bronchi or the lungs, making respiration very difficult.

Hepatitis B virus

Hepatitis B virus (HBV) is one of the five currently recognized members of the human hepatitis virus family. These viruses are responsible for five clinically defined diseases known as hepatitis A, B, C, delta, and E. HBV is a relatively small virus (diameter 42 nanometers) and has a structure similar to that of human papillomavirus (Fig. 17.8).

Hepatitis B virus (HBV) can be transmitted sexually, perinatally (during birth), horizontally (person to person), or parenterally. In the infected person, the virus is

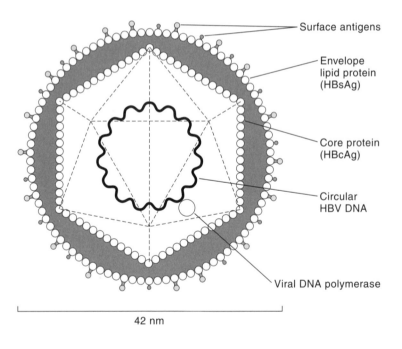

Surface antigens

Envelope
lipid protein
(HBsAg)

Core protein
(HBcAg)

Circular
HBV DNA

Viral DNA polymerase

42 nm

Figure 17.8
Structure of hepatitis B virus (HBV) (cross-sectional view). Infection by HBV is recognized by the presence of antibodies generated against surface antigens attached to lipid proteins (HBsAg) in the virus envelope. An icosohedral capsid made up of core proteins (HBcAg) houses the DNA core and a viral enzyme, DNA polymerase, which is required for viral DNA replication.

present in the saliva, semen, vaginal fluids, and most other body fluids. Approximately half of HBV infections are transmitted sexually, and these account for 50,000 to 100,000 new cases per year (Table 17.2). Sexual transmission is highly correlated with number of sexual partners and frequency of anal intercourse. Perinatal transmission is very common in Southeast Asia and Africa, where it is estimated that over 90 percent of the population is likely to have been infected in the recent past. Infants who are infected this way become carriers without developing any symptoms, but as adults they may develop chronic hepatitis. Horizontal transmission can occur by nonsexual contact with people who have been infected with HBV. Parenteral transmission occurs primarily among drug users who share needles.

The target tissue of HBV is the liver, and the main complications of HBV infections are chronic hepatitis, cirrhosis, and liver cancer. HBV is the only one of several hepatitis-producing viruses for which sexual transmission is the major means of infection. The first symptoms appear 3 to 5 months after infection, although as many as 40 percent of infections can be asymptomatic. The first symptoms may be very general: fever, rash, or painful joints. Later these can be followed by more serious symptoms—nausea, loss of appetite, fatigue, and pain in the area of the liver, jaundice (yellow eyes and skin), and even dark urine. At this stage, liver damage has begun. There is no effective therapy after an infection has begun. Surprisingly, despite the severity of the symptoms, which may require hospitalization, some 90 percent or more of patients recover completely within 2 months. Most of these (90 percent) are not infective, and are immune against reinfection. The remaining 10 percent continue to support virus replication. They are asymptomatic, but remain infective. A small fraction of infected individuals (less than 5 percent) do not recover. About 1 in 1000 infected individuals dies of acute liver failure. Others develop chronic hepatitis and eventually succumb to cirrhosis or liver cancer.

HBV is the only one of the viral STDs for which a reasonably effective vaccine is available. The vaccine provides protection for 2 to 5 years for 50–80 percent of vaccinated individuals. The vaccine is highly recommended for those in high-risk groups, or for those traveling into areas with very high rates of HBV infection. Unfortunately, the high cost of the vaccine and its multidose regimen have limited its widespread use. In some cases, people have resisted being vaccinated because of fears (unfounded) that the vaccine may be contaminated with HIV virus.

HIV/AIDS

Acquired immune deficiency disease (AIDS) is one of the most recently recognized STDs. The primary agent responsible for AIDS is human immunodeficiency virus (HIV), of which there are two types, HIV-1 and HIV-2. HIV is a complex particle consisting of an RNA core embedded in a capsid and envelope made up of several different proteins (Fig. 17.9). HIV is a member of the large *retroviral* family of viruses responsible for many different types of diseases in humans and animals.

HIV-1 was recognized as a disease-causing agent in the 1980s. According to current studies, HIV-1 strains responsible for AIDS originated in central Africa in the 1930s. Nonvirulent variants of HIV-1 may have existed in that region for hundreds of years before. The conditions that led to the emergence of the highly virulent HIV-1 strain and to the AIDS epidemic are not well understood. Since its emergence, however, HIV-1, transmitted by sexual contact, has spread all over the world. HIV-2, found primarily in West Africa, has not yet spread as widely as HIV-1.

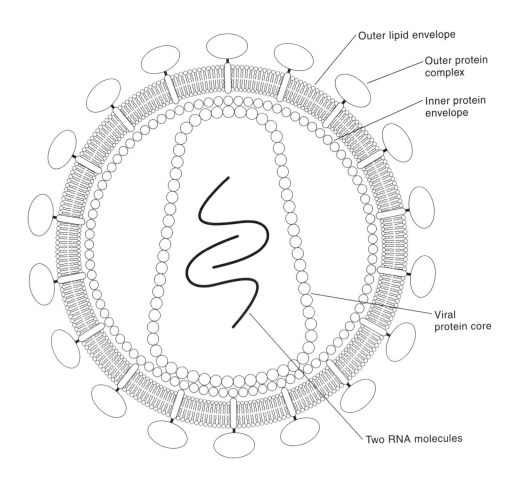

Figure 17.9
Structure of HIV (cross-sectional view). An outer lipid envelope is studded with outer protein complexes, each consisting of two different proteins. An inner protein envelope contains a viral protein core housing two RNA molecules and the enzyme reverse transcriptase, which is required for viral RNA replication.

The global estimates of the HIV/AIDS epidemic at the end of 1999 are given in Tables 17.5–17.8. In Central Africa and some countries of Southeast Asia, AIDS is an epidemic of major proportions (Tables 17.6–17.8). Originally, the focus of infection was the male homosexual communities, but in most countries of the developing world, most transmission of HIV-1 is by heterosexuals. In the United States, despite its relatively low incidence compared with other STDs, AIDS contributes significantly to STD-related deaths. Since 1982 there has been a steady increase in the number of HIV-related deaths in men 25–44 years of age, and since 1992 HIV-related deaths were the leading cause of death in men in this age range (Fig. 17.10). In women the death rate has also increased steadily, but not as rapidly as in men. Since 1995, HIV-related deaths have become the third leading cause of death in females 25–44 years.

Table 17.5
Global Summary of the HIV/AIDS Epidemic, December 1999

People newly infected with HIV in 1999	Total	**5.6 million**
	Adults	5 million
	Women	2.3 million
	Children <15 years	570,000
Number of people living with HIV/AIDS	Total	**33.6 million**
	Adults	32.4 million
	Women	14.8 million
	Children <15 years	1.2 million
AIDS deaths in 1999	Total	**2.6 million**
	Adults	2.1 million
	Women	1.1 million
	Children <15 years	470,000
Total number of AIDS deaths since the beginning of the epidemic	Total	**16.3 million**
	Adults	12.7 million
	Women	6.2 million
	Children <15 years	3.6 million

Reference: *AIDS Epidemic Update: December 1999*. Joint United Nations Programme on HIV/AIDS (UNAIDS), World Health Organization.

Table 17.6
Geographical Distribution of New HIV Infections (Adults and Children) During 1999

Region	Number	Percentage
North America	44,000	0.8
Caribbean	57,000	1.0
South America	150,000	2.7
Western Europe	30,000	0.5
Eastern Europe & Central Asia	95,000	1.7
East Asia & Pacific	120,000	2.1
South & Southeast Asia	1,300,000	23.2
North Africa & Middle East	19,000	0.3
Sub-Saharan Africa	3,800,000	67.7
Australia & New Zealand	500	0.01

Reference: *AIDS Epidemic Update: December 1999*. Joint United Nations Programme on HIV/AIDS (UNAIDS), World Health Organization.

Table 17.7
Geographical Distribution of Adults and Children Living with HIV/AIDS at the End of 1999

North America	920,000
Caribbean	360,000
South America	1,300,000
Western Europe	520,000
Eastern Europe & Central Asia	360,000
East Asia & Pacific	530,000
South & Southeast Asia	6,000,000
North Africa & Middle East	220,000
Sub-Saharan Africa	23,300,000
Australia & New Zealand	12,000

Reference: *AIDS Epidemic Update: December 1999.* Joint United Nations Programme on HIV/AIDS (UNAIDS), World Health Organization.

Table 17.8
Geographical Distribution of Adult and Child Deaths Due to HIV/AIDS from the Beginning of the Epidemic to the End of 1999

North America	450,000
Caribbean	160,000
South America	520,000
Western Europe	210,000
Eastern Europe & Central Asia	17,000
East Asia & Pacific	40,000
South & Southeast Asia	1,100,000
North Africa & Middle East	70,000
Sub-Saharan Africa	13,700,000
Australia & New Zealand	8,000

Reference: *AIDS Epidemic Update: December 1999.* Joint United Nations Programme on HIV/AIDS (UNAIDS), World Health Organization.

The main target of HIV-1 is cells of the immune system. HIV infections progress through three stages. The first-stage symptoms, such as sore throat, fever, headache, fatigue, and swollen lymph nodes, generally appear 2 to 4 weeks after infection and may persist for about 2 weeks. In most cases, the symptoms are not severe enough to require medical attention. After this period the virus becomes latent, and the symptoms disappear. The asymptomatic period can last 5 to 8 years. The second stage, known as AIDS-related complex (ARC), is characterized by severe fatigue, weight loss, chronic headaches, high fever, and, in many cases, memory loss, personality or mood changes, and progressive cognitive loss. The final and fatal stage, AIDS, is the result of a progressive and irreversible destruction of the immune system beginning 7 to 15 years after the initial infection. The patient succumbs to opportunistic infections or cancers.

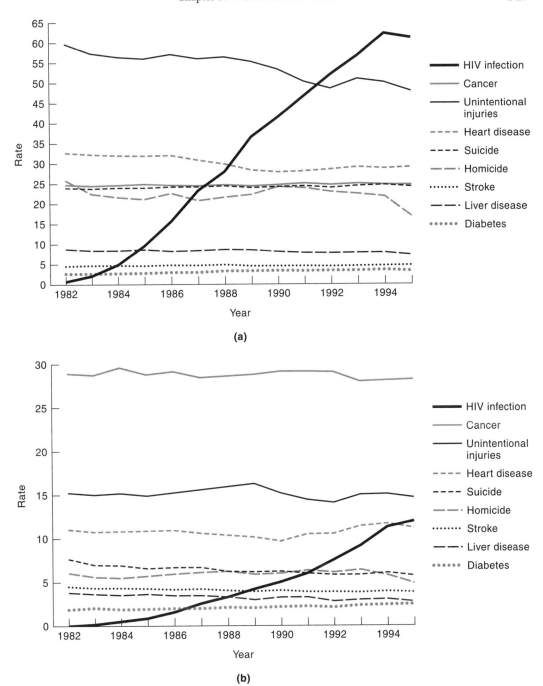

Figure 17.10
Leading causes of death among men (**a**) and women (**b**) 25 to 44 years of age in the United
States, by year, from 1982 to 1995. A steady rise in HIV-related deaths in both sexes is
apparent. (Adapted from J. A. Levy, 1998.)

Table 17.9
Major Risk Factors for HIV Transmission

Population	Risk factors
Everyone	Multiple sex partners, receipt of unscreened blood or blood products
Homosexual men	Receptive anal intercourse, recent partner with AIDS, genital herpes, syphilis
Heterosexual men	Prostitute contact, genital ulcers, uncircumcised
Heterosexual women	Genital ulcers, chlamydial or gonorrheal infection, trichomonal vaginitis, receptive anal intercourse
Intravenous drug users	Needle sharing
Health care workers	Needle stick injuries

Adapted with permission from *Research Issues in Human Behavior and STDs in the AIDS Era.* Copyright 1991 by the American Society for Microbiology.

Progression of an HIV infection is highly variable in terms of the intensity of the symptoms and the length of the asymptomatic period. The major risk factors for HIV transmission are listed in Table 17.9. For adults, infection requires sexual contact with an infected person or exposure to infected blood. Precise estimates for the likelihood of infections in a single sexual encounter are difficult to obtain, but it is probably low. Multiple sexual encounters are generally considered to be required for infection. In mother-to-infant transmission, 22 to 39 percent of infants born to infected mothers are infected. Most of these cases are diagnosed within the first 2 years, and the infants generally die within the next 2 years.

The high mortality and rapid worldwide spread of AIDS has prompted a huge effort in the United States and European countries to develop anti-AIDS therapies and AIDS vaccines to prevent infections. Some partial successes have been reported with some anti-AIDS drugs, particularly with those that use a combination of compounds that inhibit virus replication and interfere with virus infection. Promotional efforts by researchers and pharmaceutical companies, anecdotal reports of AIDS patients literally rising from their deathbeds after taking the anti-AIDS cocktails, and overly optimistic reports about progress in the development of an AIDS vaccine have contributed to the sense in many public sectors that AIDS is treatable, or maybe preventable. The real picture, however, is much grimmer. The anti-AIDS drugs are far from being as effective as their publicity initially suggested. Even if the anti-AIDS drugs were really effective, their cost of $15,000 per patient per year makes them available only to a select few. No vaccine is yet in sight. At the 1998 World AIDS Conference in Geneva, Switzerland, reports of new problems with the new anti-AIDS drugs and difficulties in developing effective vaccines have left a sense of pessimism about the research programs focused on developing anti-AIDS therapies. Public health officials continue to argue, as they have for many years, that the most effective way to contain the AIDS epidemic is prevention. However, at the present time 100 times more money is being devoted to therapeutics than to the development of prevention strategies. Sex education, needle-exchange programs, and free or inexpensive condom distribution could save millions of people, but these are the programs that appear the least likely to be supported by most governments.

SUMMARY

Genital herpes and papillomavirus infections, characterized by the formation of ulcers and warts in the external genitalia, respectively, are by far the most prevalent STDs in the United States. College-age individuals constitute the primary reservoir of infectious individuals for both STDs. Both types of infections go through latent phases where symptoms disappear, but individuals may remain infectious. No curative therapies are available. Hepatitis B is an unusual STD in that the liver rather than the anal-genital tissues is the target of infections. About half of hepatitis B infections are transmitted sexually. A hepatitis B vaccine is available, but it provides protection for a short period (2 to 5 years) in only 50 to 80 percent of vaccinated individuals. HIV/AIDS is probably the least common STD in the United States, but it is the most lethal. The target of HIV is the immune system, and lethality associated with HIV infections is due to the gradual and progressive destruction of the host's immune system. Although the focus of HIV infections in the 1980s was the homosexual communities, at present most transmission of HIV is by heterosexuals. HIV/AIDS has become a worldwide epidemic of devastating proportions. Sub-Saharan Africa and Southeast Asia account for close to 90 percent of infected individuals. Effective anti-HIV therapy is not yet available to contain the worldwide epidemic.

THE "SILENT" STD EPIDEMIC

There is general agreement among public health officials that the end of the last century and beginning of the new one have been marked by a worldwide STD epidemic. Although the epidemic has spread to both the industrialized and developing countries, the epidemic is having devastating consequences for the peoples of the underdeveloped world. The social, health, and economic costs of this epidemic are probably incalculable.

Scope of the epidemic

The primary victims of STDs worldwide are people between the ages of 15 and 44, economically the most productive. In 1993, the World Bank estimated that STDs were the second leading cause of mortality in women between the ages of 15 and 44 in the developing countries. The World Health Organization (WHO) estimated that there were 333 million new cases of four STDs—gonorrhea, syphilis, chlamydial infections, and trichomoniasis—worldwide during 1995 among people from 15 to 44 years of age. The United States has not escaped this STD epidemic; in fact, it contributes significantly to it. For example, although gonorrhea and syphilis appear to be declining slowly in the United States, the rates still far exceed those in other developed countries (Table 17.10). For example, in 1995, the reported incidence of gonorrhea in the United States was 149.5 cases per 100,000 people, but only 3 per 100,000 in Sweden and 18.6 per 100,000 in Canada. Indeed, the incidence of both gonorrhea and syphilis in the United States is higher than in many developing countries.

The multidimensional scope of the United States epidemic, which affects all socioeconomic groups, can be appreciated by considering its characteristics.

1. More than 12 million people, 3 million of whom are teenagers, are infected with STDs each year.

2. In 1995, STDs accounted for 87 percent of all cases reported among the top ten most frequently reported diseases.

3. Eight new STDs have been identified since 1980.

Table 17.10
Reported Cases of Gonorrhea and Syphilis in the
United States and Other Developed Countries
(Per 100,000 Persons Per Year)

Country	Gonorrhea	Syphilis
United States	149.5	6.3
England	34.1	1.0
Canada	18.6	0.4
Denmark	5.5	0.4
Germany	4.9	1.4
Sweden	3.0	0.8

Adapted with permission from *The Hidden Epidemic: Confronting STDs.* Copyright 1997 by the National Academy of Sciences. Courtesy of the National Academy Press, Washington, D.C.

4. Women and adolescents are the most frequently affected by STDs.

5. STDs increase the risk of HIV transmission.

6. $10 billion is spent annually on STDs other than AIDS.

7. Cancers, infertility, ectopic pregnancy, spontaneous abortion, stillbirth, low birth weight, neurologic damage, and death are the serious complications of STDs.

A number of factors are considered to contribute to the alarming spread of STDs.

1. There has been a tremendous increase in unprotected sexual activity in the population at large and among the young in particular. It is estimated that 70 percent of 12th-grade students have had sexual intercourse, and about 30 percent have had sex with four or more partners. A discussion of the reasons for this increase in sexual activity is beyond the scope of this book. Nevertheless, this phenomenon signals the fact that the spread of STDs depends not only on the biology of the infecting organism, but on complex behavioral and social factors that are not easy to delineate.

2. There has been a consistent lack of useful and clear information about the nature and types of STDs. Surveys of women between 18 and 60 years of age reveal that over two-thirds know nothing of STDs other than AIDS. There is no main network television programming that provides adequate information about STDs other than AIDS. Since only about 10 percent of teenagers surveyed have received information about STDs from their families, a large segment of the population at prime risk for STDs receives very little information about these diseases. Given these circumstances, schools would appear to be the best place to provide information about STDs, but so-called "sex education" in the schools is fraught with controversy. In most school districts, the information about appropriate protective sexual behavior is limited or nonexistent. Sexual taboos prevent effective strategies against STDs at all levels, from the development of widespread STD education and prevention programs to the

provisions of coherent and integrated STD-related services, with adequately trained health care professionals.

3. The nature of STD infections has tended to obscure the link between STD infections and their health consequences. First, many STDs are asymptomatic, and hence remain undetected. Second, the time interval between the initial infection and the full manifestation of serious disease can be many years, so that the connection between the two remains unclear in public awareness. The only exception is HIV and AIDS, where public education programs have been quite successful in developing the necessary public awareness required for successful prevention programs.

4. The widespread lack of adequate medical insurance or health care coverage means that many cases of symptomatic STD infections are not diagnosed early enough to prevent the disease from progressing to its more serious consequences. Open and calm discussion about sexuality and STDs remains extremely difficult in the United States. A strong stigma is associated with STDs, so that many infected individuals may choose not to seek medical attention at the time when it would be most beneficial.

Contraception and STDs

STD risk is affected by contraceptive methods (Table 17.11). Condoms, for example, protect very effectively against both viral and bacterial STDs. Spermicides appear to be effective in protecting against chlamydia and gonorrhea, but their value against viral STDs has not been confirmed. IUDs and hormonal methods do not provide protection against viral STDs. COCs have been associated with an increase in chlamydia detected in the cervix.

SUMMARY

Several factors have conspired to make the STD epidemic a rather "silent" one, but the economic and human cost of the epidemic is immense. The dollar cost of STDs other

Table 17.11
Contraceptives and Bacterial and Viral STDs

Method	Bacterial STDs	Viral STDs
Condoms	Protective	Protective
Spermicides	Protective against chlamydia and gonorrhea	Not known
Diaphragms	Associated with anaerobic bacterial growth	Protection against cervical infection
Hormonal	May promote cervical chlamydia	No protection
IUD	Associated with PID in first month after insertion	No protection

Reference: Hatcher *et al.*, 1994.

than AIDS is estimated at about $10 billion annually, but the cost in suffering to the millions of affected individuals cannot be calculated. The serious health consequences of STDs (again with the exception of AIDS), which include cancer, infertility, ectopic pregnancies, spontaneous abortions, stillbirth, premature births, low birth weight, neurologic damage, and death, have remained largely beyond public awareness (Table 17.12).

QUESTIONS

1. Why has it been so difficult to control the spread of chlamydial infections? What do you think would be required to control the epidemic?

Table 17.12
Major Complications of STDs

Complication	Women	Men	Infants
Cancer	Cervical cancer Vulval cancer Vaginal cancer Anal cancer Liver cancer Kaposi's sarcoma Leukemia	Penile cancer Anal cancer Liver cancer Leukemia Kaposi's sarcoma	
Impaired reproductive function	Pelvic inflammatory disease Infertility Ectopic pregnancy Spontaneous abortion	Epididymitis Prostatis Infertility	
Problems during pregnancy	Prematurity Infection during labor Postnatal infection		Stillbirth Low birth weight Pneumonia Conjunctivitis Neonatal infection Hepatitis Birth defects
Neurologic problems	Neurosyphilis	Neurosyphilis	Meningitis Neurological damage
Other	Chronic liver disease Cirrhosis	Chronic liver disease Cirrhosis	Liver disease Cirrhosis

Adapted with permission from *The Hidden Epidemic: Confronting STDs*. Copyright 1997 by the National Academy of Sciences. Courtesy of the National Academy Press, Washington, D.C.

2. Distinguish between the progression of herpes and papilloma infections.

3. Why are STDs considered to be a major contributor to the increase in ectopic pregnancies?

4. What type of information program might be the most effective in bringing out the serious nature of STDs? At what age do you think it should begin?

5. Library project:
   a. Do animals suffer from STDs?
   b. What is known about the mechanism of carcinogenesis by papillomavirus?
   c. What is the current status in the development of vaccines for viral STDs, other than hepatitis B?

SUPPLEMENTARY READING

Centers for Disease Control and Prevention. *1998 Guidelines for Treatment of Sexually Transmitted Diseases*. MMWR 1998, **47** (No. RR-1).

Eng, T. R., and W. T. Butler (Eds.). 1997. *The Hidden Epidemic: Confronting STDs*. Institute of Medicine, Division of Health Promotion and Disease Prevention, National Academy Press, Washington, D.C.

Levy, J. A. 1998. *HIV and the Pathogenesis of AIDS*. American Society for Microbiology Press, Washington, D.C.

Rosebury, T. 1971. *Microbes and Morals*. Viking Press, New York, NY.

Waugh, M. A. 1990. History of clinical developments in sexually transmitted diseases. In *Sexually Transmitted Diseases*, 2nd ed. Chapter 1. K. K. Holmes *et al.*, Eds. McGraw-Hill, Inc., New York, NY.

INTERNET SITES

Internet sites with information about STDs

Centers for Disease Control and Prevention: www.cdc.gov
Joint United Nations Programme: www.unaids.org
World Health Organization: www.who.org

ADVANCED TOPICS

Hatcher, R. A., *et al.* 1994. *Contraceptive Technology*, 16th revised ed. Ardent Media, Inc., New York, NY.

Jones, R. B., and J. N. Wasserheit. 1991. Introduction to the biology and natural history of STDs. In *Research Issues in Human Behavior and STDs in the AIDS Era*. J. N. Wasserheit, S. O. Aral, and K. K. Holmes, Eds. American Society for Microbiology, Washington, D.C.

Morse, S. A., A. A. Moreland, and S. E. Thompson (Eds.). 1990. *Atlas of Sexually Transmitted Diseases*. J. B. Lippincott Co., Philadelphia, PA, p. vii.

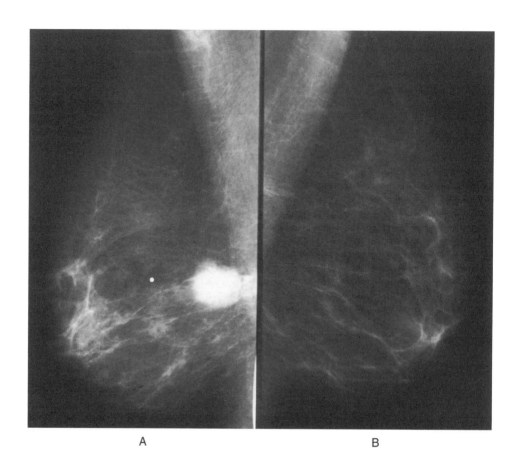

A   B

Mammogram comparing the cancerous (A) and normal (B) breast of a 54-year-old woman. The tumor is a 2.5-cm infiltrating palpable ductal carcinoma in the right breast. (Courtesy of Dr. Linda Olson, UCSD Medical Center.)

CHAPTER 18

Cancers of the Reproductive Tissues

*Since many of the known causes of cancer are avoidable,
it is possible to reduce the incidence rates of many types of cancer.*

—B. N. Ames, L. S. Gold, and W. C. Willett. 1995

The causes and prevention of cancer. *Proceedings of the
National Academy of Sciences (USA)* 92, 5258–5265.

*What we know about prevention of breast cancer
is related to lifetime estrogen exposure and exposures
to specific estrogens at vulnerable periods of life.*

—L. Kohlmeier and M. Mendez. 1997

Controversies surrounding diet and breast cancer.
Proceedings of the Nutrition Society 56(1B), 369–382.

CANCER IS THE SECOND leading cause of death in the United States. Cancers of the reproductive tissues, particularly—breast, uterine, and ovarian cancers in females and prostate cancer in males—represent a very significant fraction of all types of cancers. Cancers of the reproductive tissues are not fundamentally different from cancers of the nonreproductive tissues, but much accumulating evidence indicates that the steroid sex hormone milieu in the individual plays an important, perhaps even decisive, role in the pathogenesis of reproductive cancers. Our objective in this chapter is to summarize some of this evidence, indicating the important risk factors, possible mechanisms, and emerging suggestions for preventive strategies.

INCIDENCE AND MORTALITY

In 2000, an estimated 552,200 Americans were expected to die of cancer—more than 1500 each day. The distribution of new cases and mortality rates in males and females in 2000 for a few of the most common cancers is shown in Fig. 18.1. In females, cancers of the breast, ovary, and uterus together account for about 40 percent of new cases, and 24 percent of deaths due to cancer. These percentages translate into 256,200 new cases of breast, ovarian, and uterine cancers, as well as 64,800 deaths due to these can-

455

456

Estimated New Cancer Cases*
10 Leading Sites by Gender, U.S., 2000

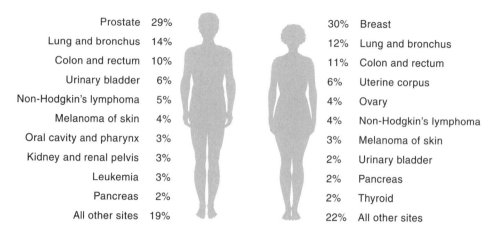

Prostate	29%		30%	Breast
Lung and bronchus	14%		12%	Lung and bronchus
Colon and rectum	10%		11%	Colon and rectum
Urinary bladder	6%		6%	Uterine corpus
Non-Hodgkin's lymphoma	5%		4%	Ovary
Melanoma of skin	4%		4%	Non-Hodgkin's lymphoma
Oral cavity and pharynx	3%		3%	Melanoma of skin
Kidney and renal pelvis	3%		2%	Urinary bladder
Leukemia	3%		2%	Pancreas
Pancreas	2%		2%	Thyroid
All other sites	19%		22%	All other sites

*Excludes basal and squamous cell skin cancers and *in situ* carcinomas except urinary bladder.
Percentages may not total 100% due to rounding.

(a)

Estimated Cancer Deaths*
10 Leading Sites by Gender, U.S., 2000

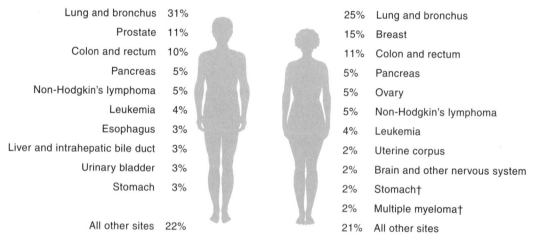

Lung and bronchus	31%		25%	Lung and bronchus
Prostate	11%		15%	Breast
Colon and rectum	10%		11%	Colon and rectum
Pancreas	5%		5%	Pancreas
Non-Hodgkin's lymphoma	5%		5%	Ovary
Leukemia	4%		5%	Non-Hodgkin's lymphoma
Esophagus	3%		4%	Leukemia
Liver and intrahepatic bile duct	3%		2%	Uterine corpus
Urinary bladder	3%		2%	Brain and other nervous system
Stomach	3%		2%	Stomach†
			2%	Multiple myeloma†
All other sites	22%		21%	All other sites

*Excludes *in situ* carcinomas except urinary bladder.
†These two cancers both received a ranking of 10; they have the same projected number of deaths
and contribute the same percentage.
Percentages may not total 100% due to rounding.

(b)

Figure 18.1
Estimated cancer cases (**a**) and cancer deaths (**b**) in the United States in 2000. The percentages indicate the estimated proportion of newly diagnosed cancers and cancer deaths for the 10 leading cancer sites. (Adapted from R. T. Greenlee *et al.*, 2000.)

Table 18.1
Estimated Number of New Selected Cancer Cases and Deaths in the United States in 2000

Cancer	New cases	Deaths
Prostate	180,400	31,900
Breast	184,200	41,200
Lung	171,600	158,900
Female	74,600	67,600
Male	89,500	89,300
Colon/rectum	130,200	55,300
Lymphoma	62,300	27,500
Uterus/cervix	48,900	9,600
Skin	56,900	16,500
Ovarian	23,100	14,000
Testis	6,900	300

Reference: Greenlee *et al.*, 2000.

cers (Table 18.1). In males, cancer of the prostate accounted for 29 percent of new cancer cases (180,400 new cases of prostate cancer), and 13 percent of cancer deaths (31,900 deaths). Lung and colon cancer, respectively, are the next two most important cancers in terms of incidence. Lung cancer for both males and females is the greatest killer—in females, the incidence is 13 percent, but it accounts for 25 percent of the deaths, while in males, the incidence is 15 percent, but it accounts for 31 percent of the deaths. In 2000, it was estimated that about 158,900 people in the United States would die of lung cancer.

The risk of developing cancer increases with age, the first significant rise occurring in the fifth decade of life. The lifetime risk of cancer for American males is one in two, and one in three for women (Table 18.2). For breast cancer, the risk is 1 in 231 for fe-

Table 18.2
Percentage of the Population Developing Cancers at Certain Ages in the United States, 1994–1996

Cancers		Birth to 39 years	40–59 years	60–79 years	Birth to death
All sites	Male	1.61	8.17	33.65	43.56
	Female	1.94	9.23	22.27	38.11
Breast	Female	0.43	4.06	6.88	12.56
Prostate	Male	<.01	1.90	13.69	15.91
Colon/rectum	Male	0.06	0.85	3.97	5.64
	Female	0.05	0.67	3.06	5.55
Lung	Male	0.04	1.29	6.35	8.11
	Female	0.03	0.94	3.98	5.69

Note the large increase in prostate cancer beginning during the fourth decade of life. (Reference: NCI Surveillance, Epidemiology and End Results Program, 2000.)

males up to 39 years, but increases to 1 in 25 between 40 and 59 years, and eventually reaches a lifetime risk of 1 in 8. In males, the risk of prostate cancer increases dramatically from 1 in 10,000 up to age 39 to 1 in 55 from 40 to 59 years. From any perspective, these are sobering numbers.

Despite the many billions of dollars that have been spent in the "cancer wars" (the first War on Cancer was declared about 25 years ago), we are still far from vanquishing the enemy. Some progress has been made, however. Except for breast and lung cancer in females and prostate cancer in males, the incidence of most cancers has remained relatively stable over the last 50 years. The huge increase in prostate cancer beginning in the late 1980s is attributed primarily to better detection. Prostate cancer appears to have peaked about 1993, and the incidence appears to have decreased significantly since then (Figs. 18.2 and 18.3).

Over the last 25 years, cancer death rates have stabilized or decreased for females and males for most cancers. However, the breast cancer death rate has not changed significantly over the last 60 years. The lung cancer death rate has increased steadily since the 1960s in females and since the 1930s in males. But even here there are some signs

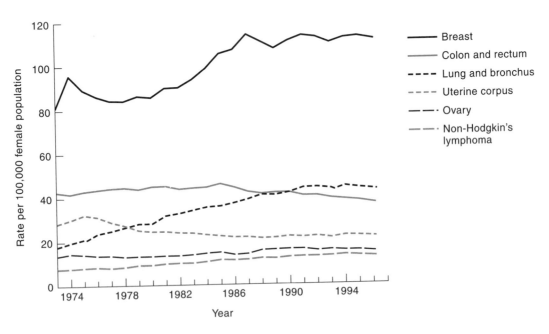

Figure 18.2
The age-adjusted incidence rates for some cancers in females in the United States, 1973–1996. The incidence of breast cancer rose during the 1980s and may have stabilized. The incidence of lung cancer increased continually since the 1960s, but it too appears to have stabilized. (Adapted from R. T. Greenlee *et al.*, 2000.)

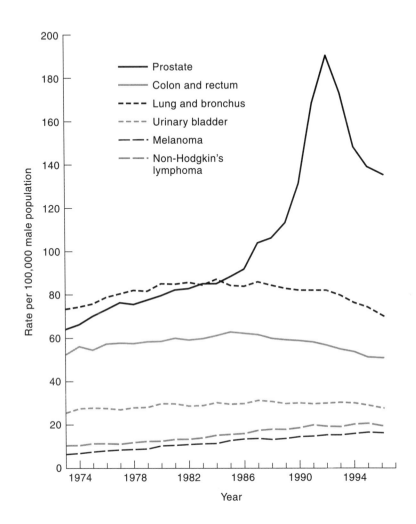

Figure 18.3
The age-adjusted incidence rates for some cancers in males in the United States, 1973–1996. The incidence of prostate cancer rose precipitously beginning about 1985, but this rise is considered to reflect primarily more aggressive efforts to detect it. Incidence peaked in 1993 and has been followed by a decrease. Incidence of other important types of cancers has stabilized or decreased. (Adapted from R. T. Greenlee *et al.*, 2000.)

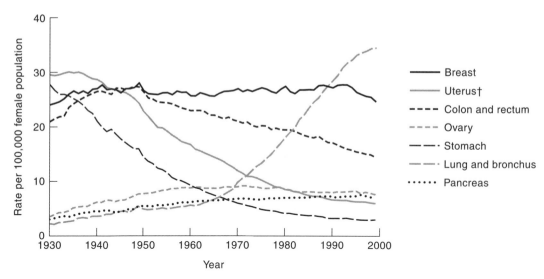

Figure 18.4
The age-adjusted cancer death rates in females by site in the United States between the years 1930 and 1996. Note that the breast cancer death rate remained constant for over 60 years but is decreasing slightly since about 1993. The lung cancer death rate increased steadily since the 1960s, but there are indications that the rate has begun to decrease. The ovarian cancer death rate increased somewhat between 1930 and 1960, but has since stabilized. (Adapted from R. T. Greenlee *et al.*, 2000.)

that the lung cancer death rate may be stabilizing and decreasing in males (Figs. 18.4 and Fig. 18.5). Most experts agree that notwithstanding the advances in our understanding of the molecular biology of cancer, as well as improvements in therapy, the most effective ways to reduce the cancer rate will be in prevention. In fact, perhaps the most significant change in the way we look at cancer in the last 10 or 15 years has been the realization that a large fraction of perhaps most cancers may be preventable.

Prevention depends first on identifying the important **risk factors** for specific cancers, that is, the conditions that are correlated with specific cancers, and second, on undertaking appropriate measures either individually or collectively to reduce the risk. Perhaps the most dramatic of the connections between a risk factor and a cancer is that of smoking and lung cancer. Tobacco use accounts for about 90 percent of lung cancers, and it contributes to cancers of the mouth, stomach, kidney, pancreas, and maybe colon. There is no doubt that a significant reduction in tobacco use would have dramatic effects on the incidence and mortality of lung cancer, and other cancers as well.

The risk factors for other types of cancers are emerging slowly from many different types of epidemiological and laboratory studies. The epidemiological studies have shown that there are significant differences in incidence and death rates in different countries for specific cancers. Japan and China, for example, have the lowest death rates for prostate cancer, but have among the highest for stomach cancer. The United States and Australia have very low rates for stomach cancer, but the United States has

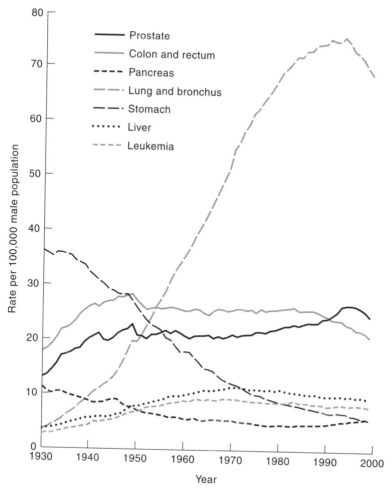

Figure 18.5
The age-adjusted cancer death rates in males by site in the United States between 1930 and 1996. Lung cancer death rates increased from the 1930s to about 1990, when the rate began to decrease. The prostate cancer death rate has been relatively constant since about 1940. It is still too early to tell whether the decrease noted in the early 1990s will continue. (Adapted from R. T. Greenlee *et al.*, 2000.)

the highest lung cancer death rate in the world for women. The United Kingdom has the highest breast cancer death rate in the world, but one of the lowest for uterine cancer. In general, cancer incidence and death rates are higher in the developed nations of the world. According to the World Health Organization (WHO), in 1996 cancer accounted for 21 percent of the deaths in the developed countries, but only 9 percent in the underdeveloped countries, indicating that cancer is in many ways a disease of affluence. Such differences in cancer rates among different countries suggest that environmental factors must be important. And indeed, a large body of work suggests that for many cancers, lifestyle factors may be the most important variables in determining cancer incidence and mortality.

REPRODUCTIVE CANCERS IN THE FEMALE

Two classes of factors have been associated with the high incidence of breast, endometrial, and ovarian cancers in the industrialized world. The first, *reproductive risk factors*, refer to parameters that characterize the reproductive life of the female; the second, *nonreproductive risk factors*, generally refer to dietary differences between women in the industrialized and nonindustrialized world. The relative importance of these two classes of risk factors is still being debated.

Reproductive risk factors in an evolutionary context

Epidemiological studies in human populations indicate that late menarche, early first birth, high parity (several births), early menopause, and lactation reduce the risk of breast, endometrial, and ovarian cancers. The converse of these—*early menarche, delayed first birth, low parity, late menopause, and lack of lactation*—are the reproductive risk factors for these cancers. It is interesting to note that these risk factors reflect the reproductive profile of women in industrialized countries. This profile is quite different from that of women in preagricultural (hunter-gatherer) societies. The identified risk factors represent deviations from those of existing foraging peoples, and presumably from our ancestral human experience. It is intriguing to note also that reproductive cancers now common in women are rarely seen in our closest nonhuman primate relatives, chimpanzees and gorillas. Under free-living conditions both primate species have reproductive profiles that are similar to those of human hunter-gatherer societies. The modern reproductive profile is a new development in our history. Human societies are perhaps 150,000 years old. The anthropological evidence available indicates that for most of our history, our way of life—nutrition, physical activity, reproductive habits—changed very little. Significant changes in these parameters began to occur only within the period of recorded history, and even then, they have occurred very unevenly, so that even now, there are still a few societies that retain the characteristics of our hunter-gatherer ancestors.

Recent epidemiological and anthropological studies have provided information for the comparison of the reproductive cancer risk in women in industrialized nations with estimates of the risk experienced by women in preagricultural societies. These latter estimates have been obtained from observation of women in existing foraging societies. Probably no society can be said to be completely untouched by modern technological civilization. However, a few still preserve in relative terms a way of life that approximates the hunter-gatherer cultural mode that characterized human societies for most of human history. The societies included in the new study were the !Kung-San from the Kalahari Desert in southwest Africa, the Aché from Paraguay, the Agta, traditional Eskimos, Australian aborigines, the Hadza from Tanzania, the Hiwi, and the Efe. These societies afford us perhaps the best window to a reproductive profile relatively free from the effects of the industrialized societies. Anthropological studies have provided the mean estimates of the various reproductive parameters shown in Table 18.3.

Reproductive profile in hunter-gatherer females

Menarche occurs at a mean age of 16.1 years in the hunter-gatherer societies. Interestingly, this is very similar to the age of menarche in nineteenth-century European nations, and to that of the Japanese (15.3 years) and Chinese (17.1 years) as late as the

Table 18.3
Reproductive Profile of Women in Existing Hunter-Gatherer Societies and in the United States

	Hunter-gatherers	U.S.
Age at menarche	16.1 years	12.5 years
Age at first birth	19.5 years	24.0 (all women)
		26.5 (educated*)
Menarche to first birth	3.4 years	11.5 (all women)
		14.0 (educated*)
Duration of lactation	2.9 years	3 months
Completed family size	5.9	1.8
Age at menopause	47	50.5
Number of ovulations	160	450**

*Women with at least some education beyond high school. **For women who have not used oral contraceptives. (Adapted with permission from Eaton *et al.* [1994] *Quarterly Review of Biology* 69[3], 353–367. Copyright © 1999 by the University of Chicago. All rights reserved.)

1940s. Mean age at first birth is 19.5 years, giving an interval between menarche and first birth of 3.4 years.

Infants are breast-fed frequently and for a relatively long time (2.9 years). For example, !Kung-San babies nurse four times an hour during the day, and several times at night until the child is 2 years old. Breast milk provides the major source of nutrients until 15 to 18 months, and continues to be a significant part of the child's nutrition until well in the third year of life. The frequency and duration of breast-feeding is sufficient to suppress ovulation for 24 to 30 months. This accounts for the long intervals between births (about 3 years) in societies that do not practice any other form of fertility control, and results in about 5.9 live births per woman who survives past menopause.

Age at menopause is difficult to gauge accurately, but estimates suggest that it occurs around 47 years. With this information, it is possible to estimate the number of lifetime ovulations. Assuming 13 ovulations per year, the total possible number is $(47 - 16) \times 13 = 403$. Correcting for the periods of anovulation (9 months pregnancy + 24 months breast-feeding per child), plus anovulatory cycles during the climacteric, results in about 160 lifetime ovulations. Hence, the reproductive profile of the hunter-gatherer female results in a large reduction (from the total number possible) in the number of ovulations that she will experience during her lifetime.

Reproductive profile in the industrialized countries

Menarche occurs at a much younger age, about 12.5 years (Table 18.3). The interval between menarche and first birth is also relatively long. For example, age at first birth for women born after 1955 is 24 years for all women in the United States. However, for women with a post–high school education, the age at first birth is 26.5, and it is over 27 for women with a college degree. The interval between menarche and first birth is between 11.5 and 14 years, rather than the 3.4 years seen in their hunter-gatherer counterparts. Also, a larger proportion of women remain childless now (15–20 percent in the 1980s compared to 10 percent in the early 1900s).

Duration of lactation is also short. In 1989, only 52 percent of American infants were breast-fed, and only 7 percent were breast-fed for as long as 12 months. Fertility in the United States is at about 1.8 births per female, similar to that of other industrialized nations. Age at menopause is 50.5 years, and it appears to have increased from the 47.5 years estimated for English and Scottish women in the mid-1800s. These differences mean that the number of lifetime ovulations will be higher in women in the industrialized countries. If oral contraceptives are not used, females in industrialized countries experience about 450 ovulations during their lifetimes. The use of oral contraceptives reduces the number of ovulations, but it is not possible to estimate by how much because use of oral contraceptives is not uniform. Nevertheless, the use of oral contraceptives is an important factor in estimating a woman's cancer risk (see below).

Significance of reproductive risk factors

Why should late menarche, early first birth, high parity, early menopause, and long lactation periods reduce the risk of breast cancer? The answer appears to be that all these factors reduce the exposure of breast tissue to circulating estrogens. Laboratory studies indicate that the susceptibility of breast tissue to cancer is directly related to the rate of epithelial cell proliferation. Breast epithelial tissue proliferates at a high rate in postpubertal females, particularly during the first 5 years after menarche. Estrogen is considered to be one of the main factors necessary for epithelial proliferation. A relationship between estrogen levels and breast cancer rates has long been noted. Asian women in general have lower blood and urinary estrogen levels, and also significantly lowered rates of breast cancer. Japanese- and Chinese-American women have higher estrogen levels and correspondingly higher breast cancer rates. Moreover, the increased breast cancer risk in obese females has been considered to reflect the increased conversion of androgen to estrogen that takes place in adipose (fat) cells.

Breast tissue appears to be particularly sensitive to the initiation of a cancer between puberty and the first pregnancy. Pregnancy and then lactation lower the risk because they induce the terminal differentiation of the epithelial tissue into the milk secretory lobules necessary for lactation. The terminally differentiated tissue is much less susceptible to the induction of cancer by carcinogenic agents to which the female may be exposed.

This estrogen-excess hypothesis also appears to account for the relationship between endometrial cancer and the reproductive risk factors. As you may recall, estrogen stimulates endometrial proliferation, and women with endometrial cancer typically have higher estrogen levels than controls. Obese women again are at higher risk for endometrial cancer. Progesterone, because it opposes the proliferative effects of estrogen, exerts a protective action. Recall that progesterone supplementation of ERT reduces the risk of breast cancer in postmenopausal women.

Similar arguments apply to the case of ovarian cancer. In addition, some investigators consider that the localized trauma induced at each ovulation exposes the ovarian epithelium to the high follicular estrogen levels. Pregnancy and the use of oral contraceptives reduce the number of ovulations, and hence reduce the overall risk of ovarian cancer.

Changes in the reproductive profile of women in the transition from the hunter-gatherer to the modern lifestyle have resulted in a significant increase in the risk for breast, endometrial, and ovarian cancers. Table 18.4 illustrates recent estimates on the magnitude of the increase in the relative cancer risk. It should be kept in mind that given the limited data on cancer incidence of hunter-gatherer females, these estimates

Table 18.4
Estimated Cumulative Relative Cancer Risk to Age 60

Population	Breast	Endometrial	Ovarian
Hunter-gatherer	1	1	1
Americans with 10 years COC	114	75	6.9
Americans who have not used COCs	114	240	24

The cancer risk is set at 1 for hunter-gatherers. COC stands for the combined oral pill type of contraceptive. Use of COCs does not appear to confer a protective effect with respect to breast cancer, but does with respect to endometrial and ovarian cancers. (Adapted with permission from Eaton *et al.* [1994] *Quarterly Review of Biology* 69[3], 353–367. Copyright © 1999 by the University of Chicago. All rights reserved.)

are not very precise. Medical anthropologists have found little evidence of reproductive cancers in present-day hunter-gatherer societies, and paleopathological studies suggest a low incidence of cancers in our remote ancestors. Nevertheless, the available data does suggest that the incidence of women's reproductive cancers today is much higher than for our ancestors, perhaps between 10-fold and 200-fold. The incidence of these cancers in two current low-risk populations (Chinese and Japanese) appears to be intermediate between the high fully Westernized rate and low ancestral rate.

On the other hand, there is substantial evidence that oral contraceptive use reduces the risk of ovarian and uterine cancers (Table 18.4).

What preventive measures are possible?

It seems unlikely that we could return to a hunter-gatherer lifestyle, no matter how desirable it might be from the point of view of reducing the risk of cancer. For example, the effects of an early first birth or an increased number of children per woman would have unacceptable sociodemographic repercussions. An important question is whether key features of the hunter-gatherer hormonal condition can be approximated without changing the birth patterns. Three possibilities have been suggested:

1. **Delay puberty.** By administering long-acting GnRH agonists puberty could be postponed or delayed until the late teens. We saw in Chapter 14 that the use of such agents is quite successful as therapy for central precocious puberty.

2. **Generate a false pregnancy.** Experiments with animals have shown that the hormonal environment of pregnancy can be mimicked by administration of high doses of estrogen and progesterone to induce the terminal maturation of breast epithelial tissue. This has been accomplished without affecting subsequent reproduction and/or lactation. It seems reasonable to suppose that these studies could be extended to humans.

3. **Lower serum estrogen to even lower levels with a new fertility control regimen.** Exploratory clinical trials indicate that this can be done in a two-step process: first, suppress ovarian function by administering GnRH agonists, and second, maintain serum estrogen at hunter-gatherer levels with a new generation of synthetic estrogens. This would assure that the levels of estrogen necessary to maintain estrogen-dependent tissues would be sustained.

The exogenous hormone therapy proposed by these suggestions sounds extremely radical, especially possibilities 1 and 2. It seems unlikely that they would be widely accepted. On the other hand, possibility 3 entails an extension of an oral contraceptive regimen already in use and widely accepted. It should be kept in mind that the use of oral contraceptives was quite radical 40 years ago.

Nonreproductive risk factors

The importance of high-fat and low-fiber diets typical of most industrialized countries has been the focus of most studies on the development of reproductive cancers in the female. The complex history of these studies is summarized below.

Dietary fat and dietary fiber: an unexpected finding

Numerous epidemiological surveys have found positive correlations between the risk of breast cancer and high dietary fat and low dietary fiber intake. In experimental studies, a high-fat diet in mice significantly increases the incidence of mammary gland tumors at all caloric levels. The increased consumption of dietary fat and decreased consumption of dietary fiber in modern times is a reflection, in part, of the transition from a preagricultural to a postagricultural lifestyle, particularly the increased consumption of meat and so-called "fast foods." The hunter-gatherer diet provided typically about 20 percent of the caloric intake as fat. During the 1970s, 41 percent of the caloric intake of American females was in fat. In the 1980s this went down to 36 percent, quite likely in response to widespread efforts to appraise people of the benefits of reducing fat intake. Women in Japan, where fat constitutes about 19 percent of the caloric intake, have a fivefold less breast cancer risk. The typical American diet provides about 15 grams/day of fiber, while it is estimated that the hunter-gatherer diet provided about 100 grams/day of fiber.

The direct relationship between fat and breast cancer and the inverse relationship between breast cancer and dietary fiber have been attributed to an increase in estrogen levels. In the case of fat, adipose tissue is known to contribute to the total estrogen levels. It is not yet clear how to explain the protective effects of a high-fiber diet. One possibility is that some components of a high-fiber diet act as antiestrogens, that is, behave as estrogen antagonists; another possibility is that they may reduce the biologically active concentrations of estrogens in the blood. Hence, for both high fat and low fiber, there seems to be a natural explanation for their contribution to the breast cancer risk.

Unexpectedly and perhaps counterintuitively, the most definitive and reliable epidemiological data available today indicates no causative association between saturated, monounsaturated, or polyunsaturated fat and the risk of breast cancer in pre- or postmenopausal women. The relative risk for various levels of energy from fat in a comprehensive study involving 337,819 women is shown in Fig. 18.6. A relative risk of 1.00 means that no increase in breast cancer incidence was observed at a given fat category, compared to that of 30 to 35 percent of total energy from fat. Hence, for example, in diets characterized by less than 20 percent energy from fat, the relative risk was about 1.05. Even at the highest levels of energy from fat, the relative risk is not significantly different from 1.00. The interpretation of this result is that in Western societies, lowering intake of fat is unlikely to reduce breast cancer risk appreciably. Similarly, another recent comprehensive study has shown that there is no relationship between dietary fiber and breast cancer.

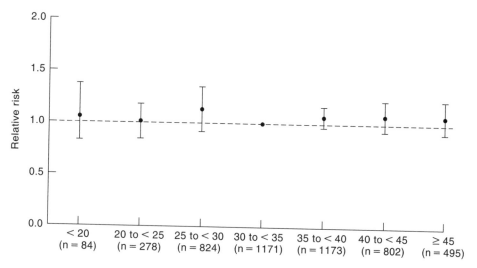

Figure 18.6
Relationship between dietary fat intake and breast cancer risk. The intake fat category of 30 to 35 percent of total energy from fat is used as the reference category. The risk at other fat intake levels is normalized to the reference category. The result from this comprehensive study indicates that there is no causative relationship between fat intake and the incidence of breast cancer. (Adapted from D. J. Hunter *et al.* [1996] *New England Journal of Medicine* 334[6], 356–361. Copyright 1996, Massachusetts Medical Society. All rights reserved.)

These two results run counter to the prevailing view of the causative relationships between cancer and dietary factors. Many investigators still consider that diet has to be an important factor in the risk equation. Perhaps the strongest evidence of a diet-related effect is the consistent increase in breast cancer in women who are tall or obese. *If it is not fat or lack of fiber, then what is the critical factor?* At present, we can only speculate. One suggested possibility is that the high caloric diets in the industrialized countries, whether from fat or carbohydrates, result in an energy imbalance, in that the diet provides much more energy than is used. A few studies are beginning to suggest that the effects of an energy-rich diet may be most acute during the prepubertal and early postpubertal periods of a female's life. One emerging hypothesis is that a higher energy intake during childhood, the prepubertal period, and adolescence may be an important predisposing factor in the development of breast cancer. None of the studies to date have examined this question properly. Only with future studies will we be able to determine the influence of the childhood diet on breast cancer risk.

Alcohol

Alcohol is emerging as an important risk factor for a number of cancers, particularly mouth, pharynx, esophagus, and liver cancer. There has also been a consistently positive association between alcohol consumption and breast cancer. As we saw in Chapter 11, the most definitive evidence has come from the critical role of alcohol in increasing the risk of breast cancer in postmenopausal women undergoing ERT. The

Table 18.5
Relative Breast Cancer Risk and Alcohol Intake

Alcohol intake (grams/day)	Relative risk
Nondrinkers	1.00
0–1.5	1.07
1.5–5	0.99
5–15	1.06
15–30	1.16
30–60	1.41

In the United States, mean alcohol content is 13.3 grams per bottle or can of beer, 10.8 grams for a glass of wine, and 15.1 grams for a shot of hard liquor. (Adapted with permission from Smith-Warner, et al. [1998] *Journal of the American Medical Association* 279[7], 535–540. Copyright 1998, American Medical Association.)

interpretation of that effect is that alcohol increases estrogen levels by about 300 percent, permitting estrogen levels to reach the cancer threshold. The most recent and methodologically reliable studies of cancers in *premenopausal* women indicate that the breast cancer incidence increases linearly with alcohol consumption. A summary of the latest comprehensive study involving 322,647 women who were followed for 11 years is given in Table 18.5. Women consuming 30 to 60 grams/day of alcohol (approximately 2.3–4.5 bottles of beer, 2.8–5.6 glasses of wine, or 2.0–4.0 shots of hard liquor) had a 41 percent higher risk of breast cancer than nondrinkers (Table 18.5). This result suggests that decreasing alcohol intake among women who consume alcohol regularly can potentially reduce breast cancer risk.

Other dietary factors

Much attention has been focused on the beneficial effects of fruit, vegetable, and/ or cereal product consumption with regard to cancer risk. Certainly, many epidemiological studies suggest an inverse relationship between the total cancer risk and consumption of fruits and vegetables. The protective effects of these dietary factors have been attributed to their antioxidant activity. However, the relationship between antioxidant activity and breast cancer specifically is more tenuous. Long-term studies designed to test the relationship between different dietary components and breast cancer risk are under way, but results are not expected for a number of years.

SUMMARY

There is growing consensus that the high incidence of breast, ovarian, and uterine cancers in Western societies can be attributed in part to the departure from the hunter-gatherer way of life (Table 18.3). This departure has resulted in exposure to significantly higher estrogen levels during a woman's lifetime. These higher levels may be the critical factor in triggering neoplastic growth in these tissues. Current estimates suggest that the modern reproductive profile leads to a 10- to 200-fold increase in risk compared to our ancestors. A reasonable approach for reducing the risk is to use exogenous hormonal administration to approximate the microanatomical and hormonal environment of the ancestral one. It is clear, however, that the search for preventive strategies proceeds in parallel with a full, societal debate about the need and desirability of such measures.

The relative importance of nonreproductive risk factors associated with life in the industrialized societies, such as high caloric and fat intake, low fiber intake, alcohol consumption, lack of physical activity, and diets low in fruits and vegetables, remains controversial. Current studies suggest that diets low in fat or high in fiber are not associated with lower cancer risk. On the other hand, more than moderate alcohol consumption is associated with increased breast cancer risk. High caloric diets (whether from fat or carbohydrates), especially during adolescence, have been suggested as being an important risk factor for reproductive cancers. Studies to test the importance of increased consumption of fruits and vegetables have been initiated.

REPRODUCTIVE CANCERS IN THE MALE

Prostate cancer is the most important reproductive cancer in males in the United States and most European countries (Fig. 18.1). It is the most commonly diagnosed cancer in men, and it is also the second leading cause of male cancer deaths in the United States. Its incidence appears to be increasing, although some of that increase may be due to increased ability to detect it early. Like most other cancers, prostate cancer is a disease of old age (Table 18.1).

Two forms of prostate cancer

The relationship between incidence and mortality of prostate cancer is more complex than with other cancers. From numerous autopsy studies it is known, for example, that a *latent* or *clinically insignificant form* of prostate cancer occurs in over 30 percent of men over 50 years old. This fraction increases with age, so that probably most men over 70 have developed this latent form of cancer. The incidence of this latent form of prostate cancer appears to be very similar among many different populations around the world. This suggests that the factors important in initiation of prostate cancer are probably the same in all populations. On the other hand, the mortality rates differ significantly from one country to another.

The mortality is currently attributed to the transformation of the latent form into the aggressive and clinically significant type. The factors responsible for this transformation remain poorly understood, but epidemiological studies indicate that environmental and lifestyle factors may play a very important role in the pathogenesis of prostate cancer. Comparisons of prostate cancer mortality among many countries have shown that the United States has one of the highest rates in the world, and that among men in the United States, African-American males have the highest rate. While the rate of latent prostate cancer is the same in United States and native Japanese and Chinese men, the incidence of clinically significant prostate cancer is about 100-fold higher in the United States. Japanese and Chinese males native to their countries have the lowest rates in the world. On the other hand, prostate cancer rates in Japanese and Chinese men who have immigrated to the United States approach the U.S. average within one generation. Significantly, in Japan, the rates of prostate cancer have been increasing, particularly among men who have adopted a more Westernized diet. These different studies have suggested that dietary factors associated with the Western diet may be especially important in promoting or stimulating the transformation of the latent form into the aggressive type of prostate cancer.

Dietary factors

Several dietary components have been identified in epidemiological studies. In the middle 1990s, the major focus was on fat. Many early studies showed a consistent association between fat, especially animal fat, and prostate cancer. The association seemed plausible especially in view of the observation that for the average American male, dietary fat constituted about 35 percent of the caloric intake, and only about 15 percent in Asian men. A causal connection between the two was considered likely. However, more recent epidemiological studies and analyses of the mechanistic basis for fat-mediated carcinogenesis suggest that the fat–prostate cancer association is not as compelling as previously thought. At present, therefore, while fat remains a possible prostate cancer risk, no specific fat components that may pose an important risk have been identified.

There are also some epidemiological studies that suggest an inverse relationship between dietary fiber and prostate cancer. A diet low in fiber is a risk factor for prostate cancer, while a diet high in fiber appears to confer a protective effect. For example, comparison of nonvegetarian and lactovegetarian Seventh-Day Adventists revealed that those who had a high intake of beans, lentils, and peas (good sources of fiber) had a significantly lower prostate cancer risk.

Other studies suggest a protective role for a class of compounds known as phytonutrients. These are a diverse group of compounds found in edible plants that have long formed an important part of the diet of many human societies around the world. A number of these phytonutrients may be important in the prevention of disease. One important source of a number of phytonutrients is the soy bean. Asians in general consume large quantities of soy-based products. The particular beneficial effects of the soy bean with regard to cancer have been attributed to a class of compounds known as *isoflavones*, the two most important of which are *genistein* and *daidzein*. The typical Japanese diet provides about 50 milligrams/day of isoflavones, while the typical American diet provides only about 2 to 4 milligrams/day. It should be emphasized that the presumed protective effects of phytonutrients or dietary fiber in the case of prostate cancer have not yet been established definitively.

It is very important to keep epidemiological studies in proper perspective. They can only establish correlations between certain factors and a particular disease. They do not establish causation. The correlations may in fact be misleading, as the example of dietary fat and breast cancer reviewed above has shown. The value of the epidemiological studies is that they may provide important clues about the causative factors or even mechanisms involved in the etiology (origin) and pathogenesis of a given disease. These clues can then be explored more definitively by appropriate experimental and clinical studies.

Etiology of prostate cancer

The current model of the origin of cancer proposes that cancers arise from oxidative damage to DNA. This damage results in the accumulation of mutations in genes whose normal function is to regulate cellular growth. Factors or conditions that promote oxidative damage, or that perturb mechanisms for controlling cell proliferation, are risk factors for cancer. For cancers of the breast, endometrium, and ovary, prolonged exposure to estrogen has been shown to be perhaps the most important causative factor.

The causative factors for prostate cancer are much less well understood. As indicated above, dietary components have been implicated in the aggressive forms of prostate cancer. Little is known, however, about the etiology of the latent form. However, since the prostate is an androgen-dependent tissue (recall that the growth, development, and function of the prostate are dependent on androgens, particularly DHT), some aspect of androgen metabolism has been considered to play a role in the origin of prostate cancer. Prostate cancer could arise because of high androgen levels within the prostate, perturbations in androgen-receptor function, or genetic differences in 5-alpha-reductase, the enzyme that converts testosterone to DHT. Prostate cancers are typically androgen dependent. Eunuchs and men with congenital abnormalities of androgen production do not develop prostate cancer. A long-standing therapy for certain types of prostate cancers is to administer *antiandrogens*, or in cases where the cancer is developing rapidly, to perform an orchidectomy, as a way of suppressing the growth of the cancer. However, as many investigators have noted, androgens by themselves cannot explain the origin of prostate cancers. If that were the case, we would expect prostate cancer to be more common among young men, since they have higher androgen levels and they are more responsive to androgens. But prostate cancer is relatively rare in young men. Androgens may be important for the continued proliferation of prostate cancer cells once they have been triggered to become cancerous, but it is not clear that they are responsible for the origin of the cancer.

A recent alternative hypothesis suggests that estrogens may play an important role in the origin of prostate cancer. Estrogen appears to increase as men age, in part because testicular conversion of androgens to estrogen increases with age, and also because of the increased estrogen production by peripheral tissues. Some investigators suggest that the critical parameter may not be the estrogen levels *per se*, but the estrogen/androgen ratio. Hence, an age-dependent increase in the estrogen/androgen ratio may account for the observation that the latent form of prostate cancer is very similar among different populations. The latent form may depend on age-dependent contributions to the estrogen/androgen ratio that are not affected significantly by environmental factors. Although the data on this question is still fragmentary, recent studies suggest that some breakdown products of estrogen may be causative factors in damaging DNA. Compounds that inhibit the growth of prostate cancer may do so because they lower the estrogen levels or because they counteract the DNA-damaging effects of some estrogen by-products. Obviously, many details still need to be worked out. Nevertheless, the attractive feature of this hypothesis is that it provides a unified way of understanding the origin and pathogenesis of cancers of the reproductive tissues in both females and males.

SUMMARY

Prostate cancer occurs in two forms—a latent, clinically insignificant form whose incidence increases with age and appears to be the same in populations around the world, and an aggressive, clinically significant form characterized by a high mortality rate. The high incidence of the aggressive form of prostate cancer in the United States and other industrialized countries compared to that in Japan and China has been attributed to dietary factors. Although epidemiological studies have suggested that diets high in animal fat and low in fiber and some phytonutrients found in soy products may be important risk factors in the development of the aggressive forms of prostate cancer, a convincing causal association has not yet been established.

Since most prostate cancers are androgen dependent, androgens have been considered to play a causative role in prostate cancer. However, the specific role that androgens might play has been difficult to elucidate. An alternative hypothesis suggests that estrogens may have a role in the development of the latent form of prostate cancer. The age-dependent increase in androgen to estrogen conversion, and recent findings that estrogen by-products may promote oxidative damage to DNA suggest a role for estrogen in the origin of prostate cancer. Perhaps surprisingly, prostate, breast, endometrial, and ovarian cancers may be traced to a common mechanism.

QUESTIONS

1. Death rates for some, but not all, cancers have decreased significantly during the last 20 or so years. What accounts for this change?

2. Reproductive and nonreproductive risk factors have been identified for cancers of the reproductive tissues in the female. Which appear to be the most important and why?

3. Which of the suggested preventive measures for reproductive cancers in the female do you find the most acceptable or feasible, and why?

4. Prostate cancer differs from breast cancer in that environmental factors appear to be much more important in its etiology. What type of evidence suggests this conclusion?

5. Library project:
   a. What types of therapies are available for prostate cancer, and how successful are they?
   b. Describe initial studies that suggest that high caloric diets in adolescence may be a significant risk factor for reproductive cancers in the female.

SUPPLEMENTARY READING

Eaton, S. B., *et al.* 1994. Women's reproductive cancers in evolutionary context. *Quarterly Review of Biology* 69(3), 353–367.

Garnick, M. B. 1994. The dilemma of prostate cancer. *Scientific American* 270(4), 72–81.

Garnick, M. B., and W. R. Fair. 1998. Combating prostate cancer. *Scientific American* 279(6), 75–83.

Gerber, M. 1998. Fibre and breast cancer. *European Journal of Cancer Prevention* 7 **Suppl. 2**, S63–S67.

Greenlee, R. T., *et al.* 2000. Cancer statistics, 2000. *CA Cancer Journal for Clinicians* 50, 7–33.

Haas, G. P., and W. A. Sakr. 1997. Epidemiology of prostate cancer. *CA Cancer Journal for Clinicians* 47(5), 273–287.

Kohlmeier, L., and M. Mendez. 1997. Controversies surrounding diet and breast cancer. *Proceedings of the Nutrition Society* 56(1B), 369–382.

Raloff, J. 1997. Radical prostates: Female hormones may play a pivotal role in a distinctly male epidemic. *Science News* 151, 126–127.

INTERNET SITES

Mammary Gland Biology: www.mammary.nih.org
National Cancer Institute: www.cancergenetics.org

ADVANCED TOPICS

Ames, B. N., L. S. Gold, and W. C. Willett. 1995. The causes and prevention of cancer. *Proceedings of the National Academy of Sciences (USA)* **92**, 5258–5265.

Fair, W. R., N. E. Flesher, and W. Heston. 1997. Cancer of the prostate: A nutritional disease? *Urology* **50**(6), 840–848.

Howe, G. R. 1997. Nutrition and breast cancer. In *Preventive Nutrition: The Comprehensive Guide for Health Professionals*. A. Bendich and R. J. Deckelbaum. Eds. Humana Press Inc., Totowa, NJ, pp. 97–107.

Humfrey, C. C. 1998. Phytoestrogens and human health effects: Weighing up the current evidence. *Natural Toxins* **6**(2), 51–59.

Hunter, D. J., *et al.* 1996. Cohort studies of fat intake and the risk of breast cancer—a pooled analysis. *The New England Journal of Medicine* **334**, 356–361.

Smith-Warner, S. A., *et al.* 1998. Alcohol and breast cancer in women: A pooled analysis of cohort studies. *Journal of the American Medical Association* **279**(7), 535–540.

Strauss, L., *et al.* 1998. Dietary phytoestrogens and their role in hormonally dependent disease. *Toxicology Letters* **102–103**, 349–354.

Joan Miró, *Ohne Titel* (*Without Title*), 1924.
(© 1999 Artists Rights Society [ARS], New York/ADAGP, Paris.)

CHAPTER 19

The Biology of Human Sexuality

Sexuality emerges from the interdependencies of biology,
awareness, and the facts and artifacts of public life.

—G. Berman. 1995

To speak in chords about sexuality. *Neuroscience and*
Biobehavioral Review 19(2), 343–348.

What can one make of traits that seem so evidently to defeat
the biological imperative of optimizing reproductive success?

—R. C. Pillard and J. M. Bailey. 1998

Human sexual orientation has a heritable component.
Human Biology 70(2), 347–365.

Ne lave pas, je reviens.

—Message sent by Napoleon to Josephine
before returning from one of his campaigns

Cited in I. Pollard. 1994. *A Guide to Reproduction.*
Cambridge University Press, Cambridge, UK, p. 106.

Aʟʟ ᴏғ ᴜs have an intuitive understanding of human sexuality, even if we may find it difficult to define. We may be in the same position as one of our U.S. Supreme Court justices regarding the definition of pornography: "*I may not be able to define it, but I know it when I see it.*" There is no doubt that our sexuality is an elemental force in our lives, one that has been sung and written about for thousands of years. We can distinguish some of its many faces and have given names to some of them: sexual desire, sexual feelings, sexual arousal, sexual energy, sexual drive, sexual capacity, sexual (gender) roles, sexual orientation, sexual (gender) identity, and sexual attraction. These terms describe different manifestations of an underlying sexuality that all humans share. Yet the nature of our sexuality, its origin, and its purpose remain perplexingly mysterious, still outside the pale of our understanding.

We recognize that at some fundamental level sexuality in all its complexity must have a biological substratum. In this sense, trying to understand sexuality may be no different than trying to understand other complex human traits, such as memory, intelligence, or consciousness. The study of human sexuality has been plagued by repeated attempts to extrapolate findings in experimental animals to the human situation. Especially because of the enormous range and flexibility in the expression of our sexuality, animals are unlikely to be good models for understanding human sexuality. This expansion of sexuality is presumably due to the increased complexity and organization of the human brain. Animal studies have been useful, however, in identifying physiological and hormonal parameters that might serve as starting points for human studies. It is clear that we need to understand the connection between genes and the central nervous system structures essential for its development and full expression. This is the research agenda in the current efforts to lay the foundation for the study of human sexuality. We are not anywhere near that goal yet.

Our objective in this chapter will be limited to examining how far our present knowledge of neuroendocrinology and endocrinology will take us in understanding five aspects to human sexuality: *gender traits*, *sexual drive* (libido), *sexual orientation*, *gender identity*, and *sexual attraction*.

GENDER TRAITS: FEMININE VERSUS MASCULINE

During the past 50 to 100 years, the biological and social sciences have provided alternative explanations for the origin of gender traits, what we generally refer to as masculine and feminine traits. These are characteristics that are said to distinguish the two sexes, and they include a variety of psychological and behavioral characteristics and perhaps even cognitive abilities. For example, it is common to say that males are *aggressive* while females are *caring*; females are *cooperative*, while males are *competitive*; boys prefer *outdoor* play and like *roughhousing*, while females prefer *indoor* and *less active* play; males are better at *mathematical tasks*, while females are better at *verbal tasks*; males have a better *spatial sense*, while females have a better *memory*, etc. Are these presumed differences innate, or are they fashioned by culture, by the way in which boys and girls are treated even when they are babies?

The nature hypothesis

The phrase "anatomy is destiny" captures the essence of the strict biological perspective in explaining the origin of gender traits. This perspective, the **nature hypothesis**, derives from the obvious and decisive influence that steroid sex hormones have in determining the female and male anatomy and physiology. The anatomical differences between the sexes, so compelling and unarguable, are considered to reflect a deeper underlying dichotomy that extends beyond the anatomical and physiological. According to this view, "brain sex," like phenotypic sex, develops ultimately from the action of the two principal gonadal hormones (estrogens and androgens) on critical regions of the brain. The male brain differs from the female brain because it has been "programmed" by testosterone and not by estradiol. In term of the nature hypothesis, this difference in programming accounts for the behavioral, psychological, and even cognitive differences between the two sexes.

The programming that converts a nonsexual brain to a sexual one, by analogy with the development of phenotypic sex, is considered to take place at certain critical periods during gestation, and possibly even during the first few months after birth. Hence, according to the **critical-period hypothesis**, the steroid hormones, testosterone and estradiol, "organize" certain regions of the brain responsible for sexual behavior into the masculine and feminine, respectively. Additional changes may take place as a consequence of the postnatal gonadotropin surge that is observed in both males and females. This programming may manifest itself during childhood, but the complete maturation of the sexual differentiation of the brain takes place after puberty, when the previously programmed regions are fully activated by the increased circulating levels of gonadal hormones. The organizational effects are considered to be permanent and irreversible.

The original evidence that led to the critical-period model came from experiments in the 1950s that examined the effects of injected testosterone into pregnant guinea pigs at certain times during gestation. The genetically female offspring of such mothers had male genitalia and they displayed the typically male mating behavior of "mounting." Nonmasculinized female guinea pigs, on the other hand, displayed a stereotypical mating behavior, an arching of the back, called *lordosis*. The masculinization of the mating behavior produced by testosterone was taken to mean that certain brain structures had been altered in a permanent way. Brain sex, then, was the natural consequence of the action of gonadal hormones on sensitive areas of the brain. This result was quickly extrapolated to other species, and it provided a straightforward way to explain the origin of gender traits in humans: males are masculine and females are feminine because of the way their brains have been programmed by the appropriate sex hormone. The nature hypothesis has a compelling neatness to it, and certainly reinforces the common conviction that there are essential differences between males and females.

The nurture hypothesis

An alternative way of looking at the origin of gender traits—the **nurture hypothesis**—has come from the social sciences. Comparative studies of many different societies have pointed out the multitude of ways in which society and culture mold and affect behavior, including sexual behavior. The nurture hypothesis posits that we learn or are taught, implicitly and explicitly, to behave in a manner that is appropriate for our anatomical sex. Boys even from a very young age are rewarded when they exhibit boyish behavior, and strongly discouraged from expressing what we consider girlish or "sissy" behavior. Feminine and masculine traits in this view do not arise from hormonal programming, but instead are arbitrary products of a sustained educational (some would say brainwashing) program that begins soon after birth, and that in fact continues throughout our lives. Hence, according to this hypothesis, gender traits are social, rather than biological, in origin.

The nurture hypothesis arises from the observation that socialization affects the expression of gender traits. It was originally considered to apply only to humans. However, recent studies suggest that it may have some validity even in the case of nonhuman species, even in the ones that originally provided the evidence for the nature hypothesis. For example, the sexual behavior of the adult rodent does not depend on testosterone *per se*, but on the amount of licking a pup receives in its anogenital region. Normally, the licking is done by the mother. Male pups get more attention because of the testosterone that they secrete in the urine. The licking can be mimicked by stroking the anogenital region with a paintbrush each day. Female pups stimulated in this manner after birth can develop masculine copulatory behavior. Moreover, these behavioral changes are associated with changes in pituitary function and anatomical changes in motor neurons in the lumbar spinal cord. Studies in monkeys have also shown that copulatory behavior in these species is much more plastic than originally thought. The social and physical context of the experimental situation can profoundly alter the outcome, so that both sexes are easily capable of displaying both masculine and feminine traits. These experiments question the *exclusive* role of gonadal hormone programming in establishing masculine or feminine traits.

SUMMARY

The nature vs. nurture dilemma is a false dichotomy. Gender traits are unlikely to arise in an either/or fashion. The so-called stereotypical mating behaviors so much studied in rodents may rarely be that. Both steroid hormones and socialization appear to be capable of determining and modifying gender traits. Masculine and feminine traits are not necessarily associated with anatomical sex. Hence, in any given individual, gender traits arise from the interplay of biological and cultural factors. This is certainly not a novel conclusion, but it may be more widely accepted by biological and social scientists now than in the past. Perhaps one of the most important inferences that can be derived from this point of view is that gender traits are not unidimensional. Masculine and feminine traits are not opposing poles in a one-dimensional continuum, where being more feminine invariably implies being less masculine, and vice versa. A more attractive alternative may be that gender traits are multidimensional. Neither masculinity nor femininity is defined by its opposite, nor are they determined in some irreversible way by gestational hormones. In this multidimensional world, an individual, irrespective of anatomy, could be very masculine and very feminine simultaneously, or perhaps neither.

SEXUAL DRIVE (LIBIDO)

The **libido** (Latin for *lust*) generally refers to conscious or unconscious sexual desire. The term is usually applied only with reference to human sexual desire or sexual behavior. Animals, however, do exhibit a variety of behaviors that are clearly sexually differentiated—for example, the frequency of mounting a receptive female, intromission, and ejaculation in males, and in females, receptivity to the male, and assumption of body positions that facilitate copulation. The relationship between sexual behavior and the gonads was clearly articulated as we saw in Chapters 5 and 6 by Aristotle, who noted the effects of castration on the sexual behavior of both male and female animals. In female animals, removal of the ovaries eliminates the periodic episodes of estrus, or "heat," that mark the time of ovulation. In male animals, castration either eliminates or reduces male sexual behavior. However, in male animals the length of time required for the loss of sexual behavior to be manifested fully is quite variable, ranging from a few days in rodents to months or even years in dogs, cats, and monkeys. The older the animal is at the time of castration and the more sexual experience he has had, the longer it takes to lose the sexual behavior.

Appropriate sexual behavior in castrated male animals can usually be restored by exogenous administration of androgens. The time required for restoration depends on how long the animal has been castrated and the dose and type of androgen used. Studies of this type have shown that there is a threshold androgen level that will restore the precastration sexual behavior. Increasing the dose beyond the threshold value does *not* produce any further intensification in sexual behavior.

Human males

The sexual drive in humans is more difficult to study than in animals, perhaps because it is less stereotypical, less innate, more subject to modulation by psychosocial context. Typical parameters that are evaluated in males are sexual thoughts and arousal; response to erotic stimuli; intensity of sexual feelings; frequency, duration, and magnitude of spontaneous and nocturnal erections; masturbation; and frequency of sexual intercourse, with and without ejaculation. The most informative of the studies that provide some insight into the relationship between gonadal hormones and the sexual drive have examined the effects of androgens in males.

In analogy with the animal studies, we might expect that androgens would play an important role in the sexual drive of human males, and many studies have been carried out to examine this question. Despite the complexity of the methodology of the studies and their often complex findings, one general conclusion seems possible from all of these studies: *androgens appear to be necessary, but not sufficient, for normal male sexuality.* Most of the parameters examined exhibit an androgen threshold effect: the different behaviors are impaired below a certain androgen threshold, and androgen administration can restore normal behavior. Increasing the androgen levels beyond the threshold does not produce any intensification of sexual behaviors. Likewise, no intensification of sexual behavior is achieved by administering additional androgens to men with normal androgen levels. Androgen concentrations necessary to maintain normal sexual behavior appear to be lower than those required to maintain normal spermatogenesis. However, not all aspects of the male sexual drive are androgen dependent. Erections in response to erotic films, for example, can be obtained in men who suffer from androgen deficiency. On the other hand, such men, in contrast to normal men, are less able to elicit erections when asked to fantasize. Taken together, these observations suggest that some minimal androgen level is required for most but not necessarily all types of male sexual behavior.

The relationship between androgens and sexual behavior is not as straightforward as this general conclusion may imply, however. The complexity of the relationship is suggested by studies in which a hypogonadal state is induced by the administration of GnRH antagonists. As we saw in previous chapters, such compounds block the action of endogenous GnRH, resulting ultimately in suppression of testicular function. In a recent placebo-controlled, double-blind study, men receiving the GnRH antagonist *Nal-Glu* experienced significant decreases in the frequency of sexual desire, fantasy, masturbation, and intercourse, all of which were restored with exogenous testosterone. However, one significant and perplexing observation was that although administration of Nal-Glu reduced testosterone to castration levels in less than a week, the behavioral changes did not begin to manifest themselves until after 4 to 6 weeks of treatment. The reason for this time delay remains unexplained. The temporal decoupling of sexual behavior from androgen levels does not invalidate the inference that some minimal level of androgens is required for sexual functioning, but it does suggest that neural structures necessary for the sexual drive retain functionality even in the absence of androgens.

Human females

In human females, it is even more difficult to define parameters that measure sexual desire in a reliable way. The role of gonadal hormones as determinants of female sexuality remains largely unresolved. Attempts to establish unequivocal correlations between cyclic patterns of testosterone or estrogen during the ovulatory cycle and different measures of sexual activity in the female have been unsuccessful. There appears to be no predictable relationship between sexual drive and ovarian hormones. Some investigators suggest, although without much evidence, that, as in the male, some threshold androgen level is necessary for normal sexual behavior in females. This level may be the "tonic" levels of androgens that are present throughout the cycle. That is, the tonic levels may be more important than the normal oscillations that take place during the cycle.

Societal influences may tend to mask the hormonal basis of female sexual behavior in very significant ways. All societies appear to invest more effort in controlling female sexuality than they do in controlling male sexuality. Males generally encounter a more favorable sexual environment than females. An important consequence of a more hostile sexual environment for females is that societal factors play a more important role in regulating the full expression of female sexuality. The importance of these inhibitory societal factors may diminish with age. For example, the incidence of female orgasm, which is usually taken as an important measure of female sexuality, increases with age. Hence, in the female, some aspects of sexual drive may increase with age without any concomitant increase in hormone levels.

SUMMARY

Under normal conditions, androgens provide the necessary basis for the development and maintenance of most, but not all, aspects, of the male libido, but it is also clear that androgens alone are not sufficient. The precise function of androgens remains largely unknown. It is not yet known how they program or activate the neural substrates that are necessary for the expression of the libido. The fact that the libido may be absent or greatly reduced under conditions when no other obvious abnormality is present indicates that other undefined psychosexual or psychogenic factors must play an important role in its maintenance. Indeed, the diminution of the libido that is experienced by many males as they age has been attributed at least in part to the greater role that psychosexual factors may play in older males. The role of gonadal hormones in determining or modulating female sexuality remains enigmatic. Psychosocial factors are considered to play a dominant role particularly in young females, but their importance may decrease with age.

SEXUAL ORIENTATION

Although being sexually attracted to the opposite sex (*heterosexuality*) is the norm, same-sex attraction (*homosexuality*) has been found consistently in all societies. In our time, for generally religious and political reasons, questions about the nature of sexual orientation and the extent to which it is innate or culturally determined have assumed great importance. The often vociferous debate of these questions has probably not shed much light on the origins of sexual orientation. In this section, we will review briefly the search for the biological basis of sexual orientation.

Conceptualization of homosexuality

The modern codification of the human trait that we call sexual orientation into categories such as *homosexual* and *heterosexual* is a relatively recent phenomenon, beginning perhaps 100 years ago. The term *homosexual* was only coined in 1869. A more recent term, considered to be less judgmental, is "gender atypical," which refers to behaviors that are not typical of a given sex. Before the end of the nineteenth century, homosexual acts were forbidden by both civil and church law in Europe, but the person who committed the forbidden act was not generally considered to belong to a special category of persons. A significant change in the perception of homosexual acts came with the development of psychiatry, which began to consider that a person who committed such acts was an "invert," a deviant. Only sexual activity that could be productive was seen as normal or natural, while sexual activity that precluded procreation was seen as unnatural. Homosexuality began to be seen as a pathological condition. This concept of sexuality was reinforced in the 1920s with the discovery and gradual appreciation of the role of the gonadal hormones, estrogens and androgens. Maleness and femaleness began to be considered as two dichotomous hormonal states. This view was expressed quite early by the zoologist F. R. Lillie in 1939 in the first textbook of reproductive endocrinology, *Sex and Secretions*:

> there are two sets of sex characters, so there are two sex hormones, the male hormone controlling the "dependent" male characters, and the female determining the "dependent" female characters.[1]

The newly developing science of endocrinology helped to establish the standard way of looking at human sexuality, the conviction that there are only two natural types or categories:

> one with female reproductive capacity, feminine behavior and a sexuality oriented towards men, the other with male reproductive capacity, masculine behavior, and sexuality oriented towards women.[2]

These notions were buttressed by psychoanalytic theories that emphasized the roles of the mother or father, or family dynamics in the genesis of homosexuality. However, things began to change in the 1960s when the psychoanalytical underpinnings of the prevailing views of homosexuality began to be questioned in the medical and scientific communities. Many studies of sexual behavior and psychosexual orientation have shown that neither a hostile or distant father nor an overly protective mother could be considered the cause of homosexuality. "Unnatural" acts could not be considered unnatural when large numbers of people admitted to committing them.

One important aspect about homosexuality is its historical ubiquity. A recent historical survey of homosexuality presents the following summary: "(1) same-sex eroticism has existed for thousands of years in vastly different times and cultures; (2) in some cultures, same-sex eroticism was accepted as a normal aspect of human sexuality, practiced by nearly all individuals some of the time; (3) in nearly every culture that has been examined in any detail, a few individuals seem to experience a compelling and abiding sexual orientation toward their own sex."[3]

The incidence of homosexuality has been the subject of much controversy. The more recent estimates suggest that in the United States 3–4 percent of males and 1–2

---

[1]Lillie, F. R. 1939. Biological introduction. In *Sex and Internal Secretions*, 2nd ed., E. Allen, Ed. Williams and Wilkins, Baltimore, MD. Cited in Gooren, 1995.

[2]Longino, H. 1990. *Science as Social Knowledge*. Princeton University Press, Princeton, NJ.

[3]Mondimore, F. M. 1996. *A Natural History of Homosexuality*. Johns Hopkins University Press, Baltimore, MD.

percent of females are exclusively homosexual. These surveys also suggest that the incidence of homosexuality has remained stable over several generations, and there is no compelling evidence that the incidence is increasing. The ubiquity of homosexuality has argued that there must be biological determinants, relatively insensitive to cultural influences, that account for the origin of such gender-atypical behavior.

The attitudes that shape society's view of homosexuals have changed remarkably since the 1970s. We have come a long way since a century ago when the categories of sexuality known as heterosexuality and homosexuality were first defined. Until the early 1970s, homosexuality was considered to be a mental disorder, and as such, physicians (mainly psychiatrists) had an obligation to "cure" it. Treatments that today would be considered completely unacceptable—electric shocks linked to photographs of nude males, hormone injections, castration of homosexual men and transplantation of testicular tissue from heterosexual men, removal of the hypothalamus—have been carried out at different times to cure homosexuality. The view that led to such practices came to an end with the 1973 declaration by the American Psychiatric Association that homosexuality was not, after all, a mental disorder.

Evolutionary perspective

From an evolutionary perspective, homosexuality presents an interesting problem. A trait that does not favor reproduction should have been lost in the course of human evolution. Its persistence suggests that it may have provided some evolutionary advantage at some time in the past. If homosexuality has a genetic basis, then the gene(s) responsible for it must have had a selective advantage, and this is why the trait persists. This argument presupposes that the principal purpose of human sexuality is reproduction. Not all investigators agree on this. Certainly, the long history of pregnancy prevention and induced abortions in human societies would suggest that reproduction is not the role of human sexuality. It is more likely that communication, bonding, and developing a stable social organization may be the true purposes of human sexuality. During our long prehistory, a flexible, rather than a fixed, sexual orientation may have been the norm, precisely because of its benefit in developing and increasing the socialization of our species.

The etiology of sexual orientation

Although familial and social factors/conditions may play a role in the origin of sexual orientation, the prevailing scientific view is that it is largely genetic (biological) in origin. The evidence for this view is not yet compelling, but it is consistent with modern ideas about the origin of complex behavior. Nevertheless, critics of the view that sexual orientation is primarily a biological phenomenon argue that it is too limiting, that it ignores the cognitive and contextual aspects of sexuality, which cannot be broken down into discrete biological parameters. Sexual orientation, they would argue, will always be more than the sum of its parts, because the interactions between its intrinsic and extrinsic elements will always be unpredictable. It may be somewhat ironic that the new gay and lesbian ideology, while rightly resisting the label of pathological or abnormal, has in general accepted the strict biological deterministic view. Indeed, the current tendency by both the progay and antigay groups is to polarize sexuality into antithetical categories of human beings, which is precisely where we were 100 years ago. The terms *gay* and *lesbian* do not reflect a biological essence that all gays or lesbians share, any more than we can identify any biological essence that all heterosexuals share. Such terms are more statements of political ideology than statements of biological fact. Homosexuals are as variable as heterosexuals.

The search for the biological basis of homosexuality

The search for the origin or cause of homosexuality has focused on trying to identify physiological or anatomical parameters that might distinguish the homosexual from the heterosexual. Three stages in this search can be discerned. They differ in the type of biological parameter that has been the subject of study.

The endocrine phase

An early model portrayed the male homosexual as an aberrant male, more like a female than a male. There was even a suggestion that the homosexual represented a third sex, in some way intermediate between male and female. It was imagined that male homosexuals would be characterized by "female" physical traits, such as wider hips, small genitalia, and reduced hairiness. Although this simplistic notion was quickly discarded, the idea of a female element embedded in the male homosexual persisted. With the discovery and increased understanding of the role of steroid sex hormones in the reproductive physiology of the female and the male, the female-like behavior of a male homosexual, that is, the desire for a male, was considered to arise from differences in sex hormone levels, either low androgen or high estrogen levels. Although some initial studies appeared to show differences, repeated studies have demonstrated that no differences in sex hormone synthesis or levels exist between heterosexual and homosexual males. Homosexuality, then, is not due to a difference in the sex steroid endocrinology in the adult.

The brain-anatomy phase

The next and more enduring hypothesis has been that homosexuality is due to some abnormality in brain anatomy. This idea has gone through several permutations. The original version was an extrapolation of the critical-period hypothesis developed in studies with rodents. The brain was said to be masculinized or feminized during a critical period in gestation by exposure to estrogens or androgens. Experiments to test this idea were carried out in monkeys by trying to elicit cross-sex behavior (generally measured by observing copulatory behavior) in castrated monkeys or in monkeys treated with androgens during gestation. All of these experiments have failed to provide compelling evidence that cross-sex behavior can be induced by exposure of the fetal brain to androgens or estrogens. Primate sexual behavior is significantly more plastic and less stereotypical than initially thought. But as we saw above, even in rodents copulatory behavior is not as dependent on gestational hormonal influences as the initial studies seemed to indicate.

The most recent stage in the brain-sex phase began in 1990 with reports of anatomical differences in different regions of the hypothalamus between male homosexual and heterosexuals. One concerned differences in the suprachiasmatic nucleus (SCN), and more recently, differences in the INAH3 nucleus. The INAH3, a very small cluster of cells in the hypothalamus, which earlier had been reported to be fractionally larger in men than in women, was also larger in heterosexual than in homosexual men. The analysis was based on the study of 19 brains from homosexual males who had died of AIDS, 16 brains from presumed heterosexual males, and 6 brains from heterosexual females.

This limited study was quickly interpreted as demonstrating that the brains of homosexuals were different from those of heterosexuals. However, this conclusion was heavily criticized on several important grounds. First, the INAH3 region is not a morphological structure with definable limits. Hence, it is in fact very difficult to measure its extent in an unambiguous way. Second, the range of sizes reported in homosexual men was enormous, spanning a 20-fold range. Moreover, there was significant overlap in sizes between homosexual and heterosexual men, with some homosexual men having an INAH3 size greater than heterosexual men. This means that INAH3 size is neither a unique nor an unambiguous determinant of homosexual behavior. No evidence was presented which distinguished between cause and effect. For example, it was pointed out that INAH3 size might reflect sexual activity rather than sexual orientation. In summary, a causal correlation between brain anatomy or brain function and homosexuality has not yet been unequivocally demonstrated.

The "homosexual genes" phase

The most recent attempt to identify the biological root of homosexuality has centered on the genetics of homosexuality. One significant impetus has been family studies that seem to suggest a strong genetic component to homosexuality. One such study involved comparisons of homosexuality in monozygotic and dizygotic twins. Recall from Chapter 12 that monozygotic (MZ) twins share 100 percent of their genes and a similar intrauterine environment, while dizygotic (DZ) twins share only 50 percent of their genes and a similar intrauterine environment. The simultaneous occurrence of a trait in both twins is referred to as the concordance of that trait. A trait found in both members of a twin pair is said to be 100 percent concordant. The concordance of a genetic trait would be expected to be higher in MZ compared to DZ twins, since MZ twins are more alike genetically.

The first large-scale twin studies comparing concordance rates for homosexuality in males and females were published in 1991 and 1993 (Table 19.1). For both male and female homosexuals the concordance for MZ twins is about 50 percent, significantly greater than that for DZ twins. This difference, it has been argued, reveals a genetic component. However, the concordance for adoptive sibs (that is, not genetically related at all) is only slightly lower than that for DZ twins in the case of male homosexuals, and almost three times lower for female homosexuals, but still significant. For those who had argued for a very strong genetic component to homosexuality, it was perhaps surprising that the concordance for MZ twins was not higher. A reasonable interpretation of these studies, despite a number of methodological problems, is that sexual orientation has a genetic component.

Table 19.1
Concordance Rates for Homosexuality from the Bailey-Pillard Studies (1991 and 1993)

	Monozygotic twins	Dizygotic twins	Adoptive sibs
Males	50%	24%	19%
Females	48%	16%	6%

Reference: Pillard and Bailey, 1998.

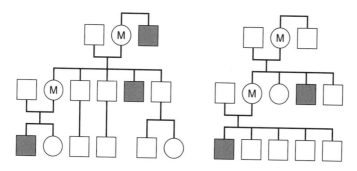

Figure 19.1
Examples of families that suggest an *X*-linked transmission pattern of male homosexuality. Homosexual men are indicated by the filled-in squares. Mothers who have homosexual brothers are indicated by the open circles marked with an M. Families of this type were used in studies by D. H. Hamer *et al.*, 1993 (see text).

A logical approach in trying to isolate the genetic component would be to examine families in which homosexuality appears to be inherited. Two examples of such families are shown in Fig. 19.1. In some of these families, male homosexuals frequently have maternal uncles who are also homosexual. Such an inheritance pattern would suggest that homosexuality could be associated with a gene on the *X* chromosome. The mothers would be carriers of the trait, but because they have another *X* chromosome, they would not manifest the trait. Some of the sons, however, might be expected to exhibit the trait. Two reports, published in 1993 and 1995, analyzed families of this type. The *X* chromosomes of 40 pairs of homosexual brothers whose family histories suggested *X*-linked inheritance of homosexuality were analyzed for DNA markers that they might have in common. The initial study found that 33 of the 40 pairs of male homosexuals shared the same DNA markers on a region (referred to as Xq28) of the tip of the long arm of the *X* chromosome. These findings were offered as evidence that at least in some males homosexuality might be associated with an *X*-linked gene. However, a recent Canadian study involving 52 pairs of homosexual brothers failed to find a linkage between homosexuality and the Xq28 region. Although the reasons for the differences between the two studies remain unexplained, it may be significant that other first reports of genetic linkage for other complex traits, such as schizophrenia, have also failed to stand up after follow-up studies.

SUMMARY

Numerous attempts to identify distinctive physiological, anatomical, or endocrinological parameters diagnostic for homosexuality have been unsuccessful. This, of course, does not mean that homosexuality does not have a biological basis, but simply that we are a long way from identifying what that basis is. The recent approaches that seek to identify genes that might play a role in sexual orientation are considered to be the most reasonable strategy in trying to decipher the elements of this complex trait. Taken at face value these still incomplete, but provocative, studies suggest that multiple genes may be involved in establishing the biological basis for sexual orientation. Although it is difficult to imagine at this stage in our understanding what these genes do, it seems reasonable to suppose that multiple genes may be involved. Sexual orientation is a complex phenomenon, varying in intensity and direction in any given individual with time, opportunity, and socialization.

GENDER IDENTITY

Gender identity, the sense of being female or being male, is, when we think about it, a difficult and mysterious concept. At present, there are no physiological parameters by which it can be measured or nuanced. In the end, we have to depend on what people say they feel about themselves. How does gender identity develop? Is it learned and acquired over time or is it hard-wired into the brain? Is it a conscious or unconscious sensation? Is gender identity a discrete property with only two possibilities or is there a continuum of identities? These questions raise profound issues that lie at the core of our sexuality. We normally assume that a person who is female anatomically will also be female psychosocially. So strong is this assumption that when surgical intervention is required in cases of congenital phenotypic sex disorders, the reconstruction is carried out to normalize the external genitalia as much as possible, and the child will be brought up in a way that is congruent with the external genitalia. We rarely question the view that psychosocial gender should be concordant with anatomical sex. It is only when we are confronted with cases in which the concordance does not hold that we begin to raise questions about the nature of sexual identity. Gender identity is also one aspect of our sexuality that is not accessible to independent confirmation by others.

Gender identity disorder (transsexuality)

Discordance between gender identity and anatomical sex is known commonly as *transsexuality*, or clinically as **gender identity disorder** (GID). Transsexuality is a rare condition. Estimates derived from recent studies in Europe suggest an incidence of about 1:20,000 males are male-to-female transsexuals and about 1:50,000 females are female-to-male transsexuals. The transsexual, anatomically of one sex, feels that she or he belongs to the other sex. For nontranssexuals, the discordance between gender identity and anatomy is impossible to conceive. The diagnosis of GID is a fairly stringent one, made only if there are no abnormalities or ambiguities in chromosomal patterns, gonads, external or internal genitalia, or circulating steroid sex hormone levels. These criteria are revealing because they suggest that gender identity is *independent* of all of the parameters by which one distinguishes the two sexes. If that were not the case, gender identity *would be* concordant with anatomy.

The transsexual does not deny his or her anatomy; rather, the anatomy is the source of psychic pain. The discordance has varying degrees of intensity, ranging from mild dissatisfaction and a vague sense of unease (a sense of being trapped in the wrong body) to a pronounced dislike, even loathing, for one's body, particularly its sexual characteristics. It is generally in these more extreme manifestations of the disassociation between identity and anatomy that surgical reconstruction is carried out. The first celebrated case of surgical reconstruction was that of George Jorgensen, a soldier in the U.S. Army, who in 1953 underwent a male-to-female reconstruction operation in Denmark to become Christine Jorgensen. At the time, this type of operation was not available in the United States. Since then, several thousand persons have undergone what are popularly called "sex-change operations" so that their anatomical sex matches their internally perceived gender. By and large, such surgery appears to reconcile body image and identity. After sex-reassignment surgery most transsexuals live as heterosexuals, find mates, and even establish families. It is also interesting to note that a small fraction establish homosexual relationships. There is a case, for example, of a man who fathered four children in a normal marriage, but finally at age 54 underwent sex-reassignment surgery because he had always thought of himself as female. After the surgery, "she" established a lesbian relationship that has lasted for over 15 years. Cases of this type show clearly that gender identity and sexual orientation are not equivalent, and can be separated.

SUMMARY

We remain abysmally ignorant of the origin of GID. It cannot correlated with any physiological, anatomical, neuroendocrinological, or genetic parameter that distinguishes the two sexes. The existence of GID raises profound questions about the nature of human sexuality. Gender identity presupposes cognition and consciousness. The question of gender identity makes no sense otherwise. This means that animal experiments cannot provide any information about the question of gender identity. Gender identity is a uniquely human problem.

SEXUAL ATTRACTION

Attraction between males and females depends on, among other things, visual, auditory, and olfactory cues. The importance of olfaction in the relationships between the sexes appears to be a constant in all societies. If olfactory cues were not important, the perfume industry would not exist. Anthropologists have long known that body scent is a sexual attractant in humans. The potent effects of body scent have been a staple of popular folklore. In nineteenth-century rural Austria, for example, girls would keep a slice of apple in their armpits during festival dances. At the end of the dance the girl would present the apple to the man of her choice who would, if the feeling was reciprocated, eat it readily. In other parts of Europe young men would woo girls by placing their handkerchief in their armpits during dances. The scented handkerchief was then offered to girls to wipe the perspiration from their face. It was considered that the axillary scent was so powerful that the girls would succumb to the young men's desires. The nature of the substances that would have such effects has long been of interest.

Pheromones

The term **pheromone** was coined in 1959 to designate compounds that insects use to communicate with each other. The term is a fusion of two Greek words, *pherein*, meaning "to transfer," and *hormon*, meaning "to arouse" or "excite." The term caught on in large part because the phenomenon it described, that is, the communication between animals by chemical signals, had been demonstrated or suspected for some time. The pheromonal concept was quickly extended from insects to mammals. Explicit in the idea of pheromones was that they were airborne (volatile) signals secreted by one individual and which elicited a behavioral response in other individuals of the same species. Such signals can be used for many different purposes (for example, a dog or a cat secretes compounds in the urine to mark territory). Pheromones, particularly in the case of mammals, have most often been considered to have a sexual or reproductive function, providing information about gender, dominance, and reproductive status, and inducing an innate pattern of behavior in an individual of the opposite sex.

In general, mammalian pheromones are found in discharges from the urogenital system, urine and feces, as well as in saliva, and secretions from the skin and hair. In some species, the pheromones are secreted by the male to elicit an appropriate sexual response in the female. In mice, for example, pheromones secreted by males can have dramatic effects on the physiology of female mice, ranging from induction of puberty to blockage of pregnancy. A pheromone secreted in the saliva of the wild boar, *androsterone*, will induce the sow to assume the position for coitus. Androsterone is structurally related to androgens, but does not function as an androgen since it does not bind to the androgen receptor. In the hamster, a polypeptide known as *aphrodisin*, secreted in the vaginal discharge of the female, has been shown to induce copulatory behavior in the male.

In a few mammalian species, specifically rodents and pigs, the existence of pheromones is considered to be well established. In humans, the existence or status of pheromones is somewhat controversial, in large part because it has been very difficult to identify human pheromones that elicit innate behavioral responses. Compounds such as androsterone and others related to it are known to be secreted by both human males and females, but their function in humans is unknown. Behavior in humans is much more complex than in other mammalian species, since it is modified extensively by learning and experience and by social contexts. It has been difficult to define unambiguously any innate behavioral traits in humans, and certainly none that can be correlated with specific chemical signals.

Two related problems will have to be addressed before the study of human pheromones achieves scientific respectability: (1) unambiguous pheromonal effects will have to be described; and (2) the compounds associated with such effects need to be identified and plausible mechanisms by which putative pheromones exert their effects will need to be identified. Recent studies suggest that the beginnings of the scientific study of pheromonal action in humans is close at hand.

Sensory systems—MOS and AOS

Classically, pheromones have been considered to be olfactory. Does this mean that pheromones are odorous compounds? Recent studies in mice suggest that pheromones need not be odorous. For example, the pheromone-accelerated puberty effect in the mouse is induced by short peptides that are nonodorous. The distinction between odorous and nonodorous compounds may be important because it suggests that pheromones may be perceived differently than odorous compounds. Mammals are known to contain two anatomically and functionally distinct olfactory systems, the **main olfactory system** (MOS) and the **accessory olfactory system** (AOS). The MOS is considered to process olfactory stimuli from odorous compounds. A moment's reflection will convince you that the MOS, able to recognize thousands of different odors, has impressive discriminatory power. It is important to realize that the perception of most such odors occurs in a conscious manner—the fragrance of a flower, a perfume, the smell of food cooking on the stove, the smoke of a fire, etc. Conscious perception

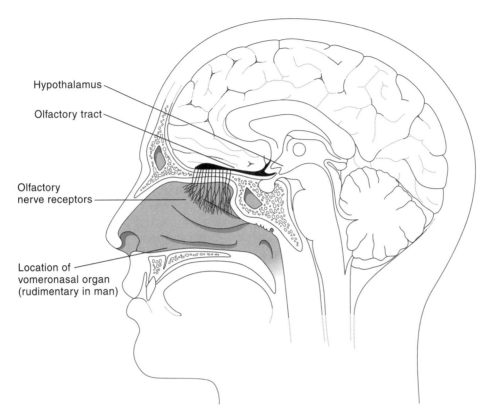

Figure 19.2
The nasopalatine cavity in humans showing the relative positions of the vomeronasal organ
(VNO) and the MOS (indicated as olfactory nerves). The VNO has direct connections to the
hypothalamus.

of odorous compounds can also elicit emotion—pleasure, sadness, anxiety—or even
memory. Simply consider the way in which a particular smell can lead us to recall a specific day or event of our childhood.

The receptor tissue of the AOS is known as the **vomeronasal organ** (VNO). It is
located above the roof of the mouth and connects to the nasal cavity or to the oral cavity via the nasopalatine canal (Fig. 19.2). The neural circuitry of the VNO differs from
that of the MOS in a number of ways. One difference is relevant to the present discussion—the MOS sends its signals to the higher cognitive centers of the CNS (which is
how we become aware of the specific odorous compound). The VNO bypasses the
higher centers. Instead, most of its connections appear to be with the hypothalamus.
Hence, it may be through its projections to the hypothalamus that stimulation of the
VNO exerts its effects on reproductive behavior. At present, the evidence that the
VNO is clearly involved in reproductive functions comes mainly from experiments
with rodents. Removal, for example, of the VNO in virgin male mice severely diminishes their sexual activity. Certain types of lesions in the VNO lead to the absence of
male-specific aggressive behaviors. In females, activation of the VNO can induce puberty in the presence of males. Pheromones are considered to be sensed by the AOS.

The VNO in humans, although present in the first few months after birth, was thought to become atretic and nonfunctional after 1 to 2 years. However, recent studies have shown that a structurally intact VNO is present in adults. Even more interesting has been the demonstration that specific neurons of the human VNO can be activated by purified components of secretions from human skin. The physiological or behavioral significance of such activation remains elusive, however.

Sources of human pheromones—apocrine glands

The human **apocrine glands**, considered to be the main source of body scent in humans, may also be the main source of human pheromones. The apocrine glands are a particularly interesting tissue. They are tubular, coiled structures that pour their secretions into *pilosebaceous canals*. Generally, a single apocrine gland is associated with each hair follicle, although in the axillae, two or three apocrine glands may be found around each hair follicle (Fig. 19.3). They are spread diffusely in human fetuses, but

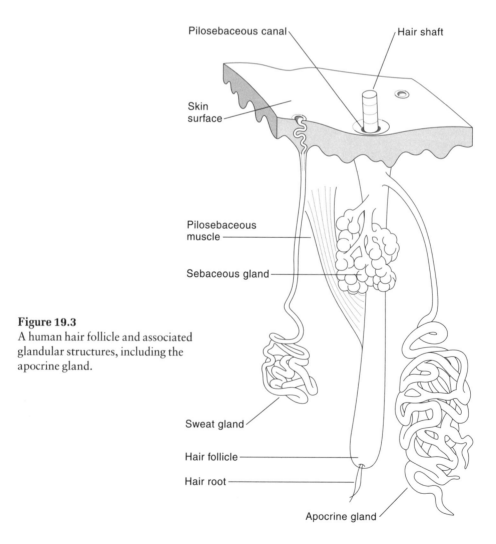

Figure 19.3
A human hair follicle and associated glandular structures, including the apocrine gland.

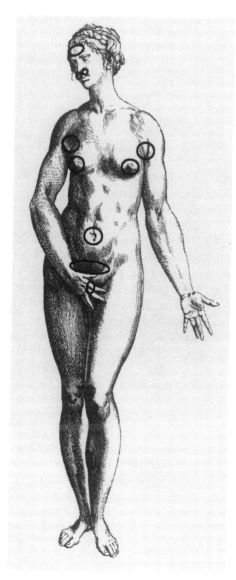

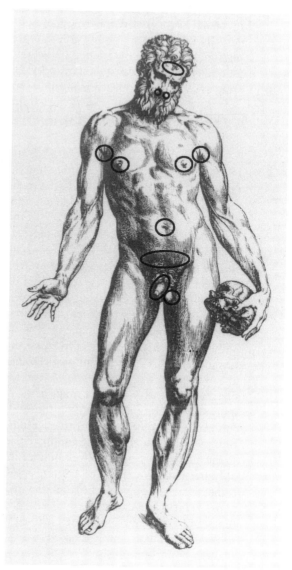

Figure 19.4
Anatomical distribution of apocrine glands in postpubertal females.

Figure 19.5
Anatomical distribution of apocrine glands in postpubertal males.

after birth are retained in only a few sites—the axillae, where they are most prevalent, the anogenital region, and the aerolae are the most prominent (Figs. 19.4 and 19.5). The glands do not mature until puberty, and it is at that time that apocrine secretions begin. Apocrine sweat by itself is odorless. The typical axillary odor is produced by the action of skin bacteria that convert sweat compounds into odorous ones. The bacterial action is also thought to produce the active states of some pheromonal compounds, and perhaps also results in the body odor that characterizes each individual. The developmental history of the apocrine glands, in particular their maturation at puberty, suggests that they may play some role in reproductive physiology, possibly as a source of pheromones that may function as sexual attractants. It has been suggested that the unique bipedal stance of humans provides an effective way of presenting and delivering axillary secretions from one individual to another.

Possible roles of pheromones

Although it is clear that communication is the principal function of pheromones in other species, it is not clear what function pheromones might serve in humans. We recognize that the sense of olfaction is important, but since communication is predominantly visual and auditory, we do not normally think of olfaction as a means of communication. Nevertheless, possible examples of olfactory communication have been suggested. For example, the infant-mother bond in the early postnatal period may depend on pheromones. The VNO is particularly well developed in infants. Studies have shown that infants have a definite preference for breast or axillary pads from their mothers, compared to those from other mothers. Anthropological evidence suggests that axillary secretion may also be important in sexual attraction. In previous times, when access to ample supplies of running water was extremely limited, whole body bathing was relatively infrequent. It was common to consider body scent sexually stimulating. Perhaps this is why, according to legend, Napoleon, when he was returning from his military campaigns to his beloved Josephine, would send a message ahead which read, "*I am coming home, don't bathe.*" The advent of frequent bathing has been accompanied by a change in the way body scent is perceived, from being sexually stimulating to its opposite. Indeed, the regions where the apocrine glands are found are often the ones most meticulously washed. Axillary hair in females is routinely shaved, which tends to eliminate bacteria and body scent for 24 hours or more. Frequent bathing effectively eliminates the pheromone secretions. However, invariably, particularly in females, bathing is followed by the application of perfumes. Perfumes are a complex mixture of scents and trace amounts of extracts of the musk organs of a few large animals. By a strange turn of events the human pheromones are replaced by animal pheromones. These examples indicate that olfactory signals may be used by humans, but it is not yet known whether these signals are processed consciously or unconsciously.

Ovulatory pheromones—the dormitory effect

The first circumstantial evidence that suggested a definitive pheromonal effect in humans was the demonstration that synchronization of menstrual cycles can take place in women living together. This synchronization was termed the **dormitory effect** because the first studies were carried out in women housed together in dormitories. These studies, reported in the 1970s, were generally greeted with some excitement by the public and with skepticism by scientists. However, the claim that the menstrual synchrony was a pheromonal effect remained unproven because other explanations for the effect (for example, visual or auditory cues, similar daily social schedules, or similar stimuli) could not be ruled out. A recent reexamination of this question in a well-controlled study has shown that humans can communicate through pheromones. The plan of the study is elegant and worth reviewing here.

Axillary secretions were collected on cotton pads taped for at least 8 hours to the underarms of "donor" females. Donor secretions were collected on a daily basis

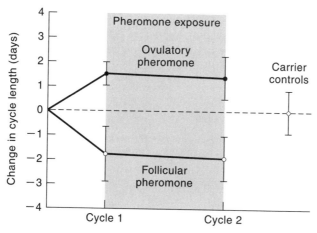

Figure 19.6

Effect of axillary pheromones from donor females on the menstrual cycles of recipient females. Axillary pheromones from the donors' follicular phase shortened the length of the recipients' cycle, while those from the ovulatory phase lengthened the recipients' cycle. Only the follicular phase of the recipients' cycle was affected by pheromones from the donors. (Adapted from K. Stern and M. K. McClintock, 1998.)

throughout the cycle of the donor females. The pads were then wiped above the upper lip (below the nose) of "recipient" females, who were asked not to wash their faces for at least 6 hours. This protocol was repeated daily for over 2 continuous menstrual cycles. The important question was *what effect do the donor secretions have on the menstrual cycles of the recipients?* The results (summarized in Fig. 19.6) demonstrated that the donor axillary secretions had two quite different effects. First, secretions taken from the donors in the follicular phase shortened the time to ovulation and the length of the cycle in the recipients. Second, secretions taken on the day the donors ovulated or during the next two days delayed ovulation and lengthened the cycle in the recipients. The results of this study indicate that the dormitory effect can be explained by pheromonal communication between females.

This study, together with our evolving understanding of apocrine gland secretions and the accessory olfactory system, suggests that pheromonal communication may be common in humans. It also raises important questions. What is the function of these ovarian-dependent pheromones in the context of everyday life? Are these pheromones only the tip of the iceberg? Are aspects of our behavior and physiology modulated by pheromonal signals that we receive from other people in the course of our social interactions? These are extremely interesting questions.

SUMMARY

Olfactory cues have long been considered to be important in sexual attraction in humans, but the molecular basis of olfactory signaling has remained relatively unexplored. Recent studies indicate that the apocrine glands secrete a class of volatile hormones called pheromones that are detected by the accessory olfactory system (AOS). Because the receptor neurons of the AOS lack connections to the higher cortical regions of the brain, detection of pheromones is unconscious. A recent reexamination of the dormitory effect has provided the first evidence for pheromonal communication in humans. This study, which showed that apocrine gland secretions from donor females influenced the length of the menstrual cycle in recipient females, suggests that pheromonal signaling may be common in humans.

QUESTIONS

1. It may be that the female brain does differ from the male brain. However, the crucial question is whether the difference is irreversible. Do you know of any evidence that would suggest that it is not?

2. Do you think the basis of the sexual drive in females differs from that in males? Explain your argument.

3. Why has it been so difficult to discover the biological basis of sexual orientation?

4. In Chapter 14, we considered several disorders of phenotypic sex. Do you think that these conditions might shed any light on the hormonal determinants of GID?

5. What are the crucial differences between olfactory molecules and pheromones?

6. What might be the biological role of the "dormitory effect"?

7. Library project:
   a. What factors might account for the difference in the U.S. and Canadian studies on the X-linked "homosexual gene"?
   b. What is known about the nature and type of apocrine gland secretions?

SUPPLEMENTARY READING

Berman, G. 1995. To speak in chords about sexuality. *Neuroscience and Biobehavioral Reviews* 19(2), 343–348.

Cohn, B. A. 1994. In search of human skin pheromones. *Archives of Dermatology* 130, 1048–1051.

Gooren, L. J. G. 1995. Biomedical concepts of homosexuality: Folk belief in a white coat. *Journal of Homosexuality* 28(3/4), 237–246.

Kinsey, A. C., W. B. Pomeroy, and C. E. Martin. 1948. *Sexual Behavior in the Human Male.* W. B. Saunders Co., Philadelphia, PA.

Mondimore, F. M. 1996. *A Natural History of Homosexuality.* Johns Hopkins University Press, Baltimore, MD.

Pillard, R. C., and J. M. Bailey. 1998. Human sexual orientation has a heritable component. *Human Biology* 70(2), 347–365.

Sell, R. L. 1997. Defining and measuring sexual orientation: A review. *Archives of Sexual Behavior* 26(6), 643–659.

Stoddart, D. M. 1990. *The Scented Ape: The Biology and Culture of Human Odour.* Cambridge University Press, Cambridge, UK.

Vines, G. 1994. *Raging Hormones: Do They Rule Our Lives?* University of California Press, Berkeley, CA.

ADVANCED TOPICS

Gooren, L. J. G. 1990. The endocrinology of transsexualism: A review and commentary. *Psychoneuroendocrinology* 15(1), 3–14.

Hamer, D. H., S. Hu, V. L. Magnuson, N. Hu, and A. M. Pattatucci. 1993. A linkage between

DNA markers on the X chromosome and male sexual orientation. *Science* **261(5119)**, 321–327.

Hutchinson, K. A. 1995. Androgens and sexuality. *The American Journal of Medicine* **98(Suppl. 1A)**, 111S–115S.

Monti-Bloch, L., C. Jennings-White, D. S. Dolberg, and D. L. Berliner. 1994. The human vomeronasal system. *Psychoneuroendocrinology* **19(5–7)**, 673–686.

Pillard, R. C., and J. M. Bailey. 1995. A biologic perspective on sexual orientation. *Psychiatric Clinics of North America* **18(1)**, 71–84.

Rice, G., C. Anderson, N. Risch, and G. Ebers. 1999. Male homosexuality: Absence of linkage to microsatellite markers at Xq28. *Science* **284**, 665–667.

Robbins, A. 1996. Androgens and male sexual behavior: From mice to men. *Trends in Endocrinology and Metabolism* **7(9)**, 345–350.

Stern, K., and M. K. McClintock. 1998. Regulation of ovulation by human pheromones. *Nature* **392**, 177–179.

Jackson Pollock, *Echo (Number 25, 1951)*, 1951.
(© 1999. Pollock-Krasner Foundation/Artists Rights Society [ARS], New York, NY.)

CHAPTER 20

Human Embryo Research

*I think it would be mind-bogglingly fascinating
to watch a younger edition of myself growing up
in the twenty-first century instead of the 1940s.*

—R. Dawkins. 1998

Cited by D. Butler and M. Wadman. In Calls for
cloning ban sell science short. *Nature* 386, 8–9.

*Cloning appalls us, unnerves us, disgusts, horrifies
and revolts us precisely because it engages our deepest
concerns about personhood, identity, life, and sex.*

—W. I. Butler. 1998

What's wrong with cloning? In *Clones and Clones:
Facts and Fantasies about Human Cloning.*
M. C. Nussbaum and C. R. Sunstein, Eds. W. W. Norton, New York, NY.

However correct scientifically, such arguments
(in support of cloning) *fail to acknowledge that the assault
on individual identity and freedom committed by cloning is
not just biological, it is psychological and social as well.*

—H. I. Kaye. 1998

Anxiety and genetic manipulation: A sociological view.
Perspectives in Biology and Medicine 41(4), 483–490.

THE EPIGRAPHS ILLUSTRATE highly divergent views expressed about the prospect of human **cloning**, a possibility which suddenly entered into our public discourse in the waning years of the twentieth century. We are not yet able to clone humans, and we may wonder whether we should ever attempt it, but there is no doubt that the possibility of doing so is close at hand. This epoch-making possibility comes as the result of research on amphibian and mammalian embryos that began some 40 years ago. A watershed in this work came with the report in February 1997 of the birth of Dolly, a lamb cloned from an adult sheep. Dolly's birth drew immediate widespread expression of concern and even indignation from politicians, religious leaders, philosophers, and ordinary citizens.

No less important than the cloning of Dolly, and more significant from a biomedical and therapeutic perspective, were the reports appearing in November and December 1998 of the successful derivation of human pluripotent **embryonic stem (ES) cells** from embryos and fetuses. The important feature of ES cells is that they retain the ability to differentiate into many of the different types of cells that make up the body. This report appears to have been received with much more equanimity than Dolly's birth, although it may be too early to draw that conclusion. It may be that the promise of beneficial results from ES cell research seems more tangible, or that ES cell technology is seen as less threatening. There is no doubt, however, that both technologies are likely to have revolutionary consequences for biology and medicine, but they also bring with them, in the view of many, unprecedented ethical dilemmas.

As a society we seem to have been caught off guard by the apparently sudden emergence of these technologies, although many scientists have been describing their development in animal studies for several years now. We are faced now with their extension and application to humans. Even if we accept what supporters of these technologies tell us of their potential for human benefit, they still arouse a deep-seated anxiety and apprehension in many of us. The reaction to Dolly is simply an expression of that unease. In the words of one critic of human cloning, our apprehension comes because *"there are certain large constraints on being human and we have certain emotions that tell us when we are pressing against those constraints in a dangerous way."*[1] For many there is a clear sense that in some fundamental way we are being forced to reexamine our concept of a human being, but we are not yet certain how to deal with these questions in the context of our cultural, ethical, and religious traditions.

One thing appears certain, however, and this may contribute to our anxiety as well: there is no way we can turn the clock back, nor can we ignore these developments. The genie has been let out of the bottle, and there is no way it can be returned. In one way or another, we will have to respond to the challenge that these technologies present. The crucial question is *How should we respond?* There is really only one effective and sensible way. First, we need to understand the science behind these new technologies. Second, with that understanding, we should be able to examine the implications of application of these technologies to humans. The most straightforward way to do this is to provide a national (perhaps even an international) framework by which we can control their further development and application. This will enable us to proceed with circumspection so that publicly and privately funded research and application might proceed with the broadest possible public understanding and acceptability. Our objectives in this chapter are to review the essentials of cloning and derivation of ES cell lines, the potential benefits of animal cloning and human ES cell research, the ethical concerns that have been raised by the application of these technologies to humans, and finally to describe briefly a format proposed by some scientists for proceeding with the development of both technologies.

SCIENTIFIC BACKGROUND

Cloning and ES cell technologies are similar in that both involve the generation and manipulation of embryos *in vitro*; they differ in that the objective of the manipulation is different. In the first case, the objective is the gestation of a new individual, while in

---

[1]Butler, W. I. What's wrong with cloning? In *Clones and Clones: Facts and Fantasies about Human Cloning.* M. C. Nussbaum and C. R. Sunstein, Eds. W. W. Norton.

the second, the objective is to develop cells from a preimplantation embryo into semi-permanent cell lines that retain ability to differentiate into derivatives of the trilaminar disc, the three cell layers that are the progenitors of all the tissues in the developing embryo (see Chapter 12). These cell lines are known as embryonic stem (ES) cell lines. There is a profound difference in the intended outcomes.

Cloning

A fundamental feature of asexual reproduction, as we saw in Chapter 2, is that progeny are genetically identical to the parent, that is, they are clones of the parent. In sexual reproduction, however, genetic identity between parent and progeny is not possible because the progeny contain genes from both parents. Nevertheless, special types of cloning do occur in sexually reproducing species under certain conditions. For example, in humans, monozygotic twinning is a type of cloning. The monozygotic twins are clones of each other because they are genetically identical to each other, but they are not clones of either parent. The possibility of generating clones of animals or humans experimentally by sidestepping sexual reproduction has probably been an old dream of humans, but until very recently, cloning was possible only in science fiction writings. Things began to change about 40 years ago when the experimental cloning of animals was first reported with the cloning of frogs in the 1960s. This work did not appear to attract the interest or attention of the general public. However, the cloning event that made history, both scientifically and publicly, was the report in February 1997 by scientists of the Roslin Institute in Edinburgh, Scotland, of the birth of Dolly, the first mammal (lamb) cloned from an adult cell.

The same Scottish research group had reported the cloning of embryonic sheep in 1995, but the birth of the two lambs, Megan and Morag, went largely unnoticed by the media and public. Hence, cloning in 1997, even cloning mammals, was not new, but Dolly was. Why was Dolly so special? The difference was that Dolly had been cloned from an adult sheep, in fact, an adult ewe that had been dead for some time. Why was this fact so important? To both appreciate the scientific importance of Dolly and understand the reason for the apprehension Dolly brought requires that we have a clearer idea of the experiments that led to Dolly.

All current cloning protocols are variations of the *nuclear transfer procedure*, which is conceptually quite straightforward. (1) An egg is obtained from a donor female by stimulating ovulation. (2) The nucleus of the egg is removed, and it is replaced with the nucleus from a cell of the individual to be cloned. (3) The newly reconstituted egg is implanted into a properly prepared uterus of a surrogate mother (Fig. 20.1). Sexual reproduction is bypassed since the nucleus in the egg is not a zygotic nucleus; that is, it is not the product of a fertilization event. If implantation in the surrogate female is successful, this procedure will yield a clone of the individual from which the nucleus was taken. It is important to keep in mind that although the new individual is not the direct result of sexual reproduction, cloning still requires an egg and a uterus. At present and for the foreseeable future, there is no way to clone an animal completely in a culture dish or a test tube. We have a long way to go before we can construct an artificial uterus to support the embryonic and fetal development of a new individual.

As indicated above, prior to Dolly, in 1995, two sheep, Meagan and Morag, had been cloned using nuclei from cells of the inner cell mass of a blastocyst embryo. In Dolly's case, however, the nucleus came from an udder cell, not an embryonic cell, but a highly differentiated, adult cell. This variation on the cloning procedure is now referred to as **somatic cell nuclear transfer** to indicate that the source of the nucleus to be

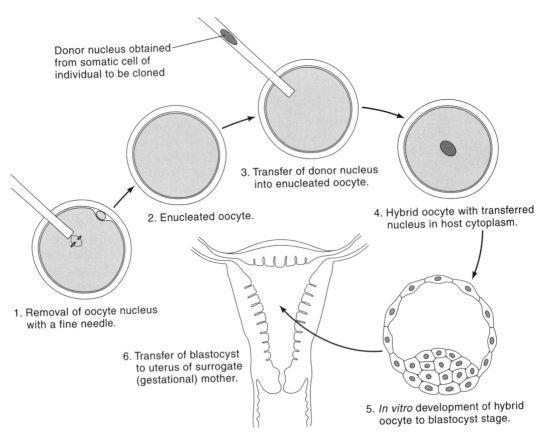

Donor nucleus obtained from somatic cell of individual to be cloned

3. Transfer of donor nucleus into enucleated oocyte.

2. Enucleated oocyte.

4. Hybrid oocyte with transferred nucleus in host cytoplasm.

1. Removal of oocyte nucleus with a fine needle.

6. Transfer of blastocyst to uterus of surrogate (gestational) mother.

5. *In vitro* development of hybrid oocyte to blastocyst stage.

Figure 20.1
Schematic illustrations of the somatic nuclear transfer protocol. This procedure has only been carried out in the mouse. In the reported sheep and cow cloning experiments, nuclear transfer is accomplished by fusing a somatic cell with an enucleated oocyte.

transferred is highly differentiated adult tissue, rather than embryonic tissue. The Dolly experiment used three strains of sheep; one was the nucleus donor, the second was the egg donor, and the third was the surrogate mother. The three strains were chosen because they were easily distinguishable by their markings, and hence, the clone would differ from both the egg donor and the surrogate mother. The nucleus was obtained from an udder cell of a Finn Dorset sheep. In fact, the ewe from which the udder cell came had been dead for a long time. An udder cell line taken from that sheep had been established and kept frozen. The nucleus from the udder cell of a Finn Dorset sheep replaced the nucleus of an *enucleated* egg (that is, an egg whose own nucleus has been removed) produced by a Poll Dorset sheep. This newly constructed egg was implanted in a properly prepared uterus of a Scottish Blackface sheep. In the experiments reported, 434 eggs were used, but only 29 successful implantations took place, and only one (Dolly) survived to birth. Dolly has developed normally, is fertile, and is now herself a mother.

From a biological point of view, the most important and surprising aspect of Dolly's cloning was that the donor nucleus came from a fully differentiated adult cell. Many previous unsuccessful attempts at cloning using somatic cell nuclear transfer had been interpreted to mean that a differentiated nucleus could not direct the development of a new organism. The success in cloning with embryonic cells was consistent with this view, since embryonic cells, because they were much less highly differentiated, were considered to have retained the potential to carry out the embryonic developmental program. Nuclei from embryonic cells were considered to be *totipotent*, or at least, totipotency could be restored easily. On the other hand, a nucleus from a highly differentiated cell had been programmed to express only a certain unique set of genes, and its developmental potential would therefore be severely restricted. The important question was: could a highly differentiated nucleus be *reprogrammed* to become a zygotic nucleus so that it could now direct the development of a new organism? This reprogramming would necessarily involve turning off the genes of the differentiated state, and reestablishing the conditions that would make the differentiated nucleus equivalent to a normal zygotic nucleus. The birth of Dolly demonstrated that at least *some* differentiated nuclei can be reprogrammed. In Dolly's case, nuclei from udder cells were capable of becoming zygotic nuclei. Nuclei from other tissues were tried but with no success. It is not yet clear whether the failure of nuclei from other tissues to reprogram is the consequence of an intrinsic biological difficulty or simply a technical problem.

Dolly is no longer the only mammal cloned by somatic cell nuclear transfer. In July 1998, a team of scientists from the University of Hawaii and the University of Tokyo reported the birth of 22 healthy, cloned female mice, using nuclei obtained from highly differentiated granulosa cells. Interestingly, cells from other tissues were also tried, but the cloning was unsuccessful. At about the same time, a group at Tokyo University of Agriculture reported the successful cloning of two calves using nuclei from uterine tube cells. Another four cows are reported to be pregnant with cloned embryos. These different experiments leave no doubt that mammalian cloning based on somatic cell nuclear transfer is an accomplished fact. The cloning frequency is still low, but undoubtedly this will improve in the next few years. Animal cloning, at least in a few species, will become a widely accessible technique.

Embryonic stem (ES) cell lines

In most tissues in an adult mammal, the cells are continually being replaced. The cells from which these new cells arise are referred to as *stem cells*. Three well-studied examples are the stem cells of the blood (hematopoietic system), skin (epidermis), and small intestine (intestinal epithelium). The regenerative activity of these tissues is impressive. For example, 10^{11} cells of the human small intestine are replaced daily. In the adult, then, the function of stem cells is to maintain established tissues and cell types under a variety of conditions.

In the very early embryo the situation is quite different. Recall from our discussion in Chapter 12 that after fertilization the development of the preimplantation embryo goes through several critical stages, in particular, from the morula to the blastocyst. From studies in the mouse, any cell from the morula is capable of becoming part of any embryonic or extraembryonic tissue. However, by the blastocyst stage, the cells of the trophoblast are irreversibly committed to trophectoderm-derived cell types. The cells of the inner cell mass (ICM) have a broader developmental potential, for they contrib-

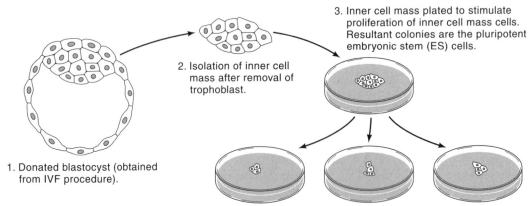

3. Inner cell mass plated to stimulate proliferation of inner cell mass cells. Resultant colonies are the pluripotent embryonic stem (ES) cells.

2. Isolation of inner cell mass after removal of trophoblast.

1. Donated blastocyst (obtained from IVF procedure).

4, 5, 6. Isolation of lineage-restricted stem cells by using specific combinations of growth factors. Each culture dish would contain a specific type of differentiated stem cell derivative to be used for transplantation therapy.

Figure 20.2
Establishment of embryonic stem (ES) cell lines and lineage-restricted derivatives. The working out of culture conditions that permit the establishment of semi-permanent pluripotent ES cell lines (Step 3) is a critical step. Realization of the potential benefits of ES cell technology will require working out the culture conditions that will establish appropriately differentiated stem cell lines (Steps 4, 5, 6, . . .).

ute to both embryonic and extraembryonic lineages. In particular, ICM cells give rise first to the bilaminar disc; cells of the epiblast give rise to the trilaminar disc, which gives rise to all of the different cell types of embryo, fetus, and eventually the adult. The extraordinary developmental potential of ICM-derived cells is under normal embryonic conditions short-lived, however. Might it be possible to maintain this developmental potential indefinitely in some way?

Experiments with mouse embryos beginning in the 1980s demonstrated that this was in fact possible. Under appropriate conditions mouse ICM cells dissociated from their blastocysts could be maintained in culture in an undifferentiated state indefinitely, and importantly, they could also be triggered to differentiate into cell types derived from the three cell lineages of the trilaminar disc. The cell lines derived in culture from these ICM cells that retained these properties were referred to as *pluripotent* embryonic stem (ES) cells (Fig. 20.2). The successful extension of these studies to primates was reported in 1995 and 1996, when ES cell lines from two primate species, the Rhesus monkey (*Macaca mulata*) and the marmoset (*Callithrix jacchus*), were established. In November 1998, the first human ES cell lines were reported. The ES cell lines were derived from ICMs that had been obtained from blastocyst embryos produced for *in vitro* fertilization (IVF) purposes and donated by couples with informed consent (Fig. 20.3).

In December 1998, the isolation of ES cell lines derived by a slightly different procedure was reported. In this case, the ES cell lines were derived from primordial germ cells (PGCs) obtained from aborted fetuses. Recall from Chapter 4 that the PGCs are the progenitors of the gametes. The PGCs, although they arise from the ICM, are

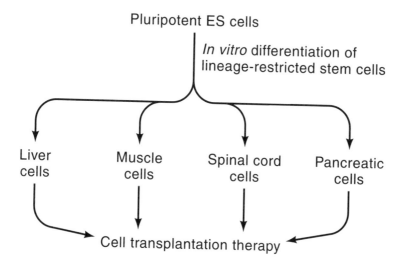

Figure 20.3
Undifferentiated human embryonic stem cell colony (20-fold magnification).
Undifferentiated human embryonic stem cells have distinct borders, a very high nucleus-to-cytoplasm ratio, and distinct nucleoli. The high nucleus-to-cytoplasm ratio is especially noted when compared to the surrounding mouse fibroblasts used as a feeder layer. The black dots seen within the nucleus are nucleoli. (Courtesy of Michelle Waknitz and Robert Becker, Wisconsin Regional Primate Research Center.)

formed during weeks 4 to 6 of gestation. The important discovery in this report is that the PGCs remain pluripotent and can generate ES cell lines.

Since U. S. federal law prohibits the use of public funds for research in which human embryos are created for research purposes or are destroyed, discarded, or subjected to greater than minimal risk, the ES cell line work was carried out in laboratories funded completely by private funds, and completely separate from government-funded laboratories. These experiments are the first step in a research program whose objective will be to realize the full potential of human ES cell lines. The most important long-range question is how to obtain the type of cells desired. Currently the ability of ES cells to differentiate into trilaminar disc derivatives cannot be controlled. It is clear that the ES cells have the potential to develop into a variety of tissues, but the conditions that will trigger differentiation into specific tissues have not yet been worked out.

SUMMARY

Cloning of animals is carried out by a nuclear transplantation protocol in which a diploid nucleus extracted from a somatic cell of tissue from one individual is transferred into an enucleated egg. The reconstituted egg is then transferred into the uterus of a surrogate mother. The resulting offspring is a clone of the donor of the somatic tissue. This procedure, which has been carried out in mice, sheep, and cows, can in principle be extended to other mammalian species including humans. Human pluripotent embryonic stem (ES) cell lines have been isolated by two different methods. Such cells have the ability to differentiate into most of the specialized cells of the human body. In one method, ES cell lines were derived by culturing cells from ICMs obtained from blastocyst embryos generated for the purposes of *in vitro* fertilization. In the second method, ES cell lines were isolated from PGCs obtained from aborted fetuses.

POTENTIAL BENEFITS OF ANIMAL CLONING AND ES CELL LINES

Animal cloning and continued research on ES cell lines in different ways offer the promise of significant advances in understanding embryonic development and in improvement in health care. The benefits from animal cloning are more likely to be realized before those of ES cell line research.

Cloning

The ability to clone animals means that animals with certain desirable traits could be generated in large numbers. For example, with farm animals such as sheep and cows, cloning could be used to ensure the quality control that is now difficult to maintain. Large numbers of sheep producing a particular type of wool could be cloned. With cows, uniformity with respect to milk or meat production could be maintained. Cloning could also be used to generate herds of cattle free of bovine spongiform encephalitis (BSE) (mad cow disease). This disease is of serious concern to many health officials because many products, including medicines, derived from cattle could be carriers of the BSE determinant, and infected animals could also infect people. Cloning of BSE-free cows could generate herds that could be used for providing ingredients for medicines certifiably free of BSE.

Cloning could be used even more beneficially to produce animals that express a gene encoding a scarce and therapeutically important human protein. This requires the production of *transgenic* clones, which is a slight variation of the standard cloning protocol. The objective here is to incorporate into the donor nucleus a segment of DNA that encodes the gene of interest. The advantage is that the cloned animal will now produce the desired protein. Consider, for example, an experiment that has already been carried out in sheep. The gene encoding the human blood-clotting protein, factor IX, which is missing in patients with hemophilia B, was incorporated into sheep embryonic cells in culture. This involves mixing the DNA containing the gene to be transferred (the *transgene*) in an appropriate solution with the desired sheep cells, and stimulating the uptake of the DNA into the cells. The cells that have incorporated the DNA into their chromosomes then serve as donor nuclei in nuclear transfer procedures. The first transgenic sheep produced in this way, Polly, was born in the summer of 1997. Polly and other transgenic clones produce factor IX protein in their milk. With improvement and refinement, these cloning techniques could make available scarce and expensive human proteins of therapeutic value to many more people.

Other possibilities include generating animals that carry genetic defects that mimic human diseases. One example that is often cited is cystic fibrosis. Sheep lungs are very similar to human lungs, and hence, the study of cystic fibrosis in sheep might be valuable in finding effective therapies for treating it in humans. A defective cystic fibrosis gene could be introduced into sheep, and their response to a variety of treatments could be studied. Animal models for other human disorders could be very useful in permitting experimentation that is not otherwise possible now, and with the hope that it might lead to the development of effective and reliable therapies for such disorders.

Cloning research in animals will likely help us understand the fundamental processes of embryogenesis. Currently, cloning in mice holds the most promise in this regard. Compared to sheep and cows, mice have a much shorter gestation time, a much better characterized genetic system, and embryos that are much easier to manipulate and study. This means that many questions about the cloning process are much more accessible by studying cloning in mice than cloning in sheep or cows. Important questions about the genetic events taking place during early embryogenesis become approachable. For example, what is the nature of reprogramming? What genes need to be turned on and when? What accounts for the low success rate of cloning? Why are some cells reprogrammable and others not? Cloning may also give us a deeper understanding of imprinted genes, that is, those genes that require a maternal and paternal imprint in order for embryonic development to take place successfully. Hence, the importance of Dolly goes much beyond economic motives that have driven much cloning work, particularly cloning in sheep and cows.

ES cell lines

The potential biomedical benefits from human ES cell lines may be more significant than from cloning. Among the possible uses for these cells are the study of abnormal development by engineering cell lines carrying specific mutations, isolation of genes involved in early embryogenesis, drug and teratogen testing, and renewable source of cells for tissue and cell replacement. From a medical perspective, this last possibility is the most important. Many human diseases arise from abnormalities in specific cell types, for example, neurodegenerative disorders such as Parkinson's disease (defective dopaminergic neurons) and juvenile-onset diabetes (defective β islet cells of the pancreas). Other serious disorders arise from injuries, such as spinal cord injuries that damage specific types of cells. ES cell line technology might provide a source of specifically differentiated cells that could be used in transplantation or replacement therapy, for example, replacing the defective islet cells of the pancreas (Fig. 20.4).

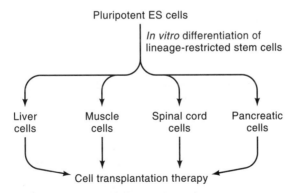

Figure 20.4
Lineage-restricted ES cell lines for cell transplantation therapy. Appropriately differentiated ES cell derivative lines could provide a limitless source of cells that could be used for treatment of diseases characterized by cell degeneration or destruction, such as insulin-dependent diabetes, neuromuscular degenerative disorders, and hepatitis.

Damaged heart muscle could be repaired by injecting cardiomyocytes developed from ES cell lines.

Human ES cells could also be used to test the efficacy of new types of drugs, as well as the effects of teratogens. Most of this testing is now carried out using animal cells or abnormal human tumor cells. ES cell lines would be an improvement because they are normal human cells. ES cell lines will probably be invaluable in studying some of the very early events in human embryogenesis and the conditions and genes that determine the differentiative events in the formation of the organ systems.

Summary

The potential benefits from animal cloning and ES cell line research may well be revolutionary, but the word "potential" must be emphasized. Most of the promised benefits are still hypothetical. A window of opportunity has been opened, however. Animal cloning work is already under way in a number of laboratories. Human ES cell line research is just beginning, and significant technical obstacles still remain before stable differentiated cell lines can be obtained and their therapeutic uses tested. The utility of ES cell lines will increase enormously once directed, lineage-restricted differentiation can be defined and controlled reasonably precisely. A tremendous amount of basic research work is still required before the theoretical potential of this technology can be realized. We are still trying to find a way to balance the ethical concerns raised by these new technologies and the need to continue to carry out the basic work that holds so much promise for society.

ETHICAL AND REGULATORY QUESTIONS

Human cloning and ES cell research are seen by many as pushing human embryo research outside the limits of acceptability. The controversy surrounding the birth of Dolly was not really about cloning sheep or other animals. We manipulate animals for so many different purposes already that cloning them cannot really be considered intrinsically more objectionable. Dolly's real importance was that it became evident that we were one step closer to the cloning of humans. Although we do not have any reliable estimate about how our society views the prospects of human cloning, there is no doubt that significant numbers of people view that prospect with some concern and apprehension.

ES cell research raises fundamental questions about the extent to which cells derived from human embryos obtained by *in vitro* fertilization or from aborted human fetuses can be used for research purposes. This dilemma is not a new one. It was first raised in a serious way by the introduction of *in vitro* fertilization and assisted reproductive techniques (ARTs). As a society we have developed an ambivalent relationship to ARTs. We permit the generation and manipulation of preimplantation human embryos to help infertile couples have children, but require that this work be carried out in privately funded laboratories and clinics. We forbid the use of federal funds for ARTs research or application because for many people research on human embryos is considered to be immoral and unacceptable. This solution has worked reasonably well so far, but an inherent problem is that the privately funded work is being carried out with little oversight. Significant risk for abuse exists. Moreover, market-driven forces are very likely to be the determinants in its development and application. The prospect of human cloning and human ES cell research have forced us to examine the ethical dilemmas presented by both, to distinguish between permissible and nonpermissible research, and to define the conditions under which the permissible research can be carried out.

Human cloning

On the same day that the cloning of Dolly was reported, President Clinton asked the U.S. National Bioethics Advisory Commission (NBAC) to *"undertake a thorough review of the legal and ethical issues associated with the use of this technology."*[2] Similar steps were taken in other countries shortly thereafter. In May 1997, the bioethics committee of the European Union decided that human cloning was unacceptable. The Church of Scotland's General Assembly went even further when they urged the British government to ban the cloning of animals used in meat and milk production. Two important questions that the U.S. NBAC faced were: what, in fact, are the ethical issues that human cloning raises, and from a policy perspective, how should the government respond to them? Over a period of 3 months, the NBAC explored the religious, ethical, and legal questions by soliciting considered opinions and testimony from representatives of different religious groups, philosophers, bioethicists, and legal experts in its efforts to provide a set of recommendations that might guide the development of government policy with respect to these two technologies. The NBAC devoted most of its attention to the issues that human cloning raises. The issues raised by ES cell research had been considered previously by the Human Embryo Panel convened in 1994 by the National Institutes of Health (NIH) and will be discussed separately below.

Religious perspectives

Religious thinkers and theologians from the major religious traditions have been wrestling with the issues presented by advances in reproductive biology for some thirty years, beginning with the introduction of oral contraceptives and more recently with the use of ARTs. Human cloning and ES cell research raise theological questions about the control of human reproduction and the nature of the human being. Although religious perspectives on these issues may arise out of particular scriptural texts or readings whose validity may not be universally accepted, many people in our society look to their religious traditions for moral and ethical guidance. Even in a pluralistic society such as ours, religious perspectives are important because they can inform and influence the development of government policy, perhaps by defining the limits of what is politically feasible. The religious responses focused on the relationship between God and humans, on God's role in procreation, and on the effects on the family. The following are some of the questions explored. *Will human beings be guilty of monumental hubris by attempting to clone a human being? Will they be guilty of "playing God"? Do we have the wisdom to literally become "co-creators," God's partners? Can the quest for knowledge be justified when we effectively begin to usurp God's control over our destiny? What is the source of a human being's dignity and intrinsic worth? Is cloning a violation of human dignity because it bypasses God's role in a human being's creation? Will cloning alter, perhaps destroy, the structure of the family, the role of the parent with respect to the child, who is a gift of God? What will be the relationship between parent and the cloned "child," and how is that role to be reconciled with God's plan?*

---

[2]NBAC, 1997.

The NBAC discovered that there is no common theological response to these and other questions about the implications of human cloning. The premises, modes of argument, and conclusions from different religious perspectives—Christian (Protestant and Roman Catholic), Jewish, and Islamic—differ considerably. For some traditions, human cloning as well as any type of human embryo research is unjustifiable, even if a "good" purpose could be imagined for it. For them the generation of a human embryo in the laboratory and its manipulation are always unacceptable. For others, however, human embryo manipulation and human cloning might be permissible under certain circumstances, but they should be strictly regulated because of the potential for abuse.

Ethical perspectives

The NBAC devoted considerable time in its deliberations to the exploration of ethical principles not arising from any particular religious tradition that might inform a reasoned governmental response to the prospects of human cloning. With the assistance of philosophers and bioethicists they examined the arguments for and against human cloning, and also tried to assess their validity. A variety of questions were explored. *What is the potential for physical and psychological harm? Would cloning inherently cause psychological harm to the cloned child? Would the cloned child suffer loss of individuality and sense of personal freedom? Would being a clone of a previously existing person imperil the right to an "open future"? To what extent would cloning restrict and limit personal choice and free will? Would the relationship between the parent and the cloned child threaten the stability of the family? What, after all, would be the purpose of cloning? Are there circumstances in which it might be acceptable to attempt cloning a human being? What principle or principles should guide us in determining when someone can be cloned? What compelling benefits, personal or societal, does cloning have that would make it worthwhile? Is cloning simply an extension of the freedom of reproductive choice? Would the banning of cloning interfere with the right to personal autonomy or with the freedom of scientific inquiry? Would cloning result in the loss of the intrinsic value of the human being? How should ethical principles influence public policy regulating human cloning?*

The NBAC was not able to come to a consensus regarding the ethics of human cloning. The NBAC felt that no compelling and sustainable arguments were presented to it that could in principle justify the complete proscription of human cloning. For example, concerns about loss of individuality and personal identity could be countered with the example of monozygotic twins, who are genetically identical, but yet retain their individuality and identity. Concerns about the effects on the family that cloning would have could be countered by asking whether they would be any greater than those already introduced by ARTs and surrogacy. Nevertheless, the NBAC clearly recognized that a much wider debate and discussion is needed before we can begin to approach anything resembling a consensus not only about the ethics of human cloning, but also about the government's role in regulating its use.

At the same time, the NBAC felt that human cloning does not offer clearly evident societal benefits that would justify attempting it. The motives generally offered, particularly the eugenic or narcissistic ones, can have only personal benefits and are precisely the ones that the public finds frivolous and objectionable. Hence, the NBAC found no compelling and sustainable arguments to proceed with human cloning in the near future. Moreover, because of safety considerations, it strongly recommended a moratorium on human cloning for an indefinite period. From a purely scientific perspective, this recommendation makes perfect sense. The preparatory research work that has made cloning possible in sheep, mice, and cows has not yet been carried out in humans. Even under the best circumstances, cloning in these animals entails the generation of many defective fetuses. While this may be acceptable with animals, it is highly unlikely that the large-scale generation of defective human fetuses would be tolerated by any society. Hence, for scientific, moral, and legal reasons, a moratorium on human cloning was the best course of action, and one that would have widespread support from all segments of society. Accordingly, the NBAC concluded that *"it is morally unacceptable for anyone in the public or private sector, whether in a research or clinical setting, to attempt to create a child using somatic cell nuclear transfer cloning."*[3] Such a moratorium could only be enforced with respect to federal funding, but the NBAC requested that the private sector comply with the intent of the federal moratorium.

Research on ES cell lines

In 1994, the Human Embryo Research Panel, consisting of scientists, physicians, lawyers, sociologists, and ethicists, was convened by the National Institutes of Health (NIH) to consider the question of the permissibility of research on preimplantation human embryos. The reticence about working with the preimplantation embryos comes from the fact that the embryos are "potential" human beings, and for those who believe that human life begins at fertilization, their use and manipulation will always be unacceptable. However, this view is by no means universal. For many, the embryo does not become a sentient being until a nervous system develops. By this criteria a preimplantation embryo is not a sentient being, and manipulation for research purposes could be permissible if a significant human benefit could be demonstrated. One of the outcomes of their deliberations was that federal funding could be used for *"research involving the development of embryonic stem cells, but only with embryos resulting from* in vitro *fertilization for infertility treatment or clinical research that have been donated with the consent of the progenitors."*[4] Despite this recommendation, the NIH decided that public funds could not be used to derive human ES cell lines. This type of work could nevertheless be conducted privately, but no enforceable guidelines of any type were imposed on privately funded work.

To a significant extent, preimplantation embryos are manipulated already in ways that implicitly assumes that they are qualitatively different from an implanted embryo and fetus. For example, in the standard ARTs protocols, more embryos are generated than are going to be used or that will eventually develop into a child. Some of the embryos that are transferred to the female will not implant; others may implant but for unknown reasons will not develop properly, and will be aborted at a later stage. Should this concern us? Is this an important ethical issue? Unused embryos are stored in a frozen state, and their status remains ambiguous. Are they considered "persons," and how do we justify keeping them in a frozen state indefinitely? Is this proper and ethical treatment? Hence, even in the context of ARTs protocols, which have support among the public, embryos are generated and manipulated with the knowledge that many will be lost or discarded. If the ends justify the means in the ARTs context, why not in generating ES cell lines, especially in view of the potential wider benefits that ES cell research might have?

[3]NBAC, 1997.

[4]National Institutes of Health Report of the Human Embryo Research Panel. 1994. U.S. Government Printing Office, Washington, D.C.

A regulatory framework

All new technologies require a period of adaptation. Eventually they become such an integral part of the society that we forget what it was like before the technology came into existence. The introduction of new technologies in biology and medicine may be more problematic, perhaps because inherently their application to humans may be seen as legitimate cause for concern. A good example of the introduction of a new technology and society's response to it was the recombinant DNA (the cloning and manipulating DNA molecules) controversy of the 1970s. At the time, because of concerns that transferring DNA between unrelated species posed a threat to public health and the environment, an initiative by scientists led to the establishment of the Recombinant DNA Advisory Committee that guided recombinant DNA research. As a forum for defining permissible and nonpermissible experiments, and for evaluating developments in recombinant DNA cloning technology, this approach was successful in showing that the original concerns were not serious. Twenty-five years later recombinant DNA technology has become so widespread and pervasive in modern biology that, except for its use in producing genetically modified foods, it is no longer news or a significant source of concern.

The questions and concerns raised by human embryo research in all its complexity, from ARTs to human cloning and human ES cell lines, are certainly as significant and substantive as, probably more so than, those raised by recombinant DNA techniques. Unfortunately, the response in the U.S. to these developments has been inconsistent and almost schizophrenic. On the one hand, human embryo research has been proscribed in the sense that no federal funds can be used for such work. On the other, such work may be carried out in the private sector with much less oversight. Market forces may govern its development even in those areas that most impinge on ethical concerns. An excellent example of the consequences of this disjointed response is the announcement in November 1998, by a biotechnology company, of the transfer of a nucleus from a human blastocyst into a cow's egg. The scientific objective of this experiment remains totally obscure, and it reinforces the call of many scientists that the most appropriate way to respond to these technologies is within a framework that provides guidelines for all human embryo research, whether publicly or privately funded.

A number of scientists have suggested that the approach that was so successful in the recombinant DNA controversy could also work once again. One of the first steps toward establishing a permanent national oversight group was taken by the director of NIH in April 1999. He convened the Working Group of the Advisory Committee to the Director, NIH (ACD) with the charge of developing guidelines governing research involving the derivation and use of human stem cell lines derived from human embryos. The Working Group was made up of individuals with varied expertise and experience, and included basic and clinical scientists, ethicists, lawyers, clinicians, patients, and patient advocates. The Working Group issued a set of proposed guidelines in December 1999 and called for the establishment of a Human Pluripotent Stem Cell Review Group (HPSCRG) to monitor all human stem cell research.

The important features of the guidelines are summarized in Table 20.1. The Working Group maintained the separation between the private and public sectors. Derivation of stem cell lines from human embryos is not permitted in federally funded laboratories, but is permitted in private laboratories. However, federally funded scientists are to be allowed to use the stem cells if they were derived according to a set of strict requirements. The requirements include the following provisions: (1) that the cell lines can only come from frozen "excess" embryos created during fertility clinics at private

Table 20.1
Proposed NIH Guidelines for Research on
Human Embryonic Stem Cells, December 1999

Deriving new cell lines from human embryos	Prohibited
Research on privately derived cell lines from human embryos	Allowed
Deriving new cell lines from human fetal tissue	Allowed
Research on cell lines from human fetal tissue	Allowed
Research that would use stem cells to create a human embryo	Prohibited
Combining human stem cells with animal embryos	Prohibited
Use of stem cells for reproductive cloning	Prohibited
Research on stem cell derived from embryos created for research purposes	Prohibited

Reference: www.nih.gov/news/stemcell/draftguidelines.htm.

clinics; (2) that strictly worded consent forms be signed by the embryo donors; and (3) that the individuals responsible for the fertility treatments and the researchers proposing to derive or use the stem cell lines not be the same persons. The guidelines prohibit any attempts to clone humans or to combine human stem cells with animal embryos, and prohibit research on cell lines obtained from human embryos created for research purposes.

The guidelines understand that some types of human embryo research are much more problematic than other types. For example, there is no doubt that a moratorium on human cloning needs to be established, and the proposed guidelines do prohibit human cloning. To be effective such a moratorium should apply to both publicly and privately funded laboratories. Although human cloning has not been explicitly prohibited in privately funded laboratories, no reputable scientist or private laboratory would be likely to undertake it. It is of course always possible to imagine that clandestine human cloning laboratories could be established and financed by extremely wealthy individuals, but with appropriate monitoring this scenario seems unlikely.

On the other hand, because of the promise that ES cell research holds, and the fact that the generation of human embryos is already permitted in privately run clinics, it seems reasonable to permit research on the establishment and manipulation of human ES cell lines. The proposed guidelines, if ratified, would provide a framework by which human embryo research would be brought under a stringent and open review process. This framework would likely be followed in the private sector also.

The government's decision, presented in an address by President George W. Bush on August 9, 2001, is summarized in Table 20.2. The president emphasized that the potential benefits to be derived from ES cell research make it important that such research be permitted. The president's decision conforms to the proposed NIH guidelines (Table 20.1), and importantly establishes a council to monitor and oversee stem cell research and recommend new guidelines and regulations. The council will also explore the biomedical and ethical questions raised by this type of research.

SUMMARY

Although neither different religious traditions nor ethical perspectives offer a uniform or universal response to the issues raised by human embryo research, a large segment

Table 20.2
Summary of U.S. Government's Decision
Regarding Stem Cell Research, August 9, 2001

1. Federal funds can be used for research with human ES cell lines generated before August 9, 2001, if the embryo from which the stem cells were derived no longer had the possibility of development, and if four additional criteria were met:
   a. The stem cells must have been derived from an embryo created for reproductive purposes.
   b. The embryo was no longer needed for these purposes.
   c. Informed consent must have been obtained for the donation of the embryo.
   d. No financial inducements were provided for donation of the embryo.
   The government identified eleven research institutions that had developed human ES cell lines that met these criteria.

2. The National Institutes of Health was directed to create a Human Embryonic Stem Cell Registry that will list the human ES cell lines that meet the eligibility criteria.

3. Investigators or institutions that generate human ES cell lines that satisfy the eligibility criteria can apply for listing in the registry.

4. Human cloning is proscribed.

5. A President's Council, consisting of scientists, doctors, ethicists, lawyers, theologians, and others, will be appointed to monitor and oversee human stem cell research, recommend guidelines and regulations, and examine the medical and ethical ramifications of biomedical innovation.

Reference: www.nih.gov/news/stemcell/index/htm.

of the public probably is uneasy about the prospects of embarking on such a path. In general, most would call for a ban or moratorium on human cloning. Opinions on the advisability of deriving human ES cell lines and their application are more varied. The U.S. government in August 2001 defined conditions under which human ES cell lines can be derived, and established a President's Council to monitor stem cell research. In providing for an open, circumspective forum for evaluating human stem cell research, the ethical concerns raised by this research are more likely to be addressed successfully. The formulation of widely accepted guidelines should minimize potential abuses, and at the same time foster public trust and acceptance, so that the potential benefits of this research can be explored in a less controversial way.

QUESTIONS

1. To date, cloning has been successful only with nuclei derived from one or two tissues, and it is not yet clear whether this is a technical or a biological problem. From a biological perspective, what may prevent nuclei from most somatic tissues from being used in cloning?

2. Some opponents of human cloning have argued that it would destroy the already delicate structure of family life. Do you agree? Why or why not?

3. Which do you think offers the best prospects for human benefit in the future: cloning or establishment of highly specialized embryonic stem cell lines?

4. What is your view about the ethical issues raised by cloning and ES cell lines? Would you ap-

prove of the generation of human embryos for well-defined and well-controlled research projects?

5. Library project: The March 2000 media report of the cloning of five pigs claimed that a new cloning method was used. How does it differ from normal cloning protocols?

SUPPLEMENTARY READING

Cole-Turner, R. (Ed.). 1998. *Human Cloning: Religious Responses.* John Knox, Westminister, UK.

Kaye, H. I. 1998. Anxiety and genetic manipulation: A sociological view. *Perspectives in Biology and Medicine* **41**(4), 483–490.

Kitcher, P. 1997. *The Lives to Come: The Genetic Revolution and Human Possibilities.* Touchstone Press, New York, NY.

Murchison, W. 1998. Can we clone souls? *Human Life Review* **24**, 7–15.

National Bioethics Advisory Commission. 1997. *Cloning Human Beings, Vol. 1 & Vol. II. Report and Recommendations of the U.S. National Bioethics Advisory Commission.* Rockville, MD.

Nussbaum, M. C., and C. R. Sunstein (Eds.). 1998. *Clones and Clones: Facts and Fantasies about Human Cloning.* W. W. Norton, New York, NY.

Pederson, R. A. 1999. Embryonic stem cells for medicine. *Scientific American* **280**(4), 68–75.

Robertson, J. A. 1998. Human cloning and the challenge of regulation. *New England Journal of Medicine* **339**(2), 119–123.

Wilmut, I. 1998. Cloning for medicine. *Scientific American* **279**(6), 58–65.

ADVANCED TOPICS

Shamblott, M. J., *et al.* 1998. Derivation of pluripotent stem cells from cultured human primordial germ cells. *Proceedings of the National Academy of Sciences (USA)* **95**(23), 13726–13731.

Thomson, J. A., *et al.* 1998. Embryonic stem cell lines derived from human blastocysts. *Science* **282**, 1145–1147.

Glossary

A

abortifacient agent that induces an abortion.

accessory olfactory system (AOS) one of two mammalian olfactory systems; its receptors, which are sensitive to pheromones, are found in the vomeronasal organ.

acrosome reaction fusion of the sperm head membrane with the outer acrosome membrane; necessary for fertilization.

adipose tissue fat-storing tissue.

adrenarche production of steroid sex hormones by adrenal glands at the onset of puberty.

agonist a compound that binds to a hormone receptor and elicits a response similar to that of the naturally occurring hormone.

allele one of several different forms of a gene.

5-alpha-reductase enzyme that converts testosterone into DHT.

alveoli milk-producing tissue in the mammary gland.

amenorrhea lack of menstruation.

amine hormones modified amino acids that function as neurotransmitters or neuroendocrine hormones; examples are dopamine, melatonin, and serotonin.

amino acid compound containing carboxyl and amino groups; building blocks of proteins.

amnion membrane that forms around developing embryo.

anabolic steroid synthetic testosterone agonist often used to build muscle mass.

androgen insensitivity syndrome (AIS) disorder caused by mutations in the gene encoding the androgen receptor protein; the consequence is the development of female external genitalia, breasts, and female body type in *XY* individual.

androgen replacement therapy administration of testosterone or testosterone agonists to restore androgen-dependent tissues and erectile function in males who have suffered testicular damage or to restore muscle mass and tone and erectile function in older men.

androgens family of steroid sex hormones; testosterone and DHT, the most important members, stimulate the male reproductive system development and maintenance, as well as secondary sex characteristics.

anencephaly failure of the brain to develop.

anestrus absence of an estrus period.

aneuploidy condition in which the chromosome number is different than the normal.

anorexia nervosa severe eating disorder characterized by amenorrhea and extreme weight loss.

anovulatory lack of ovulation.

antagonist a compound that binds to a receptor eliciting a response opposite to that of the natural hormone.

anterior lobe of pituitary one of two embryologically distinct parts of the pituitary; secretes several hormones in response to distinct hypothalamic-releasing hormones.

antioxidants compounds that neutralize free radicals.

antiparamesonephric hormone (APH) a large protein produced by the fetal testis that suppresses paramesonephric duct development; also known as anti-Mullerian hormone (AMH) or Mullerian-inhibiting substance (MIS).

antrum the fluid-filled cavity in a follicle when it reaches the antral stage.

apocrine glands exocrine tissue associated with hair follicles that secretes many different kinds of compounds; after puberty, found primarily in the axillary and genital regions.

apoptosis programmed cell death.

aromatase enzyme that converts androgens to estrogens.

asexual reproduction reproduction characterized by the division of the parent into two identical organisms.

aspermatogenic unable to produce sperm.

assisted reproductive techniques (ARTs) procedures developed to assist infertile couples to have children.

athletic amenorrhea exercise-induced amenorrhea.

514

atresia   loss of oocytes and follicles beginning during gestation and continuing through menopause.

autocrine action   local hormone action that occurs when the secretor and target cell are the same.

autosome   a chromosome that does not determine sex; humans have 22 pairs of autosomes.

azoospermic   unable to produce spermatozoa.

B

ß-endorphin   a member of the endogenous opioid family of peptides that inhibit GnRH pulse generator activity.

ß-endorphinergic system   ß-endorphin-producing neurons.

bilaminar disc   derived from the inner cell mass and consisting of two cell layers, the epiblast and hypoblast.

binary fission   a type of asexual reproduction in which the daughter cells are derived by the splitting in two of the mother cell.

biovars   different forms of an organism that cause similar types of infection; the term is applied most commonly to *Chlamydia trachomatis*.

bipotential   the capacity to develop into either of two types of cells; for example, PGCs can become either ova or sperm.

bisphosphonates   nonsteroidal compounds used in the treatment of osteoporosis.

blastocyst   preimplantation embryo consisting of the inner cell mass and trophoblast.

bone mineral density (BMD)   a measure of the mineral (mainly calcium) content of bone; bone with a low BMD fractures easily.

budding   a type of asexual reproduction in which daughter cells grow out of the mother cell.

bulbourethral glands   component of male internal genitalia that contributes secretions to the seminal fluid.

bulimia nervosa   eating disorder characterized by binge eating, followed by purging.

C

capacitation   changes in spermatozoa that take place in the uterus and that in some species are required to enable the sperm to fertilize an egg.

castration   see *gonadectomy*.

central nervous system (CNS)   the brain and spinal cord in vertebrates.

centromere   the region on a chromosome to which spindle fibers attach during cell division.

cervical crypts   regions in the cervix that serve as temporary reservoirs for spermatozoa.

chancroid   a painful ulcerative sexually transmitted disease that is often associated with enlarged and inflamed lymph nodes in the groin and is caused by the bacterium *Hemophilus ducreyi*.

chlamydia   a sexually transmitted disease caused by *Chlamydia trachomatis*; causes abnormal vaginal and urethral discharge; chlamydia also causes eye disease and pneumonia in infants born to infected women.

chorion   membrane derived from the trophoblast that surrounds the implanted embryo and fetus.

chromatids   the replicated homologs of a chromosome.

chromosomal sex   determined by the sex chromosome contributions of each parent, with *XX* being female and *XY* being male.

chromosome mutation   large-scale changes in the genome, such as gain or loss of chromosomes or rearrangements of large segments of chromosomes.

chromosomes   coiled, DNA-protein structures into which the genome is partitioned.

cirrhosis   liver disease charcterized by the scarring of the liver and ultimately by the loss of liver function.

cleavage   divisions of the zygote that lead to the formation of the morula.

climacteric   the perimenopause.

clitoris   part of the external genitalia in the female derived from the genital tubercle.

clone   cell or organism genetically identical to another cell or organism.

codon   sequence of three nucleotides that specifies an amino acid.

coitus interruptus   in sexual intercourse, withdrawal before ejaculation.

combined oral contraceptives (COCs)   combinations of estrogens and progestins taken orally that suppress ovulation.

complete (central) precocious puberty   puberty due to the premature reactivation of the GnRH pulse generator.

conceptus   a very early stage embryo.

constitutive pathway   the female development pathway of external and internal genitalia that is not dependent on ovarian hormones; also the autonomous pathway.

contraceptive   any agent that prevents fertilization.

corpus albicans   nonfunctional relic of the corpus luteum after corpus luteum involution.

corpus luteum residual cells of the follicle after expulsion of the secondary oocyte; it secretes progesterone and estrogen.

cortical granule reaction fusion of cortical granules with oocyte member immediately after fertilization to prevent polyspermy.

corticotrophs cells in the anterior lobe of the pituitary that secrete ACTH in response to CRH from the hypothalamus.

CRH 41-amino-acid hypothalamic-releasing hormone that regulates ACTH secretion by the anterior lobe of the pituitary.

critical-period hypothesis refers to the idea that gender traits develop from exposure to steroid hormones at defined times during gestation.

crossing over see *recombination*.

cryptorchidism the failure of testes to descend from the abdomen to the scrotum.

cyclicity the periodic changes in hormone secretion during the ovulatory cycle.

cytotrophoblast noninvasive part of the implanting embryo.

D

DAX1 X-linked gene that appears to function as an inhibitor of the testis development pathway.

decidua differentiated endometrial tissue that forms in response to implantation.

decidual reaction response of the endometrium to implantation.

depoprovera synthetic progesterone analog often used in COCs.

dermoid cyst see *ovarian teratoma*.

desensitization phenomenon in which a continuous exposure to a stimulus leads to a diminished response.

digyny fertilization of an unreduced (diploid) egg.

dihydrotestosterone (DHT) androgen derived from testosterone; required in the male for development of external genitalia.

dilation and evacuation surgical abortion procedure relying on removal of the fetus by mechanical means.

dimorphic occurring in two distinct forms.

diploid number ($2n$) the number of chromosomes in a somatic cell; in humans, $2n = 46$.

disperpy fertilization by two sperm simultaneously.

dizygotic twins arising from the fertilization by two different sperm of two oocytes ovulated at the same time.

DNA helical, double-stranded nucleic acid molecule that contains the genetic material of the cell.

dominant follicle the late-tertiary-stage follicle that is ovulated.

dominant mutation mutation that manifests itself in single copy in a diploid organism.

dopamine amine hormone that regulates prolactin and GnRH release.

dormitory effect synchronization of menstrual cycles in females living together.

doubling dose radiation dose that generates the number of mutations equal to those that appear spontaneously.

down-regulation in hormone action, a decrease in hormone receptor concentration in response to a given hormone.

ductus deferens see *vas deferens*.

E

ectopic pregnancy pregnancy resulting from implantation outside the uterine cavity.

eicosanoid hormone class of hormones to which the prostaglandins belong.

embryo an animal in the earliest stages of development in the uterus; in humans, from week 3 to week 8 of gestation.

embryonic stem cells cells derived by culturing cells from the inner cell mass.

embryo transfer in an assisted reproductive technique, transfer of the embryo generated *in vitro* into a receptive uterus for implantation.

endocrine action effects of hormones on tissues at a distance from the secretor.

endocrine gland ductless gland that secretes its products directly into the circulatory system.

endocrinological switch two important hormonal changes during the ovulatory cycle; one occurs when estrogen switches from negative to positive feedback at the pituitary and hypothalamus in the latter part of the follicular phase, the other when the granulosa cells change from estrogen to progesterone synthesis and secretion just prior to the LH surge.

endogenous a product made within the body.

endogenous opioid peptides family of peptides produced in the brain and placenta that bind to opioid receptors.

endometriosis disorder characterized by the growth and development of endometrial tissue in regions outside the uterus.

endometrium the outermost layer of the uterus whose periodic proliferation and destruction defines the menstrual cycle.

epiblast progenitor tissue for the embryo proper.

epididymal maturation changes in the spermatozoa as they pass through the epididymis; necessary for the spermatozoa to be able to fertilize an egg.

epididymis a single, highly convoluted tubule that carries sperm from the rete testis to the vas deferens in the male reproductive system.

erectile dysfunction inability to sustain an erection.

estrogen insensitivity syndrome (EIS) disorder due to a mutation in the gene encoding the estrogen receptor protein; only one case has been reported in humans, a male characterized by skeletal abnormalities and osteoporosis.

estrogen replacement therapy (ERT) administration of estrogen or estrogen analogs to postmenopausal women to reduce the risk of coronary heart disease and osteoporosis.

estrogens a family of steroid sex hormones involved with control of the female reproductive cycle, breast development, and bone mineralization.

estrogen surge peak in estrogen during the follicular phase that initiates the LH surge.

etiology the origin or cause of a disorder or disease.

eukaryotic a cell that has a nucleus.

eumenorrhea menstruating normally.

eutherian placental mammal.

exocrine glands glands that discharge their products through specialized ducts to the outside of the body or to other tissues and organs.

exogenous a product made outside the body.

external genitalia the external structures of the reproductive systems; in the male, they include the penis and scrotum, and in the female, they include the clitoris and labia.

extraembryonic tissues not part of the embryo proper, for example, the placenta.

F

fallopian tube see *uterine tube*.

fetal alcohol syndrome developmental disorder due to exposure of the embryo and fetus to alcohol.

feto-placental unit the coordinated activity of the placenta and the fetal adrenals and liver in producing steroid hormones in the pregnant female.

FHA functional hypothalamic amenorrhea; amenorrhea not associated with organic lesions.

flagellum whiplike organelle that provides the motive force for sperm movement.

follicle functional unit of the ovary consisting of the primary oocyte and layers of granulosa and theca cells.

follicle-stimulating hormone (FSH) hormone released from the anterior pituitary that stimulates the granulosa cells in females and the Sertoli cells in males.

follicular phase phase of ovulatory cycle beginning with the onset of menses and during which the egg is prepared for ovulation.

free radicals highly reactive metabolic compounds that generate the primary DNA lesions that may develop into mutations.

functional hypothalamic amenorrhea amenorrhea due to suppression of the GnRH pulse generator caused by different types of stress.

G

gametes the haploid sperm or egg cells of sexually reproducing organisms.

gender identity disorder (GID) disorder in which individuals of one anatomical sex feel that they are really of the opposite sex; also known as transsexualism.

gene sequence of nucleotides that encode a polypeptide.

gene locus the fixed and unique position of a gene along the length of a chromosome.

gene mutation small-scale changes in DNA nucleotide sequence, such as nucleotide substitutions.

genital ridge progenitor tissue that can develop into testes or ovaries during embryogenesis.

genital tubercle progenitor tissue for penis or clitoris.

genome the DNA in the chromosomes characteristic of each species.

genomic response response to a hormone that activates a gene or set of genes, which causes synthesis of a new protein or set of proteins; typical of steroid hormone responses.

genotype the genetic makeup of an organism.

germ cells or **germ line** special subset of cells derived from the PGCs that has the ability to undergo meiosis.

GIFT gamete intrafallopian transfer; an ART in which the oocyte obtained by ovarian stimulation and sperm are placed into the uterine tube to permit fertilization to take place in the uterine tube rather than in a culture dish.

glucocorticoids steroid hormones produced by the adrenals involved in regulating carbohydrate metabolism.

gonad   gamete-producing organ; ovary in females and testis in males.

gonadal dysgenesis   failure of the gonads to develop.

gonadal sex   the presence of either ovaries or testes.

gonadarche   development of the testis at the onset of puberty.

gonadectomy   surgical removal of the gonads.

gonadotrophs   cells in the anterior pituitary that release the gonadotropic hormones LH and FSH in response to GnRH.

gonadotropic hormones   hormones such as LH and FSH which are secreted by the pituitary gland and whose primary target is the gonads.

gonadotropin-releasing hormone (GnRH)   hormone produced by neurons in different regions of the hypothalamus; controls synthesis of LH and FSH by the pituitary.

gonadotropin-releasing hormone (GnRH) pulse generator   the set of neurons that coordinates and synchronizes their GnRH discharges in a periodic fashion (about one pulse per hour in humans).

gonorrhea   sexually transmitted disease caused by the bacterium *Neisseria gonorrhoeae* and characterized most commonly by an abnormal vaginal or urethral discharge.

Graafian follicle   the dominant follicle that has reached the final antral phase where the oocyte has moved to one pole of the follicle.

granulosa cells   the inner layer of cells surrounding the oocyte during follicular development.

gynecomastia   breast development in males caused by an increase in estrogen/androgen ratio.

H

haploid number (*n*)   the number of chromosomes in a gamete.

hepatitis B virus (HBV)   pathogenic virus that can be transmitted sexually and that can destroy liver function.

hermaphroditic   presence of both ovarian and testicular tissue in the same individual.

herpes simplex virus (HSV)   pathogenic virus responsible for the STD genital herpes.

heterozygous   having two different alleles for a given trait.

homologous chromosomes   pairs of chromosomes that are inherited from each parent and carry the same complement of genes in different allelic forms.

hormones   chemical messengers that allow the cells, tissues, and organs of an individual to communicate with each other in order to coordinate and regulate activity.

HPG axis   the core of the reproductive system; refers to the relationships between the hypothalamus, pituitary, and gonads to regulate each other's activity.

HRE   hormone recognition element; DNA sequence to which a nuclear receptor–steroid hormone complex binds to initiate transcription.

HRT   hormone replacement therapy; administration of a combination of estrogen and progestin to postmenopausal women.

human chorionic gonadotropin (hCG)   a hormone produced by the placenta that is essentially identical to LH and is one of the earliest markers of pregnancy.

human immunodeficiency virus (HIV)   pathogenic virus responsible for AIDS.

human menopausal gonadotropin (HMG)   mixture of LH and FSH found in the urine of postmenopausal women.

human papillomavirus   pathogenic virus responsible for one of the most common STDs.

human placental lactogen (hPL)   placental hormone thought to be involved in regulating energy metabolism in the pregnant female.

hydatidiform mole   see *molar pregnancy.*

hypergonadotropic hypogonadism   disorder characterized by high levels of LH and FSH and lack of gonadal function.

hyperprolactinemia   condition characterized by excessive production of prolactin.

hypoblast   one of two parts of the bilaminar disc; progenitor tissue for extraembryonic membranes.

hypogonadism   condition characterized by suppression of gonadal function.

hypogonadotropic hypogonadism   disorder characterized by low LH and FSH levels and failure of gonadal function.

hypophysectomy   surgical removal of the pituitary gland.

hypospadia   failure of the urogenital sinus to develop properly in male embryos, resulting in the external opening of the urethra being on the underside of the penis.

hypothalamic nuclei   groups of neurons in the hypothalamus that secrete a given hormone.

hypothalamus   a gland in the brain that secretes releasing factors, such as GnRH, that regulate pituitary secretions.

hysterotomy   surgical abortion method used in the third trimester; essentially equivalent to a cesarean section.

I

ICSI intracytoplasmic sperm injection; ART in which sperm head is introduced mechanically into the oocyte.

idiopathic a disorder or condition whose cause is unknown.

immunocontraception fertility control formulations that rely on antibodies against a protein necessary for fertilization or implantation.

implantation the process by which the newly formed embryo establishes contact with the uterus and initiates the development of the placenta.

imprinting process in which certain genes are marked differentially in oogenesis and spermatogenesis.

incomplete precocious puberty GnRH-independent precocious puberty.

indifferent gonad the embryonic gonad from week 3 to week 6 of gestation when the sex of the gonad cannot be determined.

induced pathway development of the internal and external genitalia in the male; dependent on fetal testicular hormones.

infecundity inability to carry a pregnancy to term.

infertility inability to conceive.

inhibin peptide hormone secreted by the Sertoli cells which has feedback control on the secretion of FSH.

initiation first stage in the development of the follicle that will be ovulated.

inner cell mass progenitor tissue in blastocyst for embryo proper.

instillation abortion methods used in the second trimester that involve injecting toxic compounds into the amniotic sac.

interception prevention of implantation.

interstitial cell stimulating hormone (ICSH) older term for LH in males.

interstitial region the outer compartment between the seminiferous tubules of the testes that contains Leydig cells, blood vessels, and nerves.

intrauterine device (IUD) plastic or copper coil placed in the uterus to function as a contraceptive device.

in utero while still in the uterus.

in vitro in a test tube; in an artificial environment.

in vivo within the living body.

ionizing radiation high-energy, penetrating electromagnetic radiation such as X-rays.

IUGR intrauterine growth retardation; alterations in embryonic or fetal development resulting in slow or abnormal growth.

L

labor process of giving birth.

lactational amenorrhea amenorrhea due to breast-feeding.

lactiferous ducts breast epithelial tissue involved in the synthesis or transport of milk.

leptin hormone produced by adipose tissue; may signal the energy status in the individual.

Leydig cells the androgen-producing cells in the testes.

LH surge the rapid increase in LH that occurs 36 hours before ovulation and that is essential for the egg to be ovulated.

libido the sexual drive or urge.

lumen interior region of the seminiferous tubules through which sperm are transported after leaving the Sertoli cells.

luteal phase the second part of the ovulatory cycle in which the site of implantation for the fertilized egg is prepared.

luteal suppression condition resulting from defects in corpus luteum function.

luteinizing hormone (LH) hormone secreted by the pituitary gland that stimulates the Leydig cells in males and theca cells in females.

luteolysis the involution and decay of the corpus luteum.

lymphogranuloma venereum (LGV) sexually transmitted disease caused by *Chlamydia trachomatis* that results in genital ulcers and lymph node enlargement.

M

main olfactory system (MOS) olfactory system responsible for sensing odorous compounds.

male factor infertility infertility due to inability to produce sperm.

mammotrophs cells in the anterior lobe of the pituitary that secrete prolactin in response to prolactin-releasing hormone.

meiosis a two-stage type of cell division in sexually reproducing organisms that results in gametes with half the chromosome number of the original cell.

melatonin hormone secreted by the pineal gland which regulates functions related to light and seasonal changes in day length.

menarche the first menstruation.

menopause the last menstrual period in a woman's life.

menses see *menstruation*.

menstruation periodic shedding of endometrial tissue and blood from the uterus through the cervix and vagina.

mesonephric ducts ducts present in the

indifferent embryo that develop into the internal genitalia in an XY embryo; also known as the Wolffian ducts.

messenger RNA   product of transcribing a gene.

metabolic stress   stress induced by alterations in nutritional/metabolic status.

mifepristone   also known as RU-486; antagonist at the progesterone receptor.

mineralocorticoids   steroid hormones produced by the adrenals; involved in regulating sodium/potassium balance in the blood.

mitosis   cell division that preserves the chromosome number.

molar pregnancy   a pregnancy in which the conceptus contains only paternally derived chromosomes; the pregnancy terminates spontaneously by around the second month of gestation.

monestrus   having only one estrus period per year.

monogenic   condition caused by only one gene.

monosexual   organisms in which male and female gametes are produced by the same individual.

monosomy   chromosome condition defined as $2n$-1.

monotreme   primitive, egg-laying mammal.

monozygotic   genetically identical twins arising from the separation of the minor cell mass.

morbidity   the condition of being diseased or otherwise unhealthy.

morula   early-stage embryo before any overt differentiation into embryonic and extraembryonic cells begins.

Mullerian ducts   see *paramesonephric ducts*.

mutagen   agent responsible for producing mutations in DNA.

mutation rate   the number of mutations produced in DNA per generation.

N

neonatal   pertaining to the first 4 weeks after birth.

neonatal testosterone surge   large increase in testosterone release in males right after birth.

neoplasia   the formation of new and abnormal growth in a tissue that may be benign or malignant.

neural tube defects   developmental defects affecting the spinal cord and brain.

neuropeptides   peptide hormones produced by neural cells.

neurotransmitters   chemical messengers released from synaptic terminals of a neuron that stimulate adjacent cells.

nondisjunction   failure of homologous chromosomes to separate during meiosis I.

nonionizing radiation   low-energy electromagnetic radiation, such as ultraviolet radiation.

nuclear receptors   hormone receptors found in the cell nucleus that bind to steroid hormones.

O

oligospermia   low sperm count.

oocyte   cell type in the first meiotic (primary oocyte) and second meiotic (secondary oocyte) divisions of oogenesis. The secondary oocyte is ovulated and is also the cell that participates in fertilization.

oogenesis   the development of ova from primordial germ cells by meiosis.

oogonia   cells that are formed from the primordial germ cells (PGCs) and that then differentiate into primary oocytes.

opiates   exogenous nonpeptide compounds such as heroin and morphine which act as agonists at the opioid receptors.

opioids   endogenous peptides that bind to the opioid receptors and are associated with a sense of well-being; the most important opioid that affects the reproductive system is ß-endorphin.

orchidectomy   surgical removal of the testes.

osteoporosis   condition characterized by loss of calcium from bone.

ovarian teratoma   spontaneous development of an oocyte in the ovary to form a disordered array of embryonic tissues whose cells contain only maternally derived chromosomes.

ovaries   the female gonads which lie in the abdominal cavity and produce ova and reproductive hormones.

oviduct   see *uterine tube*.

ovulation   the expulsion of the secondary oocyte from the ovary midway through the ovulatory cycle; oocyte enters fimbriae of the uterine tube.

oxidative damage model   model that proposes that mutations originate as oxidative alterations in DNA nucleotides produced by free radicals, which are by-products of normal cellular metabolism.

oxytocin   hormone produced by the posterior pituitary which stimulates uterine contractions during labor and milk let-down by the mammary glands.

P

paracrine action hormone action in which secretor and target cell are neighbors.

paramesonephric ducts ducts present in the indifferent embryo that develop into the internal genitalia in an XX embryo.

parenteral transmission by injection of pathogenic organisms in contaminated blood or other fluids.

parthenogenesis development of the egg without a sperm contribution.

parturition events that lead to birth.

pathogen a disease-producing organism.

PCOS see *polycystic ovarian syndrome.*

pelvic inflammatory disease (PID) infections of the uterus and uterine tubes that often lead to severe complications such as infertility and ectopic implantations.

peptide amino acid chain with fewer than 40 amino acids.

perimenopause the period, a few years in length, prior to and just after the last menstrual period in a woman's life.

perinatal pertaining to or occurring in the period shortly before and after birth.

persistent Mullerian duct syndrome presence of a uterus and uterine tubes in males due to APH deficiency.

phenotype the physical and physiological traits of an organism.

phenotypic sex the presence of either female or male internal and external genitalia.

pheromones volatile hormones secreted by one individual which elicit a response in another individual of the same species.

physical stress stress induced by intense exercise or physical activity.

physiological hyperprolactemia high levels of prolactin that are maintained in breast-feeding women and may contribute to the suppression of ovarian function.

pituitary gland found at the base of the hypothalamus in the brain; it secretes hormones such as LH and FSH that stimulate the gonads.

placenta a structure formed from extraembryonic tissues during pregnancy for nourishing the fetus.

PMDD most severe form of premenstrual dysphoria.

PMS premenstrual syndrome; see *premenstrual dysphoria.*

polar body nonfunctional product of meiosis in oogenesis.

polycystic ovarian syndrome (PCOS) anovulatory disorder characterized by the accumulation of Graafian-stage follicles in the ovary.

polygenic trait determined by more than one gene.

polyspermy fertilization by more than one sperm.

postcoital pill a form of emergency contraception taken after coitus to prevent implantation.

posterior lobe of pituitary part of the pituitary with neural connections to the hypothalamus; oxytocin and vasopressin are released through the posterior lobe.

preeclampsia premature expulsion of the placenta.

preimplantation embryo in humans, the embryo during the 5–7 days after fertilization, during which it travels from the fertilization site to the uterus.

premenstrual dysphoria disorder of differing degrees of severity in which physiological and behavioral symptoms are manifested repeatedly a few days before the onset of menstruation.

prevalence the number of persons with a disease at a specified point in time.

primary oocyte germ cell in oogenesis which undergoes the first meiotic division.

primary spermatocyte germ cell in spermatogenesis which undergoes the first meiotic division.

primordial follicle the smallest and simplest follicle consisting of the oocyte and a single layer of granulosa cells.

primordial germ cells (PGCs) the precursors of the germ line that eventually develops into ova or spermatozoa.

progestins steroid hormones that bind to the progesterone receptor.

prokaryotic a cell without a nucleus.

prolactin (PRL) a hormone produced by the anterior pituitary that stimulates the growth and development of the mammary glands and maintains lactation; also called somatomammotropin.

proliferative phase the first phase of the menstrual cycle, which corresponds to the follicular phase of the ovulatory cycle.

prostaglandins class of hormones derived from fatty acids.

protoctista primitive unicellular eukaryotes, such as slime mold and fungi.

protozoa unicellular eukaryotic organisms that range in size from microscopic to macroscopic, some of which are pathogenic; the STD organism *Trichomonas vaginalis* is a protozoan.

pseudohermaphrodite an individual in which the gonadal sex and phenotypic sex are discordant.

psychogenic amenorrhea   amenorrhea due to emotional trauma.

psychogenic stress   stress due to emotional trauma or anxiety.

puberty   the onset of endocrinological changes that allow for reproductive capability.

R

raloxifene   a SERM-replacing estrogen in ERT/HRT.

receptor   protein that mediates all hormone action.

recombination   process during meiosis in which homologs exchange segments of DNA.

recruitment stage   the second stage in the formation of the dominant follicle.

reductional division   first division of meiosis that reduces the chromosome number to half.

regeneration   the first stage in spermiogenesis in which spermatogonial stem cells are continually replenished.

relative risk   the ratio of the risk of disease or death among the exposed to the risk among the unexposed.

reproductive stressors   conditions that suppress reproductive function.

risk factor   any characteristic (environmental, behavioral, heritable) whose presence is associated with an increased probability of infection or disease development.

risk group   a group of people whose characteristics place them at risk for a specific disease (for example, adolescents are a risk group for STDs).

S

secondary oocyte   product of the first meiotic division in oogenesis; released at ovulation.

secondary sex characteristics   characteristics that develop during the endocrinological changes that occur during puberty, including appearance of pubic and facial hair and enlargement of the external genitalia in males and pubic hair, breasts, and changes in body form in females.

secondary spermatocyte   product of the first meiotic division in spermatogenesis.

secretory phase   the second phase of the menstrual cycle, which corresponds to the luteal phase of the ovulatory cycle.

secular trend   decrease in the mean age of menarche during the last 100 years.

selection stage   the third stage of the formation of the dominant follicle that occurs 14–15 days before ovulation.

selective estrogen receptor modulators   see *SERMs*.

selective serotonin reuptake inhibitors   see *SSRIs*.

seminal fluid   produced by the prostate, seminal vesicles, and the bulbourethral glands in order to nourish the sperm.

seminal vesicles   component of the male internal genitalia; contribute secretions to the seminal fluid.

seminiferous tubules   the inner compartment of the testis that contains Sertoli and germ cells.

SERMs   compounds that bind differentially to the two forms of the estrogen receptor and have agonistic or antagonistic action in different estrogen-dependent tissues.

serotonin   amine hormone with many functions.

Sertoli cell barrier   impermeable barrier formed by Sertoli cells that isolates the seminiferous tubules from the interstitial region of the testis.

Sertoli cells   cells in the testes that regulate and direct all stages of spermatogenesis.

sex chromosomes   the chromosome pair that determines the sex of an individual; a female carries two X chromosomes, and a male carries one X and one Y chromosome.

sex determination   conversion of the indifferent gonad into a definitive testis or ovary.

sex-linked genes   genes located on a sex chromosome.

sex reversal   occurs when the chromosomal sex and gonadal sex are discordant.

sexual differentiation   development of the internal and external genitalia.

sexual reproduction   reproduction in which the gametes from two parents fuse to form progeny with unique gene combinations.

sleep-entrained   biological process associated with the onset of sleep.

somatic cell   any cell in a multicellular organism except the sperm or egg.

somatic cell nuclear transfer   cloning procedure based on transfer of a nucleus from an adult somatic cell into an enucleated egg.

somatomammotropin   see *prolactin*.

spermatogenesis   the continually renewable process of spermatozoa production. The three major stages are regeneration, meiosis, and spermiogenesis.

spermatogonia   cells that are formed from primordial germ cells (PGCs) and that then differentiate into primary spermatocytes.

spermatozoon   male gamete.

spermicides   agents that kill sperm.

spermiogenesis the last stage of spermatogenesis in which the acrosome and flagella are formed.

SRY gene gene located on the Y chromosome necessary for the conversion of the indifferent gonad into a testis.

SSRIs selective serotonin reuptake inhibitors; compounds that have been found useful in the treatment of depression and PMDD.

STDs sexually transmitted diseases; diseases caused by pathogenic organisms transmitted by sexual contact.

steroid hormone receptors see *nuclear receptors*.

steroidogenic enzymes enzymes involved in the biosynthesis of steroid hormones.

steroid sex hormones hormones synthesized from cholesterol that include the progestin, androgen, and estrogen families.

stress condition associated with the release of CRH.

stroma compartment of the ovary between the follicles consisting of a network of connective tissue and a variety of cell types.

subdermal implants progestin-only contraceptives implanted into the axillary region; provide protection for 3 to 5 years.

subzonal space space between zona pellucida and oocyte membrane.

supporting cells bipotential somatic cells in the genital ridge that can develop into Sertoli or granulosa cells.

suprachiasmatic nucleus hypothalamic neurons that regulate the 24–25-hour periodicity of many hormone secretions.

syncytiotrophoblast invasive tissue of the implanting embryo.

syphilis a sexually transmitted disease characterized by initially painless genital ulcers, but which, if untreated, can result in death; caused by the bacterium *Treponema pallidum*.

systemic pertaining to or affecting the body as a whole.

T

tamoxifen a SERM-replacing estrogen in ERT/HRT.

TDF testis determining factor, now known as the SRY protein.

teratocarcinomas malignant degeneration of ovarian teratomas.

teratogen agent that can produce developmental abnormalities during embryogenesis.

teratology study of developmental disorders.

testis the male gonad in which hormones and sperm are produced.

testosterone the most abundant androgen in males.

tetraploid cells containing four complete sets of chromosomes.

theca cells the outer layer of cells surrounding the oocyte during follicular development.

thelarche initiation of breast development in puberty.

thyrotrophs cells in the anterior lobe of the pituitary that secrete thyroid-stimulating hormone (TSH) in response to thyrotropin-releasing hormone (TRH).

transcription the first step in gene expression in which the sequence of DNA bases is converted to ribonucleic acid (RNA).

translation the second step in gene expression in which the nucleotide sequence in messenger RNA is converted to an amino acid sequence; also called protein synthesis.

transmembrane receptors receptors for nonsteroid hormones residing on the plasma membrane of the target cell.

transmission (horizontal) the spread of an infectious agent from one individual to another, usually through contact with body fluids.

trichomoniasis an STD associated with vaginal discharge; caused by the protozoan *Trichomonas vaginalis*.

triploid cells containing three complete sets of chromosomes.

trisomy chromosome condition defined as $2n+1$.

trophoblast (trophectoderm) progenitor tissue in blastocyst for the placenta and other extraembryonic tissues.

true hermaphrodite disorder defined by the presence of both ovarian and testicular tissue in the same individual irrespective of the chromosomal sex.

tubal ligation sterilization involving the ligation of the uterine tubes.

tubal pregnancy implantation in the uterine tube.

two-cell model the interaction of the theca and granulosa cells in the developing follicle responsible for producing estrogen during the follicular phase of the ovulatory cycle.

U

up-regulation in hormone action, an increase in hormone receptor concentration in response to a given hormone.

urethra the duct that carries the urine from the bladder to the exterior of the body.

urogenital folds progenitor tissue for labia or scrotum.

urogenital sinus   progenitor tissue for vagina or urethra in males.

uterine tube   part of the genital tract in the female extending from the uterus; transports sperm and zygote; also known as the oviduct or fallopian tube.

uterus   the female reproductive organ in which the embryo implants and develops.

V

vacuum respiration   surgical abortion procedure for removing a very early stage embryo by suction.

vagina   a thin-walled chamber extending from the uterus to the exterior of the body; the repository for sperm and the birth canal during labor.

vas deferens   the tube that carries sperm in the male reproductive system connecting the epididymis to the urethra; also known as the ductus deferens.

vasectomy   sterilization involving ligation and excision of the vas deferens.

vasopressin   hormone produced in the posterior lobe of the pituitary whose main function is to regulate water retention by the kidney.

virus   submicroscopic infectious agent characterized by a lack of independent metabolism; dependent for replication and transmission on appropriate host cells.

viviparity   development of the fertilized egg inside the mother.

vomeronasal organ   tissue that contains the receptors of the AOS.

W

Wolffian ducts   see *mesonephric ducts*.

X

X-linked gene   gene located on the *X* chromosome.

Y

Y-linked gene   gene located on the *Y* chromosome.

yolk sac membrane   tissue derived from the hypoblast early in embryogenesis.

Z

ZIFT   zygote interfallopian transfer; ART in which the zygote formed *in vitro* is transferred into the uterine tube before cleavage divisions begin.

zona pellucida (ZP)   the glycoprotein layer between the oocyte and the inner granulosa cell layer.

zygote   the fertilized egg.

Index

525